Current Topics in
Developmental Biology

Volume 31

Cytoskeletal Mechanisms
during Animal Development

Current Topics in Developmental Biology

Volume 31

Cytoskeletal Mechanisms during Animal Development

Edited by

David G. Capco
Molecular and Cellular Biology Program
Arizona State University
Tempe, Arizona

Academic Press
San Diego   New York   Boston   London   Sydney   Tokyo   Toronto

Front cover photograph: Scanning electron micrograph showing the side view of an egg undergoing the deformation movement. (For more details see Chapter 7, Figure 6.)

This book is printed on acid-free paper. ∞

Academic Press, Inc.
A Division of Harcourt Brace & Company
525 B Street, Suite 1900, San Diego, California 92101-4495

United Kingdom Edition published by
Academic Press Limited
24-28 Oval Road, London NW1 7DX

International Standard Serial Number: 0070-2153

International Standard Book Number: 0-12-153131-7

PRINTED IN THE UNITED STATES OF AMERICA
95 96 97 98 99 00 BC 9 8 7 6 5 4 3 2 1

Contents

Preface to Section I
Nonchordates   1

Section I
Cytoskeletal Mechanisms in Nonchordate Development

1
Cytoskeleton, Cellular Signals, and Cytoplasmic Localization in *Chaetopterus* Embryos

William R. Eckberg and Winston A. Anderson

2
Cytoskeleton and Ctenophore Development

Evelyn Houliston, Danièle Carré, Patrick Chang, and Christian Sardet

3

Sea Urchin Microtubules

Kathy A. Suprenant and Melissa A. Foltz Daggett

4

Actin–Membrane Cytoskeletal Dynamics in Early Sea Urchin Development

Edward M. Bonder and Douglas J. Fishkind

5

RNA Localization and the Cytoskeleton in *Drosophila* Oocytes

Nancy Jo Pokrywka

Contents

6

Role of the Actin Cytoskeleton in Early *Drosophila* Development

Kathryn G. Miller

7

Role of the Cytoskeleton in the Generation of Spatial Patterns in *Tubifex* Eggs

Takashi Shimizu

Preface to Section II
Chordates   237

Section II
Cytoskeletal Mechanisms in Chordate Development

8

Development and Evolution of an Egg Cytoskeletal Domain in Ascidians

William R. Jeffery

9

Remodeling of the Specialized Intermediate Filament Network in Mammalian Eggs and Embryos during Development: Regulation by Protein Kinase C and Protein Kinase M

G. Ian Gallicano and David G. Capco

10

Mammalian Model Systems for Exploring Cytoskeletal Dynamics during Fertilization

Christopher S. Navara, Gwo-Jang Wu, Calvin Simerly, and Gerald Schatten

14

Intermediate Filament Organization, Reorganization, and Function in the Clawed Frog *Xenopus*

Michael W. Klymkowsky

Contributors

Numbers in parentheses indicate the pages on which the authors' contributions begin.

Winston A. Anderson Department of Biology, Howard University, Washington, District of Columbia 20059; and Marine Biological Laboratory, Woods Hole, Massachusetts 02543 (5)

Edward M. Bonder Department of Biological Sciences, Rutgers University, Newark, New Jersey 07102; e-mail: bonder@actin.rutgers.edu (101)

David G. Capco Molecular and Cellular Biology Program, Arizona State University, Tempe, Arizona 85287; e-mail: atdgc@asuvm.inre.asu.edu (277)

Danièle Carré Unité de Biologie Cellulaire Marine, (URA 671 CNRS/ Université Paris VI), Observatoire de Villefranche-sur-mer, Station Zoologique 06230 Villefranche-sur-mer, France (41)

Byeong Jik Cha Department of Biology, University of Utah, Salt Lake City, Utah 84112 (383)

Patrick Chang Unité de Biologie Cellulaire Marine, (URA 671 CNRS/ Université Paris VI), Observatoire de Villefranche-sur-mer, Station Zoologique, 06230 Villefranche-sur-mer, France; e-mail: chang@ccrv.obsvlfr.fr (41)

William R. Eckberg Department of Biology, Howard University, Washington, District of Columbia 20059; and Marine Biology Laboratory, Woods Hole, Massachusetts 02543; e-mail: eckberg@access.howard.edu (5)

Douglas J. Fishkind Department of Biological Sciences, University of Notre Dame, Notre Dame, Indiana 46556; e-mail: douglas.j.fishkind1@nd.edu (101)

Richard A. Fluck Department of Biology, Franklin & Marshall College, Lancaster, Pennsylvania 17604; e-mail: r_fluck@acad.fandom.edu (343)

Melissa A. Foltz Daggett Department of Physiology and Cell Biology, University of Kansas, Lawrence, Kansas 66045 (65)

G. Ian Gallicano Molecular and Cellular Biology Program, Arizona State University, Tempe, Arizona 85287 (277)

David L. Gard Department of Biology, University of Utah, Salt Lake City, Utah 84112; e-mail: gard@bioscience.utah.edu (383)

Nathan H. Hart Department of Biological Sciences, Rutgers University, Piscataway, New Jersey 08855 (343)

Evelyn Houliston Unité de Biologie Cellulaire Marine, (URA 671 CNRS/ Université Paris VI), Observatoire de Villefranche-sur-mer, Station Zoo-

logique, 06230 Villefranche-sur-mer, France;
e-mail: houliston@ccrv.obsvlfr.fr (41)

William R. Jeffery Section of Molecular and Cellular Biology, University of California, Davis, California 95616; e-mail: wrjeffery@ucdavis.edu (243)

Michael W. Klymkowsky Department of Molecular, Cellular, and Developmental Biology, University of Colorado, Boulder, Colorado 80309; e-mail: klym@spot.colorado.edu (455)

Carolyn A. Larabell Lawrence Berkeley National Laboratory, University of California, Berkeley, California 94720; e-mail: larabell@lbl.gov (433)

Kathryn G. Miller Department of Biology, Washington University, St. Louis, Missouri 63130; e-mail: miller@biodec.wustl.edu (167)

Christopher S. Navara Departments of Zoology and Obstetrics and Gynecology and the Wisconsin Regional Primate Research Center, University of Wisconsin, Madison, Madison, Wisconsin 53706; e-mail: csnavara@students.wisc.edu (321)

Nancy Jo Pokrywka Department of Biology, Vassar College, Poughkeepsie, New York 12601; e-mail: napokrywka@vaxsar.vassar.edu (139)

Christian Sardet Unité de Biologie Cellulaire Marine, (URA 671 CNRS/ Université Paris VI), Observatoire de Villefranche-sur-mer, Station Zoologique, 06230 Villefranche-sur-mer, France; e-mail: sardet@ccrv.obsvlfr.fr (41)

Gerald Schatten Departments of Zoology and Obstetrics and Gynecology and the Wisconsin Regional Primate Research Center, University of Wisconsin, Madison, Madison, Wisconsin 53706; e-mail: schatten@macc.wisc.edu (321)

Marianne M. Schroeder Department of Biology, University of Utah, Salt Lake City, Utah 84112 (383)

Takashi Shimizu Division of Biological Sciences, Graduate School of Science, Hokkaido University, Sapporo 060, Japan; e-mail: shitak@clu.hines.hokudai.ac.jp (197)

Calvin Simerly Departments of Zoology and Obstetrics and Gynecology and the Wisconsin Regional Primate Research Center, University of Wisconsin, Madison, Madison, Wisconsin 53706; e-mail: csimerly@macc.wisc.edu (321)

Kathy A. Suprenant Department of Physiology and Cell Biology, University of Kansas, Lawrence, Kansas 66045; e-mail: ksupre@kuttub.cc.ukans.edu (65)

Gwo-Jang Wu[1] Departments of Zoology and Obstetrics and Gynecology and the Wisconsin Regional Primate Research Center, University of Wisconsin, Madison, Madison, Wisconsin 53706 (321)

---

[1] Present Address: Department of Ob/Gyn, TSGH, Taipei, Taiwan 100, Republic of China.

Notice

The Editor and Editorial Board would like to encourage authors of topical reviews in any aspect of developmental biology to submit them for consideration for publication in *Current Topics in Developmental Biology*. Such submissions will be peer-reviewed by members of the Editorial Board or external reviewers, at the Editor's discretion. Authors with questions about this process may wish to contact the Editor(s) directly in writing or by facsimile [(415) 476-6951 (Dr. Pedersen); or (608) 262-7319 (Dr. Schatten)]. If possible, please include an abstract of the proposed manuscript in this initial correspondence, including information about manuscript length and number and type of illustrations.

Introduction

Gametes, zygotes, and blastomeres of the embryo are cells and must exhibit all of the functional characteristics of a cell in order to survive. In addition to all the requisite cell functions, gametes, zygotes, and blastomeres of the embryo face challenges posed by the developmental program that regulates these cells. Gametes, zygotes, and embryos contain adaptations that allow these specialized cells to meet and surmount the challenges posed by the developmental program. These developmental challenges are directed at the structure and function of these specialized cells, and consequently the adaptations act through specializations in the cytoskeleton.

Many of these specializations in the cytoskeleton are most clearly detectable at the time that these specialized cells undergo major remodeling of structure and function, that is, at the time of a developmental transition. Developmental transitions represent major partitions or landmarks in the developmental program where the gametes, zygote, or blastomeres of the embryo undergo a major structural and functional change. Several developmental transitions are common to (or conserved among) all classes of organisms, for example, gametogenesis, fertilization, and gastrulation. In addition, there are typically developmental transitions specialized for classes of organisms, for example, see Chapters 5, 6, 9, and 10. These transitions cause a radical change in cell function due to an underlying remodeling of intracellular structure (or in the case of the multicellular embryo both intracellular and intercellular remodeling result). This remodeling alters the engineering of the cell, and as a consequence, the function of the cell changes.

The chapters in this volume focus on the cytoskeletal specializations that allow these cells to face and surmount the special developmental problems unique to gametes, zygotes, and blastomeres of the embryo. In each of the chapters readers will identify specializations of the cytoskeleton to meet the challenges of the developmental program that exist at both conserved and specialized developmental transitions. These cytoskeletal specializations set gametes, zygotes, and blastomeres of the embryo apart from somatic cells and also demonstrate remarkable adaptability in elements of the cytoskeleton and in the elaboration of cytoskeletal structures.

Much of the current understanding of cytoskeletal organization and function comes from analysis of results obtained from studies of somatic cells.

The somatic cells employed in many of these studies were obtained either from cell lines maintained *in vitro* (e.g., 3T3 mouse fibroblasts, MDCK cells, endothelial cell) or by explant from the organism (e.g., blood platelets, macrophages, intestinal epithelium). From studies on such cells a minimum of four roles for the cytoskeleton are generally accepted: (1) The cytoskeleton provides the shape and infrastructural support for a cell as well as positioning the organelles and nucleus. (2) Elements of the cytoskeleton serve as "roadways" for the movement of cellular components, including membranous elements, through the action of molecular motors. (3) The cytoskeleton also positions both proteins and mRNA in nonrandom distributions within cells, presumably at sites where such components are necessary. (4) The cytoskeleton mediates cell motility.

The somatic cell types used to obtain the information outlined in the previous paragraph are certainly important and central to the field of cell biology. However, it must be recognized that there are limitations to the type of knowledge obtained by analysis of somatic cells that can be applied to the understanding of cells exhibiting specialized developmental roles. These limitations exist at two levels. First, not all cells will survive under *in vitro* culture conditions, and most that do lose their histotype. Even those cells that are explanted from an organism and studied immediately, such as intestinal epithelial cells, may retain their histotype, but may exhibit a wound response that modifies the action of the cytoskeleton. Thus, while results obtained from investigation of such cells certainly represent an activity of the cytoskeleton within the cell's repertoire, they may not representative of the activity of the cell in its natural location or normal histotype. Moreover, they may not be representative of cell types that cannot be maintained for *in vitro* analysis even for short-term studies. Second, these somatic cells do not face the special developmental challenges of gametes, zygotes, and blastomeres of the embryo.

What are the special developmental challenges faced by gametes, zygotes, and blastomeres and what adaptations exist to allow these special cells to overcome the challenges? The answer to that question is the subject of this volume. Some of these challenges will be common to all species, whereas other challenges will be species-specific. The chapters in this volume present these aspects for several classes of organisms. Any developmental biologist could easily conceive of some of the challenges presented by the developmental program that are conserved among different classes of organisms. A few examples follow: (1) Oocytes, eggs, and blastomeres of the early embryo contain an unusually large cytoplasmic volume compared to that of somatic cells. This can present special problems in intracellular communication when the cell must undergo a coordinated change, such as a progression through the cell cycle in the case of blastomeres or a response of the egg to the penetrating sperm. (2) The zygote is developmentally totipotent through the elaboration of its developmental program. No somatic animal cell is developmentally totipotent. (3) Fertilization requires cell fusion (i.e., between the egg and the sperm). In most species a mechanism exists to permit entry of only one sperm. Typically, somatic cells do not fuse (this statement excludes the terminal expression of a developmental program in cell types such as muscle). Even when somatic cells are induced to fuse through experimental manipulation, for example, to produce a hybridoma, a totipotent zygote is not produced. (4) Fertilization requires the restoration of ploidy through the unification of two different populations of chromosomes without the loss of a chromosome or part of a chromosome. This event occurs as pronuclear fusion or the unification of the two chromosomal populations during M phase of the cell cycle. Fusion of somatic cells through experimental manipulations usually results in the loss of one or more chromosomes from the heterokaryon. (5) Eggs and blastomeres of embryos exhibit unusual cell cycle regulation (i.e., specific cell cycle arrest points for eggs and modified cell cycles for blastomeres). Typically, a somatic cell is either progressing through the cell cycle (i.e., a stem cell) as is the case for skin epithelial cells, or it is arrested late in Gap_1 of the cell cycle in a state referred to Gap_0. In the latter case, the cell cycle arrest point differs from that of the egg, as does the mechanism of recusing the cell from Gap (e.g., the cell cycle arrest in the egg is released by fusion with the sperm). In the former case where the stem cell is progressing through the cell cycle, the amount of time spent in Gap_1, Gap_2, and the synthesis phase (DNA synthesis) for the stem cell is significantly longer than the times exhibited by blastomeres of the embryo.

Several of the conserved modifications of cytoskeletal function that have been identified in eggs, zygotes, and blastomeres address some of these developmental challenges. Some examples follow: (1) To allow for rapid, synchronized changes in large cells, such as the egg, cytoplasmic signal transduction mechanisms are responsible for the rapid remodeling events (of all parts of the egg including the cytoskeleton) at the developmental transition that converts the egg into the zygote. (2) Where examined, microtubule arrays appear to participate in the approximation of male and female pronuclei within the egg/zygote cytoplasm, permitting syngamy to occur. (3) Eggs contain extensive, cortical cytoskeletal domains that remodel as a result of fertilization and perhaps permit exocytosis of cortical granules, which provides the long-term block to polyspermy. (4) In those cases investigated, the cortical cytoskeletal domain has been shown to be associated (in some cases directly and in other cases indirectly) with components capable of influencing the developmental fate of subsequently formed blastomeres. (5) Developmental transitions are accompanied by a remodeling of both the cortical and the internal cytoskeletal components, and in those cases investigated the cytoskeletal remodeling has been shown to be regulated by cytoplasmic signal transduction mechanisms.

The occurrences outlined in the previous paragraph, and other developmental roles for the cytoskeleton, are presented in more detail in this volume. The studies in this volume demonstrate a central role for the cytoskeleton in development. Moreover, these studies demonstrate that the cytoskeleton in eggs, zygotes, and blastomeres of the embryo is a remarkably malleable structure. Even more remarkable is that the three main filament networks (i.e., networks composed of actin filaments, microtubules, and intermediate filaments) are capable of this vast array of specialized activities. To date, no new filament network has been identified in association with these special cellular functions during development, although the existing cytoskeletal networks have been identified in highly unusual aggregations and forms.

The cytoskeleton exhibits functions and activities in these specialized cells that, to date, have no parallels in somatic cells. Yet all somatic cells ultimately arise from the penetration of an egg by a sperm. Could it be that these specialized activities of the cytoskeleton are involved only during development and that once a somatic cell is formed the cytoskeleton no longer can exhibit these special roles? Or could it be that our knowledge of cytoskeletal function in somatic cells is skewed by the cell types available to cell biologists for study? Let us look and wonder together.

Preface to Section I: Nonchordates

The first section of this volume focuses on cytoskeletal mechanisms involved with early development in nonchordates. The chapters listed in parentheses denote chapters in which comparable cytoskeletal mechanisms also have been reported.

In Chapter 1 (by William Eckberg and Winston Anderson) evidence is presented to demonstrate cytoskeletal involvement with cell shape changes that accompany fertilization, and also both the localization of mRNA and the redistribution of mRNAs into specific patterns within the zygote (Chapters 3, 5, 8, and 12–14). In addition, this chapter demonstrates that the egg contains an extensive cortical cytoskeleton (referred to as cortical cytoskeletal domain), while the cytoskeleton in the egg interior is highly reduced. This chapter also considers mechanisms for regulation of cytoskeletal organization by the action of signal transduction events, specifically the level of intracellular free calcium and the action of protein kinase C (Chapters 3, 4, 7, 9, 11, 13, and 14).

Chapter 2 (by Evelyn Houliston and co-workers) examines cytoskeletal mechanisms in ctenophore development. Again, the cytoskeleton is involved in physical changes in cell shape. In addition, evidence is presented which suggests involvement of microtubules in promotion of pronuclear juxtaposition (Chapters 3 and 10) as well as in cortical rotation (similar to that found during insect oogenesis; Chapter 5) and postfertilization development (Chapters 8, 12, and 13) in some chordates that promotes axis formation. Moreover, the cytoskeleton is involved in positioning of morphogenetic determinants (Chapters 1, 5, and 7).

Two chapters examine cytoskeletal mechanisms in sea urchin eggs. The microtubule network in sea urchin eggs is the focus of Chapter 3 (by Kathy Suprenant and Melissa Foltz). Here the authors present a role for microtubules in pronuclear movement (Chapters 2 and 10) and discuss a special cortical cytoskeletal domain composed of microtubules. They review the data indicating that the dynamics of microtubule assembly is regulated by signal transducers such as kinases and phosphatases (Chapters 1, 4, 5, 7, 9, 11, and 14). Finally, they present evidence that ribosomes attach to the cytoskeleton (whereas in other systems reports demonstrate mRNA attached to the cytoskeleton; Chapters 1, 3, 5, 8, and 12–14) and that this may have a role in translational regulation. In Chapter 4 (by Bonder and

1

Fishkind) the actin cytoskeleton of the sea urchin egg is examined. Here evidence is presented to demonstrate that the cortical actin cytoskeleton undergoes extensive remodeling at the time of fertilization (Chapters 1, 2, 7, 10, 12, and 13), that the egg undergoes a cortical contraction (Chapters 1, 2, 7, and 13), and that signal transduction events regulate the reorganization of the cytoskeleton (Chapters 3, 7, 9, 11, and 14).

There are two chapters on insect development, specifically *Drosophila*. Both chapters highlight the ability to conduct experiments which manipulate the genetics of the system to reveal functional roles for the filament systems as well as for a variety of cytoskeleton-associated proteins. The first, Chapter 5 (by Nancy Jo Pokrywka), considers the involvement of the cytoskeleton during oogenesis. This chapter reviews the evidence for very distinct roles for the actin filament network and the microtubule network in translocating material between the nurse cells and the oocyte. It hints at a role for the action of kinases to regulate the microtubule network (Chapters 1, 3, 4, 7, 9, 11, and 14) and demonstrates a role for microtubules in the positioning of specific RNAs and in the repositioning of the RNA as development ensues (Chapters 1, 3, 8, and 12–14). However, here the evidence suggests a two-phase process is involved in positioning of RNA (i.e., an initial localization followed by stabilization; Chapter 12). Evidence also is considered that suggests that microtubules may be involved in the rotation of the cortical cytoplasm (referred to as ooplasmic streaming), which is somewhat similar to the cortical rotations described in other systems (Chapters 2, 8, and 12); however, here the streaming occurs prior to fertilization, whereas in the other systems it occurs after fertilization. Chapter 6 (by Kathryn Miller) focuses on the role of the actin cytoskeleton in postfertilization development. In this chapter, data are considered that present a role for actin in establishing cytoplasmic domains surrounding each embryonic nucleus. The development of this embryo as a syncytium for the first nine nuclear division cycles presents special problems in regulating chromosomal separation. The actin network serves as a highly regulated mechanism (over space and time) to isolate the genetic material into distinct cytoplasmic islands. This chapter presents a variety of mutants that are certain to provide an understanding of specific functions for the cytoskeleton.

The final chapter in this section, Chapter 7 (by Takashi Shimizu), details cytoskeletal mechanisms in early development of the freshwater oligochaete, *Tubifex*. Here evidence is presented which suggests that the actin network is involved in the shape change of the egg, and that the actin network forms a cortical cytoskeletal domain. This cortical–actin domain contracts into distinct subdomains (Chapters 1, 2, 4, 7, 12, and 13) and influences the movement of morphogenetic determinants (Chapters 1, 2, 5, and 7). A role for a specific kinase in the remodeling of the cytoskeletal network is considered (Chapters 1, 3–5, 7, 9, 11, 13, and 14). In addition, a role of centrosomal positioning in unequal cleavage divisions is considered.

Section I

Cytoskeletal Mechanisms in Nonchordate Development

1

Cytoskeleton, Cellular Signals, and Cytoplasmic Localization in *Chaetopterus* Embryos

William R. Eckberg and Winston A. Anderson
Department of Biology
Howard University
Washington, DC 20059
and
Marine Biological Laboratory
Woods Hole, Massachusetts 02543

I. Introduction

A. *Chaetopterus* as a Model System in Developmental Biology

The use of *Chaetopterus* as a system for study in developmental biology evidently began with E. B. Wilson (1882), who briefly described its early development along with that of several other annelids. Experimental analysis of the development of this organism began with Jacques Loeb (1901) in a paper in which he misdescribed "differentiation without cleavage" (Lillie, 1902) resulting from K^+ activation as true parthenogenesis. Other early studies on cytoplasmic localization and embryo organization in relation to development were performed by such legendary developmental and cellular biologists as F. R. Lillie (1906, 1909), E. B. Wilson (1929, 1930), T. H. Morgan (1910, 1937, 1938, 1939; Whitaker and Morgan, 1930; Morgan and Tyler, 1938), A. Tyler (Titlebaum, 1928; Tyler, 1930), E. B. Harvey (1939), J. Brachet (1937, 1938, and more recent work cited under Section II,C), and J. Pasteels (1935, 1950).

Experimental studies of the development of *Chaetopterus* have emphasized three areas: the regulation of the cell cycle (germinal vesicle breakdown, GVBD), mechanisms of fertilization and egg activation, and the effects of egg organization on development. Although this chapter will stress the last area, we will briefly mention the advantages of *Chaetopterus* oocytes and eggs for the first two areas. The primary advantages for the study of GVBD are that large numbers of oocytes ($>10^6$ cells or 2 ml) can be obtained from a female at one time, and that all can be induced to undergo GVBD synchronously in response to either their natural trigger (an unknown trace component in seawater) or to certain cellular agonists/antagonists of known biological activity. A further advantage is that the oocytes then arrest at metaphase I of meiosis until fertilized or artificially activated. In other words, the cells can be induced to undergo the G_2/M phase transition at will without continuing to cycle. Furthermore, these oocytes can be easily labeled with isotopic markers. The availability of large numbers of synchronized, easily labeled eggs is also an important consideration in studies of fertilization and egg organization. The unique advantages of *Chaetopterus* in studies of fertilization are that the fertilizing sperm interact with morphologically definable structures on the egg surface (Anderson and Eckberg, 1983) and that the egg is at least as metabolically active before fertilization as it is after. In fact, the unfertilized egg uses much more O_2 than does the fertilized (Whitaker, 1933) or artifically activated (Brachet, 1938) egg. The fact that the physiology of the initiation of development in *Chaetopterus* eggs differs from that of sea urchins should make them an object of more detailed study.

The unique feature which makes *Chaetopterus* of particular interest for studies of egg and embryo organization in development is the ability of the artifically activated or fertilized egg to undergo differentiation without cleavage (Lillie, 1902). Artificial activation can be induced by excess KCl (Lillie, 1902; Brachet, 1937). Differentiation without cleavage also occurs in fertilized eggs subjected to temporary cleavage inhibition by treatment with cytochalasin B (Eckberg, 1981a) and in polyspermic embryos. This interesting phenomenon will be discussed in greater detail under Section II,C.

B. Early Development

Fertilized eggs undergo typical spiral cleavage. The first five cleavages are synchronous; thereafter, cleavage becomes highly asynchronous. While the cell lineage has not been followed as completely in this organism as that in some other spiralians, there is no reason to expect that it differs significantly in *Chaetopterus* (*cf.* Henry and Martindale, 1987).

The first cleavage is unequal due in part to the presence of a small polar lobe and more importantly to asymmetric placement of the metaphase spindle. In some embryos, polar lobes do not form, but the cleavage is still unequal and subsequent development is normal. Polar lobes typically have substantial morphogenetic significance as shown by the fact that removal of the polar lobe results in severely deficient embryos, although this is less true for *Chaetoperus* than for most other lobe-bearing spiralians that have been studied.

Early investigations suggested that unequal cleavage distributes morphogenetic substances unequally to the two blastomeres (Titlebaum, 1928; Tyler, 1930). Equalization of the first cleavage by compression resulted in the development of embryos with duplication of many structures. In blastomere isolation studies of *Chaetopterus*, as in similar studies on other spiralian species, isolated AB cells formed a "swimming mass, mainly of ectodermal cells," whereas isolated CD cells developed into structures which "outwardly resembled" early trochophores and occasionally formed advanced trochophores. From this description, however, the possibility that many (or all) of these "embryos" had actually undergone differentiation without cleavage cannot be excluded, because swimming masses or structures which outwardly resemble trochophores can develop in this species in the complete absence of cell division.

In fact, the results of more recent studies suggest that *Chaetopterus* embryos differ from those of other lobe-bearing spiralians in that removal of the polar lobe has only a marginal effect on larval morphogenesis (Henry, 1986). In this study, lobeless embryos cleaved normally and formed larvae which were normal except that they lacked the ability to produce bioluminescence. Even AB and CD blastomeres developed similarly to each other and to control embryos in most respects. Further studies on equalized cleavage in these embryos (Henry and Martindale, 1987) provided a wide range of results from complete symmetric twinning through incomplete twinning to normal embryos. Normal embryos accounted for about half of the cases studied. The finding that many equally cleaving embyos developed normally is unusual for spiralia, but consistent with the results of blastomere isolation experiments. In agreement with these results, other studies in which the first cell division was equalized by another method resulted in development of embryos that were normal with no apparent doubling of posterior structures (Eckberg, 1981a). In this study, cytochalasin B (CB) was used to block first cleavage. The drug was then washed out of the embryos, which then proceded to cleave into either two or four equal-sized cells at the time controls cleaved into four cells. These last results must be interpreted with some caution, however, as the embryos were not followed to advanced stages, and CB has dramatic effects on embryonic cytoskeleton organization (see below).

II. Cytoplasmic Localization

A. Localizing Movements in Living Eggs

Chaetopterus eggs undergo a series of dramatic changes in cell shape prior to first cleavage (Fig. 1). The spherical zygote flattens slightly along the animal/vegetal axis coincidently with the formation of each polar body (Fig. 1A). About 15 min after second polar body formation, the egg becomes constricted at the animal pole into a "pear" shape (Fig. 1B). Within 5 min the elongation disappears and the cell becomes flattened at the animal pole. Coincidently with this, a constriction occurs near the vegetal pole

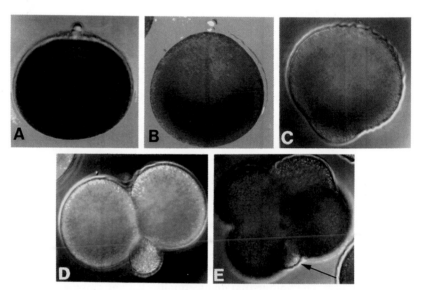

Fig. 1 Cell shape changes in living *Chaetopterus* embryos prior to first cleavage and during formation of the polar lobes. (A) Embryo flattened during second polar body formation, (B) embryo elongated ("pear-shaped") during mitosis, (C) initial formation of the first polar lobe during anaphase, (D) definitive polar lobe during first cleavage (the CD cell is on the left), and (E) second polar lobe (arrow) during second cleavage. ×320.

forming the small polar lobe (Fig. 1C). The polar lobe becomes more conspicuous as the egg constricts during cleavage (Fig. 1D). After cleavage, the polar lobe is resorbed into the CD cell. Often, a polar lobe also forms at second cleavage (Fig. 1E).

B. Localization as Seen in Cytological Preparations

Cytological preparations of early *Chaetopterus* development have been examined by both light (Lillie, 1906; Eckberg, 1981b; Jeffery and Wilson, 1983) and electron microscopy (Eckberg, 1981b). In most respects, the results of more recent studies confirm those of Lillie. Lillie's terminology recognizes ectoplasmic granules (now called cortical granules), two types of endoplasmic granules—endoplasm "a" and "b" in Lillie's terminology (yolk and lipid by electron microscopy)—but which appear by electron microscopy to colocalize throughout development, and a third, agranular, endoplasmic region which Lillie called the "residual substance" of the germinal vesicle, but, in our opinion, it is a region defined by the astral microtubules and probably glycogen aggregates (extracted by our fixation procedures) and called the "spongy" endoplasm in our studies. The cortex is also the site of the granular bodies or nuage (Eckberg, 1981b; Jeffery and Wilson, 1983; Jeffery, 1985) and polysaccharide granules (Jeffery, 1985). Importantly, the cortex contains 90–95% of the mRNA of the cell (Jeffery and Wilson, 1983). This is true for all polyadenylated RNAs, actin mRNA, and histone mRNA. It has been suggested that the granular bodies are the sites of mRNA storage (Eckberg, 1981b; Jeffery and Wilson, 1983; Jeffery, 1985), but there is no direct evidence for this. The mRNA appears to be restricted to the cortex because of its interaction with the cytoskeleton (see Section III). Interestingly, when the oocyte is centrifuged, the entire cortex, including the mRNA, dislodges as a unit and accumulates at the centrifugal pole (Jeffery, 1985).

1. Oocyte Maturation

Yolk granules of the primary oocyte appear to be distributed throughout the endoplasm (Fig. 2). The cortical granules are concentrated in about two-thirds of the egg cortex, although electron microscopy reveals that they are not completely absent from the opposite pole. Lillie (1906) claims that they are concentrated in the upper two-thirds of the cortex on the basis that he saw them moving toward their subsequent location (the vegetal pole) just after GVBD. The evidence he cites is a series of "waves" of the vitelline envelope. We have also observed these waves (or wrinklings) and have found them to be due to the extension and retraction of microvilli

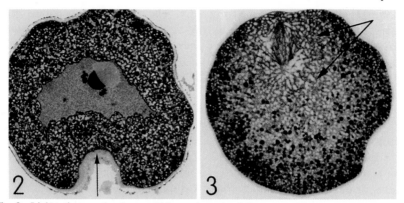

Fig. 2 Light micrograph of a semithin section of a fully grown *Chaetopterus* oocyte. The oocyte was sectioned along the animal/vegetal axis. Cortical granules are concentrated at the upper (animal) pole and rare at the vegetal pole (arrow). ×500.

Fig. 3 Light micrograph of a semithin section of a metaphase-arrested *Chaetopterus* oocyte. Note the exclusion of large cytoplasmic organelles from the spindle and aster (arrows). Also note the absence of cortical granules at the animal pole. ×500.

(Eckberg and Kang, 1981). By contrast, Jeffery *et al.* (1986) reported evidence that the cortical mRNA distribution is established early in vitellogenesis, suggesting that either the organization of some components of the cortex may be determined independently from that of others or that Lillie's observations were incorrect. The possible rearrangement of the cortex at oocyte maturation is of considerable significance to our understanding of the mechanisms by which embryonic polarity is established and therefore should be further investigated. The ability to stain components of the cortex with acridine orange (Swalla *et al.*, 1985) indicates the possibility of following its redistribution throughout oocyte maturation (and later stages as well) in living eggs.

All of the large granular components of the cortex and endoplasm are excluded from the spindle region (Fig. 3). Cortical granules are excluded from the region where the spindle anchors to the pole; all large endoplasmic granules are excluded from the spindle region of the endoplasm, perforce coming to lie in a crescent thickest at the vegetal pole. Smaller organelles, such as mitochondria, however, are not excluded from the spindle region (Eckberg, 1981b).

The cellular basis for spindle migration and cortical attachment has been investigated (Lutz *et al.*, 1988). A microneedle was used to displace the spindle from its cortical location. When the spindle was pulled toward the center of the egg, the cortex and plasma membrane above it first remained attached to the spindle and were pulled inward, forming a deep dimple which, as the spindle was pulled further, eventually detached from the spindle and recoiled to its original position. This clearly demonstrates a mechanical linkage between the spindle and the overlying cortex. The spindle failed to attach to other positions of the cortex, even when held there fore several minutes, demonstrating that the attachment point is unique. It is unknown whether this unique site is predetermined in the primary oocyte or whether it develops upon GVBD. However, the results of cytological observations (Lillie, 1906) and the fact that holding the spindle near the cortex at another location does not result in attachment suggest that it is predetermined. Successive manipulations of the same spindle demonstrate that the linkage can be broken and reformed repeatedly. When spindles were released from the microneedle, they returned to the same spot on the cortex after a pause. As they approached the cortex, they accelerated. Velocities in excess of 1 μm/sec could be achieved by the spindles. This is far in excess of the rate of spindle migration in anaphase and is of the same order of magnitude as the transport speed of cytoplasmic particles in contact with astral microtubules (Rebhun, 1972). The attractive influence of the attachment site extends up to 35 μm from the cortex. The nature of this influence is unclear. Electron microscopy of the cortex confirmed that the centriole pair is positioned approximately 3 μm from the egg surface, but revealed no distinctive or unusual structure that could be correllated with the point of spindle attachment (W. Eckberg and W. Anderson, unpublished results). Nor has it been established conclusively that there is a relationship between the unique spindle attachment site and any specific cortical domain in the prophase-arrested oocyte. If such a relationship were demonstrated, it would conclusively show that this polarity is, in fact, established in the egg prior to release from the ovary and would provide independent confirmation for Lillie's hypothesis that a preexisting polarity of the egg has developmental significance. Centrosome positions are also predetermined in cleaving embryos of *Caenorhabditis elegans* (Hyman, 1989).

2. Fertilization and Cleavage

The arrangement of the cytoplasm at metaphase I persists through the meiotic divisions. The male and female pronuclei form and closely appose one another at the time the egg becomes pear shaped. It is unclear whether they fuse or whether the chromosomes intermingle after nuclear breakdown. The fact that we have never obtained unambiguous evidence for pronuclear fusion in cytological preparations suggests that the latter possibility is most likely. As the pronuclei migrate to the center of the egg and the spindle forms, more and more endoplasmic granules are seen closer to the animal pole. These relative positions are maintained as polar lobe formation and cleavage occur, except that the cortical granules tend to be sparse in the regions where constrictions are occurring (the contractile ring and the attachment of the polar lobe to the presumptive CD cell). In contrast to Lillie's (1906) report that the polar lobe is almost entirely ectoplasmic, it contains a considerable amount of endoplasm. In fact, its cytoplasm is indistinguishable from that of the adjacent part of the CD or even the AB cell (Eckberg, 1981b).

During mitosis, a significant fraction of the mRNA leaves the cortex and associates with the periphery of the spindle (Fig. 4). At cleavage the mRNA has returned to the cortex and is partially restricted to an upper, animal domain and a lower, vegetal domain, though this restriction is not absolute (Fig. 5). The vegetal domain is centered approximately on and a substantial portion of it is included in the polar lobe, making the polar lobe especially rich in mRNA (Jeffery and Wilson, 1983). An initial investigation of mRNA distribution in the cortex did not reveal any restriction of mRNA species to different subdomains of the cortex (Jeffery and Wilson, 1983), *i.e.,* actin, histone, and all poly(A)$^+$ mRNAs were distributed similarly between the animal and vegetal domains. However, the studies performed examined only a very limited set of messages. Furthermore, Fig. 5 clearly shows that the vegetal cortical mRNA domain is not exclusively restricted to the polar lobe. In fact, it extends into the AB cell, suggesting an obvious explanation for the unusually large extent of differentiation of the isolated AB cells. Perhaps the morphogenetic determinants normally thought associated with the polar lobe are incompletely restricted to the CD cell in embryos like *Chaetopterus,* which possess small polar lobes and relatively large AB cells. The studies performed to date are not sufficient to refute the hypothesis that specific mRNAs are localized to distinct regions of the cortex, and further investigation of this possibility should be undertaken.

Subsequent cleavages follow the typical spiral pattern. The third cleavage produces a set of relatively large micromeres, and by the sixteen-cell stage and thereafter each cell of the embryo is polarized with a layer of cortical granules apically and a dense layer of endoplasmic granules basally. By the "sixty-four cell" stage (because of asynchrony, *Chaetopterus* embryos never have a typical sixty-four cell stage), the endodermal and ectodermal cell lineages are completely separated. Cortical granules are lacking in cells of the endodermal lineage. In the trochophore, yolk is absent from the ectoderm.

Thus, although the unusually wide extent of differentiation of AB cells and lobeless embryos suggests that *Chaetopterus* embryos might not be ideal candidates for studies of the localization of morphogenetic determinants to different blastomere lineages, there is very clear evidence of other kinds that suggests substances are distributed nonrandomly to blastomeres during

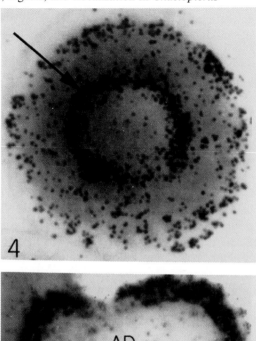

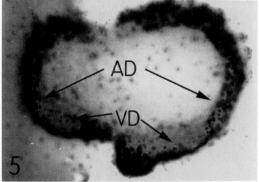

Fig. 4 *In situ* hybridization showing poly(A)$^+$ RNA localization during mitosis in a *Chaetopterus* zygote. Note that much of the mRNA is associated with the periphery of the spindle (arrow). ×600. Relabeled from Jeffery and Wilson (1983). Used by permission of Dr. William R. Jeffery and The Company of Biologists, Ltd.

Fig. 5 *In situ* hybridization showing poly(A)$^+$ RNA localization during cleavage of a *Chaetopterus* zygote. The mRNA is again associated with the cortex and is now restricted to an animal domain (AD) and a vegetal domain (VD) that includes the polar lobe. ×550. Arrows indicate the limits of the two domains. Relabeled from Jeffery and Wilson (1983). Used by permission of Dr. William R. Jeffery and The Company of Biologists, Ltd.

early development. The reorganization of the cytoplasm is a consistent process throughout early development. Localizing movements which occur during cleavage result in the distribution of ooplasmic substances (mRNA, cortical granules, and yolk) to specific regions of the embryo. As will be seen, the egg architecture is such that a comparable redistribution can occur in the absence of cell division.

C. Differentiation without Cleavage

Chaetopterus embryos which have been activated but do not cleave undergo differentiation without cleavage. This phenomenon can be induced either by eliciting parthenogenetic development—for example, by using seawater containing ≥ 60 mM K$^+$ (Loeb, 1901; Lillie, 1902), because parthenogenetic embryos of this genus never cleave—or by blocking zygotic cleavages with cytochalasin B (Eckberg, 1981a). Polyspermic embryos also undergo differentiation without cleavage, but it is impossible to achieve 100% polyspermic fertilization without resorting to other manipulations of the egg which might possibly have effects of their own on early development and differentiation without cleavage. Thus, parthenogenetic activation by elevated extracellular K$^+$ or inhibition of cleavage by cytochalasin B are the prefered methods for eliciting differentiation without cleavage.

The phenomenon has been investigated by light (Lillie, 1902, 1906; Brachet, 1937) and electron (Eckberg and Kang, 1981) microscopy. It produces unicellular, ciliated, swimming "pseudolarvae." Because their yolky endoplasm achieves an internal location (equivalent to that of the endoderm), the pseudolarvae bear a sufficient similarity to trochophore larvae that they were initially misdescribed as such (Loeb, 1901). While presumably unusual in the animal kingdom, differentiation without cleavage has also been observed in other annelids (Treadwell, 1902; Scott, 1906).

All of the localizing movements up through polar lobe formation occur normally in embryos undergoing differentiation without cleavage. Furrows may or may not form, but if they do, they regress. Coincidently with cleavages in control eggs, embryos undergoing differentiation without cleavage exhibit very strong ameboid contractions. This process is called "pseudocleavage." During pseudocleavage, DNA replication continues and either the embryo may develop a single, very large and polyploid nucleus or at least some nuclear divisions may occur with the production of a multinucleate embryo. At the same time, various cytoplasmic regions of the unfertilized egg begin to rearrange themselves. The first rearrangement observed is the segregation of the spongy layer from the granular endoplasm. Generally, the granular endoplasm will be found in the vegetal half of the "embryo" at this time. This polarization is equivalent to that in each blastomere of the sixteen-cell embryo. Subsequently, the spongy endoplasm "overflows" the granular endoplasm and comes to encircle it. This process is called "unicellular gastrulation" and results in a cell in which the dense granular endoplasm is completely surrounded by the translucent spongy endoplasm.

After unicellular gastrulation, the cell frequently differentiates cilia which may cover the entire surface. Ciliation may also be coincident with that in fertilized controls. The ciliated pseudolarva is, as noted previously, superficially similar to the trochophore. This similarity can be clearly seen in Fig. 6. The trochophore larva (Fig. 6A) exhibits a superficial layer of ectodermal nuclei embedded in the agranular endoplasm. The granular endoplasm is now restricted to the endoderm, which also has scattered nuclei. The embryo which has undergone differentiation without cleavage in response to cytochalasin B treatment (Fig. 6B) also has a ring of nuclei embedded in the agranular endoplasm, surrounding the granular endopolasm. The only obvious difference visible in sectioned material between the normal larva and that which has undergone differentiation without cleavage is the lack of nuclei in the endoplasm of the latter and the lack of organization to the cilia in the latter (not visible in these sections).

It is important to note, however, that the embryonic organ systems are only simulated by the aggregation of cytoplasmic granules characteristic of the organ system. Thus, no gut is formed, but aggregation of the granular endoplasm in the center of the pseudolarva gives the superficial appearance of a gut. The structure of these pseudolarvae is quite variable, as many of them fuse with one another and some fragment. That this fusion is the result of contact during cell motility is shown by the observation that it occurs much less frequently in embryos undergoing differentiation without cleavage in very dilute culture. Excessive embryo fusion also seems to inhibit ciliation because the larger aggregates do not develop cilia.

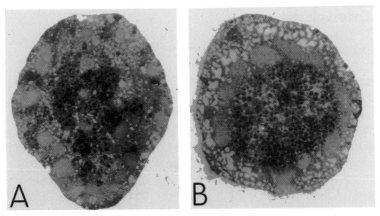

Fig. 6 Light micrographs of semithin sections through a *Chaetopterus* trochophore larva (A) and an embryo of the same age which underwent differentiation without cleavage in response to cytochalasin B treatment (B). The similarity is remarkable. (A) Longitudinal section from Eckberg (1981b); (B) from Eckberg and Kang (1981). Plane cannot be determined because such embryos lack consistent symmetry. ×475.

The fundamental question in these studiesis: does differentiation without cleavage actually involve any form of molecular differentiation, or is it strictly a rearrangement of preexisting ooplasmic components? This issue has had surprisingly little investigation, but the limited evidence suggests that it does involve at least some changes in protein components and thus some changes in gene expression. Hagiwara and Miyasaki (1977) reported that unfertilized eggs have a Ca^{2+}-dependent action potential, but that an outward K^+-dependent rectification increases dramatically during pseudo-cleavage. Furthermore, in the later pseudolarva, the action potential becomes dependent on Na^+ as well as Ca^{2+}. These results demonstrate that new K^+ and Na^+ channels appear during differentiation without cleavage at different times and thereby imply that some changes in gene expression accompany differentiation without cleavage. The only other published attempt to investigate gene expression during differentiation without cleavage involved the use of protein and RNA synthesis inhibitors (Brachet *et al.,* 1980). Neither inhibited pseudocleavage, but both inhibited segregation, unicellular gastrulation, and ciliation. Taken together, these results indicate that differentiation without cleavage is likely to involve and be dependent upon gene expression patterns which parallel those in normal embryos. More comprehensive studies of gene expression during differentiation without cleavage should be undertaken. They would be of value in identifying patterns of gene expression that do and do not depend on the cell compartmentation which occurs during cleavage.

III. Cytoskeleton

A. Isolation and Structural Analysis

Chaetopterus cytoskeletons are isolated by procedures similar to those for other cells, except that the detergent used for extraction is apparently more critical. Nonidet P-40 extraction yields typical cytoskeletons, but extraction with Triton X-100 results in degradation of the cytoskeleton (Eckberg and Langford, 1983; Jeffery, 1985; Swalla *et al.,* 1985). Cytoskeletal preparations contain approximately one-fourth of the cellular protein and rRNA and two-thirds to three-fourths of the poly(A) and actin and histone mRNA of the egg (Jeffery, 1985). Thus, as in other cells, the bulk of the mRNA is cytoskeleton associated.

In *Chaetopterus* eggs, the cytoskeleton is exclusively cortical and therefore will be referred to below as the cortical cytoskeletal domain (CCD). Its structure and composition have been examined by light and fluorescence and scanning and transmission electron microscopy (TEM). When examined by TEM (Fig. 7), the outer shell remnants of the cortical granules as well as partially extracted yolk granules and apparently intact pigment granules can be identified. Of particular interest is the large concentration of polysome-like aggregates called granular bodies (Eckberg, 1981b) or nuage (Jeffery, 1985). These are frequently associated with filaments (Fig. 8) and are likely to be the cytoplasmic organelles associated with the stored mRNA. When seen by light microscopy or scanning electron microscopy (Fig. 9), the isolated CCD is approximately 10–15 μm thick. The outer surface is relatively smooth, presumably due to the former presence of the solubilized plasma membrane. The inner surface is composed of numerous yolk-sized spherules and smaller granules embedded in a filamentous martix.

In contrast to those of unfertilized eggs (Fig. 10), cytoskeletons of cleaving zygotes are not entirely cortical. Rather, a substantial fraction of the CCD is relocalized into the endoplasm at the periphery of the spindle (Fig. 11). Previous studies of a variety of eggs have shown that various cytoplasmic particles which come in contact with astral microtubules can be transported toward the centrosome at as much as 5 μm/sec (reviewed by Rebhun, 1972). A relocalization of part of the CCD with associated cortical granules and mRNA molecules has been previously reported in intact *Chaetopterus* eggs (Jeffery and Wilson, 1983). The results reported here confirm such relocal-

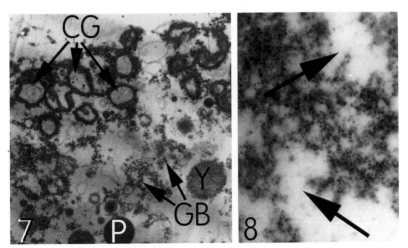

Fig. 7 Low-magnification transmission electron micrograph of a thin section of a CCD fragment. CG, partially extracted cortical granules; P, pigment granule; Y, partially solubilized yolk granule; GB, granular body, the putative mRNA storage organelle. ×5000.

Fig. 8 High-magnification transmission electron micrograph showing association of a granular body with cytoskeletal filaments (arrows). ×50,000.

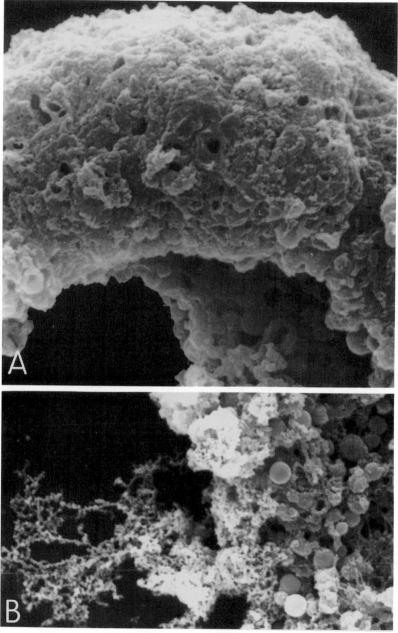

Fig. 9 Scanning electron microscope image of a CCD fragment from an unfertilized *Chaetop-terus* egg. (A) View of the outer surface. (B) View of the inner surface. The outer surface is relatively smooth, reflecting its association with the (now solubilized) plasma membrane. The inner surface consists of numerous cytoplasmic organelles embedded in a filamentous matrix. ×15,000.

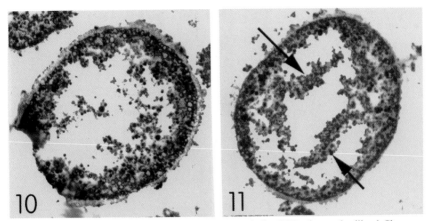

Fig. 10 Light micrograph of a semithin section through the CCD of an unfertilized *Chaetopterus* egg. The CCD is entirely cortical. ×450.

Fig. 11 Light micrograph of a semithin section through the CCD of a *Chaetopterus* zygote in mitosis. A portion of the CCD has detached from the cortex and associated with the periphery of the spindle (arrows). ×450.

ization and demonstrate that the phenomenon does not involve detachment of the CCD as a whole. They also raise the possibility that the CCD is really two or more separate structural frameworks, at least one of which can be rearranged in response to the assembly and disassembly of the mitotic spindles, and another of which is more permanently associated with the cortex. This possibility will be discussed in more detail under Section III,B.

The polypeptide composition of isolated CCDs has also been examined (Fig. 12). While the overall polypeptide composition of CCD is qualitatively similar to that of whole eggs as expected based on the presence of at least remnants of a wide spectrum of cytoplasmic organelles in the CCD, several prominent polypeptides are quantitatively depleted from the CCDs, especially a set of 26–29 kDa believed to be yolk proteins.

By contrast, certain other polypeptides are enriched in the CCD, especially a polypeptide of approximately 53 kDa. A 43-kDa polypeptide is also highly enriched, especially in the CCDs of cleaving embryos. Based on immunoblot analysis using a monoclonal antibody which recognizes cytoplasmic actin from a variety of eukaryotic species (Sigma Chemical Co.), this 43-kDa polypeptide evidently is actin (Fig. 13). There is some evidence that the cortical organization of the CCD, but not its structural integrity, depends on microfilaments. When the CCD is isolated from cytochalasin B-treated embryos, the CCD is no longer restricted to the cortex

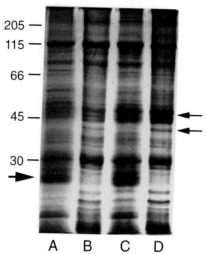

Fig. 12 SDS–PAGE analysis of polypeptide composition of *Chaetopterus* egg and cleaving embryo CCDs. (A) Unfertilized egg lysate, (B) unfertilized egg CCD, (C) cleaving embryo lysate, and (D) cleaving embryo CCD. The mobilities of molecular mass standards are given (in kDa) on the left. The large arrow on the left identifies the mobility of the yolk proteins which are depleted from the CCDs. The smaller arrows on the right indicate the 43- and 53-kDa polypeptides which are selectively enriched in the CCDs.

(Fig. 14), indicating that depletion of microfilaments does not result in the disappearance of the cytoskeletal matrix, but does prevent its restriction to the cortex.

Tubulin can also be detected in CCDs of unfertilized eggs by a monoclonal antibody from Sigma (Fig. 13); however, it apparently is not essential for the integrity of the CCD. Isolation of CCDs from colchicine-treated embryos yields CCDs which are indistinguishable from the controls (Fig. 15).

Immunoblotting using a monoclonal antibody to mammalian vimentin (Chemicon) reveals the presence of this intermediate filament protein in CCDs (Fig. 12). Polyclonal antibodies to sequences in the conserved domain of intermediate filaments (gift of Dr. Mary Lopez, Persptive Biosystems, Inc.) also reveal the presence of vimentin (and possibly other intermediate filament proteins) in CCDs (data not shown). The identities of the polypeptides other than vimentin recognized by these antibodies have not been further studied. The lack of availability of antibiotics which specifically disrupt intermediate filaments has made direct analysis of their role in CCD organization impossible. However, it seems likely that mRNA is localized to the CCD through interaction of the intermediate filament network with the 3′ end of the mRNA (Jeffery *et al.*, 1986). The experimental evidence indicating that it is the intermediate filament network of the

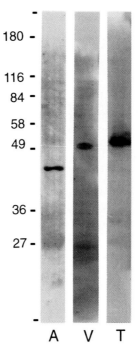

Fig. 13 Immunoblot analysis of cytoskeletal proteins in *Chaetopterus* egg CCDs. A, Monoclonal antiactin (Sigma); V, monoclonal antivimentin (Chemicon); and T, monoclonal antitubulin (Sigma). Secondary antibodies were alkaline phosphatase conjugated; the detection reagent was stabilized BCIP/NBT (Promega). The mobilities of molecular mass standards (in kDa) are given on the left.

CCD that is responsible for mRNA localization results from treatment of CCDs with high salt concentrations or microtubule or microfilament inhibitors. These treatments have no effect on mRNA binding to the CCD and therefore effectively exclude microfilaments and microtubules as being essential for mRNA binding. The evidence that mRNA interacts with intermediate filaments through its 3′ end results from experiments in which poly(U) was bound to the cytoskeletal mRNA, followed by RNase digestion. This treatment did not release the poly(U) from the CCD, indicating that the poly(A), which is resistant to RNase by reason of its hybridization to poly(U), is very closely associated with the CCD. These results suggest that the mRNA binding to the CCD is mediated by the poly(A) tail. However, it seems unlikely that such an interaction could target specific mRNAs to cytoskeletal domains. This result could also be consistent with localization being due to other sequences in the 3′ untranslated region of the mRNA as is the case for localized messages in *Drosophila* and *Xenopus* (Gavis and Lehmann, 1992; Macdonald, 1992; Mowry and Melton, 1992).

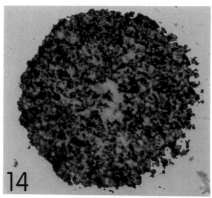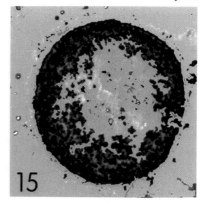

Fig. 14 Light micrograph of a semithin section of a CCD isolated from a cytochalasin B-treated embryo at the time controls were cleaving. Note that the cytoskeletal matrix is no longer restricted to the cortex. ×450.

Fig. 15 Light micrograph of a semithin section of a CCD isolated from a colchicine-treated embryo at the time controls were cleaving. The CCD cannot be distinguished from that of a normal unfertilized egg. ×450.

B. Function of the CCD in Cytoplasmic Localization

1. Experiments with Centrifugal Force

The function of the CCD in development has been examined by centrifugal displacement (Lillie, 1906, 1909; Wilson, 1929, 1930; Harvey, 1939; Jeffery, 1985; Swalla et al., 1985). Although the CCD is relatively resistant to centrifugation in comparison to the endoplasmic granules (Lillie, 1906, 1909), when dislodged, it sediments to the centrifugal pole as the densest cytoplasmic component (Lillie, 1909; Jeffery, 1985; Swalla et al., 1985). Experimental studies of the development of such centrifuged eggs have not been entirely consistent. While most authors have indicated that the cleavage pattern and development of light egg fragments—which should not contain the CCD if the centrifugal force had been sufficient to dislodge it—is essentially normal at least through cleavage (Wilson, 1929, 1930; Harvey, 1939), the structure and organization of the cortex was not examined in these early studies. Different results were obtained in more recent studies in which the distribution of the CCD in the egg fragments was known (Swalla et al., 1985). In this study, unfertilized egg fragments without the CCD failed to cleave at all, but instead underwent ameboid contractions characteristic of the pseudocleavage phase of differentiation without cleavage. By contrast, when Swalla et al. (1985) centrifuged zygotes, part of the CCD remained in the centripetal fragment, and these fragments developed normally. It is unclear why Swalla *et al.* (1985) obtained results different from those of Wilson (1929, 1930) and Harvey (1939). It seems unlikely that the CCD did not dislodge in the early studies, because it sediments to the centrifugal pole of the egg under less force than that required to fragment the egg. However, its resistance to centrifugation varies with the developmental stage of the oocyte (Wilson, 1951). Nevertheless, in Swalla *et al.*'s studies (1985), the extent of differentiation in centripetal fragments correlated positively with the presence of remnants of the CCD. The results of these recent studies suggest that the CCD must be (at least partialy) intact for normal development to occur.

The failure of normal development in centripetal fragments lacking remnants of the CCD must be accepted with some caution, as these fragments failed to cleave. Many types of nonspecific stress (*e.g.,* heat or cold shock) can cause *Chaetopterus* eggs to fail to cleave and therefore undergo differentiation without cleavage, and their figure is fully consistent with that of eggs undergoing pseudocleavage. Furthermore, cleavage stimuli are believed to be imparted by the spindle organizing an actin microfilament-based contractile ring. Pseudocleavage is, therefore, most likely due to a failure of the spindle to organize the contractile ring(s) properly, not to the absence or redistribution of an mRNA-containing intermediate filament network. Thus, incomplete development in egg fragments which fail to cleave cannot be considered conclusive evidence in support of any hypothesis.

The mechanism by which the spindle interacts with cortical microfilaments in *Chaetopterus* eggs is unknown. There may be more than one such mechanism. As noted previously, the meiotic spindle pole has a unique cortical attachment site that is outside the cortical granule-containing CCD (Lutz *et al.,* 1988). Furthermore, this attachment is not disturbed by centrifugal forces which dislodge the CCD (Lillie, 1909). This indicates that the attachment of the spindle pole to the cortex is independent of the presence of the CCD. However, at mitosis, the distal ends of the astral microtubules interact with much or all of the cortex. That this new interaction is with the CCD is demonstrated by the fact that a part of the CCD becomes displaced into the cytoplasm during mitosis. The mechanisms by which astral microtubules interact with the cell cortex are worthy of study; however, very little is known about them. Several proteins have been found to interact with both microfilaments and microtubules (Murofushi *et al.,* 1983; Itano and Hatano, 1991; Pedrotti *et al.,* 1994). The presence of cytoplasmic organelles which are visibly moved by this interaction makes *Chaetopterus* eggs a uniquely ideal system in which to study it.

2. Experiments with Cytoskeleton Antagonists

The effects of cytoskeleton antagonists have been studied on both normally fertilized eggs (Eckberg, 1981a) and artificially activated eggs undergoing differentiation without cleavage (Brachet and Donini-Denis, 1978). Both colchicine and CB inhibit cleavage in fertilized eggs, with their minimum effective concentrations ($\geq$90% inhibition of cleavage) being 10 and 0.1 μg/ml, respectively (Eckberg, 1981a). At the lowest effective concentrations, CB does not inhibit furrowing, but rather inhibits the completion of furrowing and does not inhibit either the meiotic divisions or any of the cell shape changes through polar lobe formation and furrowing. Slightly higher concentrations of CB (0.3–1 μg/ml) allow the early shape changes to occur, but not furrowing; concentrations in excess of 1 μg/ml inhibit all shape changes associated with the meiotic divisions and cleavage. Embryos which are cleavage inhibited by low concentrations of CB undergo typical differentiation without cleavage, even when the CB concentration is high enough to block furrowing. At all concentrations tested (up to 10 μg/ml), CB-treated zygotes underwent cytoplasmic reorganization characteristic of the segregation phase of differentiation without cleavage. At concentrations up to 1 μg/ml, CB-treated zygotes underwent normal differentiation without cleavage, including pseudogastrulation and ciliation. High concentrations of CB (10 μg/ml) allowed segregation but not unicellular gastrulation or ciliation (Brachet and Donini-Denis, 1978); at low concentrations, ciliation was variable (Eckberg, 1981a). These results suggest that normal microfilament organization is not essential for differentiation without cleavage.

The effects of colchicine on development and differentiation without cleavage are quite different from those of CB. Embryos develop normally in 1 μg/ml colchicine. However, at 10 μg/ml, the lowest concentration which blocks cleavage in $\geq$90% of the embryos, all subsequent cytoplasmic reorganization is blocked (Eckberg, 1981a). Even lower concentrations of colchicine block reorganization in KCl-activated eggs undergoing differentiation without cleavage (Brachet and Donini-Denis, 1978). These results demonstrate that, in contrast to normal microfilament function, normal microtubule function is absolutely required for differentiation without cleavage.

How do microtubules bring about cytoplasmic reorganization? Microscopy of CCDs isolated from cleaving zygotes and from zygotes treated with sufficient colchicine to block cleavage suggest that colchicine does not directly affect the CCD; treatment of embryos with colchicine results in CCDs which are indistinguishable from those of unfertilized eggs (Fig. 15). Thus, colchicine has no effect on the CCD *organization,* but it blocks the *reorganization* of the CCD. It may function by blocking spindle assembly and therefore by blocking effects of spindle assembly and disassembly on the CCD. This hypothesis has been tested indirectly (Brachet *et al.,* 1981; Alexandre *et al.,* 1982). Aphidicolin, a selective inhibitor of DNA synthesis, blocks the cell cycle by blocking transit from S to M. Concentrations of aphidicolin sufficient to block cell division have no effect on pseudocleavage, but they block ooplasmic segregation, pseudogastrulation, and ciliation. Since they have similar effects on differentiation without cleavage, both colchicine and aphidicolin probably block differentiation without cleavage by blocking sequential aster assembly and disassembly.

As noted previously, at mitosis a substantial fraction of the mRNA and the CCD becomes associated with the periphery of the aster (Figs. 5 and 11). This clearly indicates that the astral microtubules exert a pulling force on the CCD and implies a physical interaction between the CCD and astral microtubules. It is unknown how astral microtubules interact with peripheral structures of the cell. However, the fact that a strong aster/ cortex interaction can be followed in *Chaetopterus* embryos indicates that these might be good candidates for analysis of mechanisms by which such interactions occur. An initial search for molecules responsible for mediating the interaction between astral microtubules and the CCD should concentrate on proteins with homology to microtubule motor proteins, since these are generally involved in mediating the interaction of other cellular components with microtubules. If such proteins could be found that also interact with intermediate filaments, they would be especially good candidates for mediating this interaction *in vivo*.

The necessity for astral cycles and the interaction between microtubules and the cell cortex in cytoplasmic reorganization are not unique to *Chaetopterus*. Microtubules are apparently involved in several aspects of cytoplasmic reorganization in *Xenopus* (see Houliston and Elinson, 1992; Gard *et al.*, Chapter 12, for review).

IV. Cellular Signals Leading to Cytoskeletal Reorganization after Fertilization

A. Cellular Signals at Fertilization

Two cellular signaling molecules are known to be produced in significant amounts upon fertilization of sea urchin and frog eggs, inositol 1,4,5-trisphosphate (IP_3) and diacylglycerol (DG) (Ciapa and Whitaker, 1986; Stith *et al.*, 1993). These function, respectively, to release stored intracellular Ca^{2+} and to activate protein kinase C (PKC). The latter supposedly exerts its primary effects on egg activation at fertilization by phosphorylating and activating a Na^+/H^+ antiport carrier and thus making the cytoplasmic pH more alkaline (see Epel, 1990 for a review). However, this may not be the only role for PKC in sea urchin development. The enzyme undergoes a marked relocalization prior to mitosis (J. L. Olds, personal communication). This would indicate that it has effects after the cytoplasmic alkalization.

PKC evidently also acts after fertilization in frog and mammal eggs (Bement and Capco, 1990a, 1991b; Gallicano *et al.,* 1993).

Both of these signals may be produced before or after fertilization in *Chaetopterus* eggs. The DG content of oocytes was directly measured during maturation and after fertilization (Eckberg and Szuts, 1993). Natural stimulation of GVBD and fertilization both increased the cellular DG content. However, the cellular DG mass peaked at a value approximately three-fold higher during the initiation of GVBD than after fertilization. Nevertheless, although the mass of DG is much less after fertilization than it is before GVBD, it increases by 25 fmol/embryo during the first 5 min after fertilization, and the increase is statistically significant. This suggests that the DG/PKC pathway is not likely to be a primary signal for the reorganization that takes place after fertilization (although the rise in Ca^{2+} after fertilization could activate PKC even in the absence of a large increase in DG). It may also be involved in reorganization at GVBD, but this remains to be determined. It could be tested by examining the effects of phorbol esters and other PKC agonists on cytoskeleton organization in the primary oocytes. If PKC should prove to regulate cytoskeleton structure and function in these eggs, members of the MARCKS protein family would be likely candidates to mediate such effects (Aderem, 1992; Hartwig *et al.,* 1992; Li and Aderem, 1992).

pH changes after fertilization have not been investigated in *Chaetopterus.* However, eggs of other protostomes release acid after fertilization (Paul, 1975; Ii and Rebhun, 1979; Dubé and Guerrier, 1982), suggesting that such changes may also occur in *Chaetopterus.* The effects of PKC agonists and antagonists on cytoplasmic pH of these oocytes and zygotes should be examined.

Recent evidence strongly indicates that Ca^{2+} release is a prominant signal when the eggs are fertilized (Eckberg *et al.,* 1993). The results of these studies showed that fertilization is followed by a series of propagated waves and localized pulses of Ca^{2+} release which are initiated at multiple sites in the egg cortex. Similar waves and pulses occur upon induction of differentiation without cleavage by KCl.

It is not known as yet whether IP_3 provides the (a?) signal for this Ca^{2+} release. If all of the DG produced after fertilization resulted from phosphatidylinositol 4,5-bisphosphate (PIP_2) hydrolysis, the amount of IP_3 produced would be approximately 25 fmol/zygote. If distributed uniformly through the embryo, this would correspond to about 40 μM IP_3. If distributed only through the outer 10 μm of the egg, this concentration increases to approximately 150 μM, a value an order of magnitude greater than that reported for the frog egg cortex (Stith *et al.,* 1993) and several orders of magnitude more than is needed to activate *Spisula* eggs (Bloom *et al.,* 1988). Even if only a small fraction of the DG synthesized in *Chaetopterus* zygotes after fertilization is produced by PIP_2 hydrolysis, there should be more than enough IP_3 produced to release intracellular Ca^{2+}. We are currently investigating the mechanisms involved in Ca^{2+} release in these eggs.

B. Cellular Signals and the Cytoskeleton

1. Other Cells

The strongest evidence for regulation of cytoskeleton organization and cell polarity by Ca^{2+} has been obtained in plants and fungi. This evidence has been recently reviewed (Jackson and Heath, 1993; Kropf, 1994). Briefly, in *Fucus,* a Ca^{2+} gradient enters the rhizoid and exits through the thallus (Robinson and Jaffe, 1975; Brownlee and Wood, 1986). This gradient is essential for tip growth, because blocking it by BAPTA injection blocks tip growth (Speksnijder *et al.,* 1989b). A parallel proton gradient is also essential (Gabbon and Kropf, 1994). Microfilaments accumulate at regions of high Ca^{2+} and low pH (base of the rhizoid); microtubules accumulate at regions of low Ca^{2+} and higher pH (apex of the rhizoid) (Kropf, 1994).

Cytosolic Ca^{2+} can regulate cytoskeletal function and structure in cells through direct effects on cytoskeletal proteins. For example, Ca^{2+} can bind to and affect the conformation of G actin (Bertazzon *et al.,* 1990; Miki, 1990). However, most of the attention concerning Ca^{2+} effects on microfilaments has involved more indirect studies aimed at understanding the effects of phosphorylation of actin (Pollard, 1990) or actin-binding proteins (Vandecerchove, 1990) on cytoskeletal function.

Considerable attention has been directed to the possible involvement of calmodulin (CaM) and CaM-dependent protein kinases on cytoskeleton organization. Overexpression of CaM alters the organization of microfilaments, microtubules, and intermediate filaments (Rasmussen and Means, 1992). One potential target of CaM in regulating cytoskeleton structure and function could be a caldesmon-like protein. Caldesmon evidently regulates microfilament-associated myosin ATPase activity by interacting with tropomyosin (Matsumura and Yamashiro, 1993; Vibert *et al.,* 1993; Warren *et al.,* 1994). Alternatively, CaM has been reported to regulate nonmuscle myosin directly (Gough and Taylor, 1993; Wolenski *et al.,* 1993). CaM also regulates the phosphorylation and activity of MARCKS, which is involved in membrane–microfilament interactions (Hartwig *et al.,* 1992). CaM can also affect the structure of microtubules through activation of phosphorylation of a 62-kDa protein (Dinsmore and Sloboda, 1988). Doubtless, there will be more CaM-influenced cytoskeleton regulators discovered. In any case, the establishment of a pacemaker region from which waves of Ca^{2+} release originate suggests that directed Ca^{2+} fluxes might organize egg microfilaments and/or microtubules. The existence of localized pulses of Ca^{2+} suggests that localized Ca^{2+} increases might elict localized reorganizations of the cytoskeleton.

Cytoplasmic pH gradients or small pH changes could also alter cytoskeleton structure and function. For example, microtubule assembly is favored by alkaline pH and inhibited by acidic pH in marine invertebrate eggs both *in vivo* (Schatten *et al.,* 1985) and *in vitro* (Suprenant, 1991).

PKC evidently also elicits changes in cytoskeletal organization at fertilization in frog and mammal eggs (Bement and Capco, 1989, 1990b, 1991a; Capco *et al.,* 1992; Gallicano *et al.,* 1993). PKC can induce microfilament reorganization as indicated by cortical contractions in *Xenopus* zygotes (Bement and Capco, 1989, 1990b; Capco *et al.,* 1992). Also PKC can reorganize the intermediate filament network of mammal eggs (Gallicano and Capco, 1995).

2. *Chaetopterus*

The possible existence of Ca^{2+} and pH gradients has not been directly investigated in *Chaetopterus.* Indeed, we would not necessarily expect the same mechanism to establish polarity in embryos which do and do not grow directionally. However, one consistent and possibly relevant feature of the postfertilization waves of Ca^{2+} release in *Chaetopterus eggs* was that the last few waves always originated from the same quadrant of an egg (Fig. 16). This implies that these waves set up a kind of "pacemaker" region similar to that which is involved in the establishment of polarity in the ascidian egg (Speksnijder *et al.,* 1989a; 1990; Speksnijder, 1992). The relationship between these Ca^{2+} waves in *Chaetopterus* eggs and the polarity of the zygote has not been studied directly, but by analogy with ascidian eggs and other cells discussed previously, it seems reasonable to predict that these Ca^{2+} waves are involved in the establishment of embryonic polarity and the localization of components of the zygote cytoskeleton. Alternatively, they might also reflect preexisting embryonic, and therefore cytoskeletal, polarity. It seems unlikely that Ca^{2+} fluxes are responsible for determining the position of the spindle, because we have not detected any such pulses before or during GVBD (W. Eckberg and A. Miller, unpublished results).

Although it is indirect, there is some evidence that cytosolic Ca^{2+} regulates cytoskeletal reorganizations in *Chaetopterus* eggs through its effects on calmodulin. The calmodulin antagonists, chlorpromazine, calmidazolium, and W-7, all could initiate the ameboid contractions characteristic of differentiation without cleavage (Carroll and Eckberg, 1983). The amplitude of these contractions was dependent on the concentration of the antagonists, with the highest concentrations tested (100–300 μM) eliciting

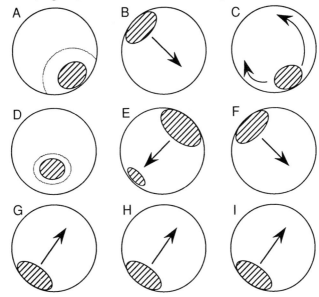

Fig. 16 Origin and direction of travel for the sequence of resolved Ca^{2+} transients in a fertilized *Chaetopterus* egg. Hatched ovals in each represent the origin of the transient. Arrows indicate the direction of propagation of waves. Straight arrows indicate waves which appeared to propagate across the entire egg; curved arrows represent waves which were definitely restricted to the cortex. The lighter lines in A and D represent the largest extent of these transients which, unlike the others, did not propagate across the egg. The smaller hatched area in E represents the position of a nonpropagating transient which began at the same time as the wave indicated by the larger hatched area.

massive contractions which eventually resulted in cytolysis. These drugs, however, also antagonize a number of other cellular processes, so the evidence that calmodulin regulates cytoskeletal reorganization requires further investigation.

V. Summary and Future Prospects

Chaetopterus eggs have received less attention than they merit, probably due to their unfamiliarity to most developmental biologists active today. We hope that this review will stimulate increased use of this resource. Their ease of handling (see Appendix), the availability of quantities of eggs sufficient for biochemical studies, and their interesting developmental program (including differentiation without cleavage) provide unique opportunities for investigation of several aspects of early development. With respect to the cytoskeleton and its organization, several areas of investigation can now be approached fruitfully, including:

1. Mechanisms by which the astral microtubules interact with the cortex and/or other cytoskeletal elements. This is important because such mechanisms are responsible for asymmetric spindle positioning in unequal cell division and may be responsible for localizing a wide variety of substances in cells. They are generally difficult to study because of the lack of measurable effect of this attachment on the cortex of most cells. The transient detachment of a part of the CCD from the cortex at mitosis provides a visible marker for this interaction and thereby could greatly facilitate its study. Furthermore, the existence of a unique, possibly predetermined, cortical attachment site for the meiotic spindle indicates the possibility of identifying such a site biochemically.

2. Role of the cytoskeleton in distributing localized substances. If the cytoskeleton is responsible for the distribution and redistribution of localized substances within the embryo, then its study in an "embryo" undergoing differentiation without cleavage—whose localization, in the absence of cytokinesis, closely parallels that of normal embryos—should provide insight into the nature of the localized substances and their mechanisms of interaction with the cytoskeleton. The careful application of cytoskeleton inhibitors in such a study should help to determine more precisely which component(s) of the cytoskeleton is responsible for cytoplasmic localization.

3. Mechanisms of establishment of the embryonic cytoskeleton and embryonic polarity. Experimental results from several systems, including *Chaetopterus*, suggest that the cytoplasmic signaling systems active at fertilization are involved in the establishment of the embryonic cytoskeleton. Preliminary correllative studies suggest the possible involvement of intracellular Ca^{2+} waves in the establishment of early embryonic polarity in *Chaetopterus*. Such a hypothesis is directly testable. Possible effects of other signals which have not yet been studied in these eggs could also easily be investigated.

4. Mechanisms of mRNA storage in the egg. Nearly all mRNA is located in the cortex, which is also the site of granular bodies/nuage which have been proposed as the mRNA storage sites. Since these can be dislodged from the cortex, they should also be amenable to isolation and characterization both biochemically and structurally. The evidence that such mRNA also interacts closely with intermediate filaments suggests specific approaches to the question of how the mRNA becomes localized in the cytoskeleton.

5. The possible rearrangement of the cortex at oocyte maturation. This is of considerable significance to our understanding of the mechanisms by which embryonic polarity is established and therefore should be further investigated. The ability to stain components of the cortex, specifically,

indicates the potential of ultimately following its redistribution in living eggs. Double staining the eggs with acridine orange and antibodies to centrosomal components could be used to establish the relationship between preexisting polarity in the germinal vesicle-intact oocyte and that of the metaphase-arrested and fertilized egg.

6. The extent to which gene expression in differentiation without cleavage parallels that of normal embryos. This has important relevance to our understanding of the role of cell compartmentation in cell differentiation.

Appendix: Methods for Obtaining and Using Eggs and Embryos in Studies of the Cytoskeleton and Early Development

A. Handling Adults

Adult *Chaetopterus pergamentaceus* live in U-shaped tubes buried in the sand. The leathery tube gives the animals one of their common names, "the parchment worm." Their other common name, "the innkeeper worm," dervies from the fact that most *Chaetopterus* share their tubes with commensal crabs. At Woods Hole, where we obtain animals from the Marine Resources Department of the Marine Biological Laboratory, their reproductive season lasts from early July to the middle of August. The reproductive season is longer in southern populations. Fertile specimens may be obtained from Cape Fear Biological Supply, P.O. Box 10218, Southport, NC 28461 from March until November. This species is also very common in Florida, but we have not investigated these specimens. Another, much smaller species, *Chaetopterus variopedatus,* is available on the Pacific coast, and (presumably) the same species is also present in the Mediterranean. When animals are first brought into the lab, we leave them in their tubes. Sex can be determined by either cutting off one end of the tube and squeezing the worm to the cut end of the tube until the gonadal parapodia are visible or alternatively by cutting a slit in the middle of the tube and viewing the parapodia through the slit. Gonaldal parapodia from fertile males will be a uniform milky white; those from fertile females will contain yellow ovaries which appear coiled in the living animal. In either case, the worms are not injured and will reseal the cut tube. If observation of the specimens is desired, they may be removed from their tubes and placed in curved glass or Tygon tubing. They will eventually secrete opaque tube material which will make them invisible, but they can be observed for weeks before that happens.

To prevent spontaneous spawning, we keep the worms at 15–18°C. This seems to have no affect on their fertility. If running seawater is available, the animals need not be fed as long as there is sufficient water flow for them to obtain and adequate supply of plankton from the water. If they are kept in closed aquaria, they can be fed commercial filter-feeder diets. If they are not fed, they begin to resorb their gametes within 2 or 3 weeks.

We have also kept fertile adults for at least 2 weeks, singly or in groups, in fingerbowls of running seawater after they are removed from their tubes. This is convenient and especially valuable for specimen conservation if the studies under way do not require a large number of eggs. Since the sperm from one parapodium are more than sufficient to fertilize even a very large number of eggs, this is especially convenient for obtaining sperm over the course of several experiments.

If the sexes are not kept segregated, the females should be briefly rinsed with distilled (or tap) water to kill any absorbed sperm before eggs are obtained.

B. Obtaining Gametes

1. Eggs

Chetopterus eggs suitable for use in studies involving fertilization or development may be obtained essentially as described (Costello and Henley, 1971). The methods below work on both species, but the adult *C. variopedatus* from the Pacific coast are much smaller than the Atlantic *C. pergamentaceus* and yield only ~10% as many eggs per adult female. Sperm and egg structure and morphology are similar in the two species, though the egg sizes differ slightly. We have never been able to force the sperm of either species to fertilize the eggs of the other.

The simplest way to obtain eggs is to cut the tips off several gonadal parapodia and leave the worm in a dish of seawater for 10–15 min. The worm will release a few hundred to a few thousand eggs into the dish. The eggs may be collected by a Pasteur pipet and washed with seawater prior to insemination. To increase yield, eggs are vigorously teased out of cut parapodia into filtered seawater and filtered through cheesecloth. These eggs will be suspended in a dense mucus which must be removed by washing. Washing can be speeded up dramatically by centrifuging the eggs briefly at low speed (<2000 rev/min). Overcentrifugation of the eggs will stratify their contents and the experimenter may find him/herself unintentionally repeating some of Lillie's, Wilson's, Morgan's, or Harvey's experiments! Usually nearly all of the mucus is lost in the first wash, so subsequent washes can be done at unit gravity. This method can be expected to yield a homogeneous population of metaphase I-arrested cells approximately 105 μm in diameter. If a significant fraction ($\geq$5%) of the eggs are smaller or their germinal vesicles do not break down within 30 min of contact with seawater, the egg batch should be discarded. Please note that oocytes of this organism normally will *not* undergo germinal vesicle breakdown in artificial seawater (Ikegami *et al.*, 1976), so it is essential, to have some natural seawater available. Once the eggs reach metaphase I, however, fertilization and development will occur as well in artificial seawater as in natural seawater.

2. Sperm

Sperm are obtained "dry" by cutting a single parapodium into a small (*ca.* 10 ml) beaker which is then filled with seawater. Contrary to some earlier workers, we have found that freshly diluted sperm of *Chaetopterus* are never motile, and that they become motile within 2–5 min after dilution. All sperm preparations should be observed by phase-contrast microscopy to ensure motility. A significant number of nongametic cells will be observed in a typical sperm suspension. They do not appear to interfere with fertilization, but since they are much larger than the sperm, they can be removed by differential centrifugation if necessary or desired.

3. Fertilization and Embryo Culture

When most of the sperm are motile, they are used to inseminate eggs. Since the eggs are resistant to polyspermy (Eckberg and Anderson, 1985), the actual amount of sperm added to the eggs is no more critical than that for the more familiar sea urchin. *Chaetopterus* have both rapid and slow blocks to polyspermy. The rapid block is probably electrical in nature (Jaffe, 1983; Eckberg and Anderson, 1985). The slow block is somewhat unique in that it results from the retraction of egg microvilli from the vitelline envelope in the absence of a cortical reaction comparable to that in many other organisms (Anderson and Eckberg, 1983). A similar mechanism also acts in another annelid, *Neanthes* (Sato and Osani, 1986).

Excess sperm are washed away by letting the eggs settle at unit gravity and decanting and replacing the seawater a couple of times, and the zygotes are cultured. Since there is no cortical reaction or vitelline envelope elevation comparable to that of the sea urchin, it is not possible to tell immediately if eggs have been fertilized. However, 15 min after fertilization, the first polar bodies form, and concomitantly the vitelline envelope begins to ruffle due to the formation and retraction of very long microvilli from the zygote surface. A thin perivitelline space forms when the microvilli retract, leaving a "fertilization envelope;" however, since exposure of the eggs to slightly hypertonic seawater can cause the same effect, we recommend that polar body formation be used as the criterion for fertilization.

The lack of a cortical reaction comparable to that of sea urchins, *etc.* may frighten investigators interested in examining the earliest phase of development. This should not be disturbing, however, as polar body formation occurs within 15 min, and a sample of a culture can be kept to ensure full activation. In this context, it should also be remembered that, although the cortical reaction in a sea urchin egg corellates in time with its initiation of development because both result from the same cell–cell interaction and probably from some of the same intracellular signals, the cortical reaction has no obligatory role in development and is, thus, not the best possible criterion for egg activation. Thus, the lack of such a response in *Chaetopterus* can be seen as an advantage.

The term "egg activation" also has a slightly different meaning in *Chaetopterus* than it does in sea urchins. When these eggs are fertilized, their oxygen consumption *decreases* (Whitaker, 1933; Brachet, 1938) as do their protein synthetic and some metabolite transport activities (W. Eckberg, unpublished data), indicating that metabolic activity is maximal prior to fertilization and may actually be depressed somewhat after fertilization.

Fertilized eggs can be cultured at room temperature and will develop normally in either monolayer or suspension culture. They can even be cultured for at least an hour at a concentration of at least 10% (v/v) without deleterious effects on development. This reduces the amount of isotopes needed for labeling studies. Development is very rapid with swimming blastulae produced in 4 or 5 hr and fully differentiated trochophore larvae produced in less than 24 hr at 22°C. Embryos do not appear to "hatch," but cilia grow out through the vitelline envelope. The trochophore is slightly atypical, lacking a prototrochal ciliary band.

4. Microscopy and Micromanipulation

Although the egg is opaque and densely pigmented, all conventional light microscopic techniques can be applied, including phase- and differential interference-contrast, epifluorescence, and polarization. However, because of the size and opacity of the egg, a significant fraction of any fluorescent signal from cytosolic molecules in parts of the egg which are distant from the objective lens will be quenched (W. Eckberg and A. L. Miller, submitted for publication). The presence of refringent granules in the egg makes visualization of spindles by polarization optics slightly more complex than in some other cells. However, this can be compensated for by slightly flattening the eggs with a coverslip (Eckberg and Palazzo, 1992) or by centrifuging them into fragments (Lutz *et al.*, 1988). The eggs can also be microinjected and impaled with electrodes (Hagiwara and Miyasaki, 1977; Hagiwara and Jaffe, 1979; Jaffe, 1983; Eckberg *et al.*, 1993).

Since neither impaling these eggs with electrodes nor microinjecting them with markers suitable for observation during subsequent development prevented their being fertilized, similar manipulations should be effective procedures for examining other phenomena.

Acknowledgments

We thank Bill Jeffery and George Langford for helpful discussions, Patricia Polk for assistance with electron microscopy, and Tracey Williams for assistance with immunoblotting. W.R.E. is currently supported by grants from the National Institutes of Health and the Council for Tobacco Research, USA, Inc. W.A.A. is currently supported by a grant from the National Instiues of Health.

References

Aderem, A. (1992). The MARCKS brothers: A family of protein kinase C substrates. *Cell* **71,** 713–716.

Alexandre, H., DePetrocellis, B., and Brachet, J. (1982). Studies on differentiation without cleavage in *Chaetopterus.* Requirement for a definite number of DNA replication cycles shown by aphidicolin pulses. *Differentiation* **22,** 132–135.

Anderson, W. A., and Eckberg, W. R. (1983). A cytological analysis of fertilization in *Chaetopterus. Biol. Bull.* **165,** 110–118.

Bement, W. M., and Capco, D. G. (1989). Activators of protein kinase C trigger cortical granule exocytosis, cortical contraction, and cleavage furrow formation in *Xenopus laevis* oocytes and eggs. *J. Cell Biol.* **108,** 885–892.

Bement, W. M., and Capco, D. G. (1990a). Protein kinase C acts downstream of calcium at entry into the first mitotic interphase of *Xenopus laevis. Cell Regul.* **1,** 315–326.

Bement, W. M., and Capco, D. G. (1990b). Transformation of the amphibian oocyte into the egg: Structural and biochemical events. *J. Electron Microsc. Technol.* **16,** 202–234.

Bement, W. M., and Capco, D. G. (1991a). Analysis of inducible contractile rings suggests a role for protein kinase C in embryonic cytokinesis and wound healing. *Cell Motil. Cytoskeleton* **20,** 145–157.

Bement, W. M., and Capco, D. G. (1991b). Parallel pathways of cell cycle control during *Xenopus* egg activation. *Proc. Natl. Acad. Sci. USA* **88,** 5172–5176.

Bertazzon, A., Tian, G. H., Lamblin, A., and Tsong, T. Y. (1990). Enthalpic and entropic contributions to actin stability: Calorimetry, circular dichroism, and fluorescent study and effects of calcium. *Biochemistry* **29,** 291–298.

Bloom, T. L., Szuts, E. Z., and Eckberg, W. R. (1988). Inositol trisphosphate, inositol phospholipid metabolism, and germinal vesicle breakdown in surf clam oocytes. *Dev. Biol.* **129,** 532–540.

Brachet, J. (1937). La différenciation sans clivage dans l'oeuf de Chétoptère envisagée aux points de vue cytologique et métabolique. *Arch. Biol. (Liege)* **48,** 561–589.

Brachet, J. (1938). The oxygen consumption of artificially activated and fertilized *Chaetopterus* eggs. *Biol. Bull.* **74,** 93–98.

Brachet, J., De Petrocellis, B., and Alexandre, H. (1981). Studies on differentiation without cleavage in *Chaetopterus:* Effects of inhibition of DNA synthesis with aphidicolin. *Differentiation* **19,** 47–54.

Brachet, J., de Petrocellis, B., Ficq, A., and Alexandre, H. (1980). Studies on differentiation without cleavage in *Chaetopterus variopedatus:* Requirements for nucleic acid and protein synthesis. *Differentiation* **17,** 63–76.

Brachet, J., and Donini-Denis, S. (1978). Studies on maturtion and differentiation without cleavage in *Chaetopterus variopedatus.* Effects of ions, ionophores, sulfhydril reagents, colchicine and cytochalasin B. *Differentiation* **11,** 19–37.

Brownlee, C., and Wood, J. W. (1986). A gradient of cytoplasmic free calcium in growing rhizoid cells of *Fucus serratus. Nature* **320,** 624–626.

Capco, D. G., Tutnick, J. M., and Bement, W. M. (1992). The role of protein kinase C in reorganization of the cortical cytoskeleton during the transition from oocyte to fertilization-competent egg. *J. Exp. Zool.* **264,** 395–405.

Carroll, A. G., and Eckberg, W. R. (1983). Possible involvement of calmodulin in maturation and activation of *Chaetopterus* eggs. *Dev. Biol.* **99,** 1–6.

Ciapa, B., and Whitaker, M. (1986). Two phases of inositol polyphosphate and diacylglycerol production at fertilisation. *FEBS Lett.* **195,** 347–351.

Costello, D. P., and Henley, C. (1971). "Methods for Obtaining and Handling Marine Eggs and Embryos." Marine Biological Laboratory, Woods Hole, Massachusetts.

Dinsmore, J. H., and Sloboda, R. E. (1988). Calcium and calmodulin-dependent phosphorylation of a 62kd protein induces microtubule depolymerization in sea urchin mitotic apparatuses. *Cell* **53,** 769–780.

Dubé, F., and Guerrier, P. (1982). Activation of *Barnea candida* (Mollusca, Pelecypoda) oocytes by sperm of KCl, but not by NH$_4$Cl, requires a calcium influx. *Dev. Biol.* **92,** 408–417.

Eckberg, W. R. (1981a). The effects of cytoskeleton inhibitors on cytoplasmic localization in *Chaetopterus pergamentaceus. Differentiation* **19,** 55–58.

Eckberg, W. R. (1981b). An ultrastructural analysis of cytoplasmic localization in *Chaetopterus pergamentaceus. Biol. Bull.* **160,** 228–239.

Eckberg, W. R., and Anderson, W. A. (1985). Blocks to polyspermy in *Chaetopterus. J. Exp. Zool.* **233,** 253–260.

Eckberg, W. R., and Kang, Y. H. (1981). A cytological analysis of differentiation without cleavage in cytochalasin B- and colchicine-treated embryos of *Chaetopterus pergamentaceus. Differentiation* **19,** 154–160.

Eckberg, W. R., and Langford, G. M. (1983). Isolation of cytoskeletons from *Chaetopterus* eggs. *Biol. Bull.* **165,** 514.

Eckberg, W. R., and Palazzo, R. E. (1992). Regulation of M-phase progression in *Chaetopterus* oocytes by protein kinase C. *Dev. Biol.* **149,** 395–405.

Eckberg, W. R., and Szuts, E. Z. (1993). Diacylglycerol content of *Chaetopterus* oocytes during maturation and fertilization. *Dev. Biol.* **159,** 732–735.

Eckberg, W. R., Miller, A. L., Short, L. G., and Jaffe, L. F. (1993). Calcium pulses during the activation of a protostome egg. *Biol. Bull.* **185,** 289–290.

Epel, D. (1990). The initiation of development at fertilization. *Cell Differ. Dev.* **29,** 1–12.

Gabbon, B. C., and Kropf, D. L. (1994). Cytosolic pH gradients associated with tip growth. *Science* **263,** 1419–1421.

Gallicano, G. I., and Capco, D. G. (1995). Remodeling of a specialized intermediate filament network in mammalian eggs and embryos during development: regulation by protein kinase M. *In Current Topics in Developmental Biology* (D. G. Capco, ed.), Vol. 31, Chapter 9. Academic Press, Orlando, Florida.

Gallicano, G. I., Schwarz, S. M., McGaughey, R. W., and Capco, D. G. (1993). Protein kinase C, a pivotal regulator of hamster egg activation, functions after elevation of intracellular free calcium. *Dev. Biol.* **156,** 94–106.

Gavis, E. R., and Lehmann, R. (1992). Localization of *nanos* RNA controls embryonic polarity. *Cell* **71,** 301–313.

Gough, A. H., and Taylor, D. L. (1993). Fluorescence anisotropy imaging microscopy maps calmodulin binding during cellular contraction and locomotion. *J. Cell Biol.* **121,** 1095–1107.

Hagiwara, S., and Jaffe, L. A. (1979). Electrical properties of egg cell membranes. *Annu. Rev. Biophys. Bioeng.* **8,** 385–416.

Hagiwara, S., and Miyasaki, S. (1977). Changes in excitability of the cell membrane during 'differentiation without cleavage' in the egg of the annelid, *Chaetopterus pergamentaceus. J. Physiol.* **272,** 197–216.

Hartwig, J. H., Thelen, M., Rosen, A., Janmey, P. A., Nairn, A. C., and Aderem, A. (1992). MARCKS is an actin filament crosslinking protein regulated by protein kinase C and calcium–calmodulin. *Nature* **356,** 618–622.

Harvey, E. B. (1939). Development of half-eggs of *Chaetopterus pergamentaceus* with special reference to parthenogenetic merogony. *Biol. Bull.* **76,** 384–404.

Henry, J. J. (1986). The role of unequal cleavage and the polar lobe in the segregation of developmental potential during first cleavage in the embryo of *Chaetopterus variopedatus. Roux' Arch. Dev. Biol.* **195,** 103–116.

Henry, J. J., and Martindale, M. Q. (1987). The organizing role of the D quadrant as revealed through the phenomenon of twinning in the polychaete *Chaetopterus variopedatus. Roux' Arch. Dev. Biol.* **196,** 499–510.

Houliston, E., and Elinson, R. P. (1992). Microtubules and cytoplasmic reorganisation in the frog egg. *Curr. Topics Dev. Biol.* **26,** 53–70.

Hyman, A. A. (1989). Centrosome movement in the early divisions of *Caenorhabditis elegans:* A cortical site determining centrosome position. *J. Cell Biol.* **109,** 1185–1193.

Ii, I., and Rebhun, L. I. (1979). Acid release following activation of surf clam (*Spisula solidissima*) eggs. *Dev. Biol.* **72,** 195–200.

Ikegami, S., Okada, T. S., and Koide, S. S. (1976). On the role of calcium ions in oocyte maturation in the polychaete *Chaetopterus pergamentaceus. Dev. Growth Differ.* **18,** 33–43.

Itano, N., and Hatano, S. (1991). F-actin bundling protein from *Physarum polycephalum:* Purification and its capacity for co-bundling of actin filaments and microtubules. *Cell Motil. Cytoskeleton* **19,** 244–254.

Jackson, S. L., and Heath, I. B. (1993). Roles of calcium ions in hyphal tip growth. *Microbiol. Rev.* **57,** 367–382.

Jaffe, L. A. (1983). Fertilization poteintials from eggs of the marine worms *Chaetopterus* and *Saccoglossus. In* "The Physiology of Excitable Cells" (V. Chang-Palay and S. L. Palay, eds.) pp. 211–218. A. R. Liss, New York.

Jeffery, W. R. (1985). The spatial distribution of maternal mRNA is determined by a cortical cytoskeletal domain in *Chaetopterus* eggs. *Dev. Biol.* **110,** 217–229.

Jeffery, W. R., Speksnijder, J. E., Swalla, B. J., and Venute, J. M. (1986). Mechanism of maternal mRNA localization in *Chaetopterus* eggs. *In* "Advances in Invertebrate Reproduction" (J.-C. Andries and A. Dhainaut, eds.), Vol. 4, pp. 229–240. Elsevier, New York.

Jeffery, W. R., and Wilson, L. J. (1983). Localization of messenger mRNA in the cortex of *Chaetopterus* eggs and early embryos. *J. Embryol. Exp. Morphol.* **75,** 225–239.

Kropf, D. L. (1994). Cytoskeletal control of cell polarity in a plant zygote. *Dev. Biol.* **165,** 361–371.

Li, J., and Aderem, A. (1992). MacMARCKS, a novel member of the MARCKS family of protein kinase C substrates. *Cell* **70,** 791–801.

Lillie, F. R. (1902). Differentiation without cleavage in the egg of the annelid *Chaetopterus pergamentaceus. Wilhelm Roux Arch. Entw. Org.* **14,** 477–499.

Lillie, F. R. (1906). Observations and experiments concerning the elementary phenomena of embryonic development in *Chaetopterus. J. Exp. Zool.* **3,** 153–268.

Lillie, F. R. (1909). Polarity and bilaterality of the annelid egg. Experiments with centrifugal force. *Biol. Bull.* **16,** 54–79.

Loeb, J. (1901). Experiments on artificial parthenogenesis in annelids (*Chaetopterus*) and the nature of the process of fertilization. *Am. J. Physiol.* **4,** 423–459.

Lutz, D. A., Hamaguchi, Y., and Inoue, S. (1988). Micromanipulation studies of the asymmetric positioning of the maturation spindle in *Chaetopterus sp.* oocytes: I. Anchorage of the spindle to the cortex and migration of a displaced spindle. *Cell Motil. Cytoskeleton* **11,** 83–96.

Macdonald, P. M. (1992). The means to the ends: Localization of maternal messenger RNAs. *Semin. Dev. Biol.* **3,** 413–424.

Matsumura, F., and Yamashiro, S. (1993). Caldesmon. *Curr. Opin. Cell Biol.* **5,** 70–76.

Miki, M. (1990). Resonance energy transfer between points in a reconstituted skeletal muscle thin filament: A conformational change of the thin filament in response to changes in Ca^{2+} concentration. *Eur. J. Biochem.* **187,** 155–162.

Morgan, T. H. (1910). Cytological studies of centrifuged eggs. *J. Exp. Zool.* **9,** 593–656.

Morgan, T. H. (1937). The factors locating the first cleavage plane in the egg of *Chaetopterus. Cytologia* Fugii Jubilee Vol., 711.

Morgan, T. H. (1938). A reconsideration of the evidence concerning a dorso-ventral preorganization of the egg of *Chaetopterus. Biol. Bull.* **74,** 395–400.

Morgan, T. H. (1939). The effects of centrifuging on the polar spindles of the egg of *Chaetopterus* and *Cumingia. Biol. Bull.* **76,** 339–358.

Morgan, T. H., and Tyler, A. (1938). The relation between entrance point of the spermatozoön and bilaterality of the egg of *Chaetopterus. Biol. Bull.* **74,** 401–402.

Mowry, K., and Melton, D. (1992). Vegetal messenger RNA localization directed by a 340 nucleotide RNA sequence element in *Xenopus* oocytes. *Science* **255,** 991–994.

Murofushi, H., Minami, Y., Matsumoto, G., and Sakai, H. (1983). Bundling of microtubules *in vitro* by a high molecular weight protein aprepared from the squid axon. *J. Biochem.* **93,** 639–650.

Pasteels, J. (1935). Recherches sur le déterminisme de l'entrée en maturation de l'oeuf chez divers invertébres marins. *Arch. Biol. (Liege)* **46,** 229–262.

Pasteels, J. (1950). Mouvements localisés et rythmiques de la membrane de fécondation chez des oeufs fécondés ou activés (*Chaetopterus, Mactra, Nereis*). *Arch. Biol. (Liege)* **61,** 197–220.

Paul, M. (1975). Release of acid and changes in light-scattering properties following fertilization of *Urechis caupo* eggs. *Dev. Biol.* **43,** 299–312.

Pedrotti, B., Colombo, R., and Islam, K. (1994). Microtubule associated parotein MAP1A is an actin-binding and crosslinking protein. *Cell Motil. Cytoskeleton* **29,** 110–116.

Pollard, T. D. (1990). Actin. *Curr. Opin. Cell Biol.* **2,** 330–340.

Rasmussen, C. D. and Means, A. R. (1992). Increased calmodulin affects cell morphology and mRNA levels of cytoskeletal protein genes. *Cell Motil. Cytoskeleton* **21,** 45–57.

Rebhun, L. I. (1972). Polarized intracellular particle transport: Saltatory movements and cytoplasmic streaming. *Int. Rev. Cytol.* **32,** 93–137.

Robinson, K. R., and Jaffe, L. F. (1975). Polarizing fucoid eggs drive a calcium current through themselves. *Science* **187,** 70–72.

Sato, M., and Osani, K. (1986). Morphological identification of sperm receptors above egg microvilli in the polychaete, *Neanthes japonica. Dev. Biol.* **113,** 263–270.

Schatten, G., Beston, T., Balczon, R., Henson, J., and Schatten, H. (1985). Intracellular pH shift leads to microtubule assembly and microtubule-mediated motility during sea urchin fertilization: Correlations between elevated internal pH and microtubule activity and depressed intracellular pH and microtubule disassembly. *Eur. J. Cell Biol.* **36,** 116–127.

Scott, J. W. (1906). Morphology of the parthenogenetic development of *Amiphitrite*. *J. Exp. Zool.* **3**, 49–97.

Speksnijder, J. E. (1992). The repetitive calcium waves in the fertilized ascidian egg are initiated near the vegetal pole by a cortical pacemaker. *Dev. Biol.* **153**, 259–271.

Speksnijder, J. E., Corson, D. W., Sardet, C., and Jaffe, L. F. (1989a). Free calcium pulses following fertilization in the ascidian egg. *Dev. Biol.* **135**, 182–190.

Speksnijder, J. E., Miller, A. L., Weisenseel, M. H., Chen, T.-H., and Jaffe, L. F. (1989b). Calcium buffer injections block fucoid egg development by facilitating calcium diffusion. *Proc. Natl. Acad. Sci. USA* **86**, 6607–6611.

Speksnijder, J. E., Sardet, C., and Jaffe, L. F. (1990). Periodic calcium waves cross ascidian eggs after fertilization. *Dev. Biol.* **142**, 246–249.

Stith, B. J., Goalstone, M., Silva, S., and Jaynes, C. (1993). Inositol 1,4,5-triphosphate mass changes from fertilization through first cleavage in *Xenopus laevis*. *Mol. Biol. Cell* **4**, 435–443.

Suprenant, K. A. (1991). Unidirectional microtubule assembly in cell-free extracts of *Spisula solidissima* oocytes is regulated by subtle changes in pH. *Cell Motil. Cytoskeleton* **19**, 207–220.

Swalla, B. J., Moon, R. T., and Jeffery, W. R. (1985). Developmental significance of a cortical cytoskeletal domain in *Chaetopterus* eggs. *Dev. Biol.* **111**, 434–450.

Titlebaum, A. (1928). Artificial production of Janus embryos. *Proc. Natl. Acad. Sci. USA* **14**, 245–247.

Treadwell, A. L. (1902). Notes on the nature of "artificial parthogenesis" in the egg of *Podarke obscura*. *Biol. Bull.* **3**, 235–240.

Tyler, A. (1930). Experimental production of double embryos in annelids and mollusks. *J. Exp. Zool.* **57**, 347–407.

Vandecerchove, J. (1990). Actin-binding proteins. *Curr. Opin. Cell Biol.* **2**, 41–50.

Vibert, P., Craig, R., and Lehman, W. (1993). Three-dimensional reconstruction of caldesmon-containing smooth muscle thin filaments. *J. Cell Biol.* **123**, 313–321.

Warren, K. S., Lin, J. L., Wamboldt, D. D., and Lin, J. J. (1994). Overexpression of human fibroblast caldesmon fragment containing actin-, Ca^{++}/calmodulin-, and tropomyosin-binding domains stabilizes endogenous tropomyosin and microfilaments. *J. Cell Biol.* **125**, 359–368.

Whitaker, D. M. (1933). On the rate of oxygen consumption of fertilized and unfertilized eggs. IV. *Chaetopterus* and *Arbacia punctulata*. *J. Gen. Physiol.* **16**, 475–495.

Whitaker, D. M., and Morgan, T. H. (1930). The cleavage of polar and anti-polar halves of the egg of *Chaetopterus*. *Biol. Bull.* **58**, 145–149.

Wilson, E. B. (1882). Observations on the developmental stages of some polychaetous annelids. *Stud. Biol. Lab. Johns Hopkins Univ.* **2**, 271–299.

Wilson, E. B. (1929). The development of egg-fragments in annelids. *Wilhelm Roux' Arch.* **117**, 180–210.

Wilson, E. B. (1930). Notes on the development of fragments of the fertilized *Chaetopterus* egg. *Biol. Bull.* **59**, 71–80.

Wilson, W. L. (1951). The rigidity of the cell cortex during cell division. *J. Cell. Comp. Physiol.* **38**, 409–415.

Wolenski, J. S., Hayden, S. M., Forscher, P., and Mooseker, M. S. (1993). Calcium–calmodulin and regulation of brush border myosin-I MgATPase and mechanochemistry. *J. Cell Biol.* **122**, 613–621.

2

Cytoskeleton and Ctenophore Development

Evelyn Houliston, Danièle Carré, Patrick Chang, and Christian Sardet
Unité de Biologie Cellulaire Marine
(URA 671 CNRS/Université Paris VI)
Observatoire de Villefranche-sur-mer
Station Zoologique
06230 Villefranche-sur-mer, France

I. Introduction to Ctenophores

Ctenophores are gelatinous marine organisms characterized by rows of beating "combs" of cilia (Fig. 1A). They are related only very distantly to other animals. Although their precise evolutionary position is controversial, it is clear that they diverged very early from other metazoans (see Willmer, 1990).

   The ctenophore egg is a beautiful eukaryotic cell containing dynamic microtubule and actin filament networks with organelles shuffling around

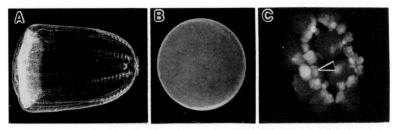

Fig. 1 (A) *Beroe ovata* juvenile, about 1 cm long in dark field. The mouth is on the left, and rows of comb plates run toward the apical organ at the opposite "aboral" pole. (B) Egg (about 1 mm in diameter) and (C) 64-cell stage embryo, both seen by natural fluorescence. Autofluorescent vesicles are found in the layer of ectoplasm in the egg, later concentrating in micromeres (e.g., at arrowhead).

them. During the cleavage stages of development, a stereotypic asymmetric cleavage pattern produces cells with distinct shapes, sizes, and composition. Associated with these cleavages are dramatic cytoplasmic reorganizations which segregate components between different cells (Fig. 1C). This is of particular interest to developmental biologists because ctenophore embryos exhibit classic mosaic development. Distinct types of blastomere from the early embryo develop autonomously to give particular adult structures, implying that different "developmental potentials" become segregated at each cleavage (see Reverberi, 1971; Ortolani, 1989; Sardet *et al.*, 1990 for reviews). The challenge is to understand the relationship between the cytoplasmic events and the attribution of developmental potential to different cells.

We have found *Beroe ovata* eggs and embryos to be ideal for detailed microscopy and have begun to investigate the cell biology of early developmental events. We have, in particular, examined the relationship between the behavior of the microtubule network and cytoplasmic relocation. In this chapter we summarize our findings and outline the possibilities for future study in the context of our embryological knowledge.

A. Organization of the Adult, Embryo, and Egg

Ctenophores have a simple body plan comprising a single axis (oral–aboral) with two perpendicular planes of symmetry (Fig. 2). These are termed the "sagittal" and "tentacular" planes (although *Beroe* species, unlike other ctenophores, have no tentacles). There is a direct correspondence between these planes of symmetry and the cleavage planes of the early embryo. The first two divisions produce four blastomeres identical in composition, which give rise to the four quadrants of the animal. The subsequent pattern

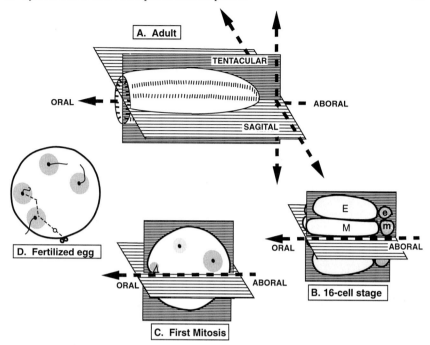

Fig. 2 Organization of the body plan in *Beroe ovata* showing the relationship of the planes of symmetry in the adult (A) to cleavage planes in the early embryo (B) and the first mitotic spindle in the egg (C). The developmental axis is not established in the egg prior to pronuclear fusion (D). The 16-cell embryo has four distinct cell types, E and M macromeres, and e and m micromeres.

of unequal divisions is unique to ctenophores. The third division divides each blastomere obliquely into external **(E)** and middle **(M)** blastomeres and the fourth produces large **E** and **M** "macromeres" and small **e** and **m** "micromeres" in each quadrant (Fig. 2B). Underlying each comb plate row are the gonads (Dunlap Pianka, 1974). *Beroe ovata,* like most but not all ctenophore species described, are hermaphroditic, with adjacent pairs of testes and ovaries.

Ctenophore egg diameters range from 100 or 200 μm (for instance, *Bolinopsis, Pleurobrachia,* and *Mnemiopsis* species) to more than a millimeter (e.g., *B. ovata*). All the early cleavage divisions are unipolar, cutting through the egg from a single starting point above superficially positioned mitotic spindles. This highlights an important feature of egg organization. All the active cytoplasm of the egg, known as "ectoplasm," forms a peripheral layer 10–30 μm thick (Fig. 1B). The center of the egg is occupied by "endoplasm," containing little cytoplasm, with densely packed yolk platelets. Each nucleus, spindle, and centrosome is thus positioned on one side of the egg, close to the cell surface, giving every cell an inbuilt asymmetry.

B. Axis Establishment

The animal pole of the ctenophore egg, marked by the position of the meiotic spindle and polar bodies, was originally thought to correspond to the oral pole of the embryo, but this is not always so (Fig. 2D). It has now been demonstrated that there is no fixed relationship between the animal pole and the position of the future body axis. In fact, the organization of the body plan cannot be reliably predicted until the first cleavage furrow starts to cut through the egg. This furrow starts at the future oral pole and passes through the future sagittal plane (Fig. 2C). Thus, the place where the zygote nucleus lies and the spindle forms corresponds to the future oral pole of the embryo (Freeman, 1977; Carré and Sardet, 1984). In *Pleurobrachia* and *Bolinopsis,* experimental displacement of the zygote nucleus reorients the future oral–aboral axis, the oral pole always corresponding to the site of cleavage initiation (Freeman, 1977). Whether the zygote nucleus exerts some influence on egg organization, or whether the spindle or cleavage furrow, which form at the same site, are responsible is not yet clear.

Since the sperm-derived male pronuclei do not move from the place where they form, the zygote nucleus forms close to the site of sperm entry, and since sperm entry is not restricted to a particular site, axis orientation can be said to be determined by the site of sperm entry. In *B. ovata* the situation is often complicated by physiological polyspermy. If several sperm enter at different sites, the migration of the female pronucleus to "choose" one of them decides where the future oral pole will lie (see Section IIIb: Carré and Sardet, 1984; Rouvière *et al.,* 1994).

The precise time at which significant asymmetries in the distribution of developmental potentials is recognizable is difficult to pinpoint. Differences *along* the future axis are not detectable until the cleavage furrow has advanced and two cells have formed (Freeman and Reynolds; 1973; Freeman, 1976a, 1977; Houliston *et al.,* 1993). Freeman (1976a, 1977), experimenting on *Mnemiopsis, Bolinopsis,* and *Pleurobrachia* species, monitored the relocation of comb plate-forming potential to the aboral pole, where comb plate-providing micromeres are fated to form. Removal of cytoplasm from the future aboral pole (opposite the nuclei) of two-cell blastomeres results in the reduction in the ability of operated embryos to form comb plates, while removal of cytoplasm from the same region of eggs undergoing first cleavage did not. This may suggest that events closely associated with cleavage are involved with setting up the axis. But removal of other cytoplasmic regions at earlier times can result in defects in embryonic organization (Fischel, 1903; Driesch and Morgan, 1895b; Yatsu, 1912; Houliston *et al.*, 1993). In particular, if cytoplasm is removed from near the site of cleavage initiation (future oral pole), larvae with reduced numbers of comb plates, or even lacking entire comb-plate rows, develop (Fischel, 1903). Eggs which are cut parallel to the presumptive oral–aboral axis from the time to mitosis onwards may also develop into defective larvae (or defective half-larvae if only one daughter nucleus is present), despite apparently normal cleavage patterns (Houliston *et al.*, 1993). These observations indicate that significant asymmetries in cytoplasmic organization may begin to be established slightly before cleavage (Freeman and Reynolds, 1973; Houliston *et al.*, 1993). Perhaps comb-plate potential first becomes grouped around the oral pole at the beginning of cleavage, before being relocated progressively by subsequent cleavages. It should not be surprising if cytoplasmic determinants do not congregate first around their final destination, particularly given the extensive cytoplasmic reorganizations that occur within the egg and early blastomeres (Spek, 1926).

It is clear that axis establishment is linked to precise restructuring of the cytoplasm in the fertilized egg, in particular at the times of first mitosis and cleavage. In this chapter we will describe how the cytoskeleton is implicated.

C. Mosaicism and Regulation

Early ctenophore embryos are "mosaic" in that the distinct fate of different blastomeres is not dependent on the presence of neighboring cells (Chun, 1880; Driesch and Morgan, 1895a,b; Fischel, 1897, 1903; see Reverberi, 1971; Freeman, 1979; Sardet *et al.*, 1990 for review). Interactions between cells, however, do occur later (Farfaglio, 1963), allowing for regeneration of missing parts in the adult (Martindale, 1986).

Blastomeres separated at the 2-cell stage each form half-embryos with four rows of comb-plate cilia, while those separated at the 4-cell stage each form quarter-embryos. At the 8-cell stage, isolated **E** blastomeres develop comb-plate cilia, while **M** blastomeres do not. This implies that the "potential" to form comb-plate cilia becomes segregated to the E lineage at the third cell division. At the 16-cell stage this potential is segregated similarly to the **e** micromeres (Fig. 2B). It is clear, however, that this mosaicism of the embryo is *not* an indication of prelocalized cytoplasmic determinants in the egg since, as discussed previously, the egg has no preestablished polarity, and developmental potentials in the egg are deemed to be evenly distributed because eggs bisected along any plane before or soon after fertilization regulate to yield small but complete embryos (Driesch and

Morgan, 1895b; Ziegler, 1898; Fischel, 1903; Yatsu, 1912; Freeman and Reynolds, 1973; Houliston et al., 1993).

These observations would seem at first glance to imply that there is a dramatic transition from regulation to mosaicism around the time of first division. This is accounted for in part by the way the terms mosaic and regulative are used. Embryos are considered to be mosaic because the fate of each *cell* is the same whether isolated or in place in the embryo. Eggs are considered regulative because any fertilized *fragment*, containing cytoplasm from any part of the egg, can give rise to a small but perfectly organized embryo. These are not equivalent tests. The egg, like any other cell of the embryo, is mosaic, since it develops with its normal fate if "isolated"! Conversely if the mosaicism of the embryo is tested by bisection of the individual cells, little prelocalization is detected. For example, localization of comb-plate potential toward the presumptive aboral pole is detectable at the two-cell stage, but is only a pale foreshadow of the dramatic segregation of developmental potential during micromere formation. When the future aboral region was removed from two-cell stage *Pleurobrachia* embryos, about 65% of derived EM isolates still developed comb plates, the result relating to the size of the region removed (Freeman, 1977). It seems that cytoplasmic asymmetries increase gradually during each cell cycle, with definitive "segregation" (as far as it can be assayed) accompanying specific cleavage divisions.

All in all, the concepts of localization or segregation of developmental potentials are fraught with difficulty. Consider the first two blastomeres in the ctenophore embryo. They are identical in composition and form. Nothing is segregated unequally between these cells during the first division. Yet the fate of each cell has become restricted to give a half-embryo, with four comb-plate rows. In contrast, the nucleated fragment of a bisected egg, which also contains half the material from the fertilized egg, develops into a small but complete embryo with eight comb-plate rows. Organizational changes must occur between these two stages. Our ultimate aim is to understand what these changes are and how they influence the development of the embryo.

II. Features of the Cytoplasm

As mentioned previously, the central endoplasm of the ctenophore egg is filled with densely packed, transparent yolk platelets, with cytoplasmic activity being confined to the peripheral ectoplasm. This arrangement provides two big advantages for the cell biologist. First, all cytoplasmic events, for example, growth of the sperm aster, nuclear migration, and mitosis, occur close to the cell surface, so they can be viewed clearly *in vivo* or in fixed preparations by interference contrast, fluorescence, and confocal microscopy. Second, the transparent yolk gives the egg an exceptional optical clarity allowing detailed light microscopy (Carré and Sardet, 1984; Carré *et al.,* 1991; Houliston *et al.,* 1993; Rouvière *et al.,* 1994). A further experimental advantage is that the egg is astonishingly elastic. Although normally adopting a spherical form, it can be squeezed, pummelled, and mounted flat between coverslips for live viewing or video recording without obvious ill effect. The egg heals rapidly and develops well following microinjection or bisection. Some details of experimental techniques we use for examination of *B. ovata* eggs, as well as some of the difficulties encountered with this material, are provided in the Appendix.

The endoplasm, and to a lesser extent the ectoplasm have an alveolar appearance due to the inclusion of the transparent yolk platelets (see Reverberi, 1971). The ectoplasm is rich in mitochondria of different shapes and sizes, as well as a variety of other granular and vesicular organelles (De Leo, 1980; Carré *et al.,* 1991; see Fig. 4). In *B. ovata,* the ectoplasm contains naturally fluorescent vesicles, which give the ectoplasm a green aspect when viewed by reflected light against a dark background. The fluorescent vesicles provide useful natural markers for the cytoplasmic reorganization, for example, gathering in the center of sperm asters and segregating into micromeres (Carré and Sardet, 1984; Figs. 1C and 3).

The cytoskeletal elements of the egg are confined mainly to the ectoplasm. Actin microfilaments are found in dynamic undulating folds of membrane ("micropapillae") extending from the cell cortex all over the egg surface. These can undergo rapid remodeling, for instance, elongating into long microvilli in local response to sperm entry (Carré *et al.,* 1991). Microtubules tend to run parallel to the plasma membrane, interacting closely with layered ER sheets (Fig. 4A; Houliston *et al.,* 1993; De Leo, 1980). No intermediate filaments have yet been identified. Microtubule reorganization in relation to the early events has been studied in most detail and will be described under Section III.

III. Cell Biology of Early Developmental Events

A. Fertilization

Oocytes are shed during the first meiotic division and complete meiosis irrespective of fertilization (Carré *et al.,* 1991). Fertilization may occur immediately or during the completion of meiosis, or even many hours later,

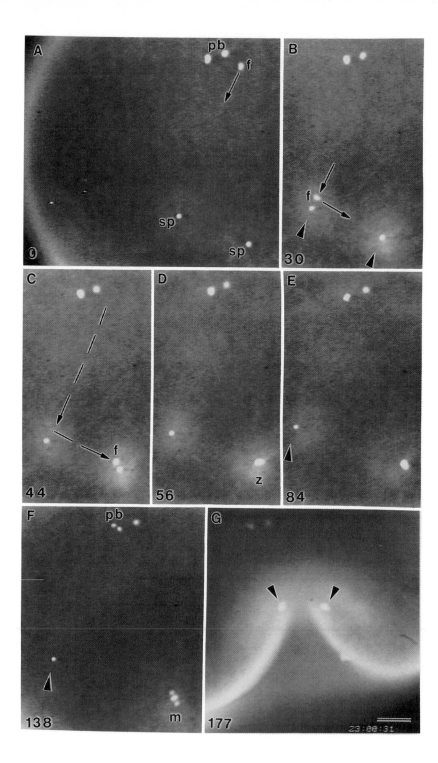

after the formation of an interphase female pronucleus. Sperm entry has been observed with exceptional clarity in *B. ovata* (Carré *et al.*, 1991). It is accompanied by a localized contraction followed by rapid changes in the form of the surrounding cell surface, the transient formation of a fertilization cone, and localized growth of microvilli around the site of sperm entry. Exocytosis occurs locally, but does not propagate, and no recognizable cortical granules are seen. Extracellular layers, including a jelly layer, and egg envelope provide at least a partial block to self-fertilization (Carré *et al.*, 1991). The structure and organization of these layers become modified first on contact with seawater and second following fertilization. The modifications following fertilization may eventually inhibit further sperm penetration, providing a slow block to polyspermy. In *B. ovata* at least, it appears there is no fast block to polyspermy. Multiple sperm entry has been observed frequently (Carré and Sardet, 1984; Carré *et al.*, 1991), each sperm triggering transient electrical changes across the membrane (Goudeau and Goudeau, 1993). Supernumarary male pronuclei do not interfere with normal development, degenerating during the first few division cycles.

In the absence of a male pronucleus, the female pronucleus remains in a decondensed interphase form for many hours, then eventually disintegrates. In fertilized eggs, the time taken to complete nuclear migration can vary widely depending on when and where sperm enter (Carré *et al.*, 1991). Entry into the embryonic cell cycle is likely to be associated somehow with the apposition or fusion of the pronuclei.

After meiosis a specialized cytoplasmic domain develops around each sperm pronucleus (Carré and Sardet, 1984; Carré *et al.*, 1991). This is in the form of "plaque" characterized by the accumulation of various organelles including mitochondria and naturally fluorescent granules, and the exclu-

---

Fig. 3 Fluorescence images from a video sequence of a fertilized *Beroe ovata* egg, covering pronuclear migration and fusion, "precleavage waves," first mitosis, and cleavage. Times in minutes are shown at bottom left. Chromatin was labeled with Hocchst 33342, and the natural fluorescence of vesicles in the ectoplasm is also visible. (A) At time 0, the female pronucleus (f) is starting to migrate in the direction of the arrow. (B) It approaches a first male pronucleus (sp) 30 min later, then moves off in a different direction. Perinuclear plaques (arrowheads) are visible due to the accumulation of fluorescent vesicles at the center of the sperm asters. (C) The female pronucleus then enters the perinuclear plaque around a second male pronucleus. The migration path is indicated. (D) After 56 min pronuclei have fused to form a zygote nucleus (z). (E) Just before prophase the precleavage waves are under way, as manifest by the disappearance of the perinuclear plaque around the zygote nucleus and partial dispersal of those around supernumary male pronuclei (arrowhead). (F) At the time of mitosis, chromatin in the supernumary male pronuclei is still condensed (arrowhead), while the polar body has divided. (G) A unipolar furrow separates the two daughters nuclei (arrowheads). m, Mitotic apparatus; pb, polar bodies; sp, male pronuclei; z, zygote nucleus. Scale bar = 100 μm.

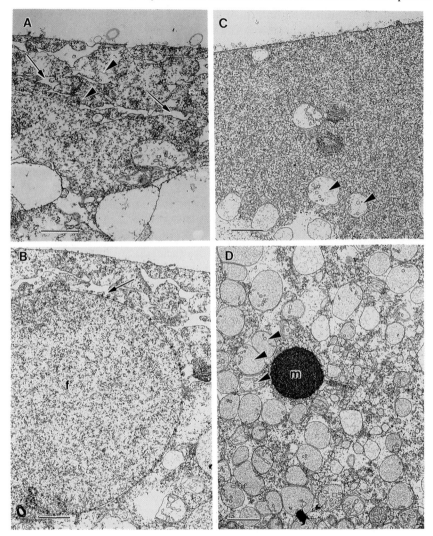

Fig. 4 Features of *Beroe* egg cytoplasm seen by transmission electron microscopy. (A) Peripheral cytoplasm of an egg fixed at the time of pronuclear migration. Microtubules (arrowheads) and ER (arrows) are organized parallel to the egg surface. (B) Migrating female pronucleus (f), showing extensive connections between the nuclear membrane and the ER (arrow). Nuclear pores are concentrated in folds at the bottom of the picture. (C) Section through a plaque of accumulated organelles overlying a male pronucleus (below the picture). (D) Mitochondria (arrowheads) gathered around a male pronucleus (m). Scale bars = 2 μm.

sion of yolk platelets (Figs. 3, 4C, and 4D). These organelle movements reflect the change in microtubule organization from a random network to the dominance of giant sperm asters (Houliston *et al.,* 1993). Centrioles have not been identified directly but the microtubule behavior implies that here, as, for instance, in amphibians and many other animals, the egg loses its centrioles during oogenesis and the fertilizing sperm provides centrioles for the zygote (see Schatten, 1994). The sperm asters are magnificent dynamic microtubule structures, lying flat in the ectoplasmic layer while sometimes attaining diameters exceeding 1 mm (in *B. ovata*). Accumulation of certain organelles at the center of these asters to create the perinuclear plaques presumably results from minus end-directed motor activity on astral microtubules. Microtubule-depolymerizing drugs prevent these domains from forming, but do not promote their dispersion (Houliston *et al.,* 1993). Such accumulation of organelles in microtubule asters is a common feature of eggs and early embryos (see Sardet *et al.,* 1994).

B. Pronuclear Migration

In eggs from many animals, the male and female pronuclei migrate simultaneously to meet in the center of the egg (for instance, in sea urchins, frogs, and mice; see Schatten, 1994). In ctenophores, however, the male pronuclei remain stationary in the ectoplasm near the sperm entry site(s) (Carré *et al.,* 1991), while the female pronucleus migrates (Fig. 3). In some polyspermic *B. ovata* eggs the female pronucleus approaches several sperm pronuclei successively, or may oscillate between nearby nuclei within an enlarged perinuclear plaque before finally choosing one to fuse with (Carré and Sardet, 1984; Carré *et al.,* 1990, 1991; Rouvière *et al.,* 1994). Whether this choice between different male pronuclei is random or has some basis remains a mystery.

In *B. ovata* eggs the female pronuclei may migrate very long distances if the sperm enter far from the site of meiosis. This has enabled a detailed investigation of the mechanism of nuclear migration (Rouvière *et al.,* 1994). The long interphase microtubules that form following the completion of meiosis provide the support for nuclear movement. In fertilized eggs, female nuclei tend to move straight toward the center of the sperm asters. In unfertilized eggs, where the microtubules form a loose random network, the nucleus follows a correspondingly random path and changes direction much more frequently. It continues to migrate at random for several hours. Experimental depolymerization of microtubules arrests nuclear migration, whereas stabilization of microtubules with taxol does not, implying a microtubule motor-based migration mechanism. Many aligned astral microtubules can be detected in association with the pronucleus during migration.

The densest microtubules overlie the migrating nucleus, and the sparser, deeper ones seem to be pushed aside, leaving a microtubule-poor trail (Fig. 5).

In both fertilized and unfertilized eggs, female pronuclei migrate at average speeds of 0.2 μm/sec. Time-lapse video recording combined with automated image analysis revealed periodic oscillations in velocity, covering distances corresponding to about three times the nuclear diameter (Rouvière et al., 1994). Characteristic deformations of the pronuclear membrane and occasional rotation of the nuclear contents accompany these oscillations. Densely packed ER is intermeshed with sperm aster microtubules and connected extensively with the nuclear membrane, with nuclear pores infolded on one side of the nucleus (Fig. 4B). These observations lead us to suggest that the female pronucleus is transported to minus end of sperm aster microtubules using motor molecules attached either to the outer nuclear membrane and/or to the network of connecting ER (Rouvière et al., 1994).

C. Precleavage Waves

Once the female pronucleus enters the zone of accumulated organelles in the center of an aster, pronuclear fusion occurs. Then, during the period between pronuclear fusion and first cleavage, waves of reorganization traverse the cortex and ectoplasm at speeds of around 10–20 μm/min (Houliston et al., 1993). These waves are manifest as the progressive displacement and dispersal of the perinuclear plaques in the cytoplasm and a series of peristaltic surface contractions. Inhibitor experiments suggest that the cytoplasmic reorganization involves microtubules. Nocodazole and taxol, which prevent microtubule dynamics by depolymarizing or stablizing them, respectively, block plaque dispersal and reduce surface movements. Actin filaments may also be involved in the surface movements: the microfilament-disrupting drug cytochalasin B does not prevent plaque dispersal but induces abnormal surface contractions.

The wave of dispersal of the perinuclear plaques reflects the progressive disassembly of the sperm asters during this period (Figs. 6A and 6B). In polyspermic eggs, the sperm asters nearest the zygote nucleus shrink first, followed by those successively further away. Such gradients of aster size could be explained by localized activation of the cell cycle following zygote nucleus formation. Enzymic changes associated with cell cycle progression modify microtubule and other cytoskeletal dynamics (e.g., Verde et al., 1990; Belmont et al., 1990) and may be activated locally within the peripheral layer of active cytoplasm in this large cell. The localized changes in microtubule dynamics can be demonstrated by treating precleavage Beroe eggs

Fig. 5 Female pronucleus (f) migrating through the microtubules of a giant sperm aster, visualized by confocal microscopy following microinjection of rhodamine–tubulin. Successive images taken with 1.5-μm steps in depth at 10-sec intervals. Scale bar = 10 μm.

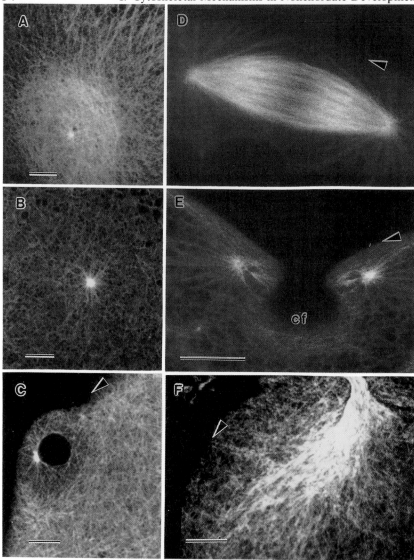

Fig. 6 Microtubule structures important in organizing the egg cytoplasm, visualized by confocal microscopy in live eggs following rhodamine–tubulin injection (A–E) or in fixed eggs by
immunofluorescence (F). (A) Sperm aster at the time of pronuclear migration; (B) sperm
aster dispersing after pronuclear fusion; (C) prophase aster; (D) anaphase spindle; (E) late
telophase from the same recording, with the unipolar cleavage furrow (cf) cutting through
the spindle; and (F) microtubules fanning out from the end of a midbody at the -2-cell stage.
Arrowheads indicate the position of the cell surface. Scale bars = 20 μm.

with taxol (Houliston *et al.*, 1993). This induces small asters around the zygote nucleus, indicating the onset of a rapid mitotic type of dynamic behavior, while stabilizing longer "interphase" microtubules elsewhere. The observed wave of organization may thus reflect the local appearance and spreading of mitotic activity as the zygote nucleus approaches mitosis.

The precleavage waves are intriguing because their direction of propagation coincides with the future oral–aboral axis of the embryo, and they occur during the period when organizational differences of significance to the embryo are first detectable with respect to this axis (see Section I,C). We must now test the hypothesis that the precleavage waves are involved in establishing cytoplasmic differences along the future body axis.

D. Mitosis and Cleavage

By the onset of mitosis most of the perinuclear plaques around supernumary sperm pronuclei have dispersed (Houliston *et al.*, 1993), and a new accumulation of mitochondria and autofluorescent vesicles reforms around the zygote nucleus (Carré *et al.*, 1991). This presumably represents accumulation of organelles on the prophase asters (Fig. 6C), while elsewhere in the egg the microtubule network is losing its distinct foci.

Mitosis itself can be followed clearly in live eggs (Figs. 6D and 6E). The opportunity this provides for examining the details of microtubule behavior in spindles, in particular with reference to the induction of the unipolar cleavage furrow in the overlying cortex, has yet to be exploited. It is interesting to note that universal cellular processes, such as mitosis and cleavage, may play a role in establishing the spatial organization of the embryo. The orientation of the spindle determines the plane of first cleavage and defines the orientation of the symmetry planes of the adult (see Section III,C). There have been hints that spindle orientation may itself be related to the direction from which the female pronucleus approaches the male prior to fusion (Carré *et al.*, 1991). Subsequent cleavages occur in fixed relation to the first cleavage plane. This probably results from a combination of centrosome positioning following each division cycle and constraint of division planes by the geometry of the cells, since cleavage planes can be altered by compressing blastomeres (Freeman, 1976b, 1979).

The unorthodox unipolar cleavages of ctenophores have long been proposed to provide a way by which cytoplasmic elements can be relocalized. The first furrow carries an accumulation of ectoplasm, with its green fluorescence (Fig. 3G), from the future oral to aboral pole (Yatsu, 1911; Spek, 1926; Carré and Sardet, 1984). The second cleavage passes along the same axis as the first, and the third obliquely in the same direction. The first three cleavages may thus be instrumental in the progressive relocation of comb-plate potential to the aboral region during this period (Freeman, 1976a). The timing as well as the orientation of cleavages is critical. For example, an oblique second cleavage induced in the same orientation as the normal third division fails to segregate comb-plate potential to the E lineage (Freeman, 1976b; see also Freeman, 1979).

The cellular basis for the acquisition of embryonic organization during cleavage, whether directly linked to organelle movement on spindle or astral microtubules or due to actin-based cortical reorganization, is unknown. De Leo (1980) noted differences in organelle distribution within the ectoplasm, the outer layer being relatively organelle free but containing multivesicular bodies, and vacuoles seeming to contribute new membrane to the furrow. Microtubules again appear good candidates as possible mediators of cytoplasmic reorganization. A prominant midbody structure, with abundant microtubules issuing from its ends, moves with the advancing furrow across the cell (Houliston *et al.,* 1993; Fig. 6E). Ectoplasm accumulates around the leading edge of the cleft and around the daughter nuclei. The continuing correspondence between microtubule structures and the organization of the ectoplasm that has been alluded to throughout this discussion is summarized in Fig. 7.

IV. Concluding Remarks

A. General Roles for the Cytoskeleton in Development

It is clear that the cytoskeleton plays a major role in organizing the cytoplasm in the ctenophore egg and developing embryo. Microtubules of the sperm asters, spindles, and midbody repeatedly gather and then redistribute cytoplasmic organelles (including ER, mitochondria, and the fluorescent vesicles), while the actin-based cortex and cleavage furrow modulate the shape of the cell and the distribution of peripheral cytoplasm. This situation is not unique to ctenophores. Sperm asters are also involved in the creation of specialized cytoplasmic domains in amphibian and ascidian eggs (see Sardet *et al.,* 1994). In these cases distinct cytoplasmic components also move into the center of the sperm aster along with the migrating female pronucleus. In the ascidian, at least, this appears developmentally significant, localizing mitochondria- and ER-rich domains to the region of the egg destined to become muscle (Sardet *et al.,* 1989; Speksnijder *et al.,* 1993).

Precleavage waves related to localized mitotic activation have also been reported in certain amphibian eggs (e.g., Sawai, 1982; Yoneda *et al.,* 1982). These eggs are of similar size to *B. ovata* eggs, with the zygote nucleus again eccentrically positioned. The "surface contraction waves" in amphibians, occurring around the time of mitosis, play a (microtubule-dependent) role in relocalizing germ plasm to one pole of the egg (Savage and Danilchik,

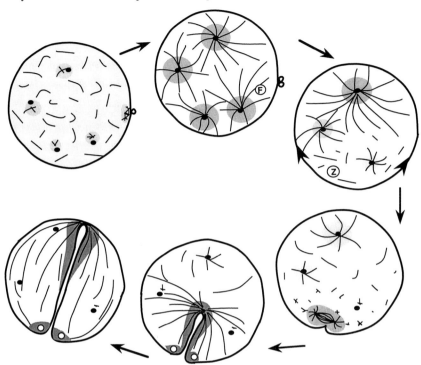

Fig. 7 Diagram summarizing microtubule reorganization and the accompanying relocations of ectoplasm in *Beroe* eggs.

1993). Amphibian eggs also show reaction of cytoplasm with advancing cleavage furrows. Superficial cytoplasm is pulled deeper into the egg with each division (Savage and Danilchik, 1993; Danilchik and Denegre, 1991).

Another general feature of egg organization which appears to be built on to organize the embryo, not only in ctenophores but also in other eggs, is the distinct nature of the egg cortex. When divided by cleavage, an inherent asymmetry is conferred to the daughter blastomeres, which now have "new" cortex on one side. This could either result directly in the inhomogenous localization of cortically attached information molecules within the blastomeres or influence cell fate indirectly though an effect on cytoskeletal organization, cleavage plane orientation, etc.

B. Perspectives for Ctenophores

Our current knowledge of ctenophore early development, outlined in this chapter, raises a number of questions which are of general interest to cell and developmental biologists. How does the meeting or fusion of the pronuclei trigger resumption of the cell cycle? How are supernumary sperm nuclei prevented from undergoing mitosis? What is the molecular and cellular basis of the precleavage waves? Do these precleavage waves influence axis establishment? What are the roles of centriole positioning and cell shape in generating the stereotypic asymmetric cleavage pattern?

The crucial embryological problem in ctenophores, concerning the nature of the segregating developmental potentials, is far from being answered. The concentration of ectoplasmic fluorescent vesicles to **e** and **m** mircromeres at the 16-cell stage is striking, but since the two types of micromere have different fates the relationship between ectoplasm and developmental potential is not simple. De Leo (1980) suggested that segregation of different types of mitochondria, perhaps with associated molecular information, may facilitate or even instruct the differentiation of different cell types from specific blastomeres. It is also conceivable that different combinations of inherited cytoplasmic elements somehow direct developmental fate. In this context it is interesting to note that although isolated ectoplasm can support the development of many larval structures, including comb-plate cilia (La Spina, 1963; Freeman and Reynolds, 1973), some of the yolky endoplasm must be present for photocytes (in *Mnemiopsis leidyi*) to develop (Freeman and Reynolds, 1973).

The evident links between mitosis, cleavage, developmental time, and the localization of developmental potential (Freeman, 1979) must have a cellular and molecular basis. All in all it may be more useful to put aside the notion of developmental potential in ctenophores for the moment and rather pose related questions in terms of cell biology: what is the difference between (the organization of) an isolated two-cell blastomere and the nucleated half of a bisected egg that can account for their different developmental fates (half larvae versus complete but half-size larvae, respectively) despite their identical constituents?

Appendix: Methods

Ctenophore embryos are not widely studied at the present time, so there is probably very limited interest in the details of experimental procedures for cell biology. Here we provide a general idea of the methods developed for *B. ovata* eggs and embryos in our laboratory. Additional details are provided in a number of published articles (e.g., Reverberi, 1971; Carré and Sardet, 1984; Carré *et al.*, 1991; Houliston *et al.*, 1993; Rouvière *et al.*, 1994). Anyone interested in exploiting *Beroe* is welcome to visit us in Villefranche. Information on other ctenophores and how to obtain their eggs can be found in Dunlap-Pianka (1974), Strathmann (1987), and Hernandez-Nicaise (1991).

A. Obtaining and Maintaining *B. ovata*

Beroe ovata are brought to the surface by upwelling currents in the Bay of Villefranche each spring. The length of the season varies from 3 to 5 months, with the most reliable period being mid-February to mid-May. Collected animals continue to spawn for 2 or 3 weeks and are kept at 15–18°C with a 12-hr light/dark cycle, with fresh seawater daily. The animals tend to spawn batches of 20–200 eggs every 1 or 2 days. Sperm are shed first and may be maintained on ice for several hours. In most cases, eggs shed subsequently by the same animal do not become fertilized. *Beroe ovata* eggs are around 1 mm in diameter, and can be picked out quite easily with the naked eye.

B. Handling and Observing Eggs

In principle, *Beroe* eggs can be fertilized in a controlled manner either by mixing eggs and sperm from different adults or by manually removing the egg envelope with fine forceps or needles to allow self-sperm access to the egg surface (Carré *et al.,* 1991). There does not appear to be an optimal moment for fertilization. Sperm will enter newly spawned, meiotic eggs or postmeiotic eggs collected many hours after spawning. Unfortunately, fertilization *in vitro* is difficult to control and does not produce highly synchronous development in egg batches. This probably reflects variability in the time taken for sperm to become activated at the egg surface and differences in nuclear migration distances. This, together with the low egg numbers and paucity of cytoplasm, does not favor biochemistry.

Once eggs are denuded, they must not be allowed to float as they may explode at air interfaces. Dishes and pipettes are coated with gelatin–form-aldehyde (GF; Zalokar and Sardet, 1984) to prevent them from sticking to the glass. Coated 50 or 100-μl capilary pipettes are convenient for handling and culturing denuded eggs.

C. Micromanipulation

Bisection of eggs and blastomere isolation can be performed while maintaining the egg envelope intact. A razor blade is simply pressed gently down in the appropriate place. This divides the envelope and leaves the developing fragments in individual culture compartments. Blastomeres from denuded embryos are easily dissociated, there being little adhesion between cells.

For microinjection we remove the egg envelope, hold the egg with forceps using one hand, and pierce the surface with a microinjection needle mounted on a micromanipulator with the other hand. Alternatively, the egg can be injected while held in a GF slide-coverslip chamber. The egg surface is deformed by the needle, so to target injected substances to the ectoplasm it is easiest to push the needle right through the center of the egg to the ectoplasm on the far side.

D. Live Observations

Denuded eggs are mounted in seawater, slightly compressed, between GF-coated slide and coverslip for live observation. Prolonged DIC or fluorescence observation and video recording are possible. Details of our microscope and video setup are found in Carré *et al.* (1991), Houliston *et al.* (1993), and Rouvière *et al.* (1994). We have also successfully followed the development of live eggs by confocal microscopy following microinjection of rhodamine-conjugated tubulin (see Houliston *et al.*, 1993; Rouvière *et al.*, 1994). Laser damage, manifest as arrest of nuclear events, can occur if observations are made too frequently, particularly in the presence of the DNA-binding dye Hoechst 33342.

E. Fixing and Staining

Ctenophore eggs are difficult to fix. They explode in almost every conventional fixative we have tried probably because of the high lipid content of the yolky endoplasm. For visualization of cytoskeletal structures by immunofluorescence we have found it best to drop denuded eggs directly into tubes of methanol or methanol/1% formaldehyde maintained in the freezer ($-20°C$). One advantage of methanol fixation is that it abolishes the natural fluorescence of the ectoplasm. When required, samples are rehydrated by underlaying the methanol with PBS containing 0.05% Triton X-100 and waiting for the eggs to sink. The endoplasm separates and can be discarded if the ectoplasm is broken and opened up into a sheet by gentle pipetting. Subsequently, embryos can be carried through primary and secondary antibodies and PBS–Triton in Eppendorf tubes in the usual sequence. To mount, we soak them in Citifluor (City University, London), then arrange them on a clean glass slide and drop on a coverslip to flatten them.

We have found it convenient to prepare isolates of the cortex and attached endoplasm for examination in detail by immunofluorescence and rapid-freeze deep-etch electron microscopy (Rouvière *et al.*, 1994). The egg surface is cleaned of jelly using EGTA-containing seawater, then stuck to polylysine-coated coverslips in an appropriate buffer. The egg is then sheared by a stream of buffer or by opening up a "sandwich" of two coverslips. In this way documented areas can be fixed and examined following live observation.

References

Belmont, L. D., Hyman, A. A., Sawin, K. E., and Mitchison, T. J. (1990). Real-time visualization of cell cycle-dependent changes in microtubule dynamics in cytoplasmic extracts. *Cell* **62**, 579–589.

Carré, D., and Sardet, C. (1984). Fertilization and early development in *Beroe ovata*. *Dev. Biol.* **105**, 188–195.

Carré, D., Sardet, C., and Rouvière, C. (1990). Fertilization in ctenophores. *In* "Mechanisms of Fertilization" (B. Dale, Ed.), pp. 626–636. Springer-Verlag, Berlin.

Carré, D., Rouvière, C., and Sardet, C. (1991). *In vitro* fertilization in ctenophores: Sperm entry, mitosis, and the establishment of bilateral symmetry in *Beroe ovata*. *Dev. Biol.* **147**, 381–391.

Chun, C (1880). Die Ctenophoren. *In* "Fauna und Flora des Golfes von Neapel," Vol. 1. Engelmann Leipzig.

Danilchik, M. V., and Denegre, J. M. (1991). Deep cytoplasmic rearrangements during early development in *Xenopus laevis*. *Development* **111**, 845–856.

De Leo, G. (1980). Fine structure of the ctenophore egg. *Acta Embryol. Morphol. Exp.* **1**, 175–198.

Driesch, H., and Morgan, T. (1895a). Zur Analyses der ersten Entwickelungsstadien des Ctenophoreneies. I Von der Entwicklung einzelner ctenophorenblastomeren. *Arch. Entwicklungsmech. Organismen.* **2**, 204–215.

Driesch, H., and Morgan, T. (1895b). Zur Analyses der ersten Entwickelungsstadien des Ctenophoreneies. II Von der Entwicklung ungefurchter Eier mit Protoplasmadefekten. *Arch. Entwicklungsmech. Organismen.* **2**, 216–224.

Dunlap Pianka, H. (1974). Ctenophora. *In* "Reproduction of Marine Invertebrates" (A. C. Giese and J. S. Pearse, eds.), Vol. 1, pp. 201–265.

Farfaglio, G. (1963). Experiments on the formation of the cilliated plates in ctenophores. *Acta Embryol. Morphol. Exp.* **6**, 191–203.

Fischel, A. (1897). Experimental Untersuchungen an ctenophorenei. I. Von der Entwicklung isolierter Eiteile. *Arch. Entwicklungsmech. Organismen.* **6**, 109–130.

Fischel, A. (1903). Entwicklung und Organdifferenzierung. *Arch. Entwicklungsmech. Organismen.* **15**, 679–750.

Freeman, G. (1976a). The role of cleavage in the localization of developmental potential in the ctenophore *Mnemiopsis leidyi*. *Dev. Biol.* **49**, 143–177.

Freeman, G. (1976b). The effects of altering the position of cleavage planes on the process of localisation of developmental potential in ctenophores. *Dev. Biol.* **51**, 332–337.

Freeman, G. (1977). The establishment of the oral–aboral axis in the ctenophores. *J. Embryol. Exp. Morphol.* **42**, 237–260.

Freeman, G. (1979). The multiple roles which cell division can play in the localization of developmental potential. *In* "Derminants of Spatial Organization" (S. Subtelny and I. R. Konigsberg, eds.), pp. 53–76. Academic Press, New York.

Freeman, G., and Reynolds, G. (1973). The development of bioluminescence in the ctenophore *Mnemiopsis leidyi*. *Dev. Biol.* **31**, 61–100.

Goudeau, M., and Goudeau, H. (1993). Successive electrical responses to insemination and concurrent sperm entries in the polyspermic egg of the ctenophore *Beroe ovata*. *Dev Biol.* **156**, 537–551.

Hernandez-Nicaise, M-L. (1991). Ctenophora. *In* "Microscopic Anatomy of Invertebrates," Vol. 2. Wiley–Liss, New York.

Houliston, E., Carré, D., Johnston, J. A., and Sardet, C. (1993). Axis establishment and microtubule-mediated waves prior to first cleavage in *Beroe ovata*. *Development* **117**, 75–87.

La Spina, R. (1963). Development of fragments of the fertilized egg of Ctenophores and their ability to form ciliated plates. *Acta Embryol. Morphol. Exp.* **6**, 204–211.

Martindale, M. Q. (1986). The ontogeny and maintenance of adult symmetry properties in the ctenophore *Mnemiopsis mccradyi*, *Dev. Biol.* **118**, 556–576.

Ortolani, G. (1989). The ctenophores: A review. *Acta. Embryol. Mophol. Exp.* **10**, 13–31.

Reverberi, G. (1971). Ctenophores. *In* "Experimental Embryology of Marine and Fresh-Water Invertebrates" (G. Reverberi, ed.), pp. 85–103. North-Holland, Amsterdam.

Rouvière, C., Houliston, E., Carré, D., Chang, P., and Sardet, C. (1994). Characteristics of pronuclear migration in *Beroe ovata*. *Cell Motil. Cytoskeleton* **29**, 301–311.

Sardet, C., McDougall, A., and Houliston, E. (1994). Cytoplasmic domains in eggs. *Trends Cell Biol.* **4**, 166–172.

Sardet, C., Speksnijder, J., Inoue, S., and Jaffe, L. F. (1989). Fertilization and ooplasmic movements in the ascidian egg. *Dev. Biol.* **105**, 237–249.

Sardet, S., Carré, D., and Rouvière, C. (1990). Reproduction and development in ctenophores. *In* "Experimental Embrylogy in Aquatic Organisms" (J. Marty, ed.), pp. 83–94. Plenum Press, New York.

Savage, R. M., and Danilchik, M. V. (1993). Dynamics of germ plasm localization and its inhibition by ultraviolet irradiation in early cleavage *Xenopus* embryos. *Dev. Biol.* **157**, 371–382.

Sawai, T. (1982). Wavelike propagation of stretching and shrinkage in the surface of the newt's egg before the first cleavage. *J. Exp. Zool.* **222**, 59–68.

Schatten, G. (1994). The centrosome and its mode of inheritance: The reduction of the centrosome during gametogesesis and its reestoration during fertilization. *Dev. Biol.* **165**, 299–335.

Spek, J. (1926). Über gesetzmässige Substanzverteilungen bei der Furchung des Ctenophoreneies und ihre Beziehungen zu den Determinations-problemen. *Arch. Entwicklungsmech. Organismen.* **107**, 54–73.

Speksnijder, J., Terasaki, M., Hage, W. J., Jaffe, L. F., and Sardet, C. (1993). Polarity and dynamics of the endoplasmic reticulum during fertilization and ooplasmic segregation in the ascidian egg. *J. Cell Biol.* **120**, 1337–1346.

Strathmann, M. F. (1987). "Reproduction and Development of Marine Invertebrates of the Northern Pacific Coast." Univ. of Washington Press, Seattle.

Verde, F., Labbé, J-C., Dorée, M., and Karsenti, E. (1990). Regulation of microtubule dynamics by cdc2 protein kinase in cell-free extracts of *Xenopus* eggs. *Nature* **343**, 233–238.

Willmer, P. (1990). "Invertebrate Relationships. Patterns in Animal Evolution." Cambridge Univ. Press, Cambridge,.

Yatsu, N. (1912). Observations and experiments on the ctenophore egg: III Experiments on germinal localization of the egg of *Beroë ovata*. *Annotnes Zool. Japon* **8**, 5–13.

Yoneda, M., Kobayakawa, Y., Kubota, H., and Sakai, M. (1982). Surface contraction waves in amphibian eggs. *J. Cell Sci.* **54,** 35–46.

Zalokar, M., and Sardet, C. (1984). Tracing of cell lineage in embryonic development of *Phallusia mammillata* (ascidia) by vital staining of mitochondria. *Dev. Biol.* **102,** 195–205.

Ziegler, H. (1898). Experimentelle Studien über die Zelltheilung. III Die Furchungszellen von *Beroë ovata. Arch. Entwicklungsmech. Organismen.* **7,** 34–64.

3

Sea Urchin Microtubules

Kathy A. Suprenant and Melissa A. Foltz Daggett
Department of Physiology and Cell Biology
University of Kansas
Lawrence, Kansas 66045

I. Introduction

From the first demonstration that the mitotic apparatus could be isolated as a discrete entity (Mazia and Dan, 1952), to the identification of the colchicine-binding protein, tubulin (Borisy and Taylor, 1967), to the purification of an mRNA–polyribosome–microtubule complex (Hamill *et al.*, 1994), sea urchin eggs and embryos have been a model system for the study of microtubule formation and function during the cell cycle. Sea urchins can be induced to spawn in the laboratory, yielding a great volume of eggs that contain a large amount of unassembled tubulin subunits and few detectable microtubules. These eggs can be fertilized *in vitro* and within a short period of time, the microtubules of the mitotic apparatus assemble and disassemble with remarkable synchrony throughout the early cleavage divisions. Thus, sea urchin eggs are viewed as an inducible model system for studying microtubule assembly and disassembly during the embryonic cleavage cycles (Schatten, 1984).

Current Topics in Developmental Biology, Vol. 31

In this chapter, attention is directed toward the molecular components that may regulate microtubule formation and function during the embryonic cell cycles. Only cytoplasmic microtubule arrays are considered. For information regarding the composition and complex organization of ciliary and flagellar microtubules, please see the following excellent reviews (Stephens, 1986; Dentler, 1987; Witman, 1990).

II. Microtubule Organization

A. Oogenesis

Oogenesis is an extraordinary process whereby a primordial germ cell transforms itself into a mature egg that contains the molecular and spatial blueprints for an entire organism. In sea urchins, oogenesis begins with the differentiation of the germinal epithelium into either accessory cells or oogonia, the latter being identified by their prominent nucleolus (Verhey and Moyer, 1967). The further maturation of oogonia into a vitellogen stage oocyte occurs during a period of intense metabolic activity that is not seen again until after 12 hr of embryonic development (Grainger et al., 1986). It is during the premeiotic stage of oogenesis that the information required for early embryonic development is positioned and stored (reviewed in Davidson, 1986).

While it has been known for some time that mature unfertilized eggs do not have recognizable microtubules (Harris et al., 1980b; Bestor and Schatten, 1981), a prominent cortical array of microtubules does exist in germinal vesicle stage oocytes of *Strongylocentrotus droebachiensis.* Using a monoclonal antibody against α-tubulin, Boyle and Ernst (1989) demonstrated that microtubules formed a basket-like network when examined in isolated cortices. In whole mounts of oocytes prepared for immunofluorescence microscopy, a radial array of microtubules is associated with a microtubule-organizing center (MTOC) situated between the cortex and the germinal vesicle. Cortical microtubules are not easily observed in whole mounts of oocytes (Boyle and Ernst, 1989). How and where the cortical microtubules arise is not known, since no connections are detected between the cortical microtubules and the MTOC. Perhaps these microtubules are nucleated by the MTOC, severed, and translocated to the cortex, a model that has been proposed for the formation of axonal and dendritic microtubules (Yu et al., 1993). When the oocyte completes the meiotic maturation divisions necessary to produce the mature haploid egg, the MTOC is no longer detected. However, shorter and fewer microtubules are still found in isolated cortices from mature eggs. Boyle and Ernst (1989) suggest that as the egg continues the maturation process, the number and length of microtubules diminish.

These cortical microtubules may help to organize cytoplasmic organelles in the oocyte. A calsequestrin-like protein immunolocalizes to the endoplasmic reticulum of the oocyte cortex as well as to microtubule arrays in isolated cortices (Henson *et al.*, 1990). This suggests that cortical microtubules may participate in the organization of the cortical endoplasmic reticulum network, as has been postulated for somatic cells (reviewed by Terasaki, 1990).

B. Fertilization and Early Cleavage

Mature sea urchin eggs prepared for conventional immunofluorescence or electron microscopy contain numerous cytoplasmic organelles, but do not appear to have any detectable microtubules (Harris *et al.*, 1980b; Bestor and Schatten, 1981). The elevation of intracellular pH from 6.9 to 7.3 following fertilization (reviewed by Whitaker and Steinhardt, 1985) leads to microtubule assembly and microtubule-dependent motile events such as pronuclear fusion (Schatten *et al.*, 1985b, 1986a). The microtubule assembly cycles following fertilization have been documented in several immunofluorescent microscopy studies (Harris *et al.*, 1980; Balczon and Schatten, 1983; Hollenbeck and Cande, 1985).

Following sperm–egg fusion and sperm incorporation, a microtubule-independent process (Longo, 1980; Schatten *et al.*, 1980), microtubules assemble in association with the sperm centrosome to form a radial array of microtubules known as the sperm aster. The sperm aster consists of a pair of centrioles surrounded by dense osmiophilic material that acts as a MTOC (Longo and Anderson, 1968, 1969). In addition to "pushing" the sperm away from the egg cortex, the sperm aster attaches tangentially to the egg pronucleus and appears to direct the movement of both pronuclei toward each other (Zimmerman and Zimmerman, 1967; Schatten and Schatten, 1981; Schatten *et al.*, 1982; Balczon and Schatten, 1983). Following pronuclear fusion and the duplication of the paternal centrosome, two radially symmetric asters are positioned at opposite poles of the zygotic nucleus. The bipolar spindle subsequently forms as the nuclear envelope disassembles and condensed chromosomes are captured and aligned at the metaphase plate (Inoue, 1981; Paweletz *et al.*, 1984, 1987a,b; Schatten *et al.*, 1986). The bipolar spindle is responsible for the accurate segregation of the chromosomes to the daughter cells (Sluder *et al.*, 1986) as well as establishing the cleavage plane axis (Rappaport, 1986). Many of the changes in microtubule organization that occur during the early cleavage cycles are illustrated in Fig. 1.

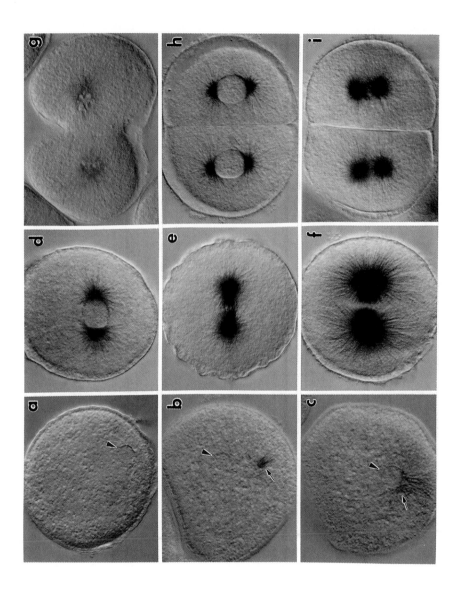

C. Gastrulation

Some of the most dramatic changes in cell shape and cell motility occur during gastrulation. The changes in microtubule distribution and orientation during the invagination of the primary mesenchyme led to a proposal that microtubules were involved in these cell shape changes (Gibbins *et al.*, 1969; Tilney and Goddard, 1970), as well as those that accompany the secondary invagination of the archenteron (Tilney and Gibbins, 1969). Recently, the role of microtubules in gastrulation was reexamined, and microtubule arrays were determined not to be essential for the rearrangement and flattening of the cells of the archenteron (Anstrom, 1989; Hardin, 1987). It is likely that the earlier results were due to the nonspecific effects of colchicine, since β-lumicolchicine, an analog of colchicine that does not bind to tubulin, also inhibits secondary invagination (Hardin, 1987). The role of microtubules in these embryonic shape changes remains unknown.

III. Biochemistry and Molecular Biology of Microtubule Components

Factors that regulate the assembly of structurally and functionally diverse microtubule arrays such as the mitotic apparatus, cortical microtubules, and interphase cytoplasmic microtubule arrays are not understood. Important players in the regulatory process may involve the synthesis of tubulin iso-

---

Fig. 1 The cycles of microtubule assembly and disassembly during the first two mitotic cleavage events in *Strongylocentrotus purpuratus* embryos are visualized by immunoperoxidase staining with a monoclonal antibody against α-tubulin, DM1A (Leslie *et al.*, 1987). Differential interference contrast light microscopy was used to visualize the nuclear membranes as well as the peroxidase staining. In the unfertilized egg (a), there are no detectable microtubules and yet, only 5 min after fertilization, microtubules of the sperm aster (arrow) are formed (b). As the microtubules of the sperm aster elongate, the male pronucleus is pushed away from the cortex as the female pronucleus (arrowhead) migrates toward the male pronucleus. At the stage just prior to pronuclear fusion (c), microtubules are broadly associated with the female pronucleus and extend outward toward the egg cortex. The centrosomes duplicate and move to opposite sides of the swelling nucleus (d). During prophase, the nuclear envelope disassembles. Chromosomes align midway between the spindle poles during metaphase (e), separate, and move toward the opposite poles during anaphase/telophase (f). During late telophase, the nuclear envelope coalesces onto chromosomes forming karyomeres as the cleavage furrow constricts to divide the zygote into two cells (g). The second cleavage division follows immediately as the centrosomes duplicate and orient themselves such that the prophase (h) and metaphase (i) mitotic apparatus forms at right angles to the first cleavage mitotic apparatus. In *S. purpuratus,* this process takes approximately 135 min at 14°C. Egg pronucleus (arrowhead). Sperm aster/pronucleus (arrow).

types, the post-translational modification of tubulins, and most likely the presence of various microtubule-associated proteins (MAPs).

A. Tubulin Synthesis and Feedback Control

1. Tubulin Genes

Sea urchins express and utilize multiple isoforms of α- and β-tubulin that belong to a multigene family (see Cleveland and Sullivan, 1985; Little and Seehaus, 1988). The tubulin gene copy number in the sea urchin genome is high, and from hybridization analyses it is estimated that there are 9–13 potential α- and β- tubulin genes (Alexandraki and Ruderman, 1981; Harlow *et al.*, 1988). Complete cDNA sequences are available for a single α-tubulin (GenBank/EMBL Data Bank accession No. X53618) and β-tubulin (GenBank/EMBL Data Bank accession No. X15389) from *Paracentrotus lividus* (Di Bernardo *et al.*, 1989; Gianguzza *et al.*, 1990). This β-tubulin is 447 amino acids long and is greater than 96% identical to human β-tubulin at the amino acid level. Similar to other tubulins, most of the amino acid substitutions are at the highly acidic C-terminus. The complete cDNA sequence for α-tubulin translates into a polypeptide of 452 amino acids, ending in a COOH-terminal tyrosine, which in most cells is enzymatically removed by a carboxypeptidase and readded by a tubulin tyrosine ligase . The functional significance of this modification is not known; however, detyrosinated tubulin is enriched in stable microtubule arrays (reviewed in Bulinski and Gunderson, 1991).

Partial cDNA sequences are available for other α-tubulin isoforms in *P. lividus* (Gianguzza *et al.*, 1989), a single *Strongylocentrotus purpuratus* β-tubulin gene (β1) (Harlow *et al.*, 1988), and several α- and β-tubulin genes of *Lytechinus pictus* (α1, α2, β1, β2, and β3) (Alexandraki and Ruderman, 1983). A comparison of the partial sequences for *L. pictus* β3 and *S. purpuratus* β1 reveals that these tubulins are highly conserved with over 95% homology at the nucleotide level (Harlow *et al.*, 1988). The nucleotide substitution rate for these β-tubulin genes is much lower than that predicted for these two species of sea urchins that diverged over 80 million years ago (Hall *et al.*, 1980). The embryonic requirement for the rapid translation of mRNAs during the early cleavage divisions may affect codon usage and bias, thus favoring certain nucleotides (Wells *et al.*, 1986; Harlow *et al.*, 1988); however, it seems more likely that the rate of nucleotide divergence reflects the intrinsic requirement for a high degree of conservation in an essential structural protein. In fact, with the exception of the C-terminal peptides, α- and β-tubulin sequences from various metazoa, protozoa, plants, algae, and fungi are all highly conserved, with minimal sequence divergence (Cleveland and Sullivan, 1985; Little and Seehaus, 1988).

The gene for γ-tubulin, a third member of the tubulin superfamily (Oakley and Oakley, 1989), has not been identified in sea urchins, although it is probably only a matter of time, since γ-tubulin is widespread in eukaryotic cells (reviewed by Oakley, 1994). γ-Tubulin is a low-abundance protein localized at MTOCs (Oakley *et al.,* 1990; Oakley, 1994) and is involved in microtubule assembly (Joshi et al., 1992).

2. Tubulin Transcripts

In addition to a stockpile of tubulin proteins (Raff *et al.,* 1971; Raff and Kaumeyer, 1973; Raff, 1975; Pfeffer *et al.,* 1976), the mature egg contains multiple α- and β-tubulin mRNAs (Alexandraki and Ruderman, 1985b; Harlow and Nemer, 1987; Gianguzza *et al.,* 1990) that are synthesized and accumulated during oogenesis (Raff *et al.,* 1972; Cogneti *et al.,* 1977). There are considerable allelic polymorphisms among the α- and β-tubulin transcripts (Alexandraki and Ruderman, 1985b). It remains unresolved as to why the number, size, and relative abundance of individual tubulin mRNAs should differ between females.

Following fertilization, these maternal tubulin mRNAs are gradually recruited into polysomes during the early cleavage divisions (Raff *et al.,* 1972; Alexandraki and Ruderman, 1985a; Harlow and Nemer, 1987). By the early blastula stage, most of the maternal tubulin mRNAs are committed to translation and new embryonic tubulin transcripts are beginning to be accumulated and translated (Alexandraki and Ruderman, 1985a; Harlow and Nemer, 1987). Individual tubulin mRNAs accumulate to different levels depending on the developmental stage and type of embryonic tissue such as ectoderm or endomesoderm (Alexandraki and Ruderman, 1985b; Harlow and Nemer, 1987). The specific constraints on microtubule function due to differences in developmental and tissue-specific expression remains to be determined.

Similar to mammalian cells (see review by Cleveland and Theodorakis, 1994), tubulin synthesis is autogenously regulated at the level of mRNA stability by the level of unpolymerized tubulin (Gong and Brandhorst, 1988a,b). Treatment of embryos with the microtubule-depolymerizing agents colchicine or nocodazole results in a rapid decline in the concentration of α- and β-tubulin mRNA. In contrast, treatment of the embryos with taxol, a microtubule-stabilizing agent, results in increased synthesis of tubulin. The decline in mRNA brought about by colchicine treatment is not due to a decrease in the transcription of tubulin genes, but rather an increase in the decay of tubulin mRNA. Gong and Brandhorst (1988c) proposed a model in which the declining levels of unpolymerized tubulin in plutei larvae account for the increased synthesis of tubulin following gastrulation.

3. Tubulin Proteins

Nearly 1% of the soluble egg protein is tubulin (Raff and Kaumeyer, 1973; Pfeffer et al., 1976; Gong and Brandhorst, 1988c) which is equivalent to a concentration of 20–25 μM tubulin dimers (Gliksman et al., 1992). Since this concentration is considerably higher than needed for microtubule assembly in vitro, the egg must have a method for maintaining the unassembled tubulin (see Section IV). In addition to comprising a large pool of microtubule precursors, the high concentration of tubulin in the unfertilized egg may also contribute to the viscosity of the cytoplasm (Sato et al., 1988). Following fertilization, this large pool of tubulin becomes incorporated into the microtubules of the first cleavage mitotic apparatus (Bibring and Baxandall, 1977) and the ciliated blastulae (Auclair and Siegel, 1966; Stephens, 1972; Bibring and Baxandall, 1981). Newly synthesized tubulin comprises no more than 0.4% of the tubulin of the first cleavage mitotic apparatus (Bibring and Baxandall, 1977) or 2–4% of the ciliary A-tubule tubulin (Bibring and Baxandall, 1981).

Egg tubulin is heterogeneous and at least two different α- and two different β-tubulin isoforms have been identified by isoelectric focusing and 2-D gel electrophoresis (Suprenant and Rebhun, 1983; Detrich and Wilson, 1983), although it is not known whether these represent different gene products or post-translational modifications. Organelle-specific tubulins can be distinguished on the basis of peptide mapping, colchicine binding, and post-translational modification (Stephens, 1978, 1992; Thompson et al., 1984; Wilson et al., 1984; Piperno and Fuller, 1985; Oka et al., 1991). Similar to other systems, the role of tubulin heterogeneity in the regulation and specification of microtubule function is not well understood (see Raff, 1994).

B. Microtubule-Associated Proteins

Microtubules are required for the organization of the mitotic apparatus and the movement of chromosomes to the daughter cells. Thus, it is likely that MAPs play an important, but as of yet, undetermined role in the formation and function of the mitotic apparatus. Several microtubule-associated proteins have been identified in sea urchin eggs and embryos (see reviews by Scholey and Leslie, 1993; Bloom and Vallee, 1989; Bloom et al., 1985b).

1. 77-kDa Echinoderm Microtubule-Associated Protein (EMAP)

In the late 1970s and early 1980s, the favored method for identifying potential microtubule-associated proteins was to purify microtubule proteins from cytosolic extracts by cycles of polymerization and depolymerization (Shelanski *et al.*, 1973; Borisy *et al.*, 1975). This method proved problematic for the purification of microtubules from sea urchin eggs (Kane, 1975) because there appeared to be a potent inhibitor of microtubule assembly (Bryan *et al.*, 1975; Naruse and Sakai, 1981; Asnes and Wilson, 1981). To overcome this problem, Keller and Rebhun (1982) succeeded in purifying microtubules from isolated mitotic apparatuses rather than from egg cytosol. An 80-kDa spindle MAP coextracted and coassembled with the mitotic apparatus tubulin through three cycles of assembly and disassembly, suggesting that this protein was a true MAP.

Subsequently, it was shown that the inhibition of microtubule assembly in egg extracts could be overcome by adding taxol (Vallee and Bloom, 1983; Scholey *et al.*, 1984; Hirokawa and Hisanaga, 1987) or dimethyl sulfoxide in a buffer at pH 7.3 (Suprenant and Marsh, 1987). A prominent microtubule-associated protein of 75–80 kDa was identified in each of these preparations. The relationship among these proteins and the 80-kDa spindle MAP is somewhat uncertain since 2-D gel electrophoresis reveals multiple 80-kDa polypeptides associated with mitotic apparatus microtubules (Leslie and Wilson, 1989). However, affinity-purified antibodies against the 77-kDa cycle-purified MAP react specifically and strongly with both the 77-kDa taxol MAP and the 80-kDa spindle MAP (Suprenant *et al.*, 1993).

Monoclonal antibodies against the 77-kDa taxol MAP (Vallee and Bloom, 1983; Bloom *et al.*, 1985a,b) and polyclonal antibodies against the cycle-purified MAP (Suprenant *et al.*, 1993) both colocalize to microtubules of the mitotic asters and central spindle, as well as microtubule arrays in adult coelomocytes. Coelomocytes are differentiated phagocytic immune cells that normally reside in the sea urchin body cavity but also flatten onto coverslips and exhibit an interphase-like microtubule array that can be easily visualized by immunofluorescence microscopy (reviewed by Edds, 1985). Both antibodies fail to recognize any polypeptides in flagellar axonemes. Because similar proteins have been identified in microtubule preparations from other echinoderms, including starfish oocytes and sand dollar eggs, we named this protein "EMAP" for echinoderm microtubule-associated protein (Suprenant *et al.*, 1993).

Recently, several cDNAs encoding the 77-kDa EMAP were isolated and sequenced (GenBank/EMBL DataBank accession No. U15551; Li and Suprenant, 1994). The deduced amino acid sequence from the cDNA clones indicates that the 77-kDa EMAP is a unique MAP with no known homologs. On the basis of charge distribution, the molecule can be divided into two distinct domains: an NH_2-terminal basic domain (aa 1–137) and a slightly acidic COOH-terminal region (aa 138–686). The COOH terminus contains 10 imperfect 40–43 amino acid repeats that end in tryptophan–aspartate (WD-40 repeats) (reviewed by van der Voorn and Ploegh, 1992; Neer *et al.,* 1994). The function of the WD-40 repeats is unknown since it is found in proteins of diverse function and subcellular location, including the β subunit of the heterotrimeric G protein β-transducin (Fong *et al.,* 1986), the tightly associated factor of RNA polymerase II transcriptional apparatus (Dynlacht *et al.,* 1993), and the yeast CDC20 gene product (Sethi *et al.,* 1991).

It is likely that the 77-kDa EMAP is of general importance to microtubules since it is found on both mitotic microtubules of the first cleavage embryo and interphase microtubules of petaloid coelomocytes (Bloom and Vallee, 1985a; Suprenant *et al.,* 1993). Recently, we suggested that EMAP may be a two-component regulator, having an effector domain to interact with and stabilize microtubules and a WD-40 repeat-containing regulatory domain (Li and Suprenant, 1994). The similarity between EMAP and the heterotrimeric G protein β-transducin suggests that EMAP might regulate microtubule formation and function through an interaction with a GTP-binding protein, perhaps even regulating the intrinsic GTPase activity of tubulin and the dynamics of microtubule assembly (Mitchison and Kirschner, 1984).

In addition to regulating microtubule formation, EMAP may be involved in the interaction of membranes with microtubules. A process which is closely linked with the formation of microtubules following fertilization is the reorganization of the endoplasmic reticulum, an organelle previously shown to be interdependent with microtubules (reviewed in Terasaki, 1990). The endoplasmic reticulum accumulates around the sperm aster and later around the spindle poles of the first cleavage mitotic apparatus (Longo and Anderson, 1968; Henson *et al.,* 1989; Terasaki and Jaffe, 1991; Jaffe and Terasaki, 1993; McDonald, 1994). It has been suggested that the ER movements are directed by minus-end motors along microtubule tracks (Terasaki and Jaffe, 1991; see also Terasaki, 1990). In this regard, the movement of a motor protein along a microtubule could be influenced by the presence of the 77-kDa EMAP.

Finally, it has been suggested that EMAP may be involved in the association of ribosomes with microtubules (Suprenant *et al.,* 1993) (see Section V). Large numbers of ribosomes are attached to EMAP-containing microtubules, but not EMAP-deficient microtubules. In addition, removal of the EMAP from microtubules by salt extraction results in the release of ribosomes from the microtubule wall, indicating that ribosome association may be mediated by the EMAP. Future studies will help to determine which of the above scenarios are most probable.

2. Buttonin

The relationship between the 77-kDa EMAP and "buttonin," a 75-kDa taxol MAP purified from the eggs of the sea urchins *Pseudocentrotus depressus* and *Hemicentrotus pulcherrimus* (Hirokawa and Hisanaga,

1987), is not clear. Buttonin is a spherical molecule, 9 nm in diameter, that stimulates the polymerization of brain tubulin at high molar ratios of buttonin:tubulin. Electron microscopy of quick-frozen and deep-etched preparations reveals that the surfaces of these microtubules are covered with hexagonally packed button-shaped molecules. Microtubules of isolated mitotic apparatus are also covered with a hexagonal array of 8 or 9-nm buttons, suggesting that buttonin is a major component of spindle microtubules (Hirokawa et al., 1985).

In addition to buttonin, larger granules (up to 26 nm in diameter) with fine tails are seen in mitotic apparatuses prepared for electron microscopy by the quick-freeze, deep-etch method (Hirokawa et al., 1985). The 26-nm granules are of the right size class to be ribosomes. RNase-sensitive ribosome-like structures associate with mitotic apparatus microtubules in vivo (Gross et al., 1958; Hartmann and Zimmerman, 1968; Goldman and Rebhun, 1969; Salmon and Segall, 1980; Silver et al., 1980; Suprenant et al., 1989). Ribosomes also copurify with egg microtubules assembled in vitro (Suprenant and Marsh, 1987; Suprenant et al., 1989, 1993; Hamill et al., 1994) (see Section V). The ribosomes are attached to the microtubule wall by an extended, slightly tapered stalk (Suprenant et al., 1989; Hamill et al., 1994).

3. Other MAPs

Several other MAPs of M_r 235, 205, 150, and 37 kDa have been identified by preparing monoclonal antibodies to less abundant polypeptides in the taxol microtubule preparations. All five of these MAPs are found in the mitotic apparatus of L. variegatus during all stages of mitosis (Vallee and Bloom, 1983; Bloom and Vallee, 1985a,b, 1989). None of these polypeptides have been detected in cilia (blastula) or flagella (sperm), indicating that they are not ciliary precursors stored in the egg (Vallee and Bloom, 1983; Bloom and Vallee, 1989). Immunofluorescence microscopy indicates that all of these proteins are found on the interphase microtubule arrays of adult, differentiated coelomocytes. It is likely that these proteins perform functions of general importance since they are found on mitotic and interphase microtubules arrays and are present in embryonic and adult cells. The 150-kDa polypeptide is unusual in that it appears to be periodically localized every 0.8 μm along microtubule bundles in coelomocytes. Both the 150- and the 235-kDa polypeptides are heat-stable MAPs (Bloom et al., 1985b), indicating that they might be related to the thermal-stable mammalian MAPs. The functions of these MAPs during mitosis and interphase are not known.

Microtubules assembled from isolated Strongylocentrotus franciscanus mitotic skeletons also contain three isoelectric variants of a 52-kDa polypeptide and a 100-kDa polypeptide with a pI of 7.0 (Leslie and Wilson, 1989).

Two additional MAPs with relative molecular masses of 24 and 34 kDa were identified by chromatographic and immunological methods (Maekawa *et al.*, 1992, 1994). Both of these proteins bind to microtubules *in vitro* and localize to the mitotic apparatus of *H. pulcherrimus* embryos *in vivo*.

In addition to the presumed structural MAPs described previously, taxol microtubules also contain a Mg–ATPase activity (Scholey *et al.*, 1984) later identified as the motor proteins cytoplasmic dynein (Porter *et al.*, 1988), kinesin (Scholey *et al.*, 1985), and a microtubule-activated 10S ATPase (Collins and Vallee, 1986), perhaps related to a dynamin-like polypeptide (Faire and Bonder, 1993; Bonder and Fishkind, Chapter 4).

Another potentially important protein identified in the mitotic apparatus is a 51-kDa basic protein (Toriyama *et al.*, 1988; Ohta *et al.*, 1988), recently identified as a GTP-binding protein that is structurally and functionally related to elongation factor-1α (EF-1α) (Ohta *et al.*, 1990; Kuriyama *et al.*, 1990b). During protein synthesis, EF-1α leads to the codon-dependent placement of the aminoacyl-tRNA at the A site of the ribosome (reviewed by Riis *et al.*, 1990). An EF-1α-like protein was identified in homogenized mitotic apparatus preparations that were capable of assembling small asters on the addition of tubulin (Toriyama *et al.*, 1988). Anti-51-kDa antibodies stained the centers of the asters and were capable of suppressing microtubule nucleation and the aster-forming activity (Ohta *et al.*, 1988). It is hard to reconcile the possible role of the EF-1α-like protein in the nucleation of microtubule assembly, especially when EF-1α purified from *Xenopus* eggs is a microtubule-severing protein (Shiina *et al.*, 1994). Both *Xenopus* EF-1α and a bacterially expressed human EF-1α rapidly sever stable microtubules when microinjected into fibroblasts. It suffices to say that EF-1α and EF-1α-like proteins may play an important role in microtubule organization; however, microtubules are not the only cytoskeletal polymer able to interact with EF-1α. EF-1α is also an actin–filament bundling protein (Yang *et al.*, 1990; Owen *et al.*, 1992). Together these results indicate that EF-1α may play a dual role in the spatial and temporal organization of the cytoskeleton and the protein-synthesizing machinery.

Finally, sea urchin mitotic spindles contain an abundant protein of 50–55 kDa that remains insoluble upon tubulin extraction (Hays and Salmon, 1983; Leslie *et al.*, 1987; Rebhun and Palazzo, 1988). These polypeptides are intermediate filament-like in their solubility and amino acid composition (Rebhun and Palazzo, 1988), yet do not cross-react with antibodies against vimentin or cytokeratin (Raymond *et al.*, 1987). In addition, there is a 52-kDa tektin-like protein in molluscan spindles that may be related to the sea urchin polypeptide (Steffen and Linck, 1992). Tektins are a family of filamentous proteins associated with the walls of axonemal microtubules of sea urchin sperm (Linck and Langevin, 1982) that have some features in common with intermediate filaments, such as amino acid composition, a high α-helical content, and similar secondary structure (reviewed in Norrander and Linck, 1994). There is some question as to whether intermediate filaments are present in sea urchin eggs because filaments have not been observed or purified; however, proteins related to intermediate filaments have been detected immunologically (Boyle and Ernst, 1989; Dufresne *et al.*, 1991; Schatten *et al.*, 1985a, 1987; St.-Pierre and Dufresne, 1990). One of the 50-kDa polypeptides coassembles with microtubules assembled in the presence of taxol, and *in vivo,* the polypeptide localizes to fibrillar material close to microtubules of the first cleavage mitotic apparatus (Raymond *et al.,* 1987). It has been suggested that the 55-kDa polypeptide provides a filamentous framework for the assembly of the mitotic apparatus (Rebhun and Palazzo, 1988; Leslie *et al.,* 1987).

IV. Regulation of Microtubule Assembly Dynamics

It is quite a puzzle as to why there are few detectable microtubules in unfertilized sea urchin eggs since the tubulin concentration in the egg is estimated at 20–25 μM (Gliksman *et al.,* 1992), a concentration that will support the self-assembly of purified egg tubulin at physiological temperatures (Keller and Rebhun, 1982; Suprenant and Marsh, 1983; Detrich and Wilson, 1983). In addition, there are several different microtubule-associated proteins found in the unfertilized egg that may promote microtubule assembly at even lower concentrations (see Section IIIB). It has been suggested that the egg contains an inhibitor(s) of microtubule assembly (Bryan *et al.,* 1975; Naruse and Sakai, 1981; Asnes and Wilson, 1981) or a dimer-sequestering mechanism (Suprenant, 1991). Previous reports have shown that sea urchin egg extracts are capable of nucleated–microtubule assembly *in vitro,* but only at the plus ends of the microtubules, suggesting an inhibitor of minus-end assembly (Simon *et al.,* 1992; Gliksman *et al.,* 1992). Recently, Spittle and Cassimeris (1994) described a dimer-binding activity that may inhibit microtubule growth specifically at the minus ends of the microtubule. This dimer-binding activity may prevent spontaneous microtubule assembly in the unfertilized egg cytoplasm such that, following fertilization, microtubule growth is nucleated by the paternal centrosome adjacent to the sperm pronucleus.

Microtubules nucleated by sperm axonemes in egg extracts *in vitro* exhibit dynamic instability (Simon *et al.,* 1992), a mechanism whereby an individual microtubule may rapidly shorten after an extended period of elongation (Mitchison and Kirschner, 1984). Egg microtubules elongate at a rate of 7.8 μm/min for approximately 1.3 min before switching to a rapid shortening or catastrophe phase (Simon *et al.,* 1992). Typically, these microtubules are rescued and begin elongating before the microtubule has completely disassembled. The kinetics of microtubule growth and shortening in unfertilized egg extracts is typical of microtubule behavior in interphase cells (Cassimeris *et al.*, 1988; Sammack and Borisy, 1988; Schultz and Kirschner, 1988). In contrast, microtubules in the sea urchin mitotic spindle are very dynamic structures (Inoue and Sato, 1967; Inoue and Ritter, 1975; reviewed by Salmon, 1989) and are capable of a rapid exchange of tubulin subunits (Salmon *et al.*, 1984; Wadsworth and Sloboda, 1983, 1984; Wadsworth and Salmon, 1986). The conversion between interphase microtubule dynamics and mitotic microtubule dynamics is thought to be brought about by the activity of the p34^{cdc2} kinase component of maturation promoting factor (Belmont *et al.*, 1990; Lamb *et al.*, 1990; Verde *et al.*, 1990, 1991, 1992) and the MAP/ERK family of kinases (Gotoh *et al.*, 1991). A general picture emerges for sea urchin extracts where protein phosphorylation promotes the formation of mitotic-like, shorter and dynamic microtubules, and dephosphorylation produces interphase-like, longer and less dynamic microtubules (Gliksman *et al.*, 1992). The target(s) for the cell cycle kinases has not been identified in sea urchins; however, it is likely that these factors regulate the frequency of rescue or catastrophe (Belmont *et al.*, 1990; Gliksman *et al.*, 1992).

Complete inhibition of protein phosphorylation in sea urchin embryos with the inhibitor 6-dimethylaminopurine (Dufresne *et al.*, 1991; Neant *et al.*, 1989), or inhibition of protein tyrosine phosphorylation with erbstatin (Wright and Schatten, 1995), blocks the formation of the sperm aster microtubules and the completion of microtubule-dependent pronuclear movements. Several phosphoproteins, including a 225-kDa polypeptide, have been identified in sea urchin centrosomes and may be targets of the p34^{cdc2} kinase (Kuriyama, 1989; Kuriyama *et al.*, 1990a). In this case, a phosphorylation event is necessary for nucleating a functional microtubule array such as the sperm aster.

Protein kinases and phosphatases copurify with isolated sea urchin mitotic apparatuses. Dinsmore and Sloboda (1988) identified a 62-kDa polypeptide that was phosphorylated by a calcium- and calmodulin-dependent kinase (CAM kinase II) in mitotic apparatuses. Under conditions in which p62 is phosphorylated, mitotic microtubules are destabilized and substantial depolymerization results. The 62-kDa polypeptide colocalizes to microtubules in isolated mitotic apparatuses (Dinsmore and Sloboda, 1988) and the level of p62 appears to be constant throughout the cell cycle (Johnston and Sloboda, 1992). Microinjection of antibodies against the 62-kDa polypeptide results in mitotic arrest and it has been suggested that phosphorylation of the p62 initiates microtubule disassembly necessary for the metaphase-to-anaphase transition (Dinsmore and Sloboda, 1989). A protein phosphatase-1 is also present in these isolated mitotic apparatus preparations which dephosphorylates the p62 in a cell cycle-dependent manner (Johnston *et al.*, 1994). It is not clear yet how p62 may promote microtubule destabilization or whether phosphorylated p62 can interact directly with microtubules.

A phosphorylation cascade initiated by the p34^{cdc2} kinase also activates a microtubule-severing activity in mitotic cells (Vale, 1991; Verde *et al.*, 1992; Shiina *et al.*, 1992; Karsenti, 1993). Stable microtubules are severed along their lengths into shorter and shorter fragments until the microtubules disassemble completely. McNally and Vale (1993) succeeded in purifying the severing protein from sea urchin egg extracts. The severing activity copurifies with katanin, a heterodimeric ATPase composed of a p81 and a p60 subunit. Severing by katanin requires Mg–ATP hydrolysis unlike the ATP-independent severing associated with p56 (Shiina *et al.*, 1992) or EF-1α in *Xenopus* (Shiina *et al.*, 1994). Severing activity may be involved in the rapid reorganization of microtubules during the interphase-to-metaphase transition or perhaps in the inhibition of microtubule growth in sea urchin eggs (Vale, 1991; Karsenti, 1993).

V. Role of Microtubules in Translational Regulation and Cytoplasmic Localization

Recent evidence suggests that microtubules and other cytoskeletal proteins are involved in mRNA localization and translational regulation (reviewed by Jeffery, 1989; Singer, 1992; Suprenant, 1993; Wilhelm and Vale, 1993; Cooley and Theurkauf, 1994). In sea urchin embryos, components of the translational machinery interact directly with microtubules *in vivo* and *in vitro*. Ribosomes are concentrated along microtubules in the mitotic apparatus (Gross *et al.*, 1958; Hartmann and Zimmerman, 1968; Goldman and Rebhun, 1969; Salmon and Segall, 1980; Silver *et al.*, 1980; Hirokawa *et al.*, 1985; Suprenant *et al.*, 1989) and on purified egg and embryo microtubules (Suprenant and Marsh, 1987; Suprenant *et al.*, 1989, 1993; Hamill *et al.*, 1994).

During the course of developing a method for the purification of egg microtubules, we noticed that the microtubules were associated with numerous, densely stained particles, approximately 24–26 nm in diameter (Suprenant and Marsh, 1987; Suprenant *et al.*, 1989). The most striking aspect of the structure was the extended and slightly curved arm that appeared to attach the particle to the microtubule wall (see Fig. 2) (Suprenant *et al.*, 1989). Based on their size, sedimentation on sucrose gradients, RNA and protein composition, RNase sensitivity, and cross-reactivity with anti-ribosome antibodies (Suprenant *et al.*, 1989; Hamill *et al.*, 1994), the microtubule-associated particles were identified as ribosomes.

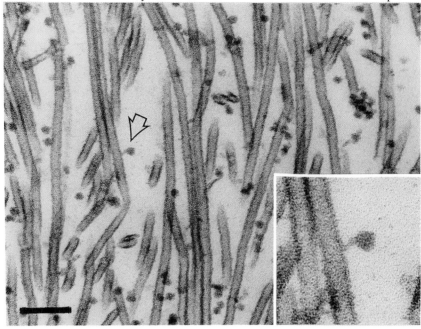

Fig. 2   Electron micrograph of a third-cycle microtubule preparation from unfertilized *S. purpuratus* eggs. The arrow points to a single ribosome attached to the microtubule by a thin stalk. An enlargement of this region is shown in the lower right. Scale bar = 0.1 μm. [Adapted from Hamill *et al.* (1994), Fig. 1; reproduced from the *Journal of Cell Biology* by copyright permission of the Rockefeller University Press.]

Ribosomes are associated with microtubules prepared from such divergent species of sea urchins as *S. purpuratus, L. pictus, L. variegatus,* and *Arbacia punctulata.* The association of ribosomes with microtubules is ionic in nature and can be disrupted in microtubule assembly buffers containing 0.45 *M* KCl. Moreover, ribosomes associate with microtubules under a variety of conditions at a constant stoichiometry of nine ribosomes per micrometer of microtubule.

The ribosomes associated with microtubules purified from unfertilized sea urchins eggs are mostly 80S monoribosomes, which is consistent with the fact that there is very little protein synthesis taking place in the unfertilized egg. In contrast, microtubules purified from two-cell embryos are covered with polyribosomes (see Fig. 3), which is consistent with the nearly 30-fold increase in protein synthesis over that observed in unfertilized eggs (Goustin and Wilt, 1981). Distinct polyribosome profiles are directly associated with the walls of the purified microtubules and, in some cases,

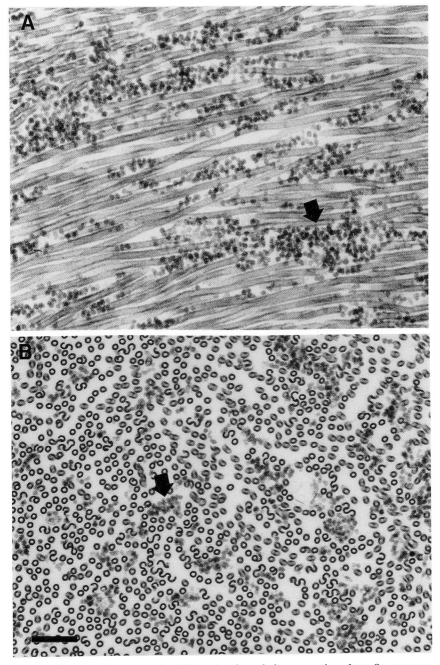

Fig. 3 Electron micrograph of a third-cycle microtubule preparations from *S. purpuratus* embryos at the two-cell stage. In longitudinal section (A) clusters of polyribosomes (arrow) are interspersed with microtubules. In cross section (B), a single polyribosome (arrow) contacts a single microtubule. Scale bar = 0.2 μm. [Adapted from Hamill *et al.* (1994), Fig. 3; reproduced from the *Journal of Cell Biology* by copyright permission of the Rockefeller University Press.]

the ribosomes are periodically associated along a microtubule with a repeat distance of 28 nm (Hamill *et al.*, 1994).

There appears to be a limiting factor for ribosome binding to egg microtubules. Only 1–3% of the embryo's ribosomes copurify with microtubules and increasing the amount of microtubules formed in the embryo extract fails to increase the amount of ribosomes that copurify. These results indicate that there may be a subset of ribosomes that bind to microtubules or that there is a limiting factor that regulates the ability of ribosomes to bind to microtubules, perhaps a ribosome-binding protein or post-translational modification (Hamill *et al.*, 1994).

In addition to ribosomes, these microtubule preparations contain two poly(A)-binding proteins of M_r 66 and 80-kDa (Drawbridge *et al.*, 1990). Both of these proteins are found in third-cycle microtubule preparations from two-cell embryos indicating that mRNA may be associated with these purified microtubules (Hamill *et al.*, 1994). A [³H]poly(U) hybridization assay indicates that 1% of the embryo's poly(A)⁺ RNA copurifies with microtubules assembled *in vitro*. On the basis of an average message size of 2000–3000 nucleotides with a 50-nucleotide poly(A) tract (Davidson, 1986), we estimate that 0.5–0.9% of the total microtubule-associated RNA is mRNA (Hamill *et al.*, 1994).

An important question is whether the microtubule-associated mRNA is translatable. To examine the translational status of the microtubule-associated RNA, microtubule-bound polyribosome complexes were added to a message-dependent rabbit reticulocyte lysate in the presence of [³⁵S]methionine. Few detectable polypeptides are translated and the microtubule preparation appears to have a general inhibitor of translation present that dramatically reduced the translation of exogenously added control RNA (Hamill *et al.*, 1994). Microtubule-associated RNA is translatable after it has been purified by phenol–chloroform extraction. Five polypeptides with relative molecular masses of 89, 77, 57, 51, and 39 kDa are synthesized well above background levels. The identities of the five messages are not known, but it is indicated that translation of specific proteins may occur on the microtubule scaffolding *in vivo*.

Without the identity of the microtubule-associated messages, we can only speculate as to the function of this association. One obvious possibility is that microtubules are involved in the localization of messages in sea urchin embryos. In the very early blastula (morula-hatching period), embryonic transcripts begin to accumulate in spatially restricted patterns (e.g., Reynolds *et al.*, 1992). Approximately 40% of these messages are progressively restricted to dividing cells, such as the oral ectoderm and endoderm of plutei larvae (Kingsley *et al.*, 1993). It is tempting to speculate that these messages are selected through an association with the mitotic apparatus microtubules (see Suprenant, 1993).

A second function for the association of polyribosomes with microtubules is in translational regulation. For example, Theodorakis and Cleveland (1992) hypothesized that tubulin synthesis may be regulated by the association of ribosomes with tubulin subunits. The autoregulation of tubulin synthesis has been described in sea urchins (Gong and Brandhorst, 1988a,b,c) and in many mammalian cells (reviewed by Cleveland and Theodorakis, 1994). In many cells, an increase in the level of unpolymerized tubulin triggers the specific degradation of tubulin mRNA. The destabilization of tubulin mRNA requires protein synthesis, specifically the synthesis of the first four amino acids of β-tubulin, Met–Arg–Glu–Ile (Pachter et al., 1987; Yen et al., 1988). How this short tubulin sequence confers mRNA instability is a matter of conjecture, but may involve the formation of a ribosome–tubulin–mRNA complex (Cleveland and Theodorakis, 1994).

VI. Prospects for the Future

Much is known regarding microtubule composition and organization in sea urchin eggs and embryos; however, we are still only speculating and forming hypotheses rather than providing detailed mechanisms for microtubule formation and function. Fundamental questions remain: How is tubulin stored in the unfertilized egg? How do MAPs function in mitosis or membrane organization? Is there a redundancy of MAP function where many different MAPs operate on the same basic principle? What is the basic principle of MAP function? During mitosis, how are catastrophe and severing factors coordinated? What are the targets for the cell cycle kinases? How is translation regulated by the cytoskeleton? We are hopeful that a biochemical and molecular approach will help identify the key players and establish potential interactions among them. In the case of microtubule-associated proteins, not only is it important to characterize their effects on microtubule dynamics, but also to identify what other proteins/macromolecules they interact with during the cell cycle. Perhaps by identifying their partners, we will understand how they function in vivo. The ability to purify large quantities of microtubule protein from a dividing cell (Vallee and Bloom, 1983; Suprenant and Marsh, 1987) and, to manipulate microtubule behavior during mitosis (Wright et al., 1993) and in cell-free extracts (Rebhun and Palazzo, 1988; Palazzo et al., 1991; Gliksman et al., 1992, 1993), makes sea urchin embryos a very useful system for studying the role of microtubules during early development.

Appendix: Methods

Most of the research in our laboratory is carried out with two species of Pacific coast sea urchins, *L. pictus* and *S. purpuratus,* although we occasionally utilize sea urchins from the Atlantic (*A. punctulata*) and Gulf coasts (*L. variegatus*). During the months of November through May when adult female *S. purpuratus* are carrying eggs, the laboratory purifies large quantities of tubulin and microtubule proteins to utilize throughout the year. Gravid *L. pictus* are usually available May through September and are utilized for smaller biochemical preparations and immunological studies.

A. Purification of Microtubules from Eggs and Embryos

One of the advantages of sea urchins is the ease in which mitotic apparatuses can be mass isolated. Since the pioneering work of Mazia and Dan (1952), several strategies have been developed for their isolation and purification (Kane, 1962; Sakai and Kuriyama, 1974; Forer and Zimmerman, 1974; Salmon and Segall, 1980; Silver *et al.,* 1980; Keller and Rebhun, 1980; Petzelt *et al.,* 1987; Rebhun and Palazzo, 1988; Dinsmore and Sloboda, 1988) and detailed methods for their isolation have been described elsewhere (Salmon, 1982; Kuriyama, 1986; Silver, 1986; Suprenant, 1986). Microtubule proteins can be purified from isolated mitotic apparatuses or mitotic cytoskeletons by cycles of temperature-dependent assembly and disassembly (Keller and Rebhun, 1980; Leslie *et al.,* 1987).

Tubulin, the major structural protein of microtubules, is easily purified from eggs and embryos by anion-exchange chromotography followed by cycles of microtubule assembly and disassembly (Kuriyama, 1977; Keller and Rebhun, 1980; Suprenant and Rebhun, 1983; Detrich and Wilson, 1983). Purified egg tubulin assembles into intact microtubules at near physiological temperatures (18°C) and at protein concentrations as low as 0.15 mg/ml (37°C). Microtubules purified in this manner are composed of greater than 95% pure tubulin and few detectable microtubule-associated proteins. Chromatography followed by polymerization in the presence of sodium L-glutamate yields highly purified tubulin with no detectable microtubule-binding proteins on silver-stained gels (Simon *et al.,* 1992). Detailed procedures for the chromatographic purification of egg tubulin are considered elsewhere (Suprenant, 1986; Detrich, 1986).

Microtubules composed of tubulin and microtubule-associated proteins can also be purified by a taxol-dependent method (Vallee and Bloom, 1983) or a pH- and temperature-dependent cycling method (Suprenant and Marsh, 1987). Both microtubule preparations contain the 77-kDa EMAP which comprises ~6 to 7% of the *S. purpuratus* microtubule mass. In general, there are more microtubule-binding proteins present in the taxol preparations than the cycle-purified microtubules. With the use of monoclonal antibodies, several of these proteins have been identified as microtubule-associated proteins with M_r of 37,000, 100,000, 150,000, 205,000, and 235,000 (Vallee and Bloom, 1983; Bloom et al., 1985a,b). A step-by-step method for the purification of microtubules from sea urchin eggs using taxol is described elsewhere (Vallee and Collins, 1986).

A significant advantage of the pH- and temperature-dependent assembly method over the taxol method is that the purified microtubule protein is capable of reversible assembly and disassembly. Thus, the microtubule protein can be used again in experiments designed to study the dynamic behavior of microtubules (e.g., Simon et al., 1992). In contrast to the taxol method, the pH- and temperature-dependent assembly method selects for microtubule-associated proteins that bind with high affinity such as the 77-kDa EMAP (Section IIIB). In addition to the 77-kDa EMAP, proteins with relative molecular masses of 39,000, 100,000, and 107,000 coassemble with these cycle-purified egg microtubules (Suprenant and Marsh, 1987; Suprenant et al., 1989, 1993). We have successfully used this procedure with eggs from the sea urchins, S. purpuratus (Suprenant and Marsh, 1987; Suprenant et al., 1989, 1993), L. pictus (Suprenant et al., 1989, 1993), L. variegatus (Suprenant et al., 1989), and A. punctulata (Suprenant et al., 1993). The greatest amounts of microtubule protein can be obtained from S. purpuratus, the Pacific coast purple urchin with the most eggs to offer. However, significant quantities of the 77-kDa EMAP can be obtained from A. punctulata in which nearly 17% of the microtubule protein by mass is the 77-kDa EMAP (Rebhun et al., 1982; Suprenant et al., 1993). In addition to sea urchins, the pH-dependent cycling method can be used to purify microtubule protein from mammalian brain and cultured cells (Tiwari and Suprenant 1993, 1994).

B. Details of the pH- and Temperature-Dependent Assembly Method

Strongylocentrotus purpuratus eggs are our major source of purified tubulin and microtubule proteins. The seasonal availability of gravid urchins has promoted the development of a scaled-up version of the method described in Suprenant and Marsh (1987) which allows us to store enough microtubule protein to work with during the "off-season."

The following method is for purifying large quantities of microtubule protein by cycles of pH- and temperature-dependent assembly and disassembly (Suprenant and Marsh, 1987, Suprenant et. al., 1989). The method does not require the housing, maintenance, and feeding of adult sea urchins and is designed to "handle" 150–300 ml of packed unfertilized sea urchin eggs; a volume of eggs, that can be obtained from one shipment of 50 very "ripe" adult *S. purpuratus*.

1. Solutions

All of our aqueous solutions are made with 18 megohm-cm type I water (Millipore Corp., Bedford, MA).

Spawning solution: 0.5 M KCl

Millipore-filtered (0.45 μm) Instant Ocean (MF-IO) (Aquarium Systems, Mentor, OH)

"19:1": 500 mM NaCl, 27 mM KCl, 2 mM EGTA, pH 7.8 Ca^{2+}-free seawater: 436 mM NaCl, 9 mM KCl, 34 mM $MgCl_2 \cdot 6H_2O$, 16 mM $MgSO_4 \cdot 7H_2O$, 1 mM EGTA, 5 mM Tris base, pH 8.0–8.3

Homogenization buffer: 100 mM PIPES–KOH, pH 7.3, 1 mM $MgSO_4$, and 4 mM EGTA

Assembly buffer: 100 mM PIPES–KOH, pH 6.9, 1 mM $MgSO_4$, and 1 mM EGTA

Dimethyl sulfoxide (DMSO): room temperature

Guanosine-5′-triphosphate (GTP), dilithium salt: 50 mM in distilled H_2O, stored at −20°C

Dithiolthreitol (DTT): 100 mM in distilled H_2O, stored at −20°C

Phenylmethylsulfonylfluoride (PMSF): 100 mM in 2-propanol, room temperature

Leupeptin: 10 mg/ml in distilled H_2O, stored at −20°C

Pepstatin: 1 mg/ml in methanol, stored at −20°C

2. Procedure

(a) In preparation for urchin arrival, the above reagents and stock solutions are made at least 1 day in advance. Fifty *S. purpuratus* sea urchins are shipped on ice via Federal Express from Marinus, Inc., (Long Beach, CA). Upon arrival, the urchins are inspected around the gonadopore region to identify any urchins which may have spontaneously spawned during shipment. If a substantial amount of spawning has occurred, the urchins are rinsed off with cold tap water before induction of spawning by the intercoelomic injection of 0.5 M KCl.

(b) Prior to spawning eggs, the styrofoam lid of the shipping container is layered with several inches of ice. Plastic tri-pour beakers (100 ml) are placed into the ice and filled with ice-cold MFIO. Urchins are injected with 5 ml of 0.5 M KCl with a 10-ml disposable syringe fitted with a 1 1/2-in. 23-guage needle; females are placed inverted on the MFIO-filled beakers. Eggs are collected on the bottom of the beakers by settling. Urchins are allowed to spawn for 30 min to 1 hr then removed from the beakers and the eggs inspected for accidental fertilization. Inspected eggs are pooled into a large glass beaker by pouring through several layers of cheesecloth to remove large debris such as spines or tubefeet. Eggs are washed twice more by settling through fresh MFIO.

(c) Washed eggs are transferred into 50-ml graduated conical-bottom plastic (polypropylene or polystyrene) centrifuge tubes (with lids) and centrifuged in a clinical centrifuge at 1000 rpm (approximate setting No. 2 on an IEC model centrifuge) for 3 min. The MFIO is aspirated away and the soft egg pellet is gently resuspended in isotonic "19:1" to remove the jelly coats. [For sea urchins other than *S. purpuratus*, the egg jelly is removed by five or six passes through a 150-μm Nitex filter (Small Parts, Inc., Miami, FL) in MFIO, not 19:1]. Eggs are once again centrifuged at 1000 rpm in a clinical centrifuge followed by one additional wash in Ca^{2+}-free artificial seawater.

(d) The volume of the packed egg pellet is estimated and an equal volume of homogenization buffer (pH 7.3), containing 2 mM DTT, 1 mM GTP, 0.2 mM PMSF, 10 μg/ml leupeptin, and 1 μg/ml pepstatin, is added. Eggs are resuspended, pooled, and homogenized in batches in an ice-cold, 55-ml Potter–Elvehjem-type tissue grinder with five to seven passes of a drill-powered pestle (Makita Cordless Driver Drill). The homogenate is transferred into 50-ml Oak Ridge polycarbonate centrifuge tubes (actual brim capacity, 43 ml) and centrifuged at 18K rpm (~39,000g) in a Beckman JA20 rotor at 4°C for 45 min. When processing greater than 320 ml of homogenate, it is necessary to use two centrifuges for the first centrifugation step.

(e) The resulting supernatants are decanted into a 500-ml or 1-liter Erlenmeyer flask at room temperature. The volume of the supernatant is estimated and a fresh aliquot of GTP is added such that the GTP concentration is increased by 1 mM. The supernatant is made 8–15% (v/v) in DMSO [In smaller microtubule preparations, we used 8% (v/v) DMSO. Scaling up the preparation sometimes requires the addition of more DMSO.] The DMSO is added in three aliquots with gentle mixing in between. Microtubule polymerization takes place for 20–30 min in a 24°C water bath. A simple water bath can be constructed from a small styrofoam container. The Erlenmeyer flask is kept from tipping over in the water bath with a lead donut around the neck of the flask. After 20 min, the polymerized microtubules in solution are poured into 50-ml Oak Ridge polycarbonate centrifuge tubes and centrifuged at 18K rpm (~39,000g) in a Beckman JA20 rotor at 24°C for 30 min. The supernatant is decanted and discarded. The lipid material on the walls of the centrifuge tubes is carefully wiped away with a Kimwipe or cotton-tipped applicator stick.

(f) First cycle microtubule pellets are suspended in ice-cold assembly buffer at pH 6.9 containing 1 mM GTP, 0.2 mM PMSF, 3 μg/ml leupeptin, and 1 μg/ml pepstatin, at 0.25× the original egg volume. The pellet is dispersed with a Teflon rod, and transferred with a pasteur pipet to a Dounce-type tissue grinder on ice. The solution is periodically homogenized over a 15- to 20-min period to disperse and depolymerize the microtubules.

(g) The depolymerized microtubule protein is transferred to 15-ml, ice-cold glass centrifuge tubes (Corex 8441) and centrifuged at 18K rpm (~39,000g) in a Beckman JA20 rotor at 4°C for 30 min The Corex 8441 tubes are used with Corning 8441 adaptors that are stored in a −20°C freezer. In this way, the microtubule solution is kept at 2–4°C. After centrifugation, the depolymerized microtubule protein in the supernatant is carefully transferred to a small Erlenmeyer flask or centrifuge tube at room temperature. DMSO is added in two aliquots until the final concentration is 10% (v/v). Microtubules are polymerized at 30°C for 20–30 min and centrifuged as described previously (step e) to obtain a second-cycle microtubule preparation.

(h) The microtubule protein is resuspended in ice-cold assembly buffer at pH 6.9 containing 1 mM GTP, 0.2 mM PMSF, 3 μg/ml leupeptin, and 1 μg/ml pepstatin at 0.1× the original egg volume. The microtubules are depolymerized and centrifuged as described in step f to obtain a second-cycle cold supernatant (C_2S). C_2S is frozen drop-wise in liquid nitrogen and stored at −80°C in disposable polypropylene centrifuge tubes.

(i) At the time of an experiment, we thaw a small number of frozen droplets of egg microtubule protein on a small weighboat. From the weight of the droplets, we estimate the volume of protein obtained. The protein is kept on ice and centrifuged briefly (10 min) in a TLA-100.3 rotor (3-ml tube size) or TLA-100 rotor (250-μl tube size) at 29,000 rpm (~39,000g) at 2°C in a tabletop ultracentrifuge (Beckman TL-100). This brief spin removes any protein that may have been denatured or aggregated upon freezing and thawing.

Acknowledgments

We thank our colleagues Jennifer Lappin, Bill Dentler, Bob Palazzo, and Dick Himes for their helpful comments on the manuscript. This work was supported by NSF MCB–9315700 and CTR 2874A.

References

Alexandraki, D., and Ruderman, J. V. (1981). Sequence heterogeneity, multiplicity, and genomic organization of α- and β-tubulin genes in sea urchins. *Mol. Cell. Biol.* **1,** 1125–1137.

Alexandraki, D., and Ruderman, J. V. (1983). Evolution of α- and β-tubulin genes as inferred by the nucleotide sequences of sea urchin cDNA clones. *J. Mol. Evol.* **19,** 397–410.

Alexandraki, D., and Ruderman, J. V. (1985a). Expression of α- and β-tubulin genes during development of sea urchin embryos. *Dev. Biol.* **109,** 436–451.

Alexandraki, D., and Ruderman, J. V. (1985b). Multiple polymorphic α- and β-tubulin mRNAs are present in sea urchin eggs. *Proc. Natl. Acad. Sci. USA* **82,** 134–138.

Anstrom, J. A. (1989). Sea urchin primary mesenchyme cells: Ingression occurs independent of microtubules. *Dev. Biol.* **131,** 269–275.

Asnes, C. F., and Wilson, L. (1981). Analysis of microtubule polymerization inhibitors in sea urchin egg extracts: Evidence for a protease. *Arch. Biochem. Biophys.* **207,** 75–80.

Auclair, W., and Siegel, B. W. (1966). Cilia regeneration in the sea urchin blastula. *Science* **154,** 913–915.

Balczon, R., and Schatten, G. (1983). Microtubule-containing detergent-extracted cytoskeletons in sea urchin eggs from fertilization through cell division: Anti-tubulin immunofluorescence microscopy. *Cell Motil.* **3,** 213–226.

Belmont, L. D., Hyman, A. A., Sawin, K. E., and Mitchison, T. J. (1990). Real-time visualization of cell cycle-dependent changes in microtubule dynamics in cytoplasmic extracts. *Cell* **62,** 579–589.

Bestor, T. H., and Schatten, G. (1981). Anti-tubulin immunofluorescence microscopy of microtubules present during the pronuclear movements of sea urchin fertilization. *Dev. Biol.* **88,** 80–91.

Bestor, T. H., and Schatten, G. (1982). Configurations of microtubules in artificially activated eggs of the sea urchin *Lytechinus variegatus. Exp. Cell Res.* **141,** 71–78.

Bibring, T., and Baxandall, J. (1977). Tubulin synthesis in sea urchin embryos: Almost all tubulin of the first cleavage mitotic apparatus derives from the unfertilized egg. *Dev. Biol.* **55,** 191–195.

Bibring, T., and Baxandall, J. (1981). Tubulin synthesis in sea urchin embryos II. Ciliary A tubulin derives from the unfertilized egg. *Dev. Biol.* **83,** 122–126.

Bloom, G. S., and Vallee, R. B. (1989). Microtubule-associated proteins in the sea-urchin egg mitotic spindle. *In* "Mitosis. Molecules and Mechanisms" J. S. Hyams and B. R. Brinkley, eds.), pp. 183–201. Academic Press, San Diego.

Bloom, G. S., Luca, F. C., Collins, C. A., and Vallee, R. B. (1985a). Use of multiple monoclonal antibodies to chracterize the major microtubule-associated protein in sea urchin eggs. *Cell Motil.* **5,** 431–446.

Bloom, G. S., Luca, F. C., Collins, C. A., and Vallee, R. B. (1985b). Isolation of mitotic microtubule-associated proteins from sea urchin eggs. *Ann. N. Y. Acad. Sci.* **466,** 328–339.

Borisy, G. G., and Taylor, E. W. (1967). The mechanism of action of colchicine. Colchicine binding to sea urchin eggs and the mitotic apparatus. *J. Cell Biol.* **34,** 535–548.

Borisy, G. G., Marcum, J. M., Olmsted, J. B., Murphy, D. B., and Johnson, K. A. (1975). Purification of tubulin and associated high molecular weight proteins from porcine brain and characterization of microtubule assembly in vitro. *Ann. N. Y. Acad. Sci.* **253,** 107–132.

Boyle, J. A., and Ernst, S. G. (1989). Sea urchin oocytes possess elaborate cortical arrays of microfilaments, microtubules and intermediate filaments. *Dev. Biol.* **134,** 72–84.

Bryan, J., Nagle, B. W., and Doenges, K. H. (1975). Inhibition of tubulin assembly by RNA and other polyanions: Evidence for a required protein. *Proc. Natl. Acad. Sci. USA* **72,** 3570–3574.

Bulinski, J. C., and Gunderson, G. G. (1991). Stabilization and post-translational modification of microtubules during cellular morphogenesis. *BioEssays* **13,** 285–293.

Cassimeris, L. U., Pryer, N. K., and Salmon, E. D. (1988). Real-time observations of microtubule dynamic instability in living cells. *J. Cell Biol.* **107,** 2223–2231.

Cleveland, D. W., and Sullivan, K. F. (1985). Molecular biology and genetics of tubulin. *Annu. Rev. Biochem.* **54,** 331–365.

Cleveland, D. W., and Theodorakis, N. G. (1994). Regulation of tubulin synthesis. *In* "Microtubules" (J. S. Hyams and C. W. Lloyd, eds.), pp. 47–58. Wiley–Liss, New York.

Cogneti, G., DiLiegro, I., and Cavarretta, F. (1977). Studies of protein synthesis during sea urchin oogenesis. II. Synthesis of tubulin. *Cell Differ.* **6,** 159–165.

Collins, C. A., and Vallee, R. B. (1986). A microtubule-activated ATPase from sea urchin eggs, distinct from cytoplasmic dynein and kinesin. *Proc. Natl. Acad. Sci. USA* **83,** 4799–4803.

Cooley, L., and Theurkauf, W. E. (1994). Cytoskeletal functions during *Drosophila* oogenesis. *Science* **266,** 590–596.

Davidson, E. H. (1986). "Gene Activity in Early Development." Academic Press, Orlando, Florida.

Dentler, W. L. (1987). Cilia and flagella. *Int. Rev. Cytol. Suppl.* **17,** 391–456.

Detrich, H. W. (1986). Isolation of sea urchin egg tubulin. *Methods Enzymol.* **134,** 128–138.

Detrich, H. W., and Wilson, L. (1983). Purification, characterization, and assembly properties of tubulin from unfertilized eggs of the sea urchin *Strongylocentrotus purpuratus. Biochemistry* **22,** 2453–2462.

Di Bernardo, M. G., Gianguzza, F., Ciaccio, M., Palla, F., Colombo, P., Di Blasi, F., Fais, M., and Spinelli, G. (1989). Nucleotide sequence of a full length cDNA clone encoding for β-tubulin of the sea urchin *Paracentrotus lividus. Nucleic Acids Res.* **17,** 5851.

Dinsmore, J. H., and Sloboda, R. D. (1988). Calcium and calmodulin-dependent phosphorylation of a 62-kDa protein induces microtubule depolymerization in sea urchin mitotic apparatuses. *Cell* **53,** 769–780.

Dinsmore, J. H., and Sloboda, R. D. (1989). Microinjection of antibodies to a 62-kDa mitotic apparatus protein arrests mitosis in dividing sea urchin embryos. *Cell* **57,** 127–134.

Drawbridge, J., Grainger, J. L., and Winkler, M. M. (1990). Identification and characterization of the poly(A)-binding proteins from the sea urchin: A quantitative analysis. *Mol. Cell. Biol.* **10,** 3994–4006.

Dufresne, L., Neant, I., St–Pierre, J., Dube, F., and Guerrier, P. (1991). Effects of 6-dimethylaminopurine on microtubules and putative intermediate filaments in sea urchin embryos. *J. Cell Sci.* **99,** 721–730.

Dynlacht, B. D., Weinzierl, R. O. J., Admon, A., and Tjian, R. (1993). The dTAF$_{II}$80 subunit of *Drosophila* TF$_{II}$D contains β-transducin repeats. *Nature* **363,** 176–179.

Edds, K. T. (1985). Morphological and cytoskeletal transformation in sea urchin coelomocytes. *In* "Blood Cells of Marine Invertebrates: Experimental Systems in Cell Biology and Comparative Physiology," pp. 53–74. A. R. Liss, New York.

Faire, K., and Bonder, E. M. (1993). Sea urchin egg 100-kDa dynamin-related protein: Identification of and localization to intracellular vesicles. *Dev. Biol.* **159,** 581–594.

Fong, H. K. W., Hurley, J. B., Hopkins, R. S., Miake-Lye, R., Johnson, M. S., Doolittle, R. F., and Simon, M. I. (1986). Repetitive segmental structure of the transducin β subunit: Homology with the CDC4 gene and identification of related mRNAs. *Proc. Natl. Acad.Sci. USA* **83,** 2162–2166.

Forer, A., and Zimmerman, A. M. (1974). Characteristics of sea urchin mitotic apparatus isolated using dimethyl sulphoxide/glycerol medium. *J. Cell Sci.* **16,** 481–493.

Gianguzza, F., Di Bernardo, M. G., Sollazzo, M., Palla, F., Ciaccio, M., Carra, E., and Spinelli, G. (1989). DNA sequence and pattern of expression of the sea urchin (*Paracentrotuslividus*) alpha-tubulin genes. *Mol. Reprod. Dev.* **1,** 170–181.

Gianguzza, F., Di Bernardo, M. G., Fais, M., Palla, F., Casao, C., Russo, R., and Spinelli, G. (1990). Sequence and expression of *Paracentrotus lividus* α tubulin gene. *Nucleic Acids Res.* **18,** 4915.

Gibbins, J. R., Tilney, L. G., and Porter, K. R. (1969). Microtubules in the formation and development of the primary mesenchyme in *Arbacia punctulata. J. Cell Biol.* **41,** 201–226.

Gliksman, N. R., Parsons, S. F., and Salmon, E. D. (1992). Okadaic acid induces interphase to mitotic-like microtubule dynamic instability by inactivating rescue. *J. Cell Biol.* **119,** 1271–1276.

Gliksman, N. R., Parsons, S. F., and Salmon, E. D. (1993). Cytoplasmic extracts from the eggs of sea urchins and clams for the study of microtubule-associated motility and bundling. *Methods Cell Biol.* **39,** 237–251.

Goldman, R. D., and Rebhun, L. I. (1969). The structure and some properties of the isolated mitotic apparatus. *J. Cell Sci.* **4,** 179–209.

Gong, Z. Y., and Brandhorst, B. (1988a). Stabilization of tubulin mRNA by inhibition of protein synthesis in sea urchin embryos. *Mol. Cell Biol.* **8,** 3518–3525.

Gong, Z. Y., and Brandhorst, B. (1988b). Autogenous regulation of tubulin synthesis via RNA stability during sea urchin embryogenesis. *Development* **102,** 31–43.

Gong, Z. Y., and Brandhorst, B. (1988c). Microtubule formation from maternal tubulins during sea urchin embryogenesis: Measurement of soluble and insoluble pools. *Mol. Reprod. Dev.* **1,** 3–9.

Gotoh, Y., Nishida, E., Matsude, S., Shiina, N., Kosako, H., Shiokawa, K., Akiyama, T., Ohta, K., and Sakai, H. (1991). *In vitro* effects on microtubule dynamics of purified *Xenopus* M phase-activated MAP kinase. *Nature* **349,** 251–254.

Goustin, A. S., and Wilt, F. W. (1981). Protein synthesis, polyribosomes, and peptide elongation in early development of *Strongylocentrotus purpuratus. Dev. Biol.* **82,** 32–40.

Grainger, J. L., von Brunn, A., and Winkler, M. M. (1986). Transient synthesis of a specific set of proteins during the rapid cleavage phase of sea urchin development. *Dev. Biol.* **114,** 403–415.

Gross, P., Philpott, D., and Nass, S. (1958). The fine-structure of the mitotic spindle in sea urchin eggs. *J. Ultrastructural Res.* **2,** 55–72.

Hall, T. J., Grula, J. W., Davidson, E. H., and Britten, R. J. (1980). Evolution of sea urchin nonrepetitive DNA. *J. Mol. Evol.* **16,** 95–110.

Hamill, D., Davis, J., Drawbridge, J., and Suprenant, K. A. (1994). Polyribosome targeting to microtubules: Enrichment of specific mRNAs in a reconstituted microtubule preparation from sea urchin embryos. *J. Cell Biol.* **127,** 973–984.

Hardin, J. (1987). Archenteron elongation in the sea urchin embryo is a microtubule-independent process. *Dev. Biol.* **121,** 253–262.

Harlow, P., and Nemer, M. (1987). Developmental and tissue-specific regulation of β-tubulin gene expression in the embryo of the sea urchin *Strongylocentrotus purpuratus*. *Genes Dev.* **1,** 147–160.

Harlow, P., Litwin, S., and Nemer, M. (1988). Synonymous nucleotide substitution rates of β-tubulin and histone genes conform to high overall genomic rates in rodents but not in sea urchins. *J. Mol. Evol.* **27,** 56–64.

Harris, P. J., Osborn, M., and Weber, K. (1980a). A spiral array of microtubules in the fertilized sea urchin egg cortex examined by indirect immunofluorescence and electron microscopy. *Exp. Cell Res.* **126,** 227–236.

Harris, P., Osborn, M., and Weber, K. (1980b). Distribution of tubulin-containing structures in the egg of the sea urchin *Strongylocentrotus purpuratus* from fertilization through first cleavage. *J. Cell Biol.* **84,** 665–679.

Hartmann, J. F., and Zimmerman, A. M. (1968). The isolated mitotic apparatus: Studies on nucleoproteins. *Exp. Cell Res.* **50,** 403–417.

Hays, T. S., and Salmon, E. D. (1983). A non-microtubular component of isolated sea urchin spindles. *J. Cell Biol.* **97,** 44a.

Henson, J. H., Begg, D. A., Beaulieu, S. M., Fishkind, D. J., Bonder, E. M., Terasaki, M., Lebeche, D., and Kaminer, B. (1989). A calsequestrin-like protein in the endoplasmic reticulum of the sea urchin: Localization and dynamics in the egg and first cell cycle embryo. *J. Cell Biol.* **109,** 149–161.

Henson, J. H., Beaulieu, S. M., Kaminer, B., and Begg, D. A. (1990). Differentiation of a calsequestrin-containing endoplasmic reticulum during sea urchin oogenesis. *Dev. Biol.* **142,** 255–269.

Hirokawa, N., and Hisanaga, S. (1987). "Buttonin," a unique button-shaped microtubule-associated protein (75-kDa) that decorates spindle microtubule surface hexagonally. *J. Cell Biol.* **104,** 1553–1561.

Hirokawa, N., Takemura, R., and Hisanaga, S. (1985). Cytoskeletal architecture of isolated mitotic spindle with special reference to microtubule-associated proteins and cytoplasmic dynein. *J. Cell Biol.* **101,** 1858–1870.

Hollenbeck, P. J., and Cande, W. Z. (1985). Microtubule distribution and reorganization in the first cell cycle of fertilized eggs of *Lytechinus pictus*. *Eur. J. Cell Biol.* **37,** 140–148.

Inoue, S. (1981). Cell division and the mitotic spindle. *J. Cell Biol.* **91,** 131s–147s.

Inoue, S., and Ritter, H., Jr. (1975). Dynamics of the mitotic spindle organization and function. *In* "Molecules and Cell Movement" (S. Inoue and R. E. Stephens, eds.), pp. 3–30. Raven Press, New York.

Inoue, S., and Sato, H. (1967). Cell motility by labile association of molecules. The nature of mitotic spindle fibers and their role in chromosome movement. *J. Gen. Physiol.* **50,** 259–292.

Jaffe, L. A., and Terasaki, M. (1993). Structural changes of the endoplasmic reticulum of sea urchin eggs during fertilization. *Dev. Biol.* **156,** 566–573.

Jeffery, W. R. (1989). Localized mRNA and the egg cytoskeleton. *Int. Rev. Cytol.* **119,** 151–195.

Johnson, J. A., and Sloboda, R. D. (1992). A 62-kDa protein required for mitotic progression is associated with the mitotic apparatus during M-phase and with the nucleus during interphase. *J. Cell Biol.* **119,** 843–854.

Johnston, J. A., Sloboda, R. D., and Silver, R. B. (1994). Phosphoprotein phosphatase 1 (PP1) is a component of the isolated sea urchin mitotic apparatus. *Cell Motil. Cytoskeleton* **29,** 280–290.

Joshi, H. C., Palacios, M. J., McNamara, L., and Cleveland, D. W. (1992). γ-tubulin is a centrosomal protein required for cell cycle-dependent microtubule nucleation. *Nature* **356,** 80–83.

Kane, R. E. (1962). The mitotic apparatus: Isolation by controlled pH. *J. Cell Biol.* **12,** 47–55.

Kane, R. E. (1975). Preparation and purification of polymerized actin from sea urchin egg extracts. *J. Cell Biol.* **66,** 305–315.

Karsenti, E. (1993). Severing microtubules in mitosis. *Curr. Biol.* **3,** 208–210.

Keller, T. C. S., and Rebhun, L. I. (1982). *Strongylocentrotus purpuratus* spindle tubulin. I. Characteristics of its polymerization and depolymerization *in vitro. J. Cell Biol.* **93,** 788–796.

Kingsley, P. D., Angerer, L. M., and Angerer, R. C. (1993). Major temporal and spatial patterns of gene expression during differentiation of the sea urchin embryo. *Dev. Biol.* **155,** 216–234.

Kuriyama, R. (1977). *In vitro* polymerization of marine egg tubulin into microtubules. *J. Biochem.* **81,** 1115–1125.

Kuriyama, R. (1986). Isolation of sea urchin spindles and cytasters. *Methods Enzymol.* **134,** 190–199.

Kuriyama, R. (1989). 225-kDa phosphoprotein associated with mitotic centrosomes in sea urchin eggs. *Cell Motil. Cytoskeleton* **12,** 90–103.

Kuriyama, R., Rao, P. N., and Borisy, G. G. (1990a). Immunocytochemical evidence for centrosomal phosphoproteins in mitotic sea urchin eggs. *Cell Structure Function* **15,** 13–20.

Kuriyama, R., Savereide, P., Lefebvre, P., and Dasgupta, S. (1990b). The predicted amino acid sequence of a centrosphere protein in dividing sea urchin eggs is similar to elongation factor (EF-1α). *J. Cell Sci.* **95,** 231–236.

Lamb, N. J., Fernandez, A., Watrin, A., Labbe, J. C., and Calvadore, J. C. (1990). Microinjection of p34[cdc2] kinase induces marked changes in cell shape, cytoskeletal organization and chromatin structure in mammalian fibroblasts. *Cell* **60,** 151–165.

Leslie, R. J., and Wilson, L. (1989). Preparation and characterization of mitotic cytoskeletons from embryos of the sea urchin *Strongylocentrotus franciscanus. Anal. Biochem.* **181,** 51–58.

Leslie, R. J., Hird, R. B., Wilson, L., McIntosh, J. R., and Scholey, J. M. (1987). Kinesin is associated with a nonmicrotubule component of sea urchin mitotic spindles. *Proc. Natl. Acad. Sci. USA* **84,** 2771–2775.

Li, Q., and Suprenant, K. A. (1994). Molecular characterization of the 77-kDa echinoderm microtubule-associated protein. *J. Biol. Chem.* **269,** 31777–31784.

Linck, R. W., and Langevin, G. L. (1982). Structure and chemical composition of insoluble filamentous components of sperm flagellar microtubules. *J. Cell Sci.* **58,** 1–22.

Little, M., and Seehaus, T. (1988). Comparative analysis of tubulin sequences. *Comp. Biochem. Physiol. B* **90,** 655–670.

Longo, F. (1980). Organization of microfilaments in sea urchin (*Arbacia punctulata*) eggs at fertilization: Effects of cytochalasin B. *Dev. Biol.* **74,** 422–431.

Longo, F. J., and Anderson, E. (1968). The fine structure of pronuclear development and fusion in the sea urchin, *Arbacia punctulata. J. Cell Biol.* **39,** 339–368.

Longo, F. J., and Anderson, E. (1969). Sperm differentiation in the sea urchins *Arbacia punctulata* and *Strongylocentrotus purpuratus. J. Ultrastructure Res.* **27,** 486–509.

Maekawa, S., Toriyama, M., and Sakai, H. (1992). A novel 24-kDa microtubule-associated protein purified from sea urchin eggs. *Eur. J. Biochem.* **205,** 1195–1200.

Maekawa, S., Mishima, M., Toriyama, M., and Sakai, H. (1994). Purification of a low molecular weight microtubule binding protein from sea urchin eggs. *Biochim. Biophys. Acta* **1207,** 194–200.

Mazia, D., and Dan, K. (1952). The isolation and biochemical characterization of the mitotic apparatus of dividing cells. *Proc. Natl. Acad. Sci. USA* **38,** 826–838.

McDonald, K. (1994). Membrane ultrastructure in early *Strongylocentrotus purpuratus* embryos: Improved resolution using high-pressure freezing and freeze substitution. *In* "Reproduction and Development of Marine Invertebrates" (W. H. Wilson, Jr., S. A. Stricker, and G. L. Shinn, eds.), pp. 50–63. The John Hopkins Univ. Press, Baltimore.

McNally, F. J., and Vale, R. D. (1993). Identification of katanin, an ATPase that severs and disassembles stable microtubules. *Cell* **75,** 419–429.

Mitchinson, T. J., and Kirschner, M. W. (1984). Dynamic instability of microtubule growth. *Nature* **312,** 237–242.

Naruse, H., and Sakai, H. (1981). Evidence that a polysaccharide from the cortex of sea urchin egg inhibits microtubule assembly through its binding to microtubule-associated proteins. *J. Biochem.* **90,** 581–587.

Neant, I., Charbonneau, M., and Guerrier, P. (1989). A requirement for protein phosphorylation in regulating the meiotic and mitotic cell cycles in echinoderms. *Dev. Biol.* **132,** 304–314.

Neer, E. J., Schmidt, C. J., Nambudripad, R., and Smith, T. F. (1994). The ancient regulatory-protein family of WD-repeat proteins. *Nature* **371,** 297–300.

Norrander, J. M., and Linck, R. W. (1994). Tektins. *In* "Microtubules" (J. S. Hyams and C. W. Lloyd, eds.), pp. 201–220. Wiley–Liss, New York.

Oakley, B. R. (1994). γ-tubulin. *In* "Microtubules" (J. S. Hyams and C. W. Lloyd, eds.), pp. 33–45. Wiley–Liss, New York.

Oakley, B. R., Oakley, C. E., Yoon, Y., and Jung, C. K. (1990). γ-tubulin is a component of the spindle-pole-body that is essential for microtubule function in *Aspergillus nidulans. Cell* **61,** 1289–1301.

Oakley, C. E., and Oakley, B. R. (1989). Identification of γ-tubulin, a new member of the tubulin superfamily encoded by mipA gene of *Aspergillus nidulans. Nature* **338,** 662–664.

Ohta, K., Toriyama, M., Endo, S., and Sakai, H. (1988). Localization of mitotic-apparatus-associated 51-kDa protein in unfertilized and fertilized sea urchin eggs. *Cell Motil. Cytoskeleton* **10,** 496–505.

Ohta, K., Toriyama, M., Miyazaki, M., Murofushi, H., Hosoda, S., Endo, S., and Sakai, H. (1990). The mitotic apparatus-associated 51-kDa protein from sea urchin eggs is a GTP-binding protein and is immunologically related to yeast polypeptide elongation factor 1α. *J. Biol. Chem.* **265,** 3240–3247.

Oka, M. T., Arai, T., and Hamaguchi, Y. (1991). Change in the heterogeneous distribution of tubulin isotypes in mitotic microtubules of the sea urchin egg by treatment with microtubule depolymerizing or stabilizing drugs. *Cell Structure Function* **16,** 125–134.

Owen, C. H., DeRosier, D. J., and Condeelis, J. (1992). Actin crosslinking protein EF-1α of *Dictyostelium discoideum* has a unique bonding rule that allows square-packed bundles. *J. Structure Biol.* **109,** 248–254.

Pachter, J. S., Yen, T. J., and Cleveland, D. W. (1987). Autoregulation of tubulin expression is achieved through specific degradation of polysomal tubulin mRNAs. *Cell* **51,** 283–292.

Palazzo, R. E., Lutz, D. A., and Rebhun, L. I. (1991). Reactivation of isolated mitotic apparatus: Metaphase versus anaphase spindles. *Cell Motil. Cytoskeleton* **18,** 304–318.

Paweletz, N., Mazia, D., and Finze, E-M. (1984). The centrosome cycle in the mitotic cycle of sea urchins. *Exp. Cell Res.* **152,** 47–65.

Paweletz, N., Mazia, D., and Finze, E-M. (1987a). Fine structural studies of the bipolarization of the mitotic apparatus in the fertilized sea urchin egg. I. The structure and behavior of centrosomes before fusion of the pronuclei. *Eur. J. Cell Biol.* **44,** 195–204.

Paweletz, N., Mazia, D., and Finze, E-M. (1987b). Fine structural studies of the bipolarization of the mitotic apparatus in the fertilized sea urchin egg. II. Bipolarization before the first mitosis. *Eur. J. Cell Biol.* **44,** 205–213.

Petzelt, C., Hafner, M., Mazia, D., and Sawin, K. W. (1987). Microtubules and Ca^{2+}-sequestering membranes in the mitotic apparatus, isolated by a new method. *Eur. J. Cell Biol.* **45,** 268–273.

Pfeffer, T. A., Asnes, C. R., and Wilson, L. (1976). Properties of tubulin in unfertilized sea urchin eggs. Quantitation and characterization by the colchicine-binding reaction. *J. Cell Biol.* **69,** 599–607.

Piperno, G., and Fuller, M. T. (1985). Monoclonal antibodies specific for an acetylated form of α-tubulin recognize the antigen in cilia and flagella from a variety of organisms. *J. Cell Biol.* **101,** 2085–2094.

Porter, M. E., Grissom, P. M., Scholey, J. M., Salmon, E. D., and McIntosh, J. R. (1988). Dynein isoforms in sea urchin eggs. *J. Biol. Chem.* **263,** 6759–6771.

Raff, E. C. (1994). The role of multiple tubulin isoforms in cellular microtubule function. *In* "Microtubules" (J. S. Hyams and C. W. Lloyd, eds.), pp. 85–109. Wiley–Liss, New York.

Raff, R. A. (1975). Regulation of microtubule synthesis and utilization during early embryonic development of the sea urchin. *Am. Zool.* **15,** 661–678.

Raff, R. A., and Kaumeyer, J. F. (1973). Soluble microtubule proteins of the sea urchin embryo: Partial characterization of the proteins and behavior of the pool in early development. *Dev. Biol.* **32,** 309–320.

Raff, R. A., Greenhouse, G., Gross, K. W., and Gross, P. R. (1971). Synthesis and storage of microtubule proteins by sea urchin embryos. *J. Cell Biol.* **50,** 516–527.

Raff, R. A., Colot, H. V., Selvig, S. E., and Gross, P. R. (1972). Oogenetic origin of messenger RNA for embryonic synthesis of microtubule proteins. *Nature* **235,** 211–214.

Rappaport, R. (1986). Establishment of the mechanism of cytokinesis in animal cells. *Int. Rev. Cytol.* **105,** 245–281.

Raymond, M-N., Foucault, G., Renner, M., and Pudles, J. (1987). Isolation of a 50-kDa polypeptide from the detergent-resistant unfertilized sea urchin egg cytomatrix and evidence for its change in organization during mitosis. *Eur. J. Cell Biol.* **45,** 302–310.

Rebhun, L. L., and Palazzo, R. E. (1988). *In vitro* reactivation of anaphase B in isolated spindles of the sea urchin egg. *Cell Motil. Cytoskeleton* **10,** 197–209.

Rebhun, L. I., Suprenant, K., Keller, T. C. S., and Folley, L. (1982). Spindle and cytoplasmic tubulins from marine eggs. *Cell Differ.* **11,** 367–371.

Reynolds, S. D., Angerer, L. M., Palis, J., Nasir, A., and Angerer, R. C. (1992). Early mRNAs, spatially restricted along the animal–vegetal axis of sea urchin embryos, include one encoding a protein related to tolloid and BMP-1. *Development* **114,** 769–786.

Riis, B., Rattan, S. I. S., Clark, B. F. C., and Merrick, W. C. (1990). Eukaryotic protein elongation factors. *TIBS* **15,** 420–425.

Sakai, H., and Kuriyama, R. (1974). The mitotic apparatus isolated in glycerol-containing medium. *Dev. Growth Differ.* **16,** 123–134.

Salmon, E. D. (1982). Mitotic spindles isolated from sea urchin eggs with EGTA–lysis buffer. *Methods Cell Biol.* **25,** 69–105.

Salmon, E. D. (1989). Microtubule dynamics and chromosome movement. *In* "Mitosis: Molecules and Mechanisms" (J. S. Hyams and B. R. Brinkley, eds.), pp. 119–181. Academic Press, San Diego.

Salmon, E. D., and Segall, R. R. (1980). Calcium–labile mitotic spindles isolated from sea urchin eggs (*Lytechinus variegatus*). *J. Cell Biol.* **86,** 355–365.

Salmon, E. D., Leslie, R. J., Saxton, W. M., Karow, M. L., and McIntosh, J. R. (1984). Spindle microtubule dynamics in sea urchin embryos: Analysis using a fluorescein-labeled tubulin and measurements of fluorescence redistribution after laser photobleaching. *J. Cell Biol.* **99,** 2165–2174.

Sammack, P. J., and Borisy, G. G. (1988). Direct observations of microtubule dynamics in living cells. *Nature (London)* **332**, 724–726.

Sato, M., Schwartz, W. H., Selden, S. C., and Pollard, T. D. (1988). Mechanical properties of brain tubulin and microtubules. *J. Cell Biol.* **106**, 1205–1211.

Schatten, G. (1984). The supramolecular organization of the cytoskeleton during fertilization. *Subcell. Biochem.* **10**, 359–453.

Schatten, G., and Schatten, H. (1981). Effects of motility inhibitors during sea urchin fertilization. *Exp. Cell Res.* **135**, 311–330.

Schatten, G., Bestor, T. H., Schatten, H., and Balczon, R. (1982). Taxol inhibits the nuclear movements during fertilization and induces asters in unfertilized sea urchin eggs. *J. Cell Biol.* **94**, 455–465.

Schatten, G., Bestor, T., Balczon, R., Henson, J., and Schatten, H. (1985a). Intracellular pH shift leads to microtubule assembly and microtubule-mediated motility during sea urchin fertilization: Correlations between elevated intracellular pH and microtubule activity and depressed intracellular pH and microtubule disassembly. *Eur. J. Cell Biol.* **36**, 116–127.

Schatten, G., Maul, G. G., Schatten, H., Chaly, N., Simerly, C., Balczon, R., and Brown, D. L. (1985b). Nuclear lamins and peripheral nuclear antigens during fertilization and embryogenesis in mice and sea urchins. *Proc. Natl. Acad. Sci. USA* **82**, 4727–4731.

Schatten, G., Bestor, T., Balczon, R., Henson, J., and Schatten, H. (1986a). Intracellular pH shift imitates microtubule-mediated motility during sea urchin fertilization. *Ann. N. Y. Acad. Sci.* **466**, 940–944.

Schatten, H., Bestor, T., and Schatten, G. (1980). Motility during fertilization: Sperm incorporation is prevented by microfilament inhibitors while pronuclear movements are prevented by microtubule inhibitors and immunofluorescence of associated microtubules. *Eur. J. Cell Biol.* **22**, 365a.

Schatten, H., Schatten, G., Mazia, D., Balczon, R., and Simerly, C. (1986b). Behavior of centrosomes during fertilization and cell division in mouse oocytes and in sea urchin eggs. *Proc. Natl. Acad. Sci. USA* **83**, 105–109.

Schatten, H., Walter, M., Mazia, D., Biessmann, H., Paweletz, N., Coffee, G., and Schatten, G. (1987). Centrosome detection in sea urchin eggs with a monoclonal antibody against *Drosophila* intermediate filament proteins: Characterization of stages of the division cycle of centrosomes. *Proc. Natl. Acad. Sci. USA* **84**, 8488–8492.

Scholey, J. M., and Leslie, R. J. (1993). Sea urchin MAPs and microtubule motors. *In* "Guidebook to the Cytoskeletal and Motor Proteins" (T. Kreis and R. Vale, eds.), pp. 120–122. Oxford Univ. Press, New York.

Scholey, J. M., Neighbors, B., McIntosh, J. R., and Salmon, E. D. (1984). Isolation of microtubules and a dynein-like MgATPase from unfertilized sea urchin eggs. *J. Biol. Chem.* **259**, 6516–6525.

Scholey, J. M., Porter, M. E., Grissom, P. M., and McIntosh, J. R. (1985). Identification of kinesin in sea urchin eggs, and evidence for its localization in the mitotic spindle. *Nature* **318**, 483–486.

Schulze, E., and Kirschner, M. (1988). New features of microtubule dynamics in living cells. *Nature (London)* **334**, 356–359.

Sethi, N., Monteagudo, M. C., Koshland, D., Hogan, E., and Burke, D. J. (1991). The CDC20 gene product of *Saccharomyces cerevisiae,* a β-transducin homolog, is required for a subset of microtubule-dependent cellular processes. *Mol. Cell. Biol.* **11**, 5592–5602.

Shelanski, M. L., Gaskin, F., and Cantor, C. R. (1973). Assembly of microtubules in the absence of added nucleotide. *Proc. Natl. Acad. Sci. USA* **70**, 765–768.

Shiina, N., Gotoh, Y., and Nishida, E. (1992). A novel homo-oligomeric protein responsible for an MPF-dependent microtubule severing activity. *EMBO J.* **11**, 4723–4731.

Shiina, N., Gotoh, Y., Kubomura, N., Iwamatsu, A., and Nishida, E. (1994). Microtubule severing by elongation factor 1α. *Science* **266**, 282–285.

Silver, R. B. (1986). Isolation of native, membrane-containing mitotic apparatus from sea urchin embryos. *Methods Enzymol.* **134**, 200–217.

Silver, R. B., Cole, R. D., and Cande, W. Z. (1980). Isolation of mitotic apparatus containing vesicles with calcium sequestering activity. *Cell* **19**, 505–516.

Simon, J. R., Parsons, S. F., and Salmon, E. D. (1992). Buffer conditions and non-tubulin factors critically affect the microtubule dynamic instability of sea urchin egg tubulin. *Cell Motil. Cytoskeleton* **21**, 1–14.

Singer, R. H. (1992). The cytoskeleton and mRNA localization. *Curr. Opin. Cell Biol.* **4**, 15–19.

Sluder, G., Miller, F. J., and Rieder, C. L. (1986). The reproduction of centrosomes: Nuclear versus cytoplasmic controls. *J. Cell Biol.* **103**, 1873–1881.

Spittle, C., and Cassimeris, L. (1994). Sea urchin eggs contain a tubulin-dimer binding activity which limits assembly to plus ends. *Mol. Biol. Cell.* **5**, 5a.

Steffen, W., and Linck, R. W. (1992). Evidence for a non-tubulin spindle matrix and for spindle components immunologically related to tektin filaments. *J. Cell Sci.* **101**, 809–822.

Stephens, R. E. (1972). Studies on the development of the sea urchin *Strongylocentrotus droebachiensis*. III. Embryonic synthesis of ciliary proteins. *Biol. Bull.* **142**, 488–504.

Stephens, R. E. (1978). Primary structural differences among tubulin subunits from flagella, cilia, and the cytoplasm. *Biochemistry* **17**, 2882–2891.

Stephens, R. E. (1986). Isolation of embryonic cilia and sperm flagella. *Methods Cell Biol.* **27**, 217–227.

Stephens, R. E. (1992). Tubulin in sea urchin embryonic cilia: Post-translational modifications during development. *J. Cell Sci.* **101**, 837–845.

St.-Pierre, J., and Dufresne, L. (1990). Identification and localization of two proteins immunologically related to intermediate filament proteins in sea urchin eggs and embryos. *Cell Motil. Cytoskeleton* **17**, 75–86.

Suprenant, K. A. (1986). Tubulin-containing structures. *Methods Cell Biol.* **27**, 189–215.

Suprenant, K. A. (1991). Unidirectional microtubule assembly in cell-free extracts of *Spisula solidissima* oocytes is regulated by subtle changes in pH. *Cell Motil. Cytoskeleton* **19**, 207–220.

Suprenant, K. A. (1993). Microtubules, ribosomes and RNA: Evidence for cytoplasmic localization and translational regulation. *Cell Motil. Cytoskeleton* **25**, 1–9.

Suprenant, K. A., and Marsh, J. C. (1987). Temperature and pH govern the self-assembly of microtubules from unfertilized sea urchin egg extracts. *J. Cell Sci.* **87**, 71–84.

Suprenant, K. A., and Rebhun, L. I. (1983). Assembly of unfertilized sea urchin egg tubulin at physiological temperatures. *J. Biol. Chem.* **258**, 4518–4525.

Suprenant, K. A., Tempero, L. B., and Hammer, L. E. (1989). Association of ribosomes with *in vitro* assembled microtubules. *Cell Motil. Cytoskeleton* **14**, 401–415.

Suprenant, K. A., Dean, K., McKee, J., and Hake, S. (1993). EMAP, an echinoderm microtubule-associated protein found in microtubule-ribosome complexes. *J. Cell Sci.* **104**, 445–456.

Terasaki, M. (1990). Recent progress on structural interactions of the endoplasmic reticulum. *Cell Motil. Cytoskeleton* **15**, 71–75.

Terasaki, M., and Jaffe, L. A. (1991). Organization of the sea urchin egg endoplasmic reticulum and its reorganization at fertilization. *J. Cell Biol.* **114**, 929–940.

Theordorakis, N. G., and Cleveland, D. W. (1992). Physical evidence for cotranslational regulation of β-tubulin mRNA degradation. *Mol. Cell Biol.* **12**, 791–799.

Thompson, W. C., Asai, D. J., and Carney, D. H. (1984). Heterogenicity among microtubules of the cytoplasmic microtubule complex detected by a monoclonal antibody to α-tubulin. *J. Cell Biol.* **98**, 1017–1025.

Tilney, L. G., and Gibbins, J. R. (1969). Microtubules and filaments in the filopodia of the secondary mesenchyme cells of *Arbacia punctulata* and *Echinarachnius parma*. *J. Cell Sci.* **5,** 195–210.

Tilney, L. G., and Goddard, J. (1970). Nucleating sites for the assembly of cytoplasmic microtubules in the ectodermal cells of blastulae of *Arbacia punctulata*. *J. Cell Biol.* **46,** 564–575.

Tiwari, S. C., and Suprenant, K. A. (1993). A pH- and temperature-dependent cycling method that doubles the yield of microtubule protein. *Anal. Biochem.* **215,** 96–103.

Tiwari, S. C., and Suprenant, K. A. (1994). pH-dependent solubility and assembly of microtubules in bovine brain extracts. *Cell Motil. Cytoskeleton* **28,** 69–78.

Toriyama, M., Ohta, K., Endo, S., and Sakai, H. (1988). 51-kDa protein, a component of microtubule-organizing granules in the mitotic apparatus involved in aster formation *in vitro*. *Cell Motil. Cytoskeleton* **9,** 117–128.

Vale, R. D. (1991). Severing of stable microtubules by a mitotically activated protein in *Xenopus* egg extracts. *Cell* **64,** 827–839.

Vallee, R. B., and Bloom, G. S. (1983). Isolation of sea urchin egg microtubules with taxol and identification of mitotic spindle microtubule-associated proteins with monoclonal antibodies. *Proc. Natl. Acad. Sci. USA* **80,** 6259–6263.

Vallee, R. B., and Collins, C. A. (1986). Purification of microtubules and microtubule-associated proteins from sea urchin eggs and cultured mammalian cells using taxol, and use of exogenous taxol-stabilized brain microtubules for purifying microtubule-associated proteins. *Methods Enzymol.* **134,** 116–127.

van der Voorn, L., and Ploegh, H. L. (1992). The WD-40 repeat. *FEBS Lett.* **307,** 131–134.

Verde, F., Labbe, J-C., Doree, M., and Karsenti, E. (1990). Regulation of microtubule dynamics by cdc2 protein kinase in cell-free extracts of *Xenopus* eggs. *Nature* **343,** 233–238.

Verde, F., Berrez, J-M., Antony, C., and Karsenti, E. (1991). Taxol-induced microtubule asters in mitotic extracts of *Xenopus* eggs: Requirement for phosphorylated factors and cytoplasmic dynein. *J. Cell Biol.* **112,** 1177–1187.

Verde, F., Dogterom, M., Stelzer, E., Karsenti, E., and Leibler, S. (1992). Control of microtubule dynamics and length by cyclin A and cyclin B dependent kinases in *Xenopus* egg extracts. *J. Cell Biol.* **118,** 1097–1108.

Verhey, C. A., and Moyer, F. H. (1972). Fine structural changes during sea urchin oogenesis. *In* "Invertebrate Oogenesis. II," pp. 194–225. MSS Information Corporation, New York.

Wadsworth, P., and Salmon, E. D. (1986). Analysis of the treadmilling model during metaphase of mitosis using fluorescence redistribution after photobleaching. *J. Cell Biol.* **102,** 1032–1038.

Wadsworth, P., and Sloboda, R. D. (1983). Microinjection of fluorescent tubulin into dividing sea urchin cells. *J. Cell Biol.* **97,** 1249–1254.

Wadsworth, P., and Sloboda, R. D. (1984). Interaction of bimane-labeled fluorescent tubulin with the isolated mitotic apparatus. *Cell Motil.* **4,** 183–196.

Wells, D., Bains, W., and Kedes, L. (1986). Codon usage in histone gene families of higher eukaryotes reflects functional rather than phylogenetic relationships. *J. Mol. Evol.* **23,** 224–241.

Whitaker, M. J., and Steinhardt, R. A. (1985). Ionic signaling in the sea urchin egg at fertilization. *In* "Biology of Fertilization" (C. B. Metz and A. Monroy, eds.), Vol. 3, pp. 168–222. Academic Press, Orlando.

Wilhelm, J. E., and Vale, R. D. (1993). RNA on the move: The mRNA localization pathway. *J. Cell Biol.* **123,** 269–274.

Wilson, L., Miller, H. P., Pfeffer, T. A., Sullivan, K. F., and Detrich, H. W. (1984). Colchicine-binding activity distinguishes sea urchin egg and outer doublet tubulins. *J. Cell Biol.* **99,** 37–41.

Witman, G. B. (1990). Introduction to cilia and flagella. *In* "Ciliary and Flagellar Membranes" (R. A. Bloodgood, ed.), pp. 1–30. Plenum Press, New York.

Wright, S. J., and Schatten, G. (1995). Protein tyrosine phosphorylation during sea urchin fertilization: Microtubule dynamics require tyrosine kinase activity. *Cell Motil. Cytoskeleton* **30,** 122–135.

Wright, B. D., Terasaki, M., and Scholey, J. M. (1993). Roles of kinesin and kinesin-like proteins in sea urchin embryonic cell division: Evaluation using antibody microinjection. *J. Cell Biol.* **123,** 681–689.

Yang, F., Demma, M., Warren, V., Dharmawardhane, S., and Condeelis, J. (1990). Identification of an actin-binding protein from *Dictyostelium* as elongation factor 1a. *Nature* **347,** 494–496.

Yen, T. J., Machlin, P. S., and Cleveland, D. W. (1988). Autoregulated instability of β-tubulin mRNAs by recognition of the nascent amino terminus of β-tubulin. *Nature* **334,** 580–585.

Yu, W., Centonze, V. R., Ahmad, F. J., and Baas, P. W. (1993). Microtubule nucleation and release from the neuronal centrosome. *J. Cell Biol.* **122,** 349–359.

Zimmerman A. M., and Zimmerman, S. (1967). Action of colcemid in sea urchin eggs. *J. Cell Biol.* **34,** 483–488.

4

Actin–Membrane Cytoskeletal Dynamics in Early Sea Urchin Development

Edward M. Bonder
Department of Biological Sciences
Rutgers University
Newark, New Jersey 07102

Douglas J. Fishkind
Department of Biological Sciences
University of Notre Dame
Notre Dame, Indiana 46556

I. Introduction
II. Sea Urchin Egg Fertilization: Physiological Activation
III. Sea Urchin Egg Fertilization: Actin Cytoskeletal Dynamics
    A. The Unfertilized Egg
    B. The Fertilized Egg
IV. Sea Urchin Egg: Actin-Binding Proteins
    A. Actin Monomer-Binding Proteins
    B. Actin Filament-Capping and -Severing Proteins
    C. Actin–Filament and Actin–Membrane Cross-Linking Proteins
    D. Other Actin Cytoskeleton-Related Proteins
V. Cortical Actin–Membrane Cytoskeletal Dynamics during Early
    Embryogenesis
    A. Cortical Membrane-Associated Actin Filament Polymerization Following
        Fertilization
    B. Formation of Elongate Microvilli after Fertilization
    C. Membrane Cytoskeletal Dynamics: Spectrin and Dynamin
VI. Concluding Remarks
    Appendix: Sea Urchin Egg and Embryo Methods
    A. Biochemical Preparation of Actin-Binding Proteins from Unfertilized Eggs
    B. Preparation of Isolated Cortical Lawns
    References

I. Introduction

For over a century, sea urchin eggs and embryos have provided a wealth of information on the cell biology of early development (Monroy, 1986; Davidson, 1986). The plethora of cell and developmental biology insights provided by the sea urchin egg/embryo model is derived from the variety of investigations and techniques applied to this experimental model. The

Current Topics in Developmental Biology, Vol. 31

diversity as an experimental system is, in part, based on the ease of obtaining large quantities of pure populations of eggs that are readily fertilized, thereby enabling the establishment of large populations of synchronous embryos. Thus, eggs and embryos have been used in a vast array of biochemical, biophysical, and cytological studies to ask and answer seminal questions on fertilization, signal transduction, cell cycle regulation, exocytosis, gene expression, cellular induction, and differentiation.

Historically, the echinoderm egg and embryo have been particularly fruitful in pushing the envelope of our understanding of the complex synergistic interactions among cell signaling events, membrane traffic, and cytoskeletal reorganizations that occur during the course of normal cellular and embryonic differentiation. Nowhere is this better exemplified than in the extraordinary works of Rappaport (1986), Hiramoto (1981), and others (see Conrad and Schroeder, 1990 for reviews) who have shown the important relationship that exists between the mitotic spindle and the cortical membrane actin cytoskeleton leading to the induction and formation of the contractile apparatus driving cytokinesis.

In this chapter, we focus our attention on the actin cytoskeleton and its interaction with egg and embryonic membranes during fertilization and morphogenesis. We highlight the known sea urchin egg actin-binding proteins and discuss their structural, functional, and regulatory cytoskeletal–membrane dynamics during development. Fertilization will serve as the starting point for reviewing the dynamic interplay between physiological activation and reorganization of the cortical actin–membrane cytoskeleton (filament assembly, cortical granule exocytosis, microvillar elongation, membrane uptake, vesicle sorting) that follows sperm–egg fusion. Our ultimate aim is to provide the reader with an appreciation of the sea urchin egg/embryo as a cell and developmental model for studying the mechanisms underlying cytoskeletal membrane reorganization and its importance to early embryonic morphogenesis and differentiation.

II. Sea Urchin Egg Fertilization: Physiological Activation

A characteristic feature of the actin cytoskeleton in sea urchin embryogenesis is its dynamic reorganization that is associated with signal transduction pathways. The pathways often rely on signaling events mediated through lipid metabolism via the phosphoinositol cycle with accompanying changes in calcium levels, phosphorylation activity, and intracellular pH. Such physiological signaling activities are known to play essential roles in regulating the organization and dynamics of the actin cytoskeleton, not only in the sea urchin egg and embryo, but in all somatic cells (Stossel, 1989; Lassing and Lindberg, 1988; Cooper, 1991). In the following section we will briefly summarize the known signaling events that play key roles in guiding the egg and embryo during development. Primary emphasis will be placed on the activation events of fertilization, since they are the best understood and most thoroughly characterized (Chandler, 1991; Epel, 1989; Jaffe and Cross, 1986; Whitaker and Swann, 1993).

The initial interaction of the sperm and egg, like many ligand–receptor interactions in other cells, triggers an initial series of ionic and metabolic changes that function to activate a cascade of subsequent events that direct the program of early development. In this instance the ligand is the sperm membrane protein, bindin, and the receptor is the egg membrane associated sperm receptor (Foltz and Lennarz, 1993). The initial egg response to the fertilization stimulus is a rapid depolarization of plasma membrane potential establishing the fast block to polyspermy (Jaffe and Cross, 1986) and activation of tyrosine kinase(s) leading to phosphorylation of sperm receptors along with several additional polypeptides (Abassi and Foltz, 1994; Ciapa and Epel, 1991; Kinsey, 1984). Tyrosine phosphorylation may serve as part of a signaling cascade that might trigger cortical actin cytoskeleton reorganization or as part of a mechanism that prevents polyspermy. In addition, the egg demonstrates a brief global contraction within the first 30 sec after fertilization (Eisen et al., 1984).

Over the same 30-sec time frame another signaling process begins relying on stimulation of phosphatidylinositol–4,5,-bisphosphate (PIP_2) metabolism, resulting in the production of inositol trisphosphate (IP_3) and diacylglycerol (DAG) (Kamel et al., 1985; Ciapa et al., 1992). Available data suggest that the initial production of IP_3 and DAG occurs via a G protein-mediated pathway leading to activation of a membrane–associated phospholipase (Turner and Jaffe, 1989). An increase in IP_3 initiates the release of Ca^{2+} from intracellular stores and produces a calcium wave, beginning ~30 sec after fertilization, that travels across the egg from the site of sperm entry to the opposite pole (Striker et al., 1992; Buck et al., 1994; Whitaker and Swann, 1993). This wave of elevated calcium propagates through the entire egg with the greatest elevation occurring in the cortical cytoplasm. Recent observations of Shen and Buck (1993) suggest that the calcium wave is preceeded by a distinct, short transient spike of calcium in the cortex which correlates to the time of sperm–egg binding. The increase in intracellular calcium concentration in turn stimulates a wave of cortical granule exocytosis, resulting in the elevation of the fertilization envelope and formation of the hyaline layer (Chandler, 1991; Eisen et al., 1984; Sardet and Chang, 1987; Vacquier, 1981). Additionally, the calcium transient is part of the triggering mechanism leading to the dynamic reorganization of the cortical actin cytoskeleton (see details below).

Interestingly, the plasma membrane of unfertilized sea urchin eggs contains an abundance of phosphatidylinositol (PI), which is converted to phosphatidylinosital monophosphate (PIP) and PIP_2 as part of the calcium response pathway (Kamel *et al.*, 1985; Ciapa *et al.*, 1992, Whitaker and Swann, 1993). Recent studies indicate that, while fertilization induces rapid conversion of PIP_2 to IP_3 and DAG, there is a parallel activation of phosphoiniositol kinase to continue production of PIP_2 (Ciapa *et al.*, 1992). Further synthesis of PIP_2 continues after the completion of the calcium transient suggesting the presence of relatively elevated levels of PIP_2 at the inner leaflet of the plasma membrane. The observation that the calcium wave produced after fertilization is not solely dependent on IP_3-mediated pathways but instead requires cyclic-ADP–ribose- and ryanodine-mediated pathways (Shen and Buck, 1993; Buck *et al.*, 1994) leaves open the possibility that the synthesis of PIP_2 could serve other functions beside fueling IP_3 and DAG synthesis.

Concomitant with the elevation of intracellular Ca^{2+} is DAG activation of a protein kinase C regulatory path that in turn activates a Na^+/H^+ exchanger (Swann and Whitaker, 1985). Activation of this exchanger presumably leads to a sustained increase in the intracellular pH of the fertilized egg that in turn functions to activate an assortment of processes driving the program of early development including, but not limited to, increased rate of protein synthesis, protein phosphorylation, initiation of DNA synthesis, increased K^+ conductance (Epel, 1989), together with major restructuring of the microtubule-based (Schatten *et al.*, 1985) and microfilament-based cytoskeletal systems (see below).

III. Sea Urchin Egg Fertilization: Actin Cytoskeletal Dynamics

Extensive remodeling of the surface of the egg occurs immediately following fertilization and is tightly linked to the ionic and metabolic changes initiated upon activation of the developmental program. Upon sperm–egg binding, the fertilization cone rapidly expands from the surface of the egg, engulfs the fertilizing sperm, and draws it into the cortical cytoplasm of the egg (Tilney and Jaffe, 1980). Meanwhile, the rest of the egg surface is rapidly transformed by the fusion and exocytosis of thousands of cortical granules (Chandler, 1984; Sardet and Chang, 1987), with subsequent formation of numerous long microvilli (Schroeder, 1978, 1979; Chandler and Heuser, 1981). Underlying these extensive changes in cell shape and topology is a dynamic reorganization of the actin-based cortical cytoskeleton which in part drives this process. In the following section, we will describe the known structural and dynamic characteristics of the unfertilized and fertilized egg's actin cytoskeleton. Correlations will be drawn between the activities of the cytoskeleton and the signaling avenues that have just been described.

A. The Unfertilized Egg

Early biochemical studies estimated that between 10 and 30% of the total pool of egg actin is positioned and associated with the cortex of the unfertilized egg (Spudich and Spudich, 1979; Otto *et al.*, 1980). Biochemical and/or cytological analyses of isolated egg cortices, whole mounts, and frozen sections have established the presence of plasma membrane associated actin filaments (Cline and Schatten 1986; Spudich *et al.*, 1988; Henson and Begg, 1988; Bonder *et al.*, 1989). One population of actin filaments coincides with the distribution of the short microvilli studding the surface of the egg, while a second population forms a thin network immediately below the microvilli (Fig. 1). Ultrastructurally there are short actin filaments within the microvilli, while longer, dispersed populations of filaments appear to be linked to the underside of the plasma membrane (Henson and Begg, 1988). The population of filaments located along the plasma membrane primarily lie between adjacent cortical granules, while the region immediately between the cortical granules and the plasma membrane appears devoid of filaments (Fig. 1; also see Bonder *et al.*, 1989). Furthermore, fluorescence cytochemical staining of frozen sections of whole eggs and ultrastructural analysis of extracted eggs detected the presence of actin filaments within the endoplasmic region of the egg (Fig 1; Foucault *et al.*, 1987; Bonder *et al.*, 1989). This endoplasmic population of actin filaments may serve as the scaffolding for the cytochalasin-sensitive saltatory movement of organelles that churn within the seemingly quiescent egg prior to fertilization (Allen *et al.*, 1992). Vesicle movements may be a consequence of actin-based motor activity (Allen *et al.*, 1992), including unconventional myosins (D'Andrea, 1993).

The observed structural organization of actin in the unfertilized egg cortex does not readily account for the amazingly high concentration of actin that coisolates with the cortex of the egg. This disparity was resolved in studies by Spudich *et al.* (1988) and Bonder *et al.* (1989) utilizing the F-actin-specific probe phalloidin coupled with immunocytochemistry using anti-actin antibodies. Both studies demonstrated that the cortical cytoplasm

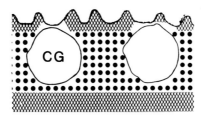

Fig. 1 Schematic representation of filamentous actin (cross-hatching) and nonfilamentous actin (solid circles) in the cortex of unfertilized sea urchin eggs.

of the unfertilized egg contains a discrete "shell" of nonfilamentous actin (Fig. 1). This shell of nonfilamentous actin exists within a 1- to 3-μm thick region that has traditionally been defined as the "cortex" of the unfertilized egg. A comparison of the two sets of data highlights another interesting observation which might be a consequence of having used either mono-clonal anti–actin antibodies (Spudich et al., 1988) or affinity-purified anti-actin antibodies (Bonder et al., 1989) that potentially demonstrate differen-tial epitope selectivity. Monoclonal anti-actin antibodies demonstrated the presence of a discrete coat of nonfilamentous actin around each cortical granule, while the affinity-purified polyclonal antibodies labeled the cyto-plasm between cortical granules. One possible molecular explanation for this observation is that the generation of nonfilamentous actin may rely on different actin-binding protein–actin complexes that individually have distinct organizational and regulatory functions.

B. The Fertilized Egg

The very earliest event of activation that might be tied to the cytoskeleton is an initial contraction of the entire egg occuring in association with sperm binding (Eisen et al., 1984). The timing of this contraction appears to coincide more directly with the depolarization of the egg, the initial transient rise in calcium, and the increase in tyrosine phosphorylation (Abassi and Foltz, 1994; Shen and Buck, 1993) rather than the calcium wave (Eisen et al., 1984).

Next is the rapid assembly and restructuring of cortically associated actin beginning at the point of sperm contact where an explosive polymerization of actin propagates as a wave across the entire surface of the egg (Cline and Schatten 1986; Hamaguchi and Mabuchi, 1988). This rapid assembly event temporally coincides with the wave of calcium passing over the egg (Yonemura and Mabuchi, 1987), and subsequently results in the generation of long surface microvilli with supporting bundles of actin filaments. The "early" polymerization step ($\sim$ 0.5–2.0 min postinsemination) leads to the assembly of loosely ordered meshworks of actin filaments within long, slender, microvillar-like processes and adjacent cortical cytoplasm (Begg et al., 1982; Carron and Longo, 1982). The reorganization also gives rise to broad, flame-like protrusions (Chandler and Heuser, 1981) that similarly contain a meshwork of actin filaments (Tilney and Jaffe, 1980; Begg et al., 1982). By 2 min after fertilization, the actin meshworks gradually trans-form into more highly ordered arrays of filament bundles (Burgess and Schroeder, 1977; Spudich and Spudich, 1979; Tilney and Jaffe, 1980; Begg et al., 1982; Carron and Longo, 1982) that progressively assemble microvilli by the membrane zippering down about the actin filament bundle (Tilney and Jaffe, 1980). This final phase of microvillar formation is completed ~ 5 min after fertilization and results in the production of thousands of surface microvilli. During this reorganization, there is a concomitant assembly of cortically associated actin filaments forming the scaffolding that supports the microvillar core bundles. By analogy, the microvillar-rich cortex and subjacent cortical cytoplasm bear striking resemblance to the brush–border organization of the actin–membrane cytoskeleton in intestinal epithelial cells.

IV. Sea Urchin Egg: Actin-Binding Proteins

The organizational structure and dynamics of the actin cytoskeleton are modulated by families of actin-binding proteins and within family members there is a high degree of evolutionary conservation (Stossel et al., 1985; Hartwig and Kwiatkowski, 1991; Cheney and Mooseker, 1992). Actin-binding proteins can be simplistically arranged into several regulatory families that (1) sequester and modulate the monomer concentration, (2) sever actin filaments and/or nucleate filament assembly, (3) crosslink and/or bundle actin filaments, (4) link actin filaments to membranes, and, (5) move actin filaments or intracellular cargo along filaments by virtue of their motor activity. In addition to the regulatory proteins, the dynamics of the actin filament (two ends with different assembly properties) and actin monomer (nucleotide exchange and hydrolysis) also play a key role in defining how actin-binding proteins function to modulate cytoarchitecture (Bonder et al., 1983; Cooper, 1991). Interestingly, the boundaries among the families are becoming less distinct since individual proteins have been identified that share functional characteristics of more than one family type. To date, proteins from all of the major classes have been identified in sea urchin eggs. What follows is a review of known egg actin-binding proteins with regard to their activity, regulation, and localization. We attempt to include the essence of each proteins molecular interaction with actin since this information aids in our understanding of cellular pathways regulating actin cytoskeletal structure and function.

A. Actin Monomer-Binding Proteins

By a number of biochemical and morphological criteria, it is well documented that the majority, estimated at between 30 and 50%, of actin in the unfertilized sea urchin egg is maintained as actin monomer or more appropriately in a "nonfilamentous form" (see above). Maintenance of actin monomer in a nonfilamentous form is believed to be a consequence of actin's interaction with sequestering proteins that bind to actin monomer limiting the amount of actin available for polymerization.

Sea urchin egg profilin was originally identified as a 13-kDa polypeptide that bound to actin with a 1 : 1 stoichiometry and decreased both the onset and the final extent of filament assembly (Mabuchi and Hosoya, 1982). Profilin has recently been cloned and its relationship to the profilin family established (Smith *et al.,* 1994). Cellular concentrations of profilin are estimated at 3×10^9 molecules/egg (Smith *et al.,* 1994) which approximately equals the number of actin molecules/egg (Otto *et al.,* 1980). However, if 30–50% of the total actin is present in a monomeric, nonfilamentous form, the total profilin concentration is in approximately twofold excess over the actin monomer pool suggesting a fairly high concentration of free profilin in unfertilized eggs. Profilins isolated and characterized from other animal sources share many common properties including (1) binding PIP_2 which inhibits actin sequestration and phospholipase C activity, (2) directing actin filament assembly to the "fast" growing or barbed end of the filament, (3) potentiating nucleotide exchange on actin monomers, and, (4) operating synergistically with thymosin-β4 to sequester monomer and promote rapid filament polymerization (see Theriot and Mitchison, 1993 for review). Detailed characterization of these properties have not been solved for egg profilin; however, it would not be unreasonable to predict similar activities given the high degree of structural and functional conservation among the members of the profilin family.

Profilin has not been localized at the cellular level in eggs or embryos, however, it has been demonstrated that egg profilin is primarily supplied from maternal stores. Interestingly, expression of profilin is elevated in certain primary and secondary mesenchymal cells that actively undergo actin filament dependent filopodial extension/retraction during gastrulation (Smith *et al.,* 1994; also see below).

The presence of thymosin-β4 in sea urchin eggs has been presented in preliminary form (reported in Nachmias, 1993) setting up the possibility for a profilin–thymosin-β4-mediated system regulating actin filament assembly in the egg and embryo.

B. Actin Filament-Capping and -Severing Proteins

While the actin monomer-sequestering proteins serve as keepers of the "fuel" that powers signal-dependent filament assembly, the barbed-end capping and filament-severing family of proteins act as factors that regulate sites where actin filaments will assemble. In general, assembly sites can be provided using several distinct pathways: generation of *de novo* polymerization sites by nucleation, uncapping of existing filament ends, or severing of existing filaments to increase the number of filaments. Barbed-end capping activity is in essence due to the actin-binding protein having a higher affinity for the barbed end of a filament than actin monomer thereby preventing further addition. This high-affinity binding to the end of a filament is fundamental for a protein's nucleation activity since these actin-binding proteins often form complexes with monomer that "possess" a free structural pointed end for monomer addition. Severing of filaments can be envisioned as a manifestation of attempted capping within a filament that subsequently destabilizes and breaks the filament into two pieces (Bonder and Mooseker, 1986; Maciever *et al.*, 1991). Not all capping proteins are capable of severing (Weeds and Maciever, 1993). In addition, many of the known actin filament-capping/severing proteins are potentially regulated by calcium and phopholipid signals (Weeds and Maciever, 1993). At least six actin filament-capping/severing proteins (depactin, actolinkin, 45 kDa, 50 kDa, gelsolin, and villin) have been characterized and their collective properties make them excellent candidates for participating in signal-induced pathways that stimulate actin polymerization at fertilization.

Actolinkin, a 20-kDa polypeptide isolated by DNAse I affinity chromatography, forms a high-affinity 1:1 complex with actin monomer (Ishidate and Mabuchi, 1988a). Unfortunately, the activity of purified actolinkin has not been adequately assessed partly because purification requires harsh treatments to dissociate the actolinkin–actin complex. Actolinkin–actin complex is capable of both capping the barbed end of actin filaments and nucleating the polymerization of *de novo* filaments. The actolinkin–actin complex does not sever or depolymerize actin filaments *in vitro* which suggests its primary role may be either nucleating or capping actin filament ends. Immunocytochemical localization of actolinkin places this actin-binding protein within the cortex of the egg both before and after fertilization, with a distribution similar to that of actin filaments (Ishidate and Mabuchi, 1988b). Interestingly, when isolated cortices, stripped of actolinkin and actin by salt extraction, are incubated with actolinkin–actin complex the cortices become competent to nucleate the polymerization of actin filaments, presumably using the complex as a membrane-associated nucleation site. Furthermore, the nucleated filaments are typically polarized with their barbed ends at the plasma membrane, the same filament polarity detected after fertilization. The *in vitro* binding properties of actolinkin with phospholipids have not been characterized and the actolinkin–actin complex does not demonstrate calcium sensitivity.

Egg depactin, a 17-kDa protein, forms a 1:1 complex with actin monomer (Hosoya *et al.*, 1982; Mabuchi, 1983) and may be related to the actophorin protein family. Depactin and depactin–actin complex exhibit the unusual property of both actively depolymerizing actin filaments while also being able to enhance polymerization. Depolymerization activity has been attributed to depactin's interaction either with the slow growing or pointed end of an actin filament (Hosoya *et al.*, 1982; Mabuchi, 1983) where depactin presumably binds ADP–actin and induces the loss of subunits as part of 1 : 1 actin–depactin complexes. Alternatively, based on the actophorin model (Maciever *et al.*, 1991), depactin may intercalate within the actin filament causing bound ADP–actin subunits to leave the filament via the 1 : 1 complex. Such intercalation would provide depactin with severing activity that generates numerous smaller filaments thereby providing greater numbers of filament ends, accounting for the reported ability of depactin to promote actin filament assembly (Hosoya *et al.*, 1982; Mabuchi, 1983). An alternate route for nucleation might rely on the type of nucleotide bound to the actin monomer. If a depactin–actin complex is formed using ATP–actin it may be able to transiently form a stable nucleus that lasts until ATP is hydrolyzed. Depactin–ADP–actin would then dissociate (depolymerize) leaving behind an oligomer capable of sustaining further polymerization. This possibility remains open for exploration. As with profilin (Theriot and Mitchison, 1993), depactin may have several different avenues for regulating actin filament assembly. To date there is little information on either the cellular or the embryonic distribution of sea urchin depactin.

Egg 45-kDa protein was simultaneously identified by Mabuchi and co-workers and Wang and Spudich either as a 1:1 complex with actin (Hosoya *et al.*, 1984) or as isolated protein (Wang and Spudich, 1984). In the absence of calcium, 45 kDa exhibited little effect on actin filament assembly, while in the presence of micromolar calcium concentrations the protein nucleated filament assembly, severed, and capped the barbed end of existing filaments (Wang and Spudich, 1984; Coluccio *et al.*, 1986). Calcium activation of 45 kDa leads to the formation of a 1:1 complex with actin which then exhibits calcium-independent capping and nucleation activity but no severing potential (Hosoya *et al.*, 1984). The binding interaction between 45 kDa and actin is not readily reversible by simple chelation of calcium. Another common feature among many of the capping/severing proteins is that while they are activated by calcium they can be inactivated by PPIs (Weeds and Maciever, 1993). Recently, Ohnuma and Mabuchi (1993) reported that PIP_2, but not PIP or PI, inhibited the calcium–mediated severing activity of 45 kDa. Of interest is the observation that in the presence of PIP_2 the 45 kDa–actin complex still formed and functioned as a nucleating and capping factor. It is not known if PIP_2 binds to both 45 kDa and the 45 kDa–actin complex and differences in these two interactions could have a dramatic impact on the mechanism of actin assembly following fertilization. Egg 45 kDa has not been localized at the cellular levels in either eggs or early embryos.

Egg 50-kDa polypeptide identified by Goldsteyn and Waisman (1989a,b) forms a 1:1 complex with actin and exhibits capping/severing activity. In preliminary studies, isolated 50 kDa protein behaves as a calcium activated nucleation, severing, and capping factor (E. M. Bonder, unpublished observations). Egg gelsolin is a 100-kDa polypeptide that exhibits calcium-activated nucleation/capping/severing activity (Hosoya et al., 1986). Unlike its gelsolin relatives (Weeds and Maciever, 1993), egg gelsolin appears to be fully reversible in its binding to actin by simple chelation of calcium. Currently, there is no available information on the cellular localization of either 50 kDa or egg gelsolin.

Utilizing an actin monomer binding method and DNAase I affinity chromatography sea urchin egg villin was isolated and characterized (Wang and Bonder, 1991). Egg villin is a 110-kDa polypeptide that has calcium-sensitive actin filament nucleation, severing, and capping activity. Capping activity is fully reversible by EGTA leading to immediate annealing of large populations of short capped filaments into a smaller number of long filaments. Additionally, in the absence of calcium, egg villin orders actin filaments into bundles through the presence of a villin "headpiece" domain that is responsible for bundling and is a structural diagnostic for the villin protein family.

Egg villin readily binds to phospholipid vesicles composed of phosphatidylcholine and PI, PIP, and PIP_2 and egg villin appears to have the greatest affinity for PIP_2 (Fig. 2). PI, PIP, and PIP_2 micelles all inhibited egg villin's calcium-mediated nucleation, capping, and severing activity with PIP_2 demonstrating the greatest effect followed by PIP and PI (Fig. 2; Wang, 1993). Preliminary studies suggest that egg villin is not localized to the cortex of the unfertilized egg; however, after fertilization villin is associated with actin filaments of the microvilli (Wang, 1993). Egg villin levels remain fairly constant during development through gastrulation and we do not yet know its embryonic localization (Wang, 1993).

C. Actin–Filament and Actin–Membrane Cross-Linking Proteins

While the sequestering and capping/severing family of proteins modulate the location and assembly of actin filaments, the cross-linking family of proteins organize the supramolecular organization of the actin cytoskeleton in cells. For brevity, we will jointly discuss the cross-linking proteins that link actin filaments to each other and to cellular membranes. Actin filament cross-linking proteins can be divided into two functional catagories—factors that link actin filaments into networks and bundling proteins that organize filaments into paracrystaline arrays of tightly packed bundles (Stossel et al., 1985; Matsudaira, 1991). Similarly, proteins that link actin to biological

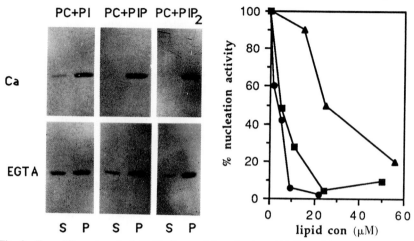

Fig. 2 Egg villin—phospholipid binding and inactivation. (A) Egg villin was mixed with phospholipid vesicles composed of PC and PI, PIP, or PIP_2, sedimented in an airfuge, and the resultant supernates (S) and pellets (P) were analyzed by SDS–PAGE. Villin bound all three types of vesicle preparations and binding was potentiated by the presence of calcium. (B) Nucleation activity of villin in the presence of calcium was assayed using pyrene-labeled actin. PIP (■) and PIP_2 micelles (●) inhibited villin's calcium-activated nucleation of actin filament assembly with approximately equal potency. PI (▲) micelles inhibited assembly, however, to a lesser degree.

membranes can be divided into proteins that link actin filaments directly to the membrane, as previously described for actolinkin, and proteins that bind actin and form ternary complexes with other membrane associated proteins (Bennett, 1990).

Fascin (58 kDa) was one of the earliest actin-binding proteins identified and characterized (Bryan and Kane, 1978). Fascin orders actin filaments into paracrystalline arrays of unidirectionally polarized bundles of actin filaments. High-resolution optical diffraction studies comparing actin filament bundles within microvilli and reconstituted *in vitro* fascin–actin bundles provided evidence for fascin's role in forming egg microvillar actin bundles (Spudich and Spudich, 1979). In addition, biochemical analysis of reconstituted actin–fascin bundles and isolated cortical preparations determined similar fascin–actin stoichiometries, 1:4.5 and 1:6.2, respectively (Bryan and Kane, 1978). Immunolocalization studies identified a cytoplasmic localization for fascin in unfertilized eggs and a microvillar localization after fertilization (Otto *et al.,* 1980). Consequently, fascin is thought to function as a key actin-bundling protein in the formation of microvilli. To date, fascin bundling activity does not appear to be responsive to either calcium or pH (Bryan and Kane, 1978). Recently, fascin has been characterized as the homolog of the *Drosophila* singed gene product (Bryan *et al.*, 1993). Comparisons of these fascins to other known actin filament cross-linking proteins suggest that they are a unique family of bundling proteins (Bryan *et al.*, 1993).

Egg α–actinin, a 95-kDa polypeptide, is a calcium regulated actin filament cross-linking protein (Mabuchi *et al.*, 1985). At calcium concentrations $<10^{-6}$ *M*, α-actinin dramatically increases the viscosity of actin filament solutions and exhibits little effect on viscosity at $>10^{-5}$ *M* calcium. Of interest is the observation that fluorescently labeled α-actinin appears not to be associated with the cortex of the unfertilized egg but instead is located in the endoplasm (Hamaguchi and Mabuchi, 1986) much the same as fascin and villin (see above). Upon fertilization, fascin is recruited to the cortex of the egg where it is localized to the submicrovillar cytoplasm and not to the actin filament bundles within the microvilli (Hamaguchi and Mabuchi, 1986). These same studies by Mabuchi *et al.* (1985) also detected a slight enrichment of α-actinin within the cortical region undergoing cytokinesis suggesting a potential role in cytokinesis. Unfortunately, the mechanism of accumulation is unknown and cleavage furrow localizations must be viewed cautiously since some cortically positioned actin binding proteins likely concentrate within the furrow as the cortical actin cytoskeleton flows from pole to equator (Bonder *et al.*, 1988; Cao and Wang, 1990). From localization and biochemical data it appears that α-actinin serves in establishing and maintaining actin filament organization in the "terminal web" region of fertilized eggs.

Several 250– to 260-kDa actin-binding proteins have been identified from various species of sea urchin and they all most probably represent members of the filamin family of cross-linking proteins (Mabuchi and Kane, 1987; Maekawa *et al.*, 1989; D. J. Fishkind, personal observations). Filamin is an elongate flexible rod that is capable of cross-linking actin filaments leading to dramatic increases in viscosity. Furthermore, egg filamin appears to be localized with actin filaments within the short microvilli of unfertilized eggs and the elongate microvilli of fertilized eggs (Maekawa *et al.*, 1989; Yoshigaki *et al.*, 1989). Electron microscopic localization of filamin places it in close association with the plasma membrane suggesting a possible role in filament–membrane linkage (Yoshigaki *et al.*, 1989). As with α-actinin, filamin was found to be enriched within the cleavage furrow region of isolated cortices.

Of all the known sea urchin egg actin-binding proteins, the most thoroughly characterized is egg spectrin (Kuramochi *et al.*, 1986; Fishkind *et al.*, 1987). Spectrin is a 237- to 234-kDa doublet that forms heterotetramers capable of cross-linking actin filaments (Fishkind *et al.*, 1987). At calcium concentrations $<10^{-6}$ *M*, spectrin cross-links actin filaments (Fishkind *et al.*, 1987) and at sufficient concentrations spectrin will even induce the formation of massive, highly stable bundles of filaments (Fishkind, 1991). Spectrin binding to and cross-linking of actin filaments decreases in the presence of $>10^{-6}$ M calcium (Fishkind *et al.*, 1987) and by calcium overlays egg spectrin appears to possess high-affinity calcium binding sites (E. M. Bonder, personal observations). Egg spectrin also demonstrates conserved binding to erythrocyte ankyrin and band 4.1 (Fishkind, 1991).

Immunolocalization studies established the colocalization of spectrin and actin within the microvilli of unfertilized eggs (Schatten *et al.*, 1986; Bonder *et al.*, 1989; Fishkind *et al.*, 1990a). Additionally, Fishkind *et al.* (1990a,b) identified that spectrin coated the surfaces of cortical granules, acidic vesicles, and yolk platelets (also see Fig. 5). The localization of spectrin on cortical granules is quite provocative since cortical granules are not associated with actin filaments but instead are embedded in the cortical domain of nonfilamentous actin. A similar spectrin–nonfilamentous actin colocalization has been observed in echinoderm sperm (D'Andrea *et al.*, 1991) in which the fertilization response leads to rapid extension of an acrosomal process by explosive polymerization of actin filaments off the nucleating actomere, a hub for actin filament assembly. Following fertilization, spectrin is associated with microvillar actin filaments and remains on the surfaces of acidic vesicle, which are translocated to the cortex (Fishkind *et al.*, 1990b).

Interestingly, the concentration of spectrin on the different membrane domains before versus after fertilization remains relatively constant suggesting that the spectrin within microvillar core bundles may be "donated" by spectrin originally located on cortical granules. Furthermore, the binding interaction of spectrin–actin and spectrin–membrane is clearly dependent on different mechanisms since spectrin can be differentially extracted from these compartments in isolated cortices (Bonder *et al.*, 1988; Fishkind *et al.*, 1990a). Through early blastula, spectrin typically lined the entire border of highly polarized blastomeres and the apically positioned acidic vesicles. Additionally, there was an accompanying loss of general cytoplasmic staining (Fishkind *et al.*, 1990b) suggesting that the cytoplasmic maternal pools of spectrin serve to maintain the membrane-associated levels of spectrin during membrane expansion that accompanies the rapid cleavage stages leading to blastulation. Beginning at late blastula stage, spectrin demonstrates enhanced cell- and tissue-specific localization within the embryo. Primary mesenchyme appear to have enhanced cellular levels of spectrin and this is coupled to specific accumulation of spectrin mRNA in cells of the vegetal plate (Wessel and Chen, 1992). During gastrulation, spectrin protein levels rise in the cells of the ingressing archenteron, while spectrin levels in primary mesenchyme fall as the cells form the subequitorial ring (Wessel and Chen, 1992). In subsequent stages, spectrin appears to be enriched within specific endodermal cells destined to form sphincters along the gut as well as a subset of secondary mesenchyme (Wessel and Chen, 1992).

D. Other Actin Cytoskeleton-Related Proteins

In addition to the three groups described there are other types of actin cytoskeleton-associated proteins in eggs and embryos. These include conventional myosin II (Mabuchi, 1973), unconventional myosins (D'Andrea, 1993), egg titin (Pudles *et al.,* 1990), tropomyosin (Maekawa *et al.,* 1989), and 53-kDa myosin-binding proteins (Walker *et al.,* 1991). Of importance to this discussion is egg tropomyosin, which binds actin filaments and has been localized to the cortex of unfertilized and fertilized eggs (Maekawa *et al.,* 1989). Interestingly, it is estimated that there is sufficient tropomyosin associated with isolated cortices to bind to every cortical actin filament (Maekawa *et al.,* 1989). This would have profound consequences on the interaction of a variety of actin-binding proteins with tropomyosin-coated filaments. For example, it is known that tropomyosin can protect actin filaments from the severing activity of villin (Bonder and Mooseker, 1982), a result that appears to hold true for other members of the severing protein family (Hartwig and Kwiatkowski, 1991).

V. Cortical Actin–Membrane Cytoskeletal Dynamics during Early Embryogenesis

As reviewed under Sections II and III, sperm–egg fusion elicits a repertoire of physiological and structural responses that appears to be finely tuned both spatially and temporally within the egg. The physiological responses include depolarization of the membrane potential, induction of phospholipid metabolism, activation of protein kinases, elevation of intracellular calcium, and an increase in intracellular pH. Overlayed upon this "physiological response" background is the explosive polymerization of short actin filaments, cortical granule exocytosis, formation of actin filament bundles within elongate microvilli, and membrane retrieval by coated vesicle endocytosis.

The early events of actin filament polymerization were shown to be dependent on elevation of intracellular calcium, while subsequent reorganization represents a separate defined transition that is dependent on elevated pH. In this context, we now consider how actin–membrane cytoskeletal reorganizations occur following fertilization via three distinct stages: (1) membrane-associated polymerization of short actin filaments fueled by the cortical pool of nonfilamentous actin, (2) filament annealing and cross-linking after completion of the calcium transient, and (3) final remodeling of the membrane surface following exo- and endocytosis. What follows is an attempt at fitting a variety of morphological, biochemical, and physiological data into a mechanistic model of how dynamic membrane cytoskeletal changes might be regulated at the cellular level (also see Stossel, 1989; Chandler, 1991; Spudich, 1992 for additional excellent reviews).

A. Cortical Membrane-Associated Actin Filament Polymerization Following Fertilization

As a point of reference, the unfertilized egg contains a reasonably precise spatial organization of actin filaments at the plasma membrane with a subjacent pool of nonfilamentous actin (Spudich *et al.*, 1988; Bonder *et al.*, 1989). This organization of spatially coupling nonfilamentous and filamentous actin is virtually a replica of an equivalent organizational strategy present in echinoderm sperm (Fig. 3). In sperm, nonfilamentous actin is stored within the profilactin cup which has an embedded bundle of short

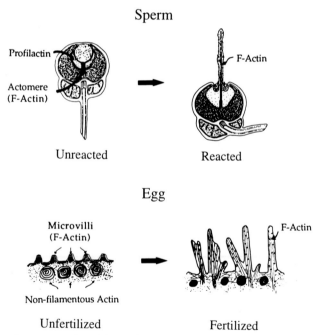

Fig. 3   Schematic representation of the spatial organization and relationship of nonfilamentous and filamentous actin in echinoderm sperm and eggs.

actin filaments called the actomere (Tilney *et al.*, 1983). During development the profilactin cup forms by the potential association of nonfilamentous actin with a specialized basal region of the acrosomal vesicle, a Golgi-derived organelle (Tilney, 1976). Interestingly, in mature sperm, spectrin is associated with nonfilamentous actin in the profilactin cup (D'Andrea *et al.*, 1991) which is equivalent in principle to the association of egg spectrin with the cortical domain of nonfilamentous actin (Bonder *et al.*, 1989). Upon elevation of intracellular calcium, a bundle of filaments is explosively produced by using the actomere as the nucleation site for polymerization, and assembly is directed toward the barbed end of the elongating filaments through the addition of profilin–actin complexes (Tilney *et al.*, 1983). The coordinated spatial organization of actin nucleation sites and nonfilamentous actin (profilactin composed of actin, profilin, and spectrin) in sperm provides an excellent paradigm with which to model early egg activation.

Simultaneous with the earliest events of fertilization is a transient decrease in the concentration of actin monomer suggesting an initial burst of filament polymerization (Dufresne *et al.*, 1987). Assuming the egg domain of nonfilamentous actin is, in part, generated by egg profilin, plasma membrane-associated profilin–actin complexes would release actin monomer upon binding to PIP and PIP_2 and the free actin monomer would rapidly assemble onto preexisting filaments at the membrane (Lassing and Lindberg, 1988; Stossel, 1989). PIP/PIP_2 molecules could also function in inhibiting the activity of capping proteins that are localized at the plasma membrane leading to additional free filament ends for polymerization. This initial burst of assembly could account for the reported transient contraction of the egg (Eisen *et al.*, 1984). Actolinkin may function in this early assembly since it localizes to the plasma membrane before fertilization.

Following this initial burst of assembly is a subsequent decrease in filamentous actin which corresponds temporally with the onset of the calcium transient (Dufresne *et al.*, 1987; Eisen *et al.*, 1984; Shen and Buck, 1993). Activation of calcium-sensitive barbed-end capping/severing proteins, such as villin, 45 kDa, 50 kDa, and gelsolin, would lead to transient and rapid depolymerization. Capping protein activity serves two purposes: it loosens the actin–membrane cytoskeleton which is believed to be a necessary step for vesicle exocytosis (Aunis and Perrin, 1984), and it provides nucleation sites for the next sustained phase of filament polymerization. Also, these same egg proteins are either known (45 kDa and villin) or thought (gelsolin and 50 kDa) to have phospholipid binding activity that can inhibit capping/severing activity. Consequently, for egg villin, this would set up a potential cycle of calcium-activated severing/capping which could be inhibited by PIP_2 binding at the membrane, thereby depositing short actin filaments with free barbed and pointed ends. The existence of free barbed ends during activation is supported by the finding that cytochalasin B, a barbed-end assembly inhibitor (Bonder and Mooseker, 1986), decreases the extent of actin filament polymerization during activation (Coffe *et al.*, 1982). Short filaments could then be linked to the membrane by egg spectrin, α-actinin, filamin, or actolinkin. A similar scenario would apply to egg 45 kDa; however, PIP_2 binding does not inhibit capping but instead could serve to target 45-kDa-capped short filaments to the membrane (Ohnuma and Mabuchi, 1993). This continued cycle of capping/severing and lipid interaction would continue as long as calcium levels were elevated, producing extensive meshworks of membrane-associated short actin filaments.

The fuel for igniting rapid polymerization is the cortically sequestered domain of nonfilamentous. Spudich *et al.* (1982) found that cortically associated actin has ADP as the bound nucleotide, suggesting the pool of cortical nonfilamentous actin is ADP–actin. In these circumstances, both profilin and thymosin-β4 would serve as sequestering proteins functioning to generate the cortical domain of nonfilamentous actin (Theriot and Mitchison, 1993). However, it is currently not known which of the two actin binding proteins is the predominant sequestering protein in unfertilized eggs. Profilin–actin could provide monomer for assembly by release of monomer following PIP/PIP_2 binding, or by direct addition of profilin–actin complexes to free barbed ends as proposed for actin filament elongation in activated sperm (Tilney *et al.*, 1983). Free profilin may also function in "desequestration" of actin from thymosin-β4–ADP–actin complexes and in facilitating nucleotide exchange to produce ATP–actin (Theriot and Mitchison, 1993). As the cycle progresses, polymer would continue to form.

B. Formation of Elongate Microvilli after Fertilization

As the calcium levels subside, calcium-dependent activation of capping/severing diminishes, slowing the production of nucleation sites followed by pH-dependent cross-linking. Additionally, PIP/PIP_2 levels remain elevated during the alkalinization period (Ciapa *et al.*, 1992) leading to further inactivation of capping/severing protein activity. With the loss of capping activity, the short filaments rapidly start annealing leading to the production of long filaments, as has been demonstrated for EGTA inactivation of egg villin (Wang and Bonder, 1991).

Cross-linking of cortical actin filaments to each other might proceed via two steps with the first step establishing "loose" cross-linked filaments organized into unidirectionally polarized assemblages and the second step involving "tight" bundling of filaments to produce paracrystalline core bundles. A two-step process for organizing actin filament bundles in microvilli is consistent with observations in intestinal epithelial cells in which villin, a weak bundling protein, localizes to the apical surface of the cell where there are loose networks prior to the positioning of the bundling protein fimbrin (Chambers and Grey, 1979; Shibayama et al., 1987; also see Louvard *et al.*, 1992). In the egg, villin, α-actinin, filamin, and spectrin have been localized to actin filament bundles after fertilization and *in vitro* they can generate loose bundles of filaments making them excellent candidates as the mediators of initial bundling. Additionally, spectrin, villin, and α-actinin are calcium regulated and their bundling activity would not begin until after the calcium wave. Spectrin is ideally suited for these early ordering events since large amounts of spectrin are localized to the cortex prior to fertilization, unlike villin and α-actinin. Completion of this filament coalescence step would then potentiate the tight, highly ordered bundling activity of fascin. Coincident with the bundling step is the membrane sheathing of the actin bundles by membrane zippering (Tilney and Jaffe, 1980). Spectrin is again suited for this purpose since it is localized to the plasma membrane, possibly through ankyrin-type binding. In the intestinal epithelial cell model, membrane zippering about actin core bundles has been attributed to the action of unconventional myosin-1 (Louvard *et al.*, 1992). Preliminary observations indicate that sea urchin eggs possess at least two unconventional myosins (D'Andrea, 1993) and if the brush-border paradigm holds they may participate in microvillar genesis as well.

C. Membrane Cytoskeletal Dynamics: Spectrin and Dynamin

1. Egg Spectrin

The synthesis of a maternal pool of spectrin appears to prime the egg with a preassembled store of membrane skeleton precursor to be utilized during the dynamic morphogenetic changes that accompany fertilization and the rapid cleavage phase of early embryogenesis.

Assembly of spectrin on cortical granules during sea urchin oogenesis is related to the concept that spectrin may facilitate the association and sequestration of nonfilamentous actin in the egg cortex during oocyte maturation. As described previously for sperm, nonfilamentous actin domains appear to become associated with a specialized region of the Golgi-derived acrosomal vesicle during spermatogenesis. In mature sperm spectrin is associated with the profilactin cup and it would not be unreasonable to propose that sperm spectrin becomes associated with the acrosomal vesicle during maturation much the same way in which egg spectrin localizes to Golgi-derived cortical granules during oogenesis (Fishkind *et al.*, 1991b). Spectrin assemblages on the surface of cortical granules may thus serve as "nucleation" centers for the deposition and accumulation of nonfilamentous actin. Presumably, such a mechanism would promote positioning of a localized pool of unassembled actin and spectrin in the egg cortex at sites where rapid actin polymerization accompanies insertion of new membrane following stimulus-induced activation and cytoskeletal transformation.

The redistribution of spectrin during sea urchin embryogenesis occurs in three sequential phases coincident with three separate developmental events. The first phase occurs after fertilization and is associated with cortical granule exocytosis and remodeling of the fertilized egg surface. Spectrin cross-bridges extend from the plasma membrane to the surface of cortical granules and acidic vesicles, as well as between adjacent vesicles (Fishkind *et al.*, 1991a). Consequently, spectrin linkers appear to tether vesicles to the overlying plasma membrane, as well as to one another. Together, the cross-linking of vesicles within the cortex and their attachment to the plasma membrane by spectrin and actin may contribute to the gel-like state of cortical cytoplasm (Hiramoto, 1981) and may function in positioning vesicles in the cortex prior to secretion, as previously described for secretory vesicles in neurons (Zagon *et al.*, 1986) and adrenal chromaffin cells (Perrin *et al.*, 1987).

The second phase of spectrin redistribution occurs within the first 20 min following fertilization when randomly dispersed, spectrin-coated acidic vesicles and pigment granules migrate to the cortex (Fishkind *et al.*, 1991b). While the mechanism responsible for vesicle movement is not understood, the actin-based cytoskeleton has been implicated since the polarized positioning and maintenance of these vesicles in the cortex is affected by cytochalasins, but not by microtubule inhibitors (Allen *et al.*, 1992). Presumably, the spectrin coat on vesicles may serve both in their targeting to the cell surface, and for anchorage within the actin–rich cortex. Such a mechanism for polarized positioning of vesicles in the cortex may function during embryogenesis when secondary waves of exocytic vesicles are delivered to the cortex for secretion (Wessel *et al.*, 1984; Alliegro and McClay, 1988).

The third phase of spectrin redistribution occurs during the early cleavage divisions, coincident with the extensive partitioning and reorganization of the embryo that results in an ~100-fold increase in the surface area to volume ratio. During this period spectrin staining in the yolk-rich cytoplasm progressively decreases, while intense staining is maintained along the plasma membrane surface of newly forming blastomeres (Fishkind *et al.*, 1991b). Furthermore, the finding that the overall content of spectrin in embryos does not change during early development suggests that the maternal pool of spectrin redistributes from cytoplasmic stores to the plasma membrane during expansion and amplification of the embryo's surface (Fishkind *et al.*, 1991b).

During late blastulation, spectrin mRNA and protein accumulates at the vegetal plate (Wessel and Chen, 1992) which is the site of epithelial–mesenchymal transformation leading to ingression of the primary mesenchymal cells (Karp and Solursh, 1985; Anstrom, 1989; Malinda and Ettensohn, 1994). This key step in embryonic differentiation relies on transformation of columnar epithelial cells by forming an apical "purse-string" contraction, coupled with broad basal protrusive activity creating a bottle cell-type morphology (Anstrom, 1992). Clearly, enhanced levels of spectrin during this stage may be a necessary prerequisite for stabilization of the basal protrusive surfaces during ingression. After ingression, the primary mesenchyme become highly motile by rapidly extending and retracting actin-dependent filopodial extensions (Karp and Solursh, 1985; Anstrom, 1989, 1992). At this stage, the cells also show enhanced levels of profilin expression (Smith *et al.,* 1994), an actin-binding protein believed to facilitate rapid filament polymerization (Buβ *et al.,* 1992; Theriot and Mitchison, 1994), which possibly participates in regulating the actin-dependent extension and retraction of filopodia (Anstrom, 1989, 1992). Thus, developmentally timed expression of profilin and spectrin may be instrumental in primary mesenchymal differentiation by stabilizing and regulating actin filament assembly and structure within filopodia.

2. Egg Dynamin

An intimate component of early activation is cortical granule exocytosis and subsequent membrane retrieval to return the surface area back to prefertilization levels (Schroeder, 1979). Fisher and Rebhun (1983) established that the initial burst of membrane retrieval occurs through the action of coated vesicle which may selectively remove specific membrane proteins, such as sperm receptors, from the cell surface. Recently, Faire and Bonder (1993) using tubulin dimer affinity chromatography isolated a 100-kDa microtubule binding protein that exhibited microtubule bundling and GTPase activity. Additionally, immunological (Faire and Bonder, 1993) and partial protein sequence data (Faire and Bonder, unpublished observations) established the identity of this protein as being a member of the dynamin protein family (Vallee, 1992; Robinson, 1994). Intense interest has been directed toward the dynamin protein family because it is now known that the dynamins function in modulating receptor-mediated endocytosis via coated vesicles (Vallee and Shpetner, 1993; Robinson, 1994). Dynamin phosphorylation/dephosphorylation has been shown to regulate GTPase activity and membrane binding and this has been linked to modulation of synaptic vesicle recycling (Robinson *et al.,* 1993; Liu *et al.,* 1994a,b).

In unfertilized sea urchin eggs, dynamin is localized to cortical granules (Fig. 4) as well as being evenly distributed within the inner cytoplasm. Immediately after fertilization, dynamin localization is greatly disrupted giving the appearance of dynamin outlining crypts formed by cortical granule exocytosis with the plasma membrane or by fusion with each other

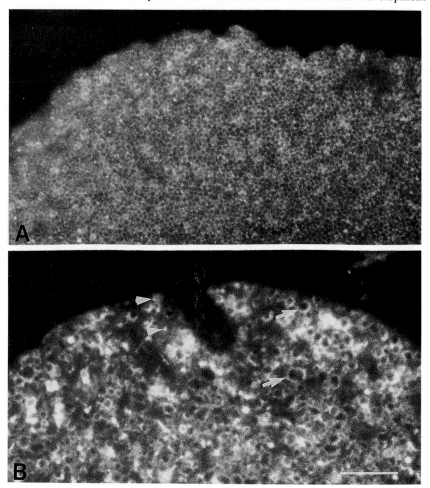

Fig. 4 Immunofluorescence localization of egg dynamin. (A) Localization of egg dynamin in an isolated cortex from an unfertilized egg. Note the well-defined outlining of individual cortical granules. In equivalent preparations in which cortical granules were removed by shearing there was no detectable staining of the plasma membrane (Faire and Bonder, 1993). (B) Localization of egg dynamin in a egg cortex isolated 1 min after fertilization. The cortical staining pattern is highly disrupted with dynamin localizing to large pockets (arrows) that possibly represent regions where cortical granules fused with the plasma membrane. Additionally, note the occasional individual bright fluorescent foci (arrowheads) that may represent dynamin associated with coated vesicles. Scale bar = 10 μm.

(Fig. 4B, arrows). Additionally, dynamin localizes to very discrete bright small spots that are estimated to be <500 nm in diameter (Fig. 4B, arrowheads). At 45 min after fertilization, dynamin is localized to the monolayer of acidic vesicles that line the cortical cytoplasm (Faire and Bonder, 1993). The localization of dynamin on cortical granules and the established role of other dynamins in coated vesicle endocytosis suggest that egg dynamin is spatially positioned to play a functional role in coated vesicle endocytosis after fertilization (Faire and Bonder, 1993).

Dynamin phosphorylation by protein kinase C and dephosphorylation by calcineurin is proposed to modulate dynamin GTPase activity which presumably regulates vesicle recycling (Robinson *et al.*, 1993; Liu *et al.*, 1994a,b). Interestingly, activation of protein kinase C by phorbol ester in unfertilized sea urchin eggs stimulates coated vesicle-mediated endocytosis in the absence of cortical granule exocytosis (Ciapa *et al.*, 1988). Thus, egg dynamin phosphorylation/dephosphorylation during membrane recycling may play an important role following fertilization and this is under investigation. Future studies directed at understanding the spatial and temporal modulation of egg dynamin are critical given the highly dynamic nature of membrane sorting and trafficking events during early embryognesis.

VI. Concluding Remarks

In this chapter, we have attempted to highlight the overall synergy that exists among physiological signals, cytoskeletal modifications, and membrane transformations vital to early embryogenesis. Clearly, given the extensive literature on this subject, we have only begun to scratch the surface regarding the detailed relationship of the many molecules that must be involved in this complex process and formally apologize for any oversight that may have been made. In closing, fertilization and early embryogenesis exhibit a remarkable series of spatially and temporally regulated actin–membrane cytoskeletal reorganizations. Today the table is well set to apply the latest set of emerging cell and molecular biological technologies to determine the molecular mechanisms underlying actin–membrane cytoskeletal dynamics and their intimate link to embryogenesis.

VII. Appendix: Sea Urchin Egg and Embryo Methods

A. Biochemical Preparation of Actin-Binding Proteins from Unfertilized Eggs

The following outline presents a specific strategy that was adopted to isolate actin-binding proteins from large quantities (50–400 ml) of unfertilized sea urchin eggs. The basic tactic is to prepare cytoplasmic extracts which are then augmented either with actin filaments or actin monomer to serve as "affinity matrices" for selectively binding F-actin- or G-actin-binding proteins. By modifying the strategy in terms of buffer conditions, calcium concentration, and pH, one could differentially select for certain actin binding characteristics. In principle, these procedures should be applicable to any cellular system and preliminary observations suggest it can be utilized to identify actin binding proteins from cytoplasmic extracts of *Drosophila* embryos (E. M. Bonder, unpublished observations).

Step 1. Preparation of cytoplasmic extracts

(a) All procedures are carried out with solutions maintained at 0–4°C.

(b) *Strongylocentrotus purpuratus* or *Lytechinus pictus* eggs are collected into beakers of artificial seawater, from gravid females by intracoelomic injection of 0.5 M KCl; the eggs are washed once with artificial seawater, and twice in divalent-free, artificial seawater (Begg and Rebhun, 1979) by repeated centrifugation at ~1,500g in 250- or 500-ml centrifuge bottles. These initial steps are taken to remove the egg's jelly coat which swells after shedding into seawater.

(c) Pelleted eggs are resuspended and washed once in "homogenization buffer" (100 mM Hepes, pH 7.5, 5 mM EGTA, 0.9 M glycerol, 0.02% NaN_3), the pellet is resuspended in a four- to sixfold volume excess of homogenization buffer containing 1 mM ATP, 1 mM DTT, 0.5 mM PMSF, 2 mM TAME, 5 mM benzamidine, 100 μg/ml aprotinin, 200 μg/ml SBTI, 25 μg/ml leupeptin, and 70 μg/ml pepstatin A, and the suspension is homogenized using a tight-fitting glass dounce homogenizer. Homogenization is monitored by light microscopy to ensure complete disruption of eggs.

(d) The homogenate is centrifuged at 20,000g for 30 min to remove nuclei, yolk platelets, and membrane fragments, and the resulting supernatant is further centrifuged at 100,000g for 2 hr to obtain a clear high-speed cytoplasmic extract. Depending on the degree of yolk in the eggs, there may a yellow, lipid film floating on the clear cytoplasmic extract in which case the extract is removed carefully with a pipette.

Step 2. $(NH_4)_2SO_4$ fractionation and precipitation of actin filament-binding proteins

(a) The 100,000g extract is made 45% $(NH_4)_2SO_4$ by slow addition, while gently stirring the extract, of ice cold saturated $(NH_4)_2SO_4$ followed by a 45-min incubation on ice.

(b) The protein precipitated between 0 and 45% $(NH_4)_2SO_4$ is collected by centrifugation (25,000g for 20 min), gently resuspended in 25–50 ml of "assembly buffer" (50 mM KCl, 10 mM Tris–HCl, 1 mM EGTA, 1 mM $MgCl_2$, 0.02% NaN_3, 1 mM

ATP, 0.5 mM DTT, 0.5 mM PMSF, 5 mM benzamidine, 25 μg/ml aprotinin, 10% sucrose, pH 7.5) per 50 ml of original egg volume, and the protein solution is supplemented with muscle actin (MacLean–Fletcher and Pollard, 1980) to a final concentration of 0.2 mg/ml exogenous actin. The rationale for supplementing with exogenous actin stems from the fact that only a portion of the endogenous egg actin partitions in the 0–45% cut. Thus, the additional actin increases the polymer concentration to facilitate maximal binding of any actin-binding proteins present in the resuspended precipitate.

(c) The actin-enriched solution is dialyzed for 24–48 hr against a total of 8.0–12 liters of assembly buffer during which time a fine, lacy-like, supramolecular aggregate begins to form. This material is collected by centrifugation (25,000g for 30 min), solubilized in 5–20 ml of KI buffer (0.8 M KI, 10 mM Tris–HCl, 5 mM EGTA, 5 mM EDTA, 0.5 mM ATP, 1 mM DTT, 0.5 mM PMSF, 5 mM benzamidine, 2 mM TAME, 5 μg/ml aprotinin, 5 μg/ml SBTI, 0.02% NaN$_3$, pH 8.0), dialyzed against the same for several hours, and clarified by centrifugation prior to size exclusion chromatography.

(d) It should be noted that while many ABPs are present in the 0–45% cut, the remaining supernatant can be saved for further fractionation of additional actin-binding proteins. For example, to isolate egg villin (Wang and Bonder, 1991) a 40–80% (NH$_4$)$_2$SO$_4$ cut is collected, resuspended in a low ionic strength buffer (2 mM Tris–HCl, pH 8.0, 0.2 mM CaCl$_2$, 0.5 mM DTT, 0.2 mM ATP containing 0.1 mg/ml sucrose, 4 μg/ml aprotinin, 0.5 mM PMSF) and dialyzed overnight against several changes of the same buffer. Depending on the degree of residual myosin in the preparation, a precipitate often forms which is collected by centrifugation (10,000g for 30 min) and the resultant supernate is supplemented with 0.1 mg/ml of gel-filtered muscle actin. In this case, the hope is to potentiate the binding of actin monomer-binding proteins which can then be isolated by either conventional chromatographies or DNAase-I affinity chromatography.

Step 3. Size-exclusion chromatography

(a) A Bio-Rad A-15 (200–400 mesh) 2.5 × 90-cm column is equilibrated with 2 or 3 column volumes of high KCl buffer (KI buffer containing 0.8 M KCl rather than KI). Immediately prior to loading the ABP-enriched sample, a 100-ml KI buffer front is placed on the column.

(b) Column fractions are monitored by standard protein assay and SDS–PAGE, and fractions enriched in proteins of interest are pooled for further purification.

    (c) In the case of egg spectrin (Fishkind *et al.*, 1987) and filamin
       (Fishkind, 1991), fractions containing high M_r polypeptides
       ranging from ~240 to 260 kDa are pooled and prepared for
       purification by hydroxylapatite chromatography. Spectrin-enriched
       fractions are collected off the hydroxylapatite column and
       subjected to phosphocellulose chromatography, while filamin-
       containing fractions are applied to a DE-52 anion-exchange
       column (Fishkind, 1991).

    (d) Using this basic strategy, together with published procedures for
       purification of a host of well-characterized actin-binding proteins,
       it is possible to optimize the isolation of a number of sea urchin
       egg actin-associated proteins. Table I lists and summarizes the
       biochemical properties of the proteins present in the actin
       filament aggregates and provides preliminary assignments based
       on information and nomenclature present in the literature.

B. Preparation of Isolated Cortical Lawns

Experimentation on preparations of cortical lawns has served as a founda-
tional approach to understanding actin cytoskeletal dynamics and organization as well as membrane dynamics during early fertilization. In the following section, we describe the procedures for preparing cortical lawns for the
application of fluorescence cytochemical light microscopy and immunogold
labeling for electron microscopy. Such preparations are also readily amenable for biochemical examination by SDS–PAGE, selective extraction, and
reconstitution experimentation.

Step 1. Preparation of eggs

    Cortical lawns are isolated from unfertilized eggs and embryos
according to the procedures of Vacquier (1975) as modified by Henson
and Begg (1988).

    (a) Poly-L-lysine-coated coverslips are prepared by placing a drop of
       freshly prepared 1 mg/ml polysine-L-lysine in distilled water on
       clean coverslips for 1–5 min followed by rinsing with distilled
       water and air drying.

    (b) Eggs are prepared as described previously for biochemical studies
       and concentrated suspensions of cells are allowed to settle onto
       poly-L-lysine-coated coverslips for 1 min.

    (c) Similarly, fertilized eggs can also be attached to coverslips. Eggs
       are fertilized with a dilute suspension of sperm. Once the
       fertilization envelopes start to elevate the eggs are briefly
       suspended in 1 M urea/5 mM Hepes (pH 8.0), passed through a

Table I Proteins which Coprecipitate with Actin Filaments

Protein	M_r (k_D)	Isoelectric point (P_i)	Stokes radius (A)	Relative quantity[a]	References[b]
Filamin	~260	ND	130	+	1,2
Spectrin	240/235	5.0	120–190	+++	3,4
?	215	5.7	110	+	
Myosin	205	4.5–5.0	120	++	5,6
?	165	4.6	V_o	+	
?	160	5.5	100	+	
?	115	4.7	ND	+	
α-Actinin	100	5.0–5.5	78	+++	7
?	97.4	5.1	ND	+	
?	97	6.1–6.5[c]	108	++	
?	88	5.5–5.9[c]	ND	+	
?	80	5.4–5.8[c]	ND	+	
?	75	5.3–5.8	75	+++	
?	70	5.7–6.2[c]	V_o	+	
?	68	5.3	ND	+	
?	65	4.5	~65	+	
?	64	5.3	~60	+	
Fascin	58	ND	~60	+++	8,9
45 and 50K	45–50	5.3–5.8	ND	+	10–13
Actin	45	5.4	ND	+++	14
Tropomyosin	33	4.6	ND	++	15
Actolinkin	20	4.5	ND	ND	16

[a] Relative quantity, taken from densitometry analysis of SDS–PAGE gels. (+) 1–3%, (++) 3–5%, and (+++) 5–10% of total protein in precipitate.

[b] References: 1, Mabuchi and Kane (1987); 2, Maekawa et al. (1987); 3, Kuramochi et al. (1986); 4, Fishkind et al. (1987); 5, Mabuchi (1973); 6, Kane (1980); 7, Mabuchi et al. (1985); 8, Bryan and Kane (1978); 9, Otto et al. (1980); 10, Golsteyn and Waisman (1989a,b); 11, Wang and Spudich (1984); 12, Hosoya et al. (1984); 13, Coluccio et al. (1986); 14, Kane (1975); 15, Maekawa et al. (1989); 16, Ishidate and Mabuchi (1988a,b).

[c] Potential polyphosphorylated proteins.

60- to 100-mm mesh nitex filter (mesh size is species dependent) by aspiration using a Buchler funnel, cultured in CaFSW (Burgess and Schroeder, 1977), and settled onto coverslips at the desired stage.

Step 2. Shearing and fixing cortical lawns

(a) Nonadherent cells are rinsed free with CaFSW. Attached eggs/embryos are sheared by gentle pipetting with a stream of cold (0–4°C) cortex-isolation buffer (CIB; see Fig. 5A). Shearing can be accomplished using a pasteur pipette or a syringe with an 18- to 20-gauge needle. For our purposes we use two basic isolation

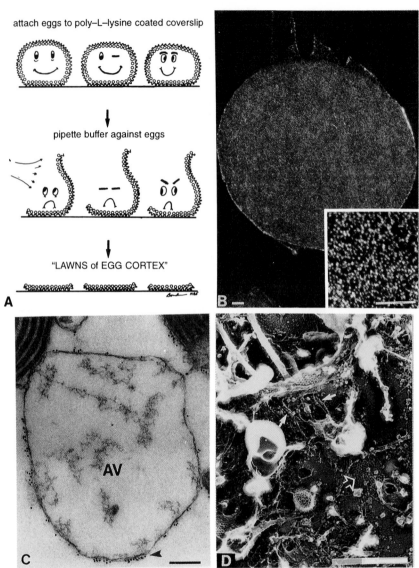

Fig. 5 Summary of cortex isolation. (A) Schematic of basic protocol for isolating sea urchin egg cortices. (B) Filamentous actin localization by rhodamine phalloidin staining of an isolated unfertilized egg cortex. At higher magnification (see inset) it is possible to detect individual brightly staining microvilli. Scale bar = 5 μm. (C) Thin section of an isolated cortex from an unfertilized egg labeled with immunogold to localize sea urchin egg spectrin (also see Fishkind *et al.*, 1990a). This electron micrograph clearly demonstrates the association of egg spectrin with the membrane of acidic vesicles. Note the linear cluster (arrowhead) of gold particles lining the acidic vesicle membrane. Scale bar = 0.1 μm. (D) Isolated fertilized egg cortex examined by quick-freeze, deep-etch, and rotary shadowing (see Bonder *et al.*, 1988; Henson and Begg, 1988). In these preparations, bundles of actin filaments are observed exiting the base of microvilli (arrows) and coated vesicles can be observed at different stages of formation (open arrow). Scale bar = 1 μm.

media that differ in ionic strength and composition. CIB I
(300 mM glycine, 250 mM K$^+$–gluconate, 20 mM PIPES, and
20 mM EGTA, pH 6.8), which was originally reported by Mazia
and colleagues (Schatten and Mazia, 1976), is particularly useful
in preserving membrane structures, while CIB II (0.6 M mannitol,
50 mM Hepes, 50 mM PIPES, 20 mM EGTA, 2.5 mM MgCl$_2$, pH
6.8) appears more effective in stabilizing microfilament–associated
structures (Henson and Begg, 1988; Fishkind *et al.,* 1990a).

(b) Care needs to be taken during the shearing process since the vigor
of shearing determines the quality of the generated lawn. This is
especially important if preparations with attached cortical granules
are required for the planned experiment.

(c) Immediately following shearing, coverslips are fixed in CIB
containing 1–3% formaldehyde/0.1% glutaraldehyde for 15–30 min
at 4°C, rinsed 3 × with cold CIB, and blocked for 30 min with
CIB containing 1% BSA.

(d) The choice of fixative is somewhat dependent on the type of
staining that will be attempted, and this may need to be
determined empirically. However, the above fixative serves as a
reliable starting point.

Step 3. Indirect Immunofluorescence staining of cortical lawns

(a) Fixed cortices are incubated with appropriate dilutions of primary
antibodies diluted in the appropriate CIB containing 1% BSA for
16 hr at 4°C. Antibody dilutions will need to be determined
empirically. Note, if other CIBs are used it is important to include
an osmoticum, such as 0.6 M sucrose, to prevent vesicles from
lysing since they can remain osmotically active after fixation.

(b) Following incubation in primary antibody, cortices are rinsed 3×
with CIB/BSA and primary antibodies are detected by incubation
with a 1:100–1:1000 dilution (depending on the source of
secondary) of affinity-purified fluorescent-secondary antibodies for
2–4 hr at 22°C.

(c) After incubation, coverslips are rinsed 3× with CIB, mounted in
an antiphotobleaching agent, and examined by fluorescence
microscopy.

(d) If fluorescence staining of actin filaments is desired, cortical lawns
are prepared as described previously and incubated in 30–50 nM
rhodamine-labeled phalloidin in CIB. Following fluorescence
labeling the coverslips are rinsed, mounted, and observed by
fluorescence microscopy (see Fig. 5B).

Step 4. Immunogold electron microscopy on cortical lawns

(a) Coverslips with freshly prepared cortices are placed in one of two
aldehyde-based fixatives for 1 hr. CIB I cortices are fixed in

400 mM NaCl, 50 mM NaPO$_4$ (NaCl buffer), 1% formaldehyde, and 0.5% glutaraldehyde at pH 7.0, and CIB II cortices are fixed in CIB II containing 1% formaldehyde and 0.5% glutaraldehyde at pH 7.0. The choice of buffers and fixative conditions are determined by the structures being analyzed and the fixative sensitivity of the primary antibodies.

(b) Cortices are rinsed 3× with the appropriate buffer (without aldehydes), and then incubated in either NaCl buffer containing 50 mM glycine/1% BSA (for CIB I cortices) or CIB II supplemented with 1% BSA (for CIB II cortices) for 4–6 hr to block reactive aldehyde groups.

(c) Coverslips are incubated overnight at 4°C with an appropriate dilution of primary antibodies in NaCl buffer–BSA for CIB I cortices or CIB II–BSA for CIB II cortices.

(d) After overnight labeling with primary antibodies, the samples are washed 3× with the appropriate blocking buffer and incubated for 4 hr at room temperature with 10 nm gold conjugated secondary antibody diluted in the appropriate blocking buffer.

(e) Immunogold-labeled cortices are washed 3× in the appropriate blocking buffer followed by three washes in buffer without BSA, and postfixed for 1 hr in either NaCl buffer with 1% glutaraldehyde (CIB I cortices) or CIB II with 1% glutaraldehyde (CIB II cortices). Following this second fixation step, cortices are washed 3× in NaCl buffer and osmicated in NaCl buffer (pH 6.0) containing 1% OsO$_4$ for 1 hr, rinsed 3× with deionized H$_2$O, and en bloc stained in 0.5% uranyl acetate overnight. Cortices are then rinsed with deionized H$_2$O, dehydrated with a graded series of ethanol, followed by propylene oxide, and flat embedded in epon–araldite.

(f) Silver and silver–gold ultrathin sections are poststained with 1% uranyl acetate and fresh Reynold's Pb–citrate, and examined with an electron microscope (see Fig. 5C).

(g) In addition to immunogold electron microscopy, cortices can be prepared for ultrastructural analysis by conventional thin sectioning, negative staining, or quick freeze and deep etching (Fig. 5D).

Acknowledgments

We thank all of our colleagues, who over the years have constantly provided stimulating and informative insight into the workings of the egg and the actin cytoskeleton. We also thank

Mr. George Sgourdas for his technical assistance in preparing the manuscript. Funding was provided through the National Institutes of Health grant HD 24649 (E.M.B.).

References

Abassi, Y. A., and Foltz, K. R. (1994). Tyrosine phosphorylation of the egg receptor for sperm at fertilization. *Dev. Biol.* **164**, 430–443.

Allen, P. G., Baltz, J. M., and Begg, D. A. (1992). Fertilization alters the orientation of pigment granule saltations in Arbacia eggs. *Cell Motil. Cytoskeleton* **21**, 223–234.

Alliegro, M. C., and McClay, D. R. (1988). Storage and mobilization of extracellular matrix proteins during sea urchin development. *Dev. Biol.* **125**, 208–216.

Anstrom, J. A. (1989). Sea urchin primary mesenchyme cells: Ingression occurs independent of microtubules. *Dev. Biol.* **131**, 269–275.

Anstrom, J. A. (1992). Microfilaments, cell shape changes, and the formation of primary mesenchyme in sea urchin embryos. *J. Exp. Zool.* **264**, 312–322.

Aunis, D., and Perrin, D. (1984). Chromaffin granule membrane–F-actin interactions and spectrin–like protein of subcellular organelles: A possible relationship. *J. Neurochem.* **42**, 1558–1569.

Begg, D. A., and Rebhun, L. I. (1979). pH regulates the polymerization of actin in the sea urchin egg cortex. *J. Cell Biol.* **83**, 241–248.

Begg, D. A., Rebhun, L. I., and Hyatt, H. A. (1982). Structural organization of actin in the sea urchin egg cortex: Microvillar elongation in the absence of actin filament bundle formation. *J. Cell Biol.* **93**, 24–32.

Bennett, V. (1990b). Spectrin: A structural mediator between diverse plasma membrane proteins and the cytoplasm. *Curr. Opin. Cell Biol.* **2**, 51–56.

Bonder, E. M., and Mooseker, M. S. (1982). Direct electron microscopic visualization of barbed end capping and filament cutting by intestinal microvillar 95-kdalton protein (villin): A new actin assembly assay using the *Limulus* acrosomal process. *J. Cell Biol.* **96**, 1097–1107.

Bonder, E. M., and Mooseker, M. S. (1986). Cytochalasin B slows but does not prevent monomer addition at the barbed end of the actin filament. *J. Cell Biol.* **102**, 282–288.

Bonder, E. M., Fishkind, D. F., and Mooseker, M. S. (1983). Direct measurement of critical concentrations and assembly rate constants at the two ends of an actin filament. *Cell* **34**, 491–501.

Bonder, E. M., Fishkind, D. J., Henson, J. H., Cotran, N. M., and Begg, D. A. (1988). Actin in cytokinesis: Formation of the contractile apparatus. *Zool. Sci.* **5**, 699–711.

Bonder, E. M., Fishkind, D. J., Cotran, N. M., and Begg, D. A. (1989). The cortical actin-membrane cytoskeleton of unfertilized sea urchin eggs: Analysis of the spatial organization and relationship of filamentous actin, nonfilamentous actin, and egg spectrin. *Dev. Biol.* **134**, 327–341.

Bryan, J., and Kane, R. E. (1978). Separation and interaction of the major components of sea urchin actin gel. *J. Mol. Biol.* **125**, 207–224.

Bryan, J., Edwards, R., Matsudaira, P., Otto, J., and Wulfkuhle (1993). Fascin, an echinoid actin–bundling protein, is a homolog of the Drosophila singed gene product. *Proc. Natl. Acad. Sci. USA* **90**, 9115–9119.

Buß, F., Temm-Grove, C., Henning, S., and Jockusch, B. M. (1992). Distribution of profilin in fibroblasts correlates with the presence of highly dynamic actin filaments. *Cell Motil. Cytoskeleton* **22**, 51–61.

Buck, W. R., Hoffmann, E. E., Rakow, T. L., and Shen, S. S. (1994) Synergistic calcium release in the sea urchin by ryanodine and cyclic ADP ribose. *Dev. Biol.* **163**, 1–10.

Burgess, D. R., and Schroeder, T. E. (1977). Polarized bundles of actin filaments within the microvilli of fertilized sea urchin eggs. *J. Cell Biol.* **74,** 1032–1037.

Carron, C. P., and Longo, F. J. (1982). Relation of cytoplasmic alkalinization to microvillar elongation and microfilament formation in the sea urchin egg. *Dev. Biol.* **89,** 128–137.

Cao, L.-G., and Wang, Y.-L. (1990). Mechanism of the formation of contractile ring in dividing cultured animal cells. II. Cortical movement of microinjected actin filaments. *J. Cell Biol.* **111,** 1905–1911.

Chambers, C., and Grey, R. D. (1979). Development of the structural components of the brush border in absorptive cells of the chick intestine. *Cell Tissue Res.* **204,** 387–405.

Chandler, D. E. (1984). Exocytosis *in vitro*: Ultrastructure of the isolated sea urchin egg cortex as seen in platinum replicas. *J. Ultrastructure* Res. **89,** 198–211.

Chandler, D. E. (1991). Multiple intracellular signals coordinate structural dynamics in the sea urchin egg cortex at fertilization. *J. Electron Micro. Technol.* **17,** 266–293.

Chandler, D. E., and Heuser, J. (1981). Postfertilization growth of microvilli in the sea urchin egg: New views from the eggs that have been quick-frozen, freeze-fractured, and deeply etched. *Dev. Biol.* **82,** 393–400.

Cheney, R. E., and Mooseker, M. S. (1992). Unconventional myosins. *Curr. Opin. Cell Biol.* **4,** 27–35.

Ciapa, B., and Epel, D. (1991). A rapid change in phosphorylation on tyrosine accompanies fertilization of sea urchin eggs. *FEBS Lett.* **295,** 167–170.

Ciapa, B., Crossley, I., and De Renzis, G. (1988). Structural modifications induced by TPA (12-O-tetradecanoyl phorbol-13-acetate) in sea urchin eggs. *Dev. Biol.* **128,** 142–149.

Ciapa, B., Borg, B., and Whitaker, M. (1992). Polyphosphoinositide metabolism during the fertilization wave in sea urchin eggs. *Development* **115,** 187–195.

Cline, C. A., and Schatten, G. (1986). Microfilaments during sea urchin fertilization: Fluorescence detection with rhodaminyl phalloidin. *Gamete Res.* **14,** 277–291.

Coffe, G., Foucault, G., Soyer, M. O., DeBilly, F., and Pudles, J. (1982). State of actin during the cycle of cohesiveness of the cytoplasm in parthenogenetically activated sea urchin egg. *Exp. Cell Res.* **142,** 365–372.

Coluccio, L. M., Sedlar, P. A., and Bryan, J. (1986). The effects of a 45K molecular weight protein from unfertilized sea urchin eggs and its 1 : 1 actin complex on actin filaments. *J. Muscle Res. Cell Motil.* **7,** 133–141.

Conrad, G. W., and Schroeder, T. E. (eds) (1990). "Cytokinesis: Mechanisms of Furrow Formation during Cell Division," (*Ann. N.Y. Acad. Sci.,* Vol. 582) New York Academy of Sciences, New York.

Cooper, J. A. (1991). The role of actin polymerization in cell motility. *Annu. Rev. Physiol.* **53,** 585–605.

D'Andrea, L., Fishkind, D. J., Begg, D. A., and Bonder, E. M. (1991). Isolation and localization of a spectrin-like protein from echinoderm sperm. *Cell Motil. Cytoskeleton* **19,** 49–61.

D'Andrea, L. (1993). Biochemical, morphological, and structural examination of the actin-based motility in echinoderm sperm, eggs, and immune coelomocytes: Analysis of dynamic actin-membrane reorganizations involving spectrin, myosin-I and myosin-II. Ph.D. Thesis, Rutgers University.

Davidson, E. H. (1986). "Gene Activity in Early Development." Academic Press, Orlando, Florida.

Dufresne, L., Swezey, R. R., and Epel, D. (1987). Kinetics of actin assembly attending fertilization or artificial activation for sea urchin eggs. *Exp. Cell Res.* **172,** 32–42.

Eisen, A., Kiehart, D. P., Wieland, S. J., and Reynolds, G. T. (1984). Temporal sequence and spatial distribution of early events of fertilization in single sea urchin eggs. *J. Cell Biol.* **99,** 1647–1654.

Epel, D. (1989). Arousal of activity in sea urchin eggs at fertilization. *In* "The Cell Biology of Fertilization" (H. Schatten and G. Schatten, eds.), pp. 361–385. Academic Press, New York.

Faire, K., and Bonder, E. M. (1993). Sea urchin egg 100 kDa dynamin-related protein: Identification and localization to intracellular vesicles. *Dev. Biol.* **159**, 581–594.

Fisher, G. W., and Rebhun, L. I. (1983). Sea urchin egg cortical granule exocytosis is followed by a burst of membrane retrieval via uptake into coated vesicles. *Dev. Biol.* **99**, 456–472.

Fishkind, D. J., Bonder, E. M., and Begg, D. A. (1987). Isolation and characterization of sea urchin egg spectrin: Calcium modulation of the spectrin–actin interaction. *Cell Motil. Cytoskeleton* **7**, 304–314.

Fishkind, D. J., Bonder, E. M., and Begg, D. A. (1990a). Sea urchin spectrin in oogenesis and embryogenesis: A multifunctional integrator of membrane–cytoskeletal interactions. *Dev. Biol.* **142**, 453–464.

Fishkind, D. J., Bonder, E. M., and Begg, D. A. (1990b). Subcellular localization of sea urchin egg spectrin: Evidence for assembly of the membrane–skeleton on unique classes of vesicles in eggs and embryos. *Dev. Biol.* **142**, 439–452.

Fishkind, D. J. (1991). The biochemistry and cell biology of sea urchin egg spectrin: Purification and characterizarion of a calcium-modulated actin crosslinking protein, and its distribution and dynamic redistribution during oogenesis and embryogenesis. Ph.D. Thesis, Harvard University.

Foltz, K. R., and Lennarz, W. J. (1993). The molecular basis of sea urchin gamete interactions at the egg plasma membrane. *Dev. Biol.* **158**, 46–61.

Foucault, G., Raymond, M. N., and Pudles, J. (1987). Cytoskeleton of the unfertilized sea urchin egg. *Biol. Cell* **60**, 63–70.

Golsteyn, R. M., and Waisman, D. M. (1989a). The 50kDa protein–actin complex from unfertilized sea urchin (*S. purpuratus*) eggs. *Biochem. J.* **257**, 817–822.

Golsteyn, R. M., and Waisman, D. M. (1989b). The purification of a 50kDa protein–actin complex from unfertilized sea urchin (*S. purpuratus*) eggs. *Biochem. J.* **257**, 809–815.

Hamaguchi, Y., and Mabuchi, I. (1988). Accumulation of fluorescently labeled actin in the cortical layer in sea urchin eggs after fertilization. *Cell Motil. Cytoskeleton* **9**, 153–163.

Hamaguchi, Y., and Mabuchi, I. (1986). Alpha-actinin accumulation in the cortex of echinoderm eggs during fertilization. *Cell Motil. Cytoskeleton* **6**, 549–559.

Hartwig, J. H., and Kwiatkowski, D. J. (1991). Actin-binding proteins. *Curr. Opin. Cell Biol.* **3**, 87–97.

Henson, J. H., and Begg, D. A. (1988). Filamentous actin organization in the unfertilized sea urchin egg cortex. *Dev. Biol.* **127**, 338–348.

Hiramoto, Y. (1970). Rheological properties of sea urchin eggs. *Biorheology* **6**, 201–234.

Hiramoto, Y. (1981) Mechanical properties of dividing cells. *In* "Cellular Dynamics: Mitosis/Cytokinesis," (Forer, A. and Zimmerman A. M. eds). Academic Press, New York.

Hosoya, H., Mabuchi, I., and Sakai, H. (1982). Actin–modulating proteins in the sea urchin egg. I. Analysis of G-actin-binding proteins by DNaseI-affinity chromatography and purification of a 17,000 molecular-weight component. *J. Biochem.* **92**, 1853–1862.

Hosoya, H., Mabuchi, I., and Sakai, H. (1984). A 45,000-mol-wt protein–actin complex from unfertilized sea urchin egg affects assembly properties of actin. *J. Cell Biol.* **99**, 994–1001.

Hosoya, H., Mabuchi, I., and Sakai, H. (1986). An 100-kDa Ca^{2+}-sensitive actin–fragmenting protein from unfertilized sea urchin eggs. *Eur. J. Biochem.* **154**, 233–239.

Ishidate, S., and Mabuchi, I. (1988a). A novel actin filament-capping protein from sea urchin eggs: A 20,000-molecular weight protein–actin complex. *J. Biochem.* **104**, 72–80.

Ishidate, S., and Mabuchi, I. (1988b). Localization and possible function of 20kDa actin-modulating protein (actolinkin) in the sea urchin egg. *Eur. J. Cell Biol.* **46,** 275–281.

Jaffe, L. A., and Cross, N. L. (1986). Electrical regulation of sperm–egg fusion. *Annu. Rev. Physiol.* **48,** 191–200.

Kamel., C. C., Bailey, J., Schoenbaum, L., and Kinsey, W. (1985). Phosphatidylinositol metabolism during fertilization in the sea urchin egg. *Lipids* **20,** 350–356.

Kane, R. E. (1975). Preparation and purification of polymerized actin from sea urchin egg extracts. *J. Cell Biol.* **66,** 305–315.

Kane, R. E. (1980). Induction of either contractile or structural actin-based gels in sea urchin egg cytoplasmic extract. *J. Cell Biol.* **86,** 803–809.

Karp, G. C., and Solursh, M. (1985). Dynamic activity of the filopodia of sea urchin embryonic cells and their role in directed migration of the primary mesenchyme *in vitro*. *Dev. Biol.* **112,** 276–283.

Kinsey, W. H. (1984). Regulation of tyrosine-specific kinase activity at fertilization. *Dev. Biol.* **105,** 137–143.

Kuramochi, K., Mabuchi, I., and Owaribe, K. (1986). Spectrin from sea urchin eggs. *Biomed. Res.* **7,** 65–68.

Lassing, I., and Lindberg, U. (1988). Evidence that the phosphoinositol cycle is linked to cell motility. *Exp. Cell Res.* **174,** 1–15.

Liu, J.-P., Powell, K. A., Sudhof, T. C., and Robinson, P. J. (1994a). Dynamin is a calcium-sensitive phospholipid-binding protein with high affinity for protein kinase C. *J. Biol. Chem.* **269,** 21043–21050.

Liu, J.-P., Sim, A.T. R., and Robinson, P. J. (1994b). Calcineurin inhibition of dynamin I GTPase activity is coupled to nerve terminal depolarization. *Science* **265,** 970–973.

Louvard, D., Kedinger, M., and Hauri, H. P. (1992). The differentiating intestinal epithelial cell: Establishment and maintenance of functions through interactions between cellular structures. *Annu. Rev. Cell Biol.* **8,** 157–195.

Mabuchi, I. (1973). A myosin-like protein in the cortical layer of the sea urchin egg. *J. Cell Biol.* **59,** 542–547.

Mabuchi, I. (1983). An actin–depolymerizing protein (depactin) from starfish oocytes: Properties and interaction with actin. *J. Cell Biol.* **97,** 1612–1621.

Mabuchi, I., and Hosoya, H. (1982). Actin–modulating proteins in the sea urchin egg. II. Sea urchin egg profilin. *Biomed. Res.* **3,** 465–476.

Mabuchi, I., and Kane, R. E. (1987). A 250K molecular-weight actin-binding protein from the actin-based gels formed in sea urchin egg cytoplasmic extract. *J. Biochem.* **102,** 947–956.

Mabuchi, I., Hamaguchi, Y., Kobayashi, T., Hosoya, H., Tsukita, S., and Tsukita, S. (1985). Alpha-actinin from sea urchin eggs: Biochemical properties, interaction with actin, and distribution in the cell during fertilization and cleavage. *J. Cell Biol.* **100,** 375–383.

Maciever, S. K., Zot, H. G., and Pollard, T. D. (1991). Characterization of actin filament severing by actophorin from *Acanthamoeba castellanii*. *J. Cell Biol.* **115,** 1611–1620.

MacLean–Fletcher, S. D., and Pollard, T. D. (1980b). Identification of a factor in conventional muscle actin preparations which inhibits actin filament self-associations. *Biochem. Biophys. Res. Commun.* **96,** 18–27.

Maekawa, S., Endo, S., and Sakai, H. (1987). A high molecular weight actin binding protein: Its localization in the cortex of the sea urchin egg. *Exp. Cell Res.* **172,** 340–353.

Maekawa, S., Toriyama, M., and Sakai, H. (1989). Tropomyosin in the sea urchin egg cortex. *Eur. J. Biochem.* **178,** 657–662.

Malinda, K. M., and Ettensohn, C. A. (1994). Primary mesenchyme cell migration in the sea urchin embryo: Distribution of directional cues. *Dev. Biol.* **164,** 562–578.

Matsudaira, P. T. (1991). Modular organization of actin crosslinking proteins. *Trends Biochem. Sci.* **16,** 87–92.

Monroy, A. (1986). A centennial debt of developmental biology to the sea urchin. *Biol. Bull.* **171**, 509–519.

Nachmias, V. T. (1993) Small actin-binding proteins: The β-thymosin family. *Curr. Opin. Cell Biol.* **5**, 56–62.

Ohnuma, M., and Mabuchi, I. (1993). 45K actin filament-severing protein from sea urchin eggs: Interaction with phosphatidylinositol-4,5-bisphosphate. *J. Biochem.* **114**, 718–722.

Otto, J. J., Kane, R. E., and Bryan, J. (1980). Redistribution of actin and fascin in sea urchin eggs after fertilization. *Cell Motil.* **1**, 31–40.

Perrin, D., Langley, O. K., and Aunis, D. (1987). Anti-α-fodrin inhibits secretion from permiabilized chromaffin cells. *Nature (London)* **326**, 498–501.

Puldles, J., Moudjou, M., Hisanaga, S.-I., Maruyama, K., and Sakai, H. (1990). Isolation, characterization, and immunochemical properties of a giant protein from sea urchin egg cytomatrix. *Exp. Cell Res.* **189**, 253–260.

Rappaport, R. (1986). Establishment of the mechanism of cytokinesis in animal cells. *Int. Rev. Cytol.* **105**, 245–281.

Robinson, M. S. (1994). The role of clathrin, adaptors, and dynamin in endocytosis. *Curr. Opin. Cell Biol.* **6**, 538–544.

Robinson, P. J., Sontag, J.-M., Liu, J.-P., Fykse, E. M., Slaughter, C., McMahon, H., and Sudhof, T. C. (1993). Dynamin GTPase regulated by protein kinase C phosphorylation in nerve terminals. *Nature* **365**, 163–166.

Salmon, E. D. (1989). Cytokinesis in animal cells. *Curr. Opin. Cell Biol.* **1**, 541–547.

Sardet, C., and Chang, P. (1987). The egg cortex: From maturation through fertilization. *Cell Differ.* **21**, 1–19.

Schatten, G., and Mazia, D. (1976). The penetration of the spermatozoan through the sea urchin egg surface at fertilization. *Exp. Cell Res.* **98**, 325–337.

Schatten, G., Bestor, T., Balczon, R., Henson, J., and Schatten, H. (1985). Intracellular pH shifts lead to microtubule assembly and microtubule-mediated motility during sea urchin fertilization: Correlations between intracellular pH and microtubule activity and depressed intracellular pH and microtubule disassembly. *Eur. J. Cell Biol.* **36**, 116–127.

Schatten, H., Cheney, R., Balczon, R., Willard, M., Cline, C., Simerly, C., and Schatten, G. (1986). Localization of fodrin during fertilization and early development of the sea urchins and mice. *Dev. Biol.* **118**, 457–466.

Schroeder, T. E. (1978). Microvilli on sea urchin eggs: A second burst of elongation. *Dev. Biol.* **64**, 342–346.

Schroeder, T. E. (1979). Surface area change at fertilization: Resorption of the mosaic membrane. *Dev. Biol.* **70**, 306–326.

Shen, S. S., and Buck, W. R. (1993). Sources of calcium in sea urchin eggs during the fertilization response. *Dev. Biol.* **157**, 157–169.

Shibayama, T., Carboni, J. M., and Mooseker, M. S. (1987). Assembly of the intestinal brush border: Appearance and distribution of microvillar core proteins in developing chick enterocytes. *J. Cell Biol.* **105**, 335–344.

Smith, L. C., Harrington, M. G., Britten, R. J., and Davidson, E. H. (1994). The sea urchin profilin gene is specifically expressed in mesenchyme cells during gastrulation. *Dev. Biol.* **164**, 463–474.

Spudich, A., and Spudich, J. A. (1979). Actin in triton-treated cortical preparations of unfertilized and fertilized sea urchin eggs. *J. Cell Biol.* **82**, 212–226.

Spudich, A., Giffard, R. G., and Spudich, J. A. (1982). Molecular aspects of cortical actin filament formation upon fertilization. *Cell Differ.* **11**, 281–284.

Spudich, A., Wrenn, J. T., and Wessells, N. K. (1988). Unfertilized sea urchin eggs contain a discrete cortical shell of actin that is subdivided into two organizational states. *Cell Motil. Cytoskeleton* **9**, 85–96.

Spudich, A. (1992). Actin organization in the sea urchin egg cortex. *Curr. Topics Dev. Biol.* **26,** 9–21.

Stossel, T. P. (1989). From signal to pseudopod. How cells control cytoplasmic actin assembly. *J. Biol Chem.* **264,** 18261–18264.

Stossel, T. P., Chaponnier, C., Ezzel, R. M., Hartwig, J. H., Janmey, P. A., Kwiatkowski, D. J., Lind, S. E., Southwick, F. S., Yin, H. L., and Zaner, K. S. (1985). Nonmuscle actin-binding proteins. *Annu. Rev. Cell Biol.* **1,** 353–402.

Striker, S. A., Centonze, V. E., Paddock, S. W., and Schatten, G. (1992) Confocal microscopy of fertilization-induced calcium dynamics in sea urchin eggs. *Dev. Biol.* **149,** 370–380.

Swann, K., and Whitaker, M. (1985). Stimulation of the Na/H exchanger of sea urchin eggs by phorbol ester. *Nature* **314,** 274–277.

Theriot, J. A., and Mitchison, T. J. (1993). The three faces of profilin. *Cell* **75,** 835–858.

Theriot, J. A., Rosenblatt, J., Portnoy, D. A., Goldschmidt-Clermont, P., and Mitchison, T. J. (1994). Involvment of profilin in the actin-based motility of Listeria monocytogenes in cells and in cell free extracts. *Cell* **76,** 505–517.

Tilney, L. G. (1976). The polymerization of actin II. How nonfilamentous actin becomes nonrandomly distributed in sperm: Evidence for the association of this actin with membranes. *J. Cell Biol.,* 51–71.

Tilney, L. G., and Jaffe, L. A. (1980). Actin, microvilli, and the fertilization cone of sea urchin eggs. *J. Cell Biol.* **87,** 771–782.

Tilney, L. G., Bonder, E. M., Coluccio, L. M., and Mooseker, M. S. (1983). Actin from Thyone sperm assembles on only one end of an actin filament: A behavior regulated by profilin. *J. Cell Biol.* **97,** 112–124.

Turner, P. R., and Jaffe, L. A. (1989). G-proteins and the regulation of oocyte maturation and fertilization. *In* "The Cell Biology of Fertilization" (H. Schatten and G. Schatten, eds.), pp. 297–318. Academic Press, New York.

Vacquier, V. D. (1975). The isolation of intact cortical granules from sea urchin eggs: Calcium ions trigger granule discharge. *Dev. Biol.* **43,** 62–74.

Vacquier, V. D. (1981). Dynamic changes of the egg cortex. *Dev. Biol.* **84,** 1–26.

Vallee, R. B. (1992). Dynamin: motor protein or regulatory GTPase. *J. Muscle Res. Cell Motil.* **13,** 493–496.

Vallee, R. B., and Shpetner, H. S. (1993). Dynamin in synaptic dynamics. *Nature* **365,** 107–108.

Walker, G., Yabkowitz, R., and Burgess, D. R. (1991). Mapping the binding domain of a myosin II binding protein. *Biochemistry* **42,** 10206–10210.

Wang, F.-S., and Bonder, E. M. (1991). Sea urchin egg villin: Identification of villin in a non-epithelial cell from an invertebrate species. *J. Cell Sci.* **100,** 61–71.

Wang, F.-S. (1993). Modulation of sea urchin egg cortical actin cytoskeleton: Isolation, identification, analysis, and localization of egg villin. Ph.D. Thesis, Rutgers Univ.

Wang, L. L., and Spudich, J. A. (1984). A 45,000 MW protein from unfertilized eggs severs actin filaments in a calcium dependent manner and increase the steady-state concentration of nonfilamentous actin. *J. Cell Biol.* **99,** 844–851.

Weeds, A., and Maciever, S. (1993). F-actin capping proteins. *Curr. Opin. Cell Biol.* **5,** 63–69.

Wessel, G. M., Marchase, R. B., and McClay, D. R. (1984). Ontogeny of the basal lamina in the sea urchin embryo. *Dev. Biol.* **103,** 235–245.

Wessel, G. M., and Chen, S. W. (1992). Transient, localized accumulation of alpha-spectrin during sea urchin morphogenesis. *Dev. Biol.* **154,** 671–682

Whitaker, M., and Swann, K. (1993). Lighting the fuse at fertilization. *Development* **117,** 1–12.

Yonemura, S., and Mabuchi, I. (1987). Wave of cortical actin polymerization in the sea urchin egg. *Cell Motil. Cytoskeleton* **7,** 46–53.

Yoshigaki, T., Maekawa, S., Endo, S., and Sakai, H. (1989). Localization of a high-molecular-weight actin binding protein in the sea urchin egg from fertilization through cleavage. *Cell Structure Function* **14,** 363–374.

Zagon, I. S., Higbee, R., Riederer, B. M., and Goodman, S. R. (1986). Spectrin subtypes in mammalian brain: An immunoelectron microscopic study. *J. Neurosci.* **6,** 2977–2986.

5

RNA Localization and the Cytoskeleton in *Drosophila* Oocytes

Nancy Jo Pokrywka
Department of Biology
Vassar College
Poughkeepsie, New York 12601

I. Introduction

Prevailing wisdom asserts that the cytoskeleton plays an important role in the organization and modeling of a cell's cytoplasm. The importance of subcellular organization is stunningly demonstrated in the oocyte of many animals, where localization of developmental determinants is essential for the earliest events of development. For example, the establishment of embryonic axes often occurs in the unfertilized egg via the subcellular localization of spatial determinants. Nowhere is this more evident than in *Drosophila*, where both the anterior–posterior and dorsal–ventral axes of the embryo are first specified during oogenesis. In *Drosophila*, specific mRNAs become localized within the oocyte; when localized RNAs are

Current Topics in Developmental Biology, Vol. 31

translated after fertilization, the result is locally high concentrations of protein which give a region of the embryo its identity. If the localization of these RNAs is prevented, large portions of the embryo never develop. Thus, RNA localization is absolutely essential for proper development. An understanding of the interactions required for RNA localization is central to our understanding of the earliest events of development. Thus, a major question in this field is how RNA localization is accomplished. How are the appropriate messenger RNAs recognized and localized to the proper place in the developing oocyte? What molecular associations are required for localization? How is the position of localized RNAs maintained?

Answers to some of these questions are evolving slowly. The genetic tractability of *Drosophila* has made possible the identification of a number of mutations which interfere with the proper localization of specific RNAs. The characterization of their functions promises to reveal much regarding the mechanisms of RNA localization. In addition, recent research into the organization of the oocyte cytoskeleton has highlighted the importance of the cytoskeleton in RNA localization events. It appears likely that the *Drosophila* cytoskeleton may participate in the localization of many, if not all, RNAs, and data to support a role for the cytoskeleton in these events are slowly accumulating. Although the evidence complied thus far is circumstantial, it is tantalizing for the glimpse it gives us of the potential mechanisms at work. The intention of this chapter is to briefly describe the experiments and results which have led researchers to focus on the role of the cytoskeleton in RNA localization in *Drosophila* and to propose possible mechanisms for how the cytoskeleton may participate in RNA localization.

II. Oogenesis in *Drosophila*

To understand the forces which influence patterning in the *Drosophila* egg, it is important to begin with a review of oogenesis, since the manner in which oogenesis proceeds appears to influence the organization of the resulting embryo. This includes information on the cytoskeletal organization of the egg chamber, since current data indicate the cytoskeleton plays an important role in RNA localization in *Drosophila.*

A. An Overview of *Drosophila* Oogenesis

Drosophila ovaries develop along a plan described as meroistic, meaning the germline stem cell gives rise to a cluster of nurse cells, in addition to the oocyte. Oogenesis begins when a germ cell divides mitotically four times to produce a cyst of 16 cells which remain interconnected by intercellular bridges (also called ring canals) that are the result of incomplete cytokinesis. One of the 16 cells is specified to become the oocyte, while the other 15 cells become nurse cells. Each cluster of 15 nurse cells, an oocyte, and a surrounding epithelium of somatic follicle cells is referred to as an egg chamber; each egg chamber produces one egg at the end of oogenesis. The cell that becomes the oocyte invariably occupies the posteriormost section of the developing egg chamber, and this arrangement establishes the anterior and posterior ends of the oocyte, and eventually the embryo. The growing oocyte is nurtured by 15 nurse cells, so named because they provide the oocyte with most of its cytoplasm and almost all of its organelles, proteins, and mRNA (the oocyte nucleus itself being virtually quiescent).

The development of the oocyte has been divided into 14 stages (as in King, 1970) on the basis of morphological features. Some of the key stages of oogenesis are outlined in Fig. 1. During the first 7 stages, the entire egg

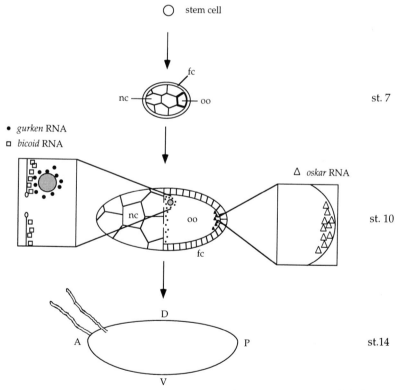

Fig. 1 A diagram of some key stages of oogenesis, with stages listed at the right (according to King, 1970). The approximate positions of *gurken, bicoid,* and *oskar* RNAs in a stage-10 oocyte are depicted. All oocytes are oriented such that anterior is left and dorsal is up.

chamber grows in volume, and the nurse cells and oocyte are of equal size. Since the oocyte is nonproductive, even at these early stages cytoplasm must flow from the nurse cells into the oocyte. Starting at stage 8, the oocyte begins to increase in volume relative to the attached nurse cells, partly due to the uptake of yolk platelets from the fly hemolymph via endocytosis. At stage 9, the oocyte occupies approximately one-third of the egg chamber, and by stage 10, this increases to one-half. At the end of stage 10, a spectacular event occurs: violent actin-based forces within the nurse cells cause contractions and a rather sudden influx of the remaining nurse cell cytoplasm into the oocyte (Gutzeit, 1986). This influx is accompanied by ooplasmic streaming which mixes the incoming material into the cytoplasm of the oocyte (Gutzeit and Koppa, 1982). This event is completed in 30 min, after which time the nurse cells degenerate, the chorion (or eggshell) is secreted, and the mature oocyte is produced.

B. The Cytoskeleton of the Developing Oocyte

In recent years, extensive studies have been undertaken to examine the distribution of microtubules and actin filaments during oogenesis. These analyses have revealed the dynamic nature of the oocyte cytoskeleton, which undergoes several reorganizations as oogenesis proceeds. Excellent reviews of the oocyte cytoskeleton have been published elsewhere (Cooley and Theurkauf, 1994; Mahajan-Miklos and Cooley, 1994; Theurkauf, 1994a). For the purposes of this discussion, only relevant information will be reviewed. In brief, even stage 1 egg chambers appear to possess a polarized microtubule array (Theurkauf et al., 1993). Most of the microtubules in the egg chamber originate in a microtubule-organizing center (MTOC), which lies at the posterior end of the oocyte. The microtubules extend from the MTOC, through the ring canals, and into the nurse cells (Theurkauf et al., 1993). Since microtubules generally extend from a MTOC with a predictable polarity, it seems likely that the microtubules are arranged with their plus ends at the farthest reaches of the nurse cells and their minus ends embedded in the MTOC (in the oocyte proper).

The MTOC in the oocyte of young egg chambers is found at the posterior edge of the oocyte periphery. This structure persists until midoogenesis (stages 7–9) when a reorganization of the microtubule network takes place. The MTOC at the posterior end of the oocyte breaks down, and microtubules begin instead to nucleate along the anterior margin of the oocyte, resulting in an anterior–posterior gradient of microtubules (Theurkauf et al., 1992). It has been proposed that the anterior oocyte margin acts as a diffuse MTOC, since mild colchicine treatments of egg chambers result in the depolymerization of most microtubules, leaving only short microtubules found at the anterior margin. The polarity of these microtubules is thought to be uniform (with the minus ends anchored at the anterior margin, and the plus ends extending toward the posterior), although this has not been tested directly (Theurkauf *et al.*, 1992). It is during these stages that the anterior–posterior and dorsoventral axes of the embryo are specified. At the end of stage 10, when large volumes of nurse cell cytoplasm enter the egg, organized ooplasmic streaming occurs. This process is distinct from the forces which result in the influx of nurse cell cytoplasm, and is preceded by the appearance of long parallel arrays of microtubules just below the surface of the oocyte (Theurkauf *et al.*, 1992). Treatment of egg chambers with microtubule inhibitors during these stages prevents ooplasmic streaming (Gutzeit and Koppa, 1982), implying the process is microtubule based.

In comparison, less is known about the role of the actin cytoskeleton during oogenesis. As the egg chamber grows, actin is concentrated at the periphery of the nurse cells and oocyte, just below the plasma membrane of each cell. The oocyte cortex appears to contain significantly higher amounts of actin, suggesting that even at early stages of oogenesis, the oocyte is specialized with regard to its cytoskeleton components. The distribution of actin remains unchanged throughout most of oogenesis until stage 10, when massive polymerization of actin occurs, and dense networks of actin filaments extend throughout the nurse cell complex (Theurkauf *et al.*, 1992; Warn *et al.*, 1985). These actin filaments appear to extend from the nurse cell nuclei to the periphery of each cell and between nurse cells via the ring canals. At this time, the nurse cells contract, squeezing their contents into the oocyte. The peripheral actin is thought to be responsible for the generation of force, while the newly created actin networks appear to anchor the nurse cell nuclei in place (Cooley *et al.*, 1992; Theurkauf *et al.*, 1992). During middle and late oogenesis the oocyte contains a dense cortical meshwork of actin filaments, but after maturation, actin filaments are also present extending throughout the deeper cytoplasm of the oocyte (Theurkauf *et al.*, 1992).

There is some evidence for the transport of macromolecules into the oocyte via an actin–based process. Using video–enhanced contrast microscopy, the preferential movement of cytoplasmic vesicles into the oocyte has been observed during stages 7–9 (Bohrmann and Biber, 1994). This movement is prevented upon treatment with cytochalasins but not microtubule inhibitors, suggesting an actin–based transport system is present in nurse cells.

III. Localization of Developmental Determinants

It is during the midstages of oogenesis that several mRNAs critical for oocyte patterning become localized. The *bicoid* locus controls the development of head and thoracic structures, and its RNA is localized to the anterior margin of the oocyte (Berleth *et al.,* 1988; Driever and Nüsslein-Volhard, 1988; Driever *et al.,* 1990). *oskar* RNA is required for polar granule formation and is a regulator of posterior development (Ephrussi *et al.,* 1991; Ephrussi and Lehmann, 1992; Smith *et al.,* 1992). *osk* RNA becomes localized to the posterior pole of the oocyte during midoogenesis. Finally, the dorsoventral polarity of the oocyte is determined during midoogenesis, largely through the localization of *gurken* RNA (and presumably, protein) to the dorso–anterior corner of the oocyte (Neuman-Silberberg and Schüpbach, 1993). These three mRNAs are clear examples of localized molecules which can act as developmental determinants, but several other RNAs become localized within the oocyte during oogenesis. In the next sections, we examine the distribution of various localized RNAs throughout oogenesis.

A. Establishing Anterior Polarity

The future anterior pole of the embryo is determined during oogenesis through the action of the *bicoid* (*bcd*) gene. Products of the *bcd* gene are responsible for the initiation of head and thoracic development (Driever and Nüsslein–Volhard, 1988; Driever *et al.,* 1990). The localization of *bcd* RNA to the anterior end of the oocyte allows for the localized translation of bcd protein during embryogenesis, which sets up a steep concentration gradient from anterior to posterior. Locally high concentrations of bcd protein in turn regulate the spatially restricted expression of downstream genes (Struhl *et al.,* 1989). Thus, the establishment of a bcd protein gradient directs the polarity and extent of anterior development. *bcd* RNA localization is absolutely required for proper anterior development, since embryos in which *bcd* RNA is not localized fail to produce anterior structures (Berleth *et al.,* 1988; Stephenson *et al.,* 1988).

bcd RNA is synthesized in nurse cells and becomes localized to the anterior pole of the oocyte during oogenesis. *bcd* RNA first appears in the egg chamber at stage 5, where it accumulates in the oocyte. As oogenesis proceeds, *bcd* RNA becomes associated with the anterior cortex. In addition to oocyte localization, *bcd* RNA is localized in the nurse cells as well, as patches concentrated on one side of each nurse cell nucleus. Localization to the anterior cortex of the oocyte persists until fertilization, when *bcd* RNA is released from the cortex and becomes localized in the deeper cytoplasm at the anterior end of the embryo (Berleth *et al.,* 1988; St Johnston *et al.,* 1989).

Several genes are thought to be directly involved in *bcd* RNA localization. *exuperantia* (*exu*) is required for the early stages of *bcd* RNA localization, as *exu* mutants have a uniform distribution of *bcd* in the nurse cells and oocyte (St. Johnston *et al.*, 1989; Pokrywka and Stephenson, 1991). *swallow* (*sww*) appears to be required for localization from stages 9 or 10 through the end of oogenesis. In *sww* mutants, early stages of localization proceed normally, but *bcd* RNA becomes unlocalized during later oogenic stages (Stephenson *et al.*, 1988). *staufen* is a gene required for localization of *bcd* RNA after fertilization and probably acts to maintain *bcd* RNA in the deep cytoplasm at the anterior end of the embryo (St Johnston *et al.*, 1991; Ferrandon *et al.*, 1994). Additionally, there are several mutations which appear to affect the organization of the oocyte cytoskeleton; these mutations also alter the localization pattern of *bcd* RNA and will be discussed later.

Although genes required for *bcd* RNA localization have been identified, little is known about how their products function to localize *bcd* RNA. A large (625 nt) cis–acting signal in the 3′ untranslated region (UTR) of the *bcd* message is known to be required for localization. The secondary structure of the UTR, but not the primary sequence, has been evolutionarily conserved in closely related species of *Drosophila*, implying the three–dimensional structure of the RNA is recognized by localization machinery (Macdonald and Struhl, 1988; Macdonald, 1990). However, at least one discrete element (termed BLE1) is required for the early stages of localization (Macdonald *et al.*, 1993). It is possible that the secondary structure of the 3′ RNA permits discrete elements to become aligned or presented to RNA-binding proteins.

Strong evidence suggests that microtubules are involved in the localization of *bcd* RNA. Treatment of egg chambers with microtubule inhibitors interferes with all aspects of *bcd* RNA localization, indicating localization relies on an intact microtubule cytoskeleton. That this connection may be direct can be inferred from experiments in which egg chambers are treated with taxol, a drug that induces microtubule stabilization and the formation of novel microtubule networks. In taxol-treated egg chambers, *bcd* RNA is no longer anteriorly localized, but becomes localized to ectopic sites in the oocyte, coincident with the formation of novel microtubule aggregates (Pokrywka and Stephenson, 1991). The sites also contain exu protein, suggesting that exu protein is required for the association of *bcd* RNA with microtubules (Wang and Hazelrigg, 1994). A direct association of *bcd* RNA with the cytoskeleton has been inferred from biochemical studies. *bcd* RNA (and other localized RNAs) are enriched in detergent extracts of oocytes. Enrichment in the detergent-insoluble fraction is dependent on the same factors required for *bcd* RNA localization *in vivo,* including microtubules, *exu,* and *sww,* indicating the procedure enriches for physiologically relevant associations. The detergent-insoluble fraction contains several localized RNAs, is depleted of other RNAs, and is thought to represent a crude preparation of the oocyte cortex (Pokrywka and Stephenson, 1994).

B. The Posterior Pole

The main determinant of the posterior system is *nanos* (Wang and Leh-mann, 1991), and *nanos* RNA becomes localized to the posterior pole of the oocyte during the last few stages of oogenesis. However, the positioning of *nanos* RNA during later stages of oogenesis is dependent on the assembly of pole plasm at the posterior pole of the oocyte during midoogenesis (Wang *et al.*, 1994). This process is initiated by the localization of *oskar* (*osk*) RNA to the posterior pole in midoogenesis. *osk* RNA is first tran-scribed during the earliest stages of oogenesis, and accumulates in the young oocyte. During midoogenesis, several interesting features of *osk* distribution become obvious. *osk* RNA is unevenly distributed in nurse cells (in a pattern similar to that described for *bcd* RNA), and as the oocyte grows larger than the attached nurse cells, *osk* RNA gradually becomes localized to the posterior tip of the oocyte (Ephrussi *et al.*, 1991; Kim-Ha *et al.*, 1991; Pokrywka and Stephenson, 1995). This localization persists through fertilization and the early stages of embryogenesis.

osk RNA localization depends on several gene products. Localization requires *orb* and *staufen* (which also participates in the last stages of *bcd* RNA localization) and is prevented in two other mutants with rather pleio-tropic effects: *cappuccino* (*capu*) and *spire* (*spir*) (Christerson and McK-earin, 1994; Ephrussi *et al.*, 1991; Kim-Ha *et al.*, 1991). In *orb* and *staufen* mutants, *osk* RNA transiently accumulates at the anterior edge of the oocyte during stages 8 and 9 but never becomes localized to the posterior pole, and even the anterior localization is lost by stage 10B. Similarly, in *capu* and *spir* mutants, *osk* RNA never becomes localized to the posterior tip of the oocyte during midoogenesis, although accumulation in the oocyte at early stages appears normal. *osk* RNA localization is also affected by mutations in *Bicaudal-D* (*Bic-D*). In these mutants, high amounts of *osk* RNA accumulate at the anterior margin of the oocyte during stages 8 and 9. However, the bulk of the *osk* message gradually shifts to the posterior pole, so that by the end of stage 10, *osk* RNA localization appears virtually normal. In addition, the oskar gene product itself is required for the mainte-nance of *osk* RNA localization during the latter stages of oogenesis (Ephrussi *et al.*, 1991; Kim-Ha *et al.*, 1991).

The cis-acting signals required for proper localization of *osk* are contained in the 3' UTR of the mRNA . Macdonald and co-workers systematically investigated the role of various 3' RNA deletions on *osk* RNA localization, and found evidence of multiple, overlapping signals, each of which affects only certain steps of localization (Kim-Ha *et al.*, 1993). Together, these studies indicate that the localization of *osk* RNA to the posterior pole is a complex multistep process.

osk RNA localization does not appear to depend on actin filaments, but exhibits a requirement for microtubules at all stages of oogenesis (Clark *et al.*, 1994; Pokrywka and Stephenson, 1995). Treatment with microtubule inhibitors, either *in vitro* or *in vivo*, results in a loss of *osk* RNA localization. The nurse cell population of *osk* RNA is most sensitive to microtubule inhibitors, and nurse cell localization is lost after even mild, 1- or 2-hr treatments. Much longer treatments are required to disrupt the localization of *osk* RNA in the oocyte, and a 2-hr taxol treatment has no effect on *osk* RNA localization. Resistance to mild drug treatments makes *osk* unique among localized RNAs tested thus far, and suggests that microtubules are required for *osk* RNA localization, but not for the maintenance of *osk* RNA at the posterior pole.

osk RNA is required for the assembly of polar granules, which are electron-dense ribonucleoprotein complexes associated with the posterior pole plasm (and eventually with germline cells). Several groups have noticed that colchicine treatments which disrupt *osk* RNA result in the accumulation of *osk* RNA-containing granules in the nurse cells and oocyte (Clark *et al.*, 1994; Pokrywka and Stephenson, 1995). However, it is uncertain whether these granules represent *osk* RNA which has been released from a localized position at the posterior pole or *osk* RNA which has yet to be localized. An obvious possibility is that the *osk* RNA particles that result from colchicine treatment may represent partially formed or fragmented polar granules.

C. Localized RNAs and Dorsoventral Patterning

The dorsal–ventral axes of the embryo can be predicted by visual inspection of the egg chamber, starting around stage 8. Two events signal the specification of the dorsal–ventral axis: (1) the oocyte nucleus migrates to the future dorsal side of the oocyte, and (2) follicle cells overlying that area of the oocyte undergo differential shape changes. Unlike the anterior–posterior axis, in which each pole is determined separately, the dorsal–ventral axis is specified as a result of a single signaling event. Initially, a signal originating in the oocyte is received by the future dorsal follicle cells, specifying their fate. By default, the follicle cells on the opposite side of the oocyte become ventralized, and ventral identity is then reinforced by the expression of ventral-specific genes. A candidate for the germline signal which directs dorsal patterning has recently been identified. This is the product of the *gurken (grk)* gene, a secreted protein of the TGF-α family. Consistent with its role in establishing dorsal fates, *grk* RNA is localized to the future dorsal side of the oocyte during midoogenesis, at the same time when dorsal–ventral patterning is initiated (Neuman-Silberberg and Schüpbach, 1993).

grk RNA is first expressed early in oogenesis and accumulates in the oocyte during stages 1–7, a pattern typical of many localized RNAs. Accumulation is seen predominantly at the posterior margin, consistent with the position of the MTOC during these stages. At stage 8, the oocyte nucleus migrates to the future dorsoanterior corner of the oocyte and becomes surrounded by microtubules. The oocyte nucleus is presumably anchored at this position by microtubules, since colchicine treatments cause displacement of the nucleus. At stage 8, *grk* RNA accumulates along the anterior margin of the oocyte. However, by the end of stage 8 *grk* RNA becomes localized around the oocyte nucleus, with highest concentrations of *grk* RNA at the anterior and dorsal margins of the nucleus. *grk* RNA localization persists until stage 10B or, occasionally, later.

Several mutations which interfere with dorsal–ventral patterning in the oocyte interfere with the localization of *grk* RNA. In *orb*, K10, *squid*, *spir*, and *capu* mutations early oocyte accumulation is normal, but later localization is disrupted. Instead of becoming localized around the oocyte nucleus, *grk* remains concentrated along the entire anterior margin (although concentrations of *grk* RNA around the oocyte nucleus are always somewhat higher) (Neuman-Silberberg and Schüpbach, 1993). *grk* is the least well-studied RNA with respect to cytoskeletal requirements for localization. To date, no investigations of the effects of cytoskeletal drugs on *grk* RNA localization have been reported. In addition, nothing is known about the *cis*-acting signals that may be required for the targeting of *grk* RNA to the dorsoanterior corner of the oocyte. Nonetheless, it seems likely that microtubules will also be involved in the localization of *grk* RNA. The position of *grk* RNA around the oocyte nucleus is coincident with a domain of high microtubule density, and a reliance on microtubules would be consistent with what is known about the localization requirements of other RNAs.

D. Early Transiently Localized (ETL) RNAs and Other Localized RNAs

In addition to those RNAs which help determine the course of embryonic patterning, there are several other RNAs which are localized during at least part of oogenesis. One class is the ETL RNAs. These RNAs have similar patterns of localization, and include K10, *Bic-D*, and *orb* (Lantz *et al.*, 1992; Prost *et al.*, 1988; Suter *et al.*, 1989). K10 and *orb* are thought to participate in the localization of *grk* RNA to the dorsoanterior corner of the oocyte in midoogenesis (Neuman-Silberberg and Schüpbach, 1993). *Bic-D* is also involved in the localization of mRNAs, but its role is unclear.

The ETL RNAs are expressed early in oogenesis and accumulate in the early oocyte. As the oocyte grows larger than the attendant nurse cells, ETL messages are concentrated at the anterior margin of the oocyte. Association with the anterior margin is coincident with the establishing of a microtubule density gradient at the anterior end of the oocyte. For ETL RNAs, localization to the anterior pole of the oocyte is transient, and by the end of stage 10, localization in the oocyte is lost.

Like the other localized RNAs discussed so far, ETL RNA localization is lost upon treatment with cytoskeletal drugs (Pokrywka and Stephenson, 1995). Localization of these RNAs is resistant to cytochalasins, but highly sensitive to microtubule inhibitors. The relative sensitivity of localized RNAs to drug treatment was determined by examining different RNAs in sequential sections from the same tissue. In this way, it was confirmed that *bcd*, *osk*, and ETL RNAs respond differently to identical drug treatments (Pokrywka and Stephenson, 1995)(see Table I). Even short treatments with microtubule drugs result in a complete loss of ETL RNA localization. In fact, ETL RNAs are the most sensitive to microtubule disruption of all localized RNAs characterized thus far. Furthermore, treatment with taxol also causes a complete loss of localization which is in direct contrast to the results obtained for *bcd* and *osk* RNAs.

The localization of K10 RNA is dependent on the genes *capu* and *spir* (which are also required for *osk* RNA localization). In *capu* and *spir* mutants, K10 transcripts accumulate in the oocyte during early stages of oogenesis, but do not become anteriorly localized (Cheung *et al.*, 1992). Nothing is known about genes which may be required for localization of other ETL RNAs. K10 RNA is also known to contain *cis*-acting sequences in the 3′ UTR which direct its localization (Cheung *et al.*, 1992).

Several other localized RNAs have been discovered which cannot be placed into existing categories. For at least one, hsp83, localization is achieved by selective degradation of message. hsp83 RNA which is associated with the posterior pole is apparently resistant to a cellular RNase,

Table I Comparing the Sensitivity of RNA Localization in Vitellogenic Egg Chambers to 2-Hr Drug Incubations *in Vitro*

Localized RNA	Sensitivity to MT inhibitors	Localization in taxol-treated cells?	Genetic requirements during oogenesis
ETL RNAs	High	No	Varies
Bicoid	Intermediate	Ectopic	*exu, sww*
oskar	Low	Yes	*capu, spir, stau, orb, osk*
gurken	?	?	*capu, spir, squid, orb,* K10

which degrades the unlocalized hsp83 RNA (Ding *et al.*, 1993a). Another RNA, *adducin-like*, is localized to the anterior pole of oocytes and embryos in a manner similar to that of *bcd* RNA. However, the mechanisms governing its localization differ from that of *bcd* RNA. *adducin-like* has no requirement for either *exu* or *staufen*, but localization does rely on sww protein (Ding *et al.*, 1993b). In *sww* mutants, *adducin-like* RNA becomes delocalized, starting at stage 8 (two full stages earlier than *bcd* RNA). The significance of this observation is unclear.

IV. The Connection between the Cytoskeleton and Positional Information

Recent reports from several groups suggest that the localization of specific RNAs relies heavily on the cytoskeletal organization of the oocyte itself. This hypothesis has recently been substantiated by several overlapping studies which examined the regulation of microtubule networks in the oocyte.

A. Mutations Which Affect Ooplasmic Streaming

cappuccino and *spire* are two mutations in which the ooplasmic streaming characteristic of stage-10 oocytes commences prematurely, during stage 7 (Theurkauf, 1994b). These mutants were originally isolated for their effects on axial patterning. Weak alleles cause a loss of posterior localization which results in posterior embryonic defects, while stronger alleles affect not only posterior patterning but dorsal–ventral patterning as well. *capu* and *spir* are the only mutations known which simultaneously affect anterior–posterior and dorsal–ventral polarity. Both *capu* and *spir* mutations prevent the posterior localization of *osk* RNA during midoogenesis, although accumulation of *osk* RNA in the oocyte during earlier stages is not affected (Kim-Ha *et al.*, 1991). Similar effects are seen on the localization of K10 and *grk* RNAs (Cheung *et al.*, 1992; Neuman-Silberberg and Schüpbach, 1993). *grk* RNA accumulates at the anterior margin of the oocyte in *capu* and *spir* mutants, but never relocalizes to the dorsoanterior corner of the oocyte, and in these mutants, the anterior localization of K10 RNA is lost. It is worth noting that the premature streaming in *capu* and *spir* mutants does not appear to affect the localization of *bcd* RNA (as implied by the observation that *capu* and *spir* mutants never have anterior patterning defects). Since *capu* and *spir* affect the localization of both *grk* and *osk*, their role in localization is likely to be more general, and may involve coordination of the cytoskeletal reorganization which occurs in midoogenesis. Some question remains as to whether the premature ooplasmic streaming directly results in an inability to localize these RNAs, perhaps by physically interfering with the localization process, or whether the streaming defect is symptomatic of more general cytoskeletal and organizational defects.

B. Mutations Which Affect Microtubule Reorganization

A group of three mutations disrupt the reorganization of the microtubule network at midoogenesis. These mutations include temperature-sensitive alleles of *Notch* and *Delta*, and mutations in the catalytic subunit of protein kinase A (PKA) (Ruohola *et al.*, 1991; Lane and Kalderon, 1994). All three have neomorphic effects on the localization of both *bcd* and *osk* RNA. The early localization of *bcd* and *osk* RNA appears normal in these mutants, but during midoogenesis, some *bcd* RNA becomes localized to the posterior pole, in addition to the *bcd* RNA anteriorly positioned. *osk* RNA is no longer found at the posterior pole, but accumulates in the center of the oocyte, as though it is excluded from both poles. Mosaic analysis has revealed that *Notch* and *Delta* function are required in the follicle cells, raising the possibility that follicle cell–oocyte signaling is required for establishing anterior–posterior polarity. PKA mutations have a similar effect on the localization of *bcd* and *osk* RNA, but PKA function is required in the oocyte, suggesting it could be part of the pathway which receives or transduces a signal from the follicle cells. Microtubule staining of PKA mutants reveals that the posteriorly located MTOC, which normally breaks down at stage 7, instead persists at the posterior pole. Under these conditions the oocyte possesses two MTOCs, one at the anterior pole and one at the posterior pole. This is clear evidence that PKA is involved in regulating the reorganization of the microtubule network, and indicates that the organization of microtubules in the oocyte is critical for proper localization of *bcd* and *osk* RNAs. One model posits that in these mutants *bcd* RNA, which may be associated with the minus ends of microtubules, becomes localized to the anterior and posterior margins of the oocyte. *osk* RNA, probably associated with the plus ends of microtubules, is shunted to the middle of the oocyte, where the plus ends now reside.

C. Microtubule Motors

In a system in which microtubules mediate the localization of individual RNAs, microtubule motors are a potential method of achieving directed transport of RNA species. As a means of investigating the possibility that motors participate in RNA localization, the distribution of two microtubule motors during oogenesis has been examined. The distribution of kinesin, a plus end-directed microtubule motor, has been indirectly investigated through the use of a kinesin : ß-gal fusion protein (Clark *et al.,* 1994). The expression and distribution of the fusion protein (and, by extension, kinesin) was monitored through the use of reagents which allow visualization of ß-gal. In oocytes, the kinesin fusion protein is localized to the posterior pole during stages 8 and 9, reminiscent of *osk* RNA localization during the same stages. Posterior localization of kinesin : ß-gal is transient, however, and is lost by the end of stage 10, just as ooplasmic streaming commences. Like *osk* RNA, localization of the fusion protein can be disrupted by treating oocytes with microtubule inhibitors and by mutations in *capu, spir, Notch,* or *Delta.* One possible interpretation of these results is that the fusion protein is a "reporter" of microtubule polarity, indicating that the plus ends of microtubules are located at the posterior end of the oocyte (consistent with descriptive studies of microtubule distributions at these stages). The fact that localization of the fusion protein is dependent on microtubules supports the model that kinesin moves along microtubules toward their plus ends until it reaches the posterior pole of the oocyte. The position of the kinesin fusion protein in *Notch* and *Delta* mutants has been interpreted to mean that, in these mutants, plus ends of microtubules are no longer located at the posterior pole.

Thus far, a fairly uniform and consistent picture of cytoskeletal organization has emerged from these studies. During early oogenesis, a single MTOC is located at the posterior end of the oocyte. It has been hypothesized that RNAs which preferentially accumulate in the pro-oocyte at this time may utilize minus end-directed motors to move from nurse cells into the oocyte toward the MTOC. During midoogenesis, from stages 8 to 10, the microtubules in the oocyte become reorganized. The posteriorly located MTOC breaks down, and instead, the entire anterior cortex gains the capacity to nucleate microtubules. Accompanying this reorganization is a reorientation of microtubule polarity, where minus ends are predicted to lie at the anterior and plus ends at the posterior pole of the oocyte. The model which has evolved from these studies is one in which posterior determinants are moved to the posterior pole by a kinesin-like microtubule motor which moves toward the plus ends of microtubules. Likewise, *bcd* RNA may be directed to the anterior pole by association with a dynein-like minus end-directed motor. This model also explains the effects of *Notch, Delta* and PKA mutations on *bcd* and *osk* RNA localization: in these mutants, microtubule minus ends are found at both poles, and their plus ends are directed toward the middle of the oocyte. *bcd* and *osk* RNAs, moving in opposite directions along disordered microtubule "highways," are ultimately mislocalized.

According to current models of cytoskeletal organization, dynein would be predicted to be found at the posterior cortex in early oocytes (stage 6) and anteriorly positioned during midoogenesis (stages 8–10). A recent report describing the subcellular distribution of dynein has confirmed some details of this model, but contradicts other details (Li *et al.,* 1994). Dynein is expressed throughout oogenesis, and at early stages is concentrated at the posterior edge of the oocyte, consistent with the predicted location of microtubule minus ends. However, during midoogenesis, dynein mirrors the distribution of a kinesin:ß-gal fusion protein; it is localized to the posterior pole of the oocyte, seemingly in conflict with the presumed orientation of microtubules at this time. There are several possible explanations for these apparently conflicting data: (1) one remote possibility is that the *Drosophila* form of dynein may actually be a plus end-directed motor. This is highly improbable, as all known dyneins are minus-end directed. (2) Perhaps the oocyte microtubule cytoskeleton is not arranged with all plus ends directed toward the posterior and all minus ends concentrated at the anterior. Instead, there may be a mixture of orientations. This further complicates models for RNA localization, which must then explain how translocating RNA particles "know" which motor (or microtubules) to become attached to. (3) Finally, it is possible that either the kinesin fusion protein or dynein (or both) do not accurately reflect microtubule polarity in the oocyte.

It is important to note that, although *bcd* RNA may become localized to the anterior margin of the oocyte via a microtubule motor, the behavior of *bcd* RNA under various conditions can just as easily be explained by proposing that *bcd* RNA has an intrinsic affinity for microtubules in general, without regard for polarity. There is some experimental precedence for the notion that *bcd* RNA has an intrinsic affinity for microtubules. In egg chambers treated with the microtubule-stabilizing drug, taxol, novel aberrant microtubule bundles are generated. *bcd* RNA appears to be associated with these bundles, as is exu protein (Pokrywka and Stephenson, 1991; Wang and Hazelrigg, 1994). A similar phenomenon may occur in the *Notch*, *Delta* and PKA mutants. Novel microtubule networks are now found at the posterior end of the oocyte, and *bcd* RNA becomes associated with them. These data must be interpreted with caution, however, because it is equally possible that *bcd* RNA associates with microtubules in general or with the minus ends of microtubules in particular. Current data cannot distinguish between either possibility.

V. Toward a General Model for RNA Localization in *Drosophila*?

The studies described previously can be viewed as a whole in order to develop a more complete picture of RNA localization in *Drosophila*. The evidence compiled on several fronts is beginning to converge, and the remaining sections explore similarities and differences among localized RNAs, with an eye toward integration of the data into a general model for RNA localization in *Drosophila.*

A. Common Features of Localized RNAs

Localized RNAs share characteristics which suggest common mechanisms of localization. Several observations suggest there are common intermediates in the localization of most or all RNAs. Most obvious is the requirement for microtubules but not actin filaments during all stages of RNA localization, implying microtubules are a universal component of RNA localization machinery (Pokrywka and Stephenson, 1991; Theurkauf *et al.*, 1993; Clark *et al.*, 1994; Pokrywka and Stephenson, 1995). Microtubules extend through the nurse cells and oocyte during the first half of oogenesis, and because they possess an intrinsic polarity, the arrangement of microtubules in the oocyte can convey directional information which may be used in localizing RNAs. Current evidence favors a model in which microtubules mediate the localization of most or all RNAs.

Consistent with this idea is the observation that all localized RNAs accumulate in the oocyte during the first seven stages of oogenesis. This characteristic appears to be unique to localized RNAs and seems to involve a common mechanism, as supported by several observations. First, oocyte accumulation of all localized RNAs which have been tested is equally sensitive to microtubule inhibitors (N. J. Pokrywka, unpublished observations). Second, RNAs which accumulate in the young oocyte are similarly enriched by a detergent-insoluble fraction of ovarian extracts, consistent with their attachment to a large, stable structure (Pokrywka and Stephenson, 1994) (see Fig. 2). Finally, many of the mutations which interfere with oocyte localization of specific RNAs at later stages have no effect on the oocyte accumulation during early oogenesis. Thus, it seems likely that RNA localization actually begins when specific RNAs are associated with the specialized cytoplasm of the oocyte cortex during early oogenesis. The early oocyte accumulation observed for most localized RNAs may in fact represent a presorting of RNAs, perhaps as a prerequisite for localization during midoogenesis.

Another feature shared by localized RNAs is an association with the anterior margin of the oocyte. This anterior localization is transient in the case of most RNAs and permanent for *bcd* and *adducin-like* RNAs. It seems likely that this association is an extension of the oocyte accumulation observed during early oogenesis. Many localized RNAs are initially routed to the anterior margin of the oocyte during midoogenesis, prior to their final localization within the oocyte (see Table II). Some, like *orb*, *Bic-D*, and K10, are never rerouted to another site in the oocyte, and localization

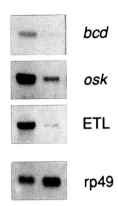

P S

bcd

osk

ETL

rp49

Fig. 2   Northern analysis of localized RNAs from stages 1–7 egg chambers. Localized RNAs, which accumulate in the oocyte at these stages (*bcd, osk,* and ETL), are enriched in a detergent-insoluble fraction (P). Most RNAs in the cell (as represented by ribosomal protein 49 RNA) are recovered in the supernatant (S). RNA from equivalent numbers of oocytes were loaded in each lane.

is lost by the end of stage 10. Other RNAs are transiently localized to the anterior end of the oocyte, and then become localized to other domains of the oocyte. That anterior localization is a discrete step in RNA localiza-

Table II   Comparison of Steps Involved in Localization of Specific RNAs

Localized RNA	Localization pathway	Reference
K10 *Bic-D* *orb*	Initial localization to anterior margin, no rerouting	Haenlin *et al.* (1987); Suter *et al.* (1989); Lantz *et al.* (1992)
bcd *adducin-like*	Initial anterior localization, followed by rerouting to the anterior cortex	Berleth *et al.* (1988); Pokrywka and Stephenson (1991); St. Johnston *et al.* (1989); Ding *et al.* (1993)
grk	Initial anterior localization, followed by rerouting to the dorsoanterior corner	Neuman-Silberberg and Schüpbach (1993)
osk	Initial anterior localization, followed by rerouting to the posterior pole	Ephrussi *et al.* (1991); Kim-Ha *et al.* (1991)

tion seems evident from several lines of research. Mutations which affect RNA localization often preserve a transient anterior localization but prevent subsequent localization to other oocyte domains. Examples include the effect of *orb* and *staufen* mutations on *osk* RNA localization, and the effect of *orb, squid,* and K10 on the localization of *grk* RNA (Christerson and McKearin, 1994; Kim-Ha *et al.,* 1991; Neuman-Silberberg and Schüpbach, 1993). Although *bcd* RNA is always found at the anterior margin of the oocyte, mutations in *exu* cause a transient localization of *bcd* RNA which is quickly lost by stage 9 (St. Johnston *et al.,* 1989).

Additional support for the idea that anterior localization is a distinct step in the localization of RNAs comes from detailed analyses of the *cis* sequences required for the localization of *bcd* and *osk* RNA. Both *osk* and *bcd* RNA contain discrete 3' RNA sequences which direct only the early stages of RNA localization (Kim-Ha *et al.,* 1993; Macdonald *et al.,* 1993). In the case of *osk* RNA, discrete elements can direct the accumulation of *osk* RNA in previtellogenic oocytes, the transient localization of *osk* RNA during midoogenesis, and finally the localization of *osk* RNA to the posterior pole. For *bcd* RNA localization, a discrete element, BLE1, is sufficient for localization to the anterior margin of the oocyte during early oogenesis, but the localization conferred by this element is transient and is quickly lost as oogenesis continues.

Another feature common to all localized RNAs is their attachment to some portion of the oocyte periphery, whether it be to anterior, posterior, or dorsal regions of the oocyte. This implies the oocyte periphery may be specialized with respect to RNA localization. The oocyte periphery contains a region of morphologically distinctive cytoplasm termed the oocyte cortex, which is enriched in actin filaments, microtubules, and other cytoskeletal elements. Two observations suggest the oocyte cortex may be a densely packed, highly stable cytoplasmic domain. First, yolk platelets and other cytoplasmic vesicles are excluded from 5 μm closest to the plasma membrane (i.e., the cortex). Second, this portion of the oocyte cytoplasm remains immobile during the rapid ooplasmic streaming which occurs at stage 10B. orb protein (a putative RNA-binding protein which is required for the localization of *osk* and *grk* RNAs) is also localized to the cortex of the oocyte during midoogenesis (Lantz *et al.,* 1994). The localization of at least one RNA, *bcd,* overlaps with this structure: *bcd* RNA is confined to a 5-μm thick layer of cytoplasm which lies just below the cell membrane (Stephenson and Pokrywka, 1992).

The notion of an oocyte domain specialized for the anchoring and stabilizing localized RNAs is further supported by two separate experiments. First, *bcd* RNA localization has been examined using reversible microtubule inhibitors with interesting results. Treatment of egg chambers with the inhibitors causes a release of *bcd* RNA from the anterior margin. When the drug is removed, *bcd* RNA reassociates with the oocyte periphery (Pokrywka and Stephenson, 1991). One interpretation of these experiments is that the cortex contains those complexes required for stabilizing localized RNAs. Second, fractionation studies demonstrate that all localized RNAs tested (*bcd, osk, Bic-D, orb*, and K10) are enriched in a detergent-insoluble fraction of ovary extracts, beginning from the time of their first appearance in the young oocyte (Pokrywka and Stephenson, 1994). RNA found in the nurse cell complex does not appear to be enriched by this procedure, indicating these associations are specific for the oocyte.

These observations can be integrated into the model shown in Fig. 3, in which RNA localization involves one or more steps that are common to all RNAs, as well as steps unique to each RNA which allow delivery to a specific position in the oocyte. The observed similarities among localized RNAs are consistent with a scenario in which RNAs which will become localized in the oocyte are "presorted" during early oogenesis by accumulation in the oocyte. Oocyte accumulation is likely to be mediated by transport along a polarized microtubule system toward the minus ends of microtubules located in the oocyte proper. This presorting is manifested as transient anterior localization during the middle stages of oogenesis, when the minus ends of microtubules are presumably found at the anterior margin of the oocyte. Once the cytoskeleton of the oocyte has reorganized (around stage 8) the RNAs are routed to the appropriate destination. The final stage of

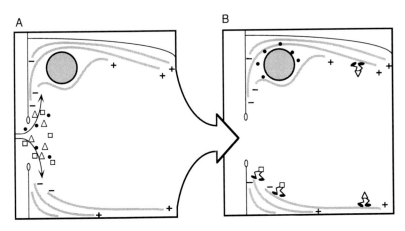

Step 1: Transport to the oocyte   Step 2: Localization

Fig. 3 A general model for RNA localization. Localized RNAs (△, •, □) are delivered to the anterior end of the oocyte by a common mechanism (which may be microtubule based). In a second step, RNAs are then sorted and routed to their final destinations within the oocyte proper. As shown here, the second step of localization may involve movement along microtubules with the aid of microtubule motors.

localization requires attachment of RNAs to specific domains of the oocyte cortex, and it is at this step that localization factors unique to each localized RNA function.

B. Differences among Localized RNAs

Despite the similarities among localized RNAs, the simple observation that different RNAs are ultimately localized to different subcellular locations suggests that some aspects of RNA localization are unique. Evidence from a variety of sources points to differences among localized RNAs that may reflect differences in mechanisms of RNA localization. These differences arise primarily in the last steps of localization, once an RNA has entered the oocyte.

Although all localized RNAs rely on microtubules, their sensitivity to microtubule drugs varies (as shown in Table I), suggesting a previously unsuspected complexity in RNA localization (Pokrywka and Stephenson, 1995). By hybridizing different probes to successive sections of a drug-treated egg chamber, it is possible to monitor the response of several RNAs to the same treatment. ETL RNAs are highly sensitive to even mild treatments with microtubule inhibitors, and taxol treatment causes a rapid loss of ETL RNA localization. In contrast, the same treatments have almost no effect on the localization of *osk* RNA. *bcd* RNA localization is intermediate in its response to microtubule inhibitors, with some *bcd* RNA remaining localized after treatments which completely disrupt ETL RNA localization. In addition, *bcd* RNA responds dramatically to taxol treatment. In taxol-treated egg chambers, *bcd* RNA becomes ectopically localized to novel sites in the oocyte which correspond to sites of aberrant microtubule formation (Pokrywka and Stephenson, 1991). These sites also contain exu protein, implying that *bcd* RNA is attached to microtubules in a process which is mediated by exu (Wang and Hazelrigg, 1994). Such variation in drug sensitivity would not be expected for RNAs whose only difference in localization was in the adapter proteins required for association with microtubules. These different responses to drug treatment may therefore indicate qualitative differences in mechanisms of RNA localization.

These data can be explained by positing that initial localization to the anterior margin requires microtubules and is easily disrupted by microtubule inhibitors (perhaps because microtubules are solely responsible for RNA position at this step). One possibility is that the same mechanism is responsible for the initial delivery of various localized RNAs into the oocyte proper, where anterior localization occurs (perhaps as a by-product of the transport or unloading process). Rerouting may require other cytoskeletal systems or protein complexes in addition to microtubules, thereby rendering these steps variably more resistant to microtubule disruption. Thus, differences in subsequent steps (rerouting) may account for experimentally observed differences in microtubule inhibitor sensitivity.

Genetic requirements for RNA localization also point to differences among the localized RNAs and support the model outlined above. The observation that no two localized RNAs share identical genetic requirements points to differences in RNA localization. In addition, several of the mutations known to interfere with RNA localization seem to affect oocyte localization, but not delivery to the anterior margin of the oocyte (as described previously). For example, mutations in *staufen* and *Bic-D* prevent the localization of *osk* RNA to the posterior pole, but do not affect the transient anterior localization which precedes it. In fact, *Bic-D* mutations actually prolong the transient anterior localization. In the case of *bcd* RNA localization, *exu* mutations sometimes result in a transient localization of *bcd* RNA to the anterior end, which is lost at later stages. *bcd* RNA may be delivered to the oocyte anterior margin by an *exu*-independent common mechanism, but in *exu* mutants, it fails to become anchored. Likewise, in K10, *squid*, and *orb* mutants, *grk* RNA accumulates along the entire anterior margin, but never becomes localized to the dorsal–anterior corner of the oocyte.

Finally, it seems likely that the process of localizing an RNA is distinct from the need to maintain that RNA in a specific position. That is, localization and stabilization are two separate functions. Current evidence suggests that the maintenance function is carried out by elements other than microtubules. The ability of microtubule inhibitors to disrupt localization seems to decrease with the age of the oocyte, so that after stage 10, some RNAs are more resistant to microtubule inhibitors than they were at earlier stages (N. J. Pokrywka, unpublished observations). *bcd* RNA is somewhat resistant to drugs, such as colchicine, and *osk* RNA localization is highly resistant. Furthermore, the localization of kinesin fusion protein, dynein, *grk* RNA, and ETL RNAs is lost after stage 10. This is the same stage of oogenesis at which many localized RNAs become dispersed throughout the oocyte, and is likely to coincide with the rearrangement of an organized microtubule array (Theurkauf *et al.*, 1992). It is possible that these molecules are associated with microtubule populations during midoogenesis, but never become anchored to more stable structures in the oocyte.

VI. Conclusions

There is substantial evidence that the cytoskeleton plays a major role in organizing positional information in *Drosophila* eggs. However, it is clear that there are holes in our knowledge of how a molecule becomes localized to a specific subdomain of the egg. What one can take from the evidence accumulated thus far is the idea that the organization of the cytoskeleton itself is of paramount importance for the localization of developmentally significant molecules. This answer is somewhat unsatisfactory, however, because it merely begs the question; instead, we must now ask, "How does the oocyte cytoskeleton become organized?" How are individual RNAs recognized and associated with the cytoskeleton? What other intermediates are involved? How is the position of a localized RNA maintained? These questions highlight those areas of the field in which our knowledge is still woefully inadequate. The hunt is now on in two distinct yet connected areas: (1) what directs the organization of the cytoskeleton during oogenesis? and (2) by what mechanisms do localized RNAs interact with the cytoskeleton? The high degree of interest in this field ensures that progress will be made in years to come.

VII. Appendix: Methods

A. Biochemical Enrichment of Localized RNAs

As an aid in the investigation of molecular associations required for RNA localization, a crude fractionation procedure which preferentially enriches for localized RNAs has been developed. The research which led to the formulation of this protocol was originally conducted in the lab of Ed Stephenson (University of Alabama) and is based on work performed by Yisraeli *et al.* (1990). The protocol is based on the classical observation that the cytoskeleton of most cells is resistant to detergent extraction. This characteristic was exploited as a first step toward the eventual biochemical purification of localization complexes (defined as that combination of molecules required for attachment of an RNA to the cytoskeleton).

Ovaries and oocytes are obtained by hand dissection of adult females. In order to collect mixed-stage oocytes, uncrowded, well-fed flies are used. If the flies are transferred to fresh media once a day, ovaries will contain a variety of stages. For the analysis of specific stages, ovaries are hand dissected and sorted. To obtain ovaries enriched for stage 14 (mature) oocytes, females are separated from males and fattened under very crowded conditions (approximately 100 females per vial) for 4 or 5 days. Ovaries from these females will contain predominantly mature oocytes, and contaminating younger stages are relatively easy to remove.

Briefly, the oocytes are homogenized in the presence of nonionic detergents and then centrifuged to separate soluble from insoluble components. The fractionation procedure solubilizes most cellular proteins: about 80–90% of total protein is recovered in the supernatant. Likewise, approximately 80–90% of cytoplasmic RNA is solubilized by this treatment and recovered in the supernatant. Results of a typical fractionation procedure are shown in Fig. 4. If equivalent amounts of oocytes are loaded in each pellet and supernatant lane, the relative amount of a specific RNA (or protein) recovered from each fraction can be compared. One convenient technique is to scan underexposed autoradiograms with a laser densitometer to determine relative percentages of a specific RNA in pellet and supernatant fractions. Since oocyte equivalents are electrophoresed, pellet and supernatant lanes can be directly compared for any given trial. For example, if four times as much of an RNA appears in the supernatant compared to the pellet lane, then it can be said that 20% of that particular RNA in a given

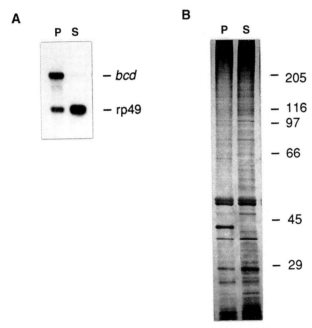

Fig. 4 RNA and protein analysis of fractionation products. (A) After fractionation of stage-14 oocytes homogenized in a nonionic detergent buffer (see text for details), RNA was extracted from the pellet (P) and supernatant (S) and analyzed by Northern hybridization. rp49 is representative of most RNAs, which are found in the supernatant. Many localized RNAs (in this figure, *bcd* RNA) are enriched in the pellet fraction by this procedure. RNA from equivalent numbers of oocytes were loaded in each lane. (B) Proteins from a routine fractionation were analyzed on a silver-stained 10% SDS–polyacrylamide gel. Most proteins are solubilized by this procedure and recovered in the supernatant fraction. At least one protein of 42 kDa is enriched in the pellet. The relative size of molecular weight markers is shown on the right (in kDa).

oocyte (or batch of oocytes) is pellet associated, while 80% is recovered in the supernatant. Thus, the relative ratio of a protein in the pellet:supernatant represents its actual distribution between the two fractions.

A variety of unlocalized mRNAs have been examined (including actin, glucose-3-phosphate dehydrogenase, *swallow*, rRNA, and rp49) and are recovered predominantly in the supernatant in amounts which correspond to the distribution of total RNA in the two fractions. Therefore, in a given experiment, if 85% of nucleic acids partition to the supernatant (as assayed by absorbance at 260 nm), then in this trial an individual, unlocalized RNA will distribute to the same degree. In theory, any RNA which is known to be unlocalized in the oocyte can serve as an appropriate control. The purpose of the control RNA is twofold: (1) to demonstrate, in a representative manner, the distribution of most cellular (i.e., nonlocalized) RNAs between the two fractions, and (2) to allow easy visual comaprison of an unlocalized RNA and a test RNA in the same sample. rp49 mRNA is used as a representative RNA in my lab for several reasons. First, it is uniformly distributed in all tissues in which it is expressed, fitting the criteria for an unlocalized mRNA. Second, it is highly abundant and therefore easy to detect by Northern blot. Finally, it is smaller than a typical *Drosophila* mRNA (rp49 RNA is ~600 bp in length) making it possible to simultaneously hybridize a filter with probe to rp49 mRNA and to an RNA of interest. This allows a direct comparison of RNA distribution within an experiment and obviates the need to control for loading or gel-running errors.

The distribution of specific proteins and mRNAs is very reproducible from experiment to experiment. In general, RNAs which are uniformly distributed in the oocyte (that is, most RNAs) are largely recovered in the supernatant. However, the fractionation behavior of localized RNAs is clearly atypical. As shown in Fig. 4, greater than 99% of all *bcd* RNA in the oocyte is pellet associated, while most RNAs are found largely in the supernatant. The ability of an RNA to become enriched in the pellet appears to be directly related to its ability to become localized in the oocyte. For *bcd* RNA and several other localized RNAs, enrichment in the pellet is dependent on the same criteria required for localization *in vivo* (that is, both processes have the same genetic and cytoskeletal requirements). It is for this reason that the procedure is likely to be useful in analyzing the factors participating in RNA localization.

What follows is the fractionation protocol routinely used in my laboratory.

1. Dissect out or otherwise collect oocytes in DEPC-treated phosphate-buffered saline (PBS). To obtain ovaries containing only stage-14 oocytes, fatten flies in very crowded conditions for ~5

days. In order to obtain enough RNA to detect a moderately abundant message, 30–50 pairs of ovaries should be considered minimum starting material.

2. Transfer oocytes to a Dounce homogenizer and remove as much PBS as possible. Add 350 μl of fractionation buffer and homogenize at least 20 strokes with a tight-fitting pestle (clearance, 0.03–0.08 mm).

3. Transfer contents to a 1.5-ml centrifuge tube and incubate at room temperature for 5 min. Spin 5 min at room temperature in a microcentrifuge (approximately 13,000g).

4. The detergent-insoluble pellet can be separated from the supernatant by removal with a wire loop. Continue with RNA extraction or resuspend each fraction in protein gel sample buffer such that final volumes of the two fractions are equal.

5. RNA extraction: to the supernatant, add 4 vol of RNA extraction buffer. Resuspend the pellet in 350 μl RNA extraction buffer. To each sample add proteinase K to a final concentration of 400 μg/ml and incubate at 37°C for 30 min. This step digests many of the yolk proteins and substantially increases RNA yield from the supernatant.

6. Continue with RNA purification according to routine protocols for phenol:chloroform extractions, followed by an ethanol precipitation.

7. In order to ensure that equivalent amounts of oocytes are analyzed when quantitative measurements are made, the supernatant and pellet fractions should be resuspended in equal volumes. This assumes that recovery from each fraction is equally efficient, which appears to be the case when using this protocol (N. J. Pokrywka, unpublished results).

Notes

- All stock solutions except for 1,4-piperazinebisethanesulfonic acid (PIPES), vanadate ribonucleoside complexes (VRC), and tRNA should be DEPC treated and autoclaved. All glassware should be baked overnight at 180°C.
- If the pellet and supernatant are to be analyzed for proteins, the VRC can probably be omitted and replaced by a proteinase inhibitor. I have routinely used 1 mM PMSF as a proteinase inhibitor, although a proteinase inhibitor cocktail of aprotonin, leupeptin, and pepstatin can also be used.
- Total yields of RNA should be >4–6 μg per ovary pair.
- This protocol was optimized for the enrichment of complexes containing *bcd* RNA, and it is possible that a variation of conditions may be necessary for the analysis of other RNAs.

Solutions

PBS
0.13 *M* NaCl
10 m*M* NaPO$_4$, pH 7.2

Fractionation buffer (made fresh)
0.5% Triton X-100
10 m*M* PIPES, pH 6.8
300 m*M* KCl
0.5 m*M* EDTA
0.5 m*M* EGTA
20 m*M* VRC (vanadate ribonucleoside complexes)
20 μg/ml tRNA

RNA extraction buffer
0.5% SDS
10 m*M* EDTA
50 m*M* Tris, pH 7.0
0.1 *M* NaCl

Acknowledgments

The author thanks Ed Stephenson and Kate Susman for critical reading of the manuscript. Data presented in this chapter were the results of work performed at the University of Rochester, and the work was supported by a Messersmith Fellowship from the University of Rochester.

References

Berleth, T., Burri, M., Thoma, G., Bopp, D., Richstein, S., Frigerio, G., Noll, M., and Nüsslein-Volhard, C. (1988). The role of localization of *bicoid* RNA in organizing the anterior pattern of the *Drosophila* embryo. *EMBO J.* **7,** 1749–1756.
Bohrmann, J., and Biber, K. (1994). Cytoskeleton-dependent transport of cytoplasmic particles in previtellogenic to mid-vitellogenic ovarian follicles of *Drosophila:* Time-lapse analysis using video-enhanced contrast microscopy. *J. Cell Sci.* **107,** 849–858.
Cheung, H.-K., Serano, T. L., and Cohen, R. S. (1992). Evidence for a highly selective RNA transport system and its role in establishing the dorsoventral axis of the Drosophila egg. *Development* **114,** 653–661.
Christerson, L. B., and McKearin, D. M. (1994). *orb* is required for anteroposterior and dorsoventral patterning during *Drosophila* oogenesis. *Genes Dev.* **8,** 614–628.
Clark, I., Giniger, E., Ruohola-Baker, H., Jan, L. Y., and Jan, Y. N. (1994). Transient posterior localization of a kinesin fusion protein reflects anteroposterior polarity of the Drosophila oocyte. *Curr. Biol.* **4,** 289–300.
Cooley, L., and Theurkauf, W. E. (1994). Cytoskeletal functions during *Drosophila* oogenesis. *Science* **266,** 590–596.

Cooley, L., Verheyen, E., and Ayers, K. (1992). *chickadee* encodes a profilin required for intercellular cytoplasm transport during *Drosophila* oogenesis. *Cell* **69**, 173–184.

Ding, D., Parkhurst, S. M., Halsell, S. R., and Lipshitz, H. D. (1993a). Dynamic Hsp83 RNA localization during *Drosophila* oogenesis and embryogenesis. *Mol. Cell. Biol.* **13**, 3773–3781.

Ding, D., Parkhurst, S. M., and Lipshitz, H. D. (1993b). Different genetic requirements for anterior RNA localization revealed by the distribution of Adducin-like transcripts during Drosophila oogenesis. *Proc. Natl. Acad. Sci. USA* **90**, 2512–2516.

Driever, W., and Nüsslein-Volhard, C. (1988). The bicoid protein determines position in the *Drosophila* embryo in a concentration-dependent manner. *Cell* **54**, 95–104.

Driever, W., Siegel, V., and Nüsslein-Volhard, C. (1990). Autonomous determination of anterior structures in the early *Drosophila* embryo by the *bicoid* morphogen. *Development* **109**, 811–820.

Ephrussi, A., Dickinson, L. K., and Lehmann, R. (1991). oskar organizes the germ plasm and directs localization of the posterior determinant nanos. *Cell* **66**, 37–50.

Ephrussi, A., and Lehmann, R. (1992). Induction of germ cell formation by oskar. *Nature* **358**, 387–392.

Ferrandon, D., Elphick, L., Nüsslein-Volhard, C., and Johnston, D. S. (1994). staufen protein associates with the 3′ UTR of *bicoid* mRNA to form particles that move in a microtubule-dependent manner. *Cell* **79**, 1221–1232.

Gutzeit, H. O. (1986). The role of microfilaments in cytoplasmic streaming in *Drosophila* follicles. *J. Cell Sci.* **80**, 159–169.

Gutzeit, H. O., and Koppa, R. (1982). Time-lapse film analysis of cytoplasmic streaming during late oogenesis of *Drosophila*. *J. Embry. Exp. Morphol.* **67**, 101–111.

Kim-Ha, J., Smith, J. L., and Macdonald, P. M. (1991). oskar mRNA is localized to the posterior pole of the Drosophila oocyte. *Cell* **66**, 23–35.

Kim-Ha, J., Webster, P. J., Smith, J. L., and Macdonald, P. M. (1993). Multiple RNA regulatory elements mediate distinct steps in localization of *oskar* mRNA. *Development* **119**, 169–178.

King, R. C. (1970). "Ovarian Development in Drosophila Melanogaster." Academic Press, New York.

Lane, M. E., and Kalderon, D. (1994). RNA localization along the anteroposterior axis of the *Drosophila* oocyte requires PKA-mediated signal transduction to direct normal microtubule organization. *Genes Dev.* **8**, 2986–2995.

Lantz, V., Ambrosio, L., and Schedl, P. (1992). The Drosophila orb gene is predicted to encode sex-specific germline RNA-binding proteins and has localized transcripts in ovaries and early embryos. *Development* **115**, 75–88.

Lantz, V., Chang, J. S., Horabn, J. I., Bopp, D., and Schedl, P. (1994). The *Drosophila orb* RNA-binding protein is required for the formation of the egg chamber and establishment of polarity. *Genes Dev.* **8**, 598–613.

Li, M. G., McGrail, Serr, M., and Hays, T. S. (1994). *Drosophila* cytoplasmic dynein, a microtubule motor that is asymmetrically localized in the oocyte. *J. Cell Biol.* **126**, 1475–1494.

Macdonald, P. M. (1990). *bicoid* mRNA localization signal: Phylogenetic conservation of function and RNA secondary structure. *Development* **110**, 161–171.

Macdonald, P. M., Kerr, K., Smith, J. L., and Leask, A. (1993). RNA regulatory element BLE1 directs the early steps of *bicoid* mRNA localization. *Development* **118**, 1233–1243.

Macdonald, P. M., and Struhl, G. (1988). Cis-acting sequences responsible for anterior localization of *bicoid* mRNA in *Drosophila* embryos. *Nature* **336**, 595–598.

Mahajan-Miklos, S., and Cooley, L. (1994). Intercellular cytoplasm transport during *Drosophila* oogenesis. *Dev. Biol.* **165**, 336–351.

Neuman-Silberberg, F. S., and Schüpbach, T. (1993). The Drosophila dorsoventral patterning gene gurken produces a dorsally localized RNA and encodes a TGFa-like protein. *Cell* **75,** 165–174.

Pokrywka, N. J., and Stephenson, E. C. (1991). Microtubules mediate the localization of *bicoid* RNA during *Drosophila* oogenesis. *Development* **113,** 55–66.

Pokrywka, N. J., and Stephenson, E. C. (1994). Localized RNAs are enriched in cytoskeletal extracts of *Drosophila* oocytes. *Dev. Biol.* **166,** 210–219.

Pokrywka, N. J., and Stephenson, E. C. (1995). Microtubules are a general component of RNA localization systems in *Drosophila* oocytes. *Dev. Biol.* **167,** 363–370.

Prost, E., Deryckere, F., Roos, C., Haenlin, M., Pantesco, V., and Mohier, E. (1988). Role of the oocyte nucleus in determination of the dorsoventral polarity of Drosophila as revealed by molecular analysis of the K10 gene. *Genes Dev.* **2,** 891–900.

Ruohola, H., Bremer, K. A., Baker, D., Swedlow, J. R., Jan, L. Y., and Jan, Y. N. (1991). Role of neurogenic genes in establishment of follicle cell fate and oocyte polarity during oogenesis in *Drosophila. Cell* **66,** 433–449.

St. Johnston, D., Beuchle, D., and Nüsslein-Volhard, C. (1991). *staufen,* a gene required to localize maternal RNAs in the *Drosophila* oocyte. *Cell* **66,** 51–63.

St. Johnston, D., Driever, W., Berleth, T., Richstein, S., and Nüsslein-Volhard, C. (1989). Multiple steps in the localization of bicoid RNA to the anterior pole of the Drosophila oocyte. *Development (Suppl.)* **107,** 3–19.

Smith, J. L., Wilson, J. E., and Macdonald, P. M. (1992). Overexpression of *oskar* directs ectopic activation of *nanos* and presumptive pole cell formation in *Drosophila* embryos. *Cell* **70,** 849–859.

Stephenson, E. C., Chao, Y. C., and Fackenthal, J. D. (1988). Molecular analysis of the *swallow* gene of *Drosophila* melanogaster. *Genes Dev.* **2,** 1655–1665.

Stephenson, E. C., and Pokrywka, N. J. (1992). Localization of bicoid message during Drosophila oogenesis. *In* "Current Topics in Developmental Biology" (E. Bearer, ed.), Vol. 26, pp. 23–34. Academic Press, Orlando.

Struhl, G., Struhl, K., and Macdonald, P. M. (1989). The gradient morphogen *bicoid* is a concentration–dependent transcriptional activator. *Cell* **57,** 1259–1273.

Suter, B., Romberg, L. M., and Steward, R. (1989). Bicaudal-D a Drosophila gene involved in developmental asymmetry: Localized transcript accumulation in ovaries and sequence similarity to myosin heavy chain tail domains. *Genes Dev.* **3,** 1957–1968.

Theurkauf, W. E. (1994a). Microtubules and cytoplasm organization during *Drosophila* oogenesis. *Dev. Biol.* **165,** 352–360.

Theurkauf, W. E. (1994b). Premature microtubule-dependent cytoplasmic streaming in *cappuccino* and *spire* mutant oocytes. *Science* **265,** 2093–2096.

Theurkauf, W. E., Alberts, B. M., Jan, Y. N., and Jongens, T. A. (1993). A central role for microtubules in the differentiation of Drosophila oocytes. *Development* **118,** 1169–1180.

Theurkauf, W. E., Smiley, S., Wong, M. L., and Alberts, B. M. (1992). Reorganization of the cytoskeleton during Drosophila oogenesis—implications for axis specification and intercellular transport. *Development* **115,** 923–936.

Wang, C., Dickinson, L. K., and Lehmann, R. (1994). Genetics of nanos localization in Drosophila. *Dev. Dynamics* **199,** 103–115.

Wang, C., and Lehmann, R. (1991). nanos is the localized posterior determinant in Drosophila. *Cell* **66,** 637–647.

Wang, S., and Hazelrigg, T. (1994). Implications for bcd mRNA localization from spatial distribution of exu protein in *Drosophila* oogenesis. *Nature* **369,** 400–403.

Warn, R. M., Gutzeit, H. O., Smith, L., and Warn, A. (1985). F-actin rings are associated with the ring canals of the *Drosophila* egg chamber. *Exp. Cell Res.* **157,** 355–363.

Yisraeli, J. K., Sokol, S., and Melton, D. A. (1990). A two-step model for the localization of maternal mRNA in *Xenopus* oocytes: Involvement of microtubules and microfilaments in the translocation and anchoring of Vg1 mRNA. *Development* **108,** 289–298.

6

Role of the Actin Cytoskeleton in Early *Drosophila* Development

Kathryn G. Miller
Department of Biology
Washington University
St. Louis, Missouri 63130

I. Introduction and Overview

Newly fertilized *Drosophila* embryos undergo 14 nuclear divisions without cytokinesis, forming a large, highly organized multinucleate cell (Fig. 1). For the first 7 divisions, the nuclei divide in the yolky internal cytoplasm (Figs. 1A and 1B); during the eighth and ninth nuclear division cycles, most of the nuclei migrate to the cortex, forming an evenly spaced monolayer beneath the plasma membrane at cycle 10 (Figs. 1C and 1D). In this syncytial blastoderm embryo, the nuclei divide four more times before being segregated into individual cells by a process known as cellularization. The cellular blastoderm begins gastrulation immediately upon completion of cellularization.

The syncytial nature of the early embryo presents some special problems for organization and function of cellular components. The presence of many nuclei within a common cytoplasm requires special organizational mechanisms. Nuclear position during both the internal and the syncytial blastoderm divisions is highly regular. Movement and positioning of nuclei within the syncytial embryo is critical for later developmental events. Since

Current Topics in Developmental Biology, Vol. 31

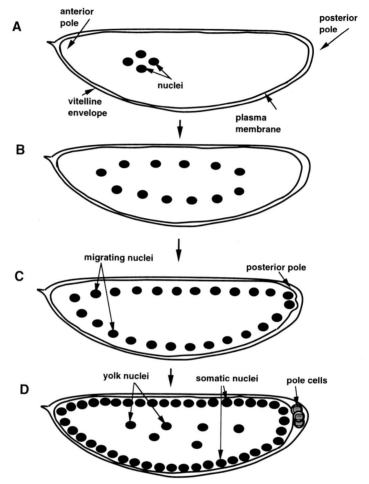

Fig. 1 Schematic diagram of early nuclear divisions in *Drosophila*. (A) During the first four divisions, nuclei divide in the internal yolk cytoplasm and are arranged equidistant from each other in a sphere centered approximately one-third of egg length from the anterior. (B) During cycles 4–6, nuclei expand axially so that their distribution mimics the embryo's shape. (C) Several nuclei reach the posterior pole of the embryo during cycle 9, prior to the arrival of most of the migrating nuclei. These nuclei become incorporated into cells prior to cellularization of the blastoderm. (D) The somatic nuclei reach the cortex at the beginning of cycle 10, forming an evenly spaced monolayer.

many centrosomes (associated with each nucleus) are present within this one cell, ensuring that nuclear division occurs with the proper fidelity requires that spindles form only between centrosomes associated with a single nucleus (sister centrosomes) and not between centrosomes from neighboring nuclei, which, at some stages (see below), are in close physical proximity. In addition, it is during this period of syncytial development that nuclei begin expressing genes involved in pattern formation in a position-specific manner (Hafen *et al.*, 1984), so that by the time that cellularization is complete, the cells are committed to their segmental fate (Simcox and Sang, 1983). Since the cascade of gene expression that leads to this commitment occurs in the syncytium, organization is likely to be very important for the orderly expression of the proper pattern.

The actin and microtubule cytoskeletons play critical roles in embryo organization. Studies on the cytoskeleton of the syncytial embryo have begun to dissect the roles that different cytoskeletal components play in this organization. Actin and tubulin localization both in fixed preparations (Warn *et al.*, 1984; Karr and Alberts, 1986; Warn and Warn, 1986; Baker *et al.*, 1993) and in living embryos (Warn *et al.*, 1987; Kellogg *et al.*, 1988) have revealed a highly organized array of cytoskeletal polymers. Actin- and microtubule-associated proteins have been identified and their distribution has been investigated (Kiehart and Feghali, 1986; Whitfield *et al.*, 1988; Miller *et al.*, 1989; Kellogg *et al.*, 1989; Pesacreta *et al.*, 1989; Young *et al.*, 1991). Mutations that affect the cytoskeletal structures present are being studied (Postner *et al.*, 1992; Sullivan *et al.*, 1993; Fogarty *et al.*, 1994). Antibodies are being used to disrupt the function of molecules important for cytoskeletal function (Kiehart *et al.*, 1990; Mermall *et al.*, 1994; Mermall and Miller, 1995). Drugs that disrupt the cytoskeleton have been used to investigate the function of the different polymer systems (Zalokar and Erk, 1976; Foe and Alberts, 1983; Edgar *et al.*, 1987). Although studies of cytoskeletal function in this organism are still in their early stages, much has already been learned about the role of specific cytoskeletal components in various aspects of organization and the dynamic process that are required to generate a normal *Drosophila* embryo. In addition, it seems clear that while the cytoskeleton is functionally specialized for this mode of syncytial development, many of the lessons that we can learn by studying the regulation and function of the cytoskeleton in this embryo will be applicable to other embryos and to eukaryotic cells in general (see below).

In this chapter, the primary focus will be the actin cytoskeleton. In particular, studies of the syncytial blastoderm stage and the function of actin and actin-associated proteins during this period will be highlighted. Under Sections I,A and I,B, a general introduction to the early *Drosophila* embryo and its cytoskeleton is presented. Under Section II, several different approaches to studying cytoskeletal function that are being used in this system are outlined. Work from several laboratories that suggests functions of particular proteins associated with the actin cytoskeleton is described in some detail. Recent reviews of the cytoskeleton in *Drosophila* early embryonic development (Schejter and Weischaus, 1993b; Fyrberg, 1995) and syncytial blastoderm development in general (Foe *et al.,* 1993) highlight additional information not dealt with here.

A. Organization of the Cytoskeleton in Early Embryos

During the very early divisions (and presumably immediately after fertilization, although this has not been documented), F-actin is highly concentrated in the cortical cytoplasm in punctate aggregates and filamentous structures visible at the light microscope level using fluorescence (Warn *et al.,* 1984; Karr and Alberts, 1986; Hatanaka and Okada, 1991; von Dassow and Schubiger, 1994). Since individual actin filaments are not resolved at this level of resolution, the filamentous-appearing material is thought to be bundles of actin filaments. The structure of punctate aggregates is unknown, but since rhodamine–phalloidin labels these structures, they must be composed of filamentous actin. F-actin is also present throughout the internal cytoplasm at much lower concentrations and is also organized into bundles of filaments and punctate material (Hatanaka and Okada, 1991; von Dassow and Schubiger, 1994). There is a focus of granular or punctate-appearing F-actin in the center of embryo, internal to the dividing nuclei (Hatanaka and Okada, 1991; von Dassow and Schubiger, 1994). In contrast to this vast network of actin throughout the embryo, microtubules are restricted in their distribution. An extensive microtubule aster is associated with the centrosome of the sperm, with some microtubules extending outward from the microtubule-organizing center (MTOC) to the cortex. Microtubules are associated with the meiotic female pronuclei, which complete meiosis after fertilization. Few other microtubule structures are visible at this stage. While a continuous layer of short microtubules organized into foci in the cortical cytoplasm has been reported when taxol (a microtubule-stabilizing drug) is present during fixation (Karr and Alberts, 1986), these microtubules are not seen in other circumstances (Warn and Warn, 1986; Kellogg *et al.,* 1988).

During the first nuclear divisions microtubules are associated with the internally dividing nuclei in a typical cell cycle-regulated manner (Karr and Alberts, 1986; Baker *et al.,* 1993). During interphase, duplicated centrosomes lie near each nucleus and nucleate the assembly of asters that appear to define a region of cytoplasm around each nucleus free of large inclusions and enriched in F-actin (von Dassow and Schubiger, 1994). F-actin fibers and punctate material are visible within these cytoplasmic islands. Actin aggregates and fibers appear somewhat enriched around the centrosome (MTOC), forming a cap that is oriented apically (toward the embryo surface). Similar-appearing actin fibers and punctate aggregates are also visible throughout the cytoplasm. Fibers appear to encircle many of the large organelles and sometimes radiate from the punctate structures. During mitosis, microtubules form typical spindles with small asters. The actin caps persist during metaphase, located in the region of the asters, and become more prominent during ana- and telophase. Actin fibers can be seen radiating from the region of the spindle poles toward the metaphase plate and into the surrounding granule-rich cytoplasm.

As their numbers increase, the nuclei form a spherical shell approximately equidistant from each other (Fig. 1A). This equidistant spacing of nuclei has been hypothesized to result from the action of microtubule-dependent motors on the microtubule arrays associated with each nucleus. The motors associated with each array are thought to push against overlapping microtubules from neighboring arrays during anaphase. The sphere of nuclei is initially centered about one-third of egg length from the anterior pole of the embryo. During nuclear cycles 4–6, the sphere expands axially so that it becomes more oblate and nuclei lie in a hollow shell equidistant from the egg surface (Fig. 1B). This axial expansion of nuclei is critical, since nuclei will reach the cortex in the mid- and anterior regions of the embryo long before they reach the posterior region if this rearrangement does not occur (Hatanaka and Okada, 1991). Normally, however, the posterior pole is the first region of the cortex to become populated with nuclei (i.e., during cycle 9; Fig. 1C). These early migrating nuclei are segregated into cells during the next cycle, significantly before cellularization of the blastoderm (Fig. 1D). These cells, called pole cells, give rise to the germline stem cells. If nuclei are delayed in their arrival in the posterior cortex, pole cells will not form (Okada, 1982). In embryos from mothers mutant at any one of several maternal effect loci, this axial expansion does not occur and nuclei do reach the cortex asynchronously in different parts of the embryo (Hantanaka and Okada, 1991). Pole cells do not form in embryos from animals mutant for these loci; thus, these animals are sterile.

Actin is required for axial expansion of the nuclear sphere into an ellipsoid (Zalokar and Erk, 1976; Hatanaka and Okada, 1991; von Dassow and Schubiger 1994). Nuclear position is disrupted when embryos are treated with cytochalasin D, a drug that prevents actin polymerization. Examination of the movement of nuclei during axial expansion has demonstrated that nuclear movement occurs during inter- and prophase, as cytoplasmic streaming of the internal cytoplasm toward the embryo poles is occurring (von Dassow and Schubiger, 1994). The nuclei are carried axially by this cytoplasmic flow. The actin meshwork that is present in the region of the sphere of nuclei appears to disassemble during interphase and prophase, as cytoplasmic streaming is occurring. It is hypothesized that the actin network is under tension. The local dissolution of the actin meshwork, and subsequent contraction of the rest of the meshwork due to the release of tension, is suggested to be the cause of the cytoplasmic streaming.

Most of the nuclei migrate to the cortex during cycles 7–9 (Figs. 1C and 1D), with microtubule-mediated movement outward occurring in ana- and telophase (Foe and Alberts, 1983). The migrating nuclei reach the cortex just as interphase of nuclear cycle 10 begins, forming the syncytial blastoderm (Fig. 1D). These nuclei become incorporated into cells after four more divisions. These blastoderm cells will become the somatic cells of the organism. A few nuclei appear to fall into the center of the embryo as the outward migration proceeds (Foe et al., 1993) and become polyploid. These nuclei remain a part of the large single yolk cell that comprises the interior yolky cytoplasm of the embryo after cellularization of the blastoderm during nuclear cycle 14.

At the syncytial blastoderm stage, nuclei with their associated centrosomes apically situated form an evenly spaced monolayer in the cortical cytoplasm (Figs. 1D and 2). The duplicated centrosomes lie at either side of each nucleus where they nucleate asters that radiate both apically and basally. The basally radiating microtubules surround the nucleus and extend into the cytoplasm basal to it. These microtubules are thought to be involved in clearing cytoplasmic enclusions (yolk granules and lipid droplets) from the cortex. The apically radiating microtubules interact with the cortex and are thought to position sites of actin accumulation (see below).

As nuclei reach the cortex, actin in the cortex rearranges into domains that associate with each nucleus (Fig. 2). Above each nucleus, actin concen-

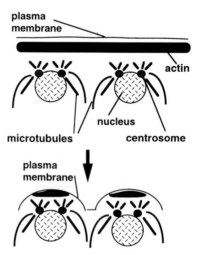

Fig. 2 Schematic diagram of nuclei migrating to the embryo cortex during late cycle 9 and early cycle 10. A cross section through a small region of the cortex is shown. Actin that is highly concentrated in the cortex in an even layer prior to nuclear migration rearranges so that it is concentrated in caps above each nucleus.

trates into a patch tightly associated with the plasma membrane, called an actin cap (Warn *et al.,* 1984; Karr and Alberts, 1986). The plasma membrane bulges outward above each nucleus, so that the actin cap forms a dome-shaped structure. Centrosomes that become separated from nuclei have been shown to induce the reorganization of actin into caps (Raff and Glover, 1989; Yasuda *et al.,* 1991). It seems likely that the interaction between centrosomes and actin in the cortex is mediated by microtubules contacting the cortex, but direct data are not available. Injection of antibodies to tubulin causes collapse of spindles and subsequent loss of nuclei from the cortex (Warn *et al.,* 1987). Actin structures are not apparent in regions where nuclei have disappeared, indicating that the continuing interactions between centrosomes/microtubules and the cortex are required for maintenance of actin structures as the cell cycle proceeds. However, detailed observations of the dynamics of loss of actin structures when microtubules are depolymerized have not been reported.

During mitosis, the actin caps flatten as the actin rearranges, and actin, presumably through the action of actin-binding proteins (ABPs), mediates the formation of transient membrane invaginations, called pseudocleavage or metaphase furrows, that divide neighboring mitotic apparati (Fig. 3). The positioning of the sites of furrow formation also depends on the microtubule

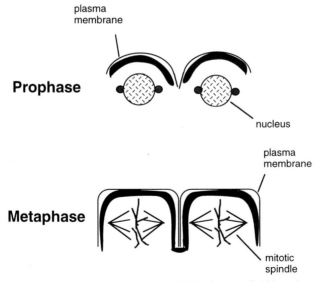

Fig. 3 Progression from prophase to metaphase during the syncytial blastoderm divisions is depicted in cross section. Actin is shown in solid black shading. As centrosome migrate to opposite sides of the nucleus, actin rearranges and mediates the formation of transient invaginations that divide neighboring spindles.

cytoskeleton. Sister centrosomes are normally situated apically at either side of the nucleus in interphase. As prophase begins they migrate to diametrically opposed positions (Fig. 3). Metaphase furrows normally form during prophase between centrosomes of neighboring nuclei, but not between sister centrosomes. In the maternal effect mutant, *daughterless-abolike* (*dal*), some centrosomes do not separate or migrate to opposite sides of the nucleus (Sullivan *et al.*, 1990). Thus, at some sites where metaphase furrows should form, centrosomes from neighboring nuclei are not juxtaposed. At these sites, metaphase furrows fail to form (Sullivan *et al.*, 1993). Where defects in furrows occur, microtubules from neighboring spindles interact and defects in division occur. It is not clear how the juxtaposition of centrosomes from neighboring nuclei creates the signal for formation of metaphase furrows at the appropriate site. Other mutations affect syncytial blastoderm organization through effects on the actin cytoskeleton (Sullivan *et al.*, 1993; Postner *et al.*, 1992; see below). Thus, proper interaction between the microtubule cytoskeleton and actin cytoskeleton is important for normal morphology of the syncytial blastoderm.

As suggested previously, the actin cytoskeleton plays a number of important roles in organizing the syncytial embryo. It is important for proper positioning of the migrating nuclei (see above). It also has important functions during the syncytial blastoderm stage. The actin caps are required to keep the nuclei in their regularly spaced array (Postner *et al.*, 1992). The apical orientation of the centrosomes (Postner *et al.*, 1992) and the arrangement of spindles parallel to the plane of the embryo surface during mitosis are likely to be a result of interactions between actin and microtubules (reviewed in Foe *et al.*, 1993). During mitosis, the transient invaginations of membrane that surround each spindle depend on actin polymerization (Zalokar and Erk, 1976). In their absence, neighboring mitotic apparati interact so that aberrant divisions result (Zalokar and Erk, 1976; Sullivan *et al.*, 1990; Postner *et al.*, 1992; Sullivan *et al.*, 1993; Mermall and Miller, 1995).

B. Some Important Questions about Cytoskeletal Function That Can Be Investigated in *Drosophila*

As described previously, in the syncytial blastoderm embryo, the spatial arrangement of actin and microtubule structures is highly interdependent. The position of actin caps and metaphase furrows depends on centrosome position. Conversely, the formation of proper microtubule arrays depends on the presence of actin caps and furrows. Little is known about the mechanism of this interaction. This type of interaction between cytoskeletal systems is not peculiar to *Drosophila* syncytial blastoderms. All cells must coordinate the distribution and morphology of actin and microtubule structures. For example, assuming an appropriate cell shape requires that both actin and microtubules be organized properly.

During the syncytial stage of *Drosophila* development, actin and microtubule arrays undergo coordinated rearrangements that are regulated both developmentally and by the cell cycle. Again, the ability to coordinately rearrange cytoskeletal organization is not specific for syncytial blastoderms. Most cells are capable of reorganizing their cytoskeletons in coordinated ways in response to intracellular and extracellular signals, including developmental cues. Such rearrangements and the cell's ability to coordinate them are perhaps most obvious in dividing cells. The actin-based contractile ring must form oriented such that division bisects a cell perpendicular to the spindle axis and at the spindle equator, so that cytoplasmic and nuclear contents are evenly distributed to daughters. During development, division plane orientation and location is often controlled very precisely relative to the overall organization of the embryo. This places a further level of regulation on the positioning of cytoskeletal systems. In addition, as cells differentiate they also often rearrange both actin and microtubule cytoskeletons so that they can perform a specialized function. The molecular mechanisms that control positioning and interactions among cytoskeletal systems remain poorly understood.

The *Drosophila* syncytial embryo has a very consistent and highly repeated array of cytoskeletal structures. This repeated, regularly spaced, and uniformly structured array makes it easy to detect even minor alterations in the organization of the cytoskeleton. These cytoskeletal structures are required for the organization of many cellular components. In particular, nuclear distribution is highly dependent on both actin and microtubule cytoskeletal function. Because nuclei are normally very evenly spaced in the syncytial blastoderm, many perturbations in the function of proteins associated with the cytoskeleton or that control cytoskeleton arrangements can be detected as a disruption of the regular array of nuclei. Disruptions in cytoskeleton function have major consequences for development, since perturbation of the organization of the embryo can cause developmental arrest. The early *Drosophila* embryo can be manipulated using injection of proteins and RNAs, antibodies and drugs, by genetic methods, and through expression of manipulated genes introduced by transformation. This system is ideal for defining functions of proteins involved in generating the proper cytoskeletal structures and in developmental and cell cycle-controlled changes in cytoskeleton distribution.

II. Studies of Actin-Binding Protein Function

It is clear that actin filaments alone are not sufficient to perform the multitude of tasks required of the actin cytoskeleton. Many proteins have been identified that bind to and affect actin filament dynamics and structure (Pollard and Cooper, 1986). Actin-binding proteins are required to mediate actin filament interactions with other cellular components. They regulate actin filament assembly through actin monomer binding and filament end capping. They determine the morphology of actin filament structures through activities such as cross-linking and bundling. Thus, much of our ability to understand actin cytoskeleton function rests on our understanding of actin-binding protein function. Although we know that actin-binding proteins perform important functions, largely through *in vitro* assays of activities, we have little evidence, in most cases, for these roles *in vivo*. To understand how the actin cytoskeleton participates in early *Drosophila* development, it is critical to determine what actin-binding proteins are present, where they are located, and their particular roles in actin cytoskeleton organization and function.

A. Genetic Strategies for the Study of Cytoskeletal Protein Function

Traditional (forward) genetic strategies in *Drosophila* are yielding information on actin cytoskeletal protein function. Mutations that affect cytoskeletal proteins have been identified (Karess *et al.*, 1991; Cooley *et al.*, 1992; Cant *et al.*, 1994; Mahajan-Miklos and Cooley, 1994) using screens for particular phenotypic effects. Using this approach, the goal was not necessarily to identify mutations specifically affecting the cytoskeleton, per se. Instead, information on the process affected in the mutant was being sought. However, once protein sequences encoded by these loci were obtained, homology searches identified them as homologs of cytoskeletal proteins identified in other systems. Detailed phenotypic analyses have been directed toward examination of the cytoskeleton and are yielding significant new information on *in vivo* functions of some actin-binding proteins that are well-studied *in vitro*.

One protein whose role *in vivo* is being clarified as a result of this approach is profilin. Profilin had traditionally been thought of as a monomer-binding protein, regulating polymerization of actin by sequestering monomers (Pollard and Cooper, 1986). Recently, *in vitro* data indicated that profilin can bind phosphoinositides with high affinity and when bound to them cannot bind actin (Goldschmidt-Clermont *et al.*, 1990). In addition, profilin can promote assembly under some conditions *in vitro* (Goldschmidt-Clermont *et al.*, 1991; Pantaloni and Carlier, 1993). These results suggest a more complex role for profilin in regulating actin polymerization in response to signaling. *In vivo* data demonstrating its role in actin polymerization are important to understanding profilin's function.

The gene encoding profilin in *Drosophila* was identified as a result of a screen for mutants that affect a specific part of oogenesis, the dumping of nurse cell cytoplasm into the oocyte. One mutation that stops this process (*chickadee*) affects a germline-specific profilin transcript (Cooley *et al.,* 1992). Examination of the actin cytoskeleton in egg chambers reveals that the prominent actin cables that normally form in the nurse cell cytoplasm just prior to cytoplasmic dumping do not form in the egg chambers of mutant animals. The other actin structures present (the actin meshwork in the cortex of the nurse cells and the ring canals that connect the nurse cells to each other and the oocyte) appear unaffected. The fact that loss of profilin function in this cell leads to a lack of a specific filamentous actin structure suggests that, *in vivo,* profilin normally has a positive effect on the assembly of at least some actin structures. Additional loss-of-function alleles that affect the profilin transcript expressed at other times during the fly life cycle have subsequently been identified (Verheyen and Cooley, 1994). Some mutant alleles are lethal, suggesting that profilin function is critical at other times in development. Examination of the phenotypic effects of these mutations suggests that profilin may have different roles in different cells. For example, excess numbers of actin bundles form in profilin mutant bristle cells. Since lack of profilin leads to excess polymerization, this phenotype suggests that profilin normally sequesters monomers in these cells. Thus, in some situations, profilin's sequestering role seems to predominate, while in others, its assembly promotion appears most important. Since alleles that permit survival to the adult stage cause female sterility, and lack of germline-specific transcript causes defects in oogenesis, information on the role of profilin in the syncytial blastoderm cannot be obtained until conditional alleles are available or other methods are applied (see below).

Other mutations that affect the same cytoplasmic dumping process have identified genes for the *Drosophila* homolog of fascin (*singed;* Cant *et al.,* 1994) and an oogenesis-specific villin/gelsolin family member (*quail;* Mahajan-Miklos and Cooley, 1994). In both cases, since oogenesis is affected, no information on function during syncytial stages has yet been obtained.

In a more direct approach toward understanding cytoskeletal function, *Drosophila* homologs of actin-binding proteins that have been identified and studied in other systems have been sought in order to take advantage of the *Drosophila* system to understand their functions. The gene that encodes a protein of interest is first identified through sequence homology with genes from other organisms or using an antibody specific for the protein of interest to screen an expression library. Once a clone is available it is mapped onto the polytene chromosome map and known mutations and deficiencies in the region are identified. Identification of an existing mutation that corresponds to the gene of interest or designing a mutant screen is the next step (see below). Phenotypic analysis can then reveal the processes that are most affected in mutants and a variety of genetic strategies are available to further manipulate animals with loss of function. This reverse genetic approach is beginning to yield new insights into actin-binding protein function.

Particularly fruitful studies on the cytoplasmic myosin in *Drosophila* have used the reverse genetic approach. These studies have demonstrated that cytoplasmic (type II) myosin is required for morphogenesis during embryonic development (Young *et al.*, 1993). Mutations in the gene that encodes the cytoplasmic myosin heavy chain (*zipper*) cause arrest of development during dorsal closure, the migration of epithelial cells over more internal cell layers. Consistent with this phenotype, cytoplasmic myosin is concentrated in the apical cell border of the cells at the leading edge of the migrating sheet. The role of cytoplasmic myosin in the movement of the sheet is thought to be similar to the role it plays in cytokinesis, causing the contraction of the apical margin of the cell. This has an effect similar to pulling a drawstring on a purse, drawing the cells toward the dorsal midline. Several mutant alleles have a weaker phenotype, with animals surviving into larval stages. The fact that these alleles later show a lethal phase suggests that cytoplasmic myosin is required for some later events as well (Young *et al.*, 1993).

Further insight into later requirements for cytoplasmic myosin function is being obtained from studies of its regulatory light chain. This light chain has been shown to regulate myosin activity in other systems. Mutations in the gene that encode the light chain (*spaghetti-squash*) lead to defects in cell division which are particularly apparent during later development (Karess *et al.*, 1991). Imaginal disc cells, which normally proliferate during larval life, are reduced in number and become highly polyploid because cytokinesis does not occur. It is interesting that mutations in this regulatory light chain have a different phenotype than mutations in the myosin heavy chain. While the single existing mutation in the regulatory light chain does not appear to be a null, the expression level of the protein is severely reduced. It appears that with these reduced levels of the regulatory light-chain function, cytoplasmic myosin is able to function well enough that dorsal closure can occur, but cytokinesis is more sensitive. Using a heat-shock-inducible rescue construct, expression of wild-type gene product can be induced at various times during development (K. Edwards and D. Kiehart, personal communication). The imaginal disc proliferation defect can be rescued by heat-shock-induced expression of the light chain at regular intervals during larval life. Other processes that require cytoplasmic myosin's function are revealed through different heat-shock regimens. Continued heat shocks during adult life are required for continued fertility, indicating that myosin function is required during oogenesis. It might be possible to find a heat-shock regimen that would provide enough activity for oogenesis to be completed, but would provide insufficient activity for syncytial embryonic development to occur.

Mutations that affect α-spectrin cause lethality during the larval stage with defects in the organization of cells of the intestine (Lee *et al.*, 1993). Recent work demonstrated that site-directed mutations which target sites important for network formation, introduced into the lethal mutant background, create a temperature-sensitive phenotype. These mutant α-spectrins, at the permissive temperature, rescue lethal mutant animals so that they survive to adulthood (Deng *et al.*, 1995). Thus, using this temperature-sensitive phenotype, it may be possible to determine the effects of α-spectrin loss of function on syncytial embryos in the near future. The fact that, at the nonpermissive temperature, these mutations cause phenotypic effects in nonerythroid cells demonstrates that spectrin network formation is important to its role in many cell types.

Other actin-associated proteins that are known to be required for viability include capping protein (Hopmann *et al.*, manuscript in preparation), α-actinin (Fyrberg *et al.*, 1990; Roulier *et al.*, 1992), and band 4.1 (Fehon *et al.*, 1994), although models for their specific functions are not yet well developed. Again, the lethality of these mutations means that the analysis of the effect of loss of function on the syncytial blastoderm is not straightforward.

Through the genetic studies described previously, evidence for important functions for the actin cytoskeleton during development is being obtained. Many of the proteins associated with the actin cytoskeleton play important roles in development, since many of them are required for viability. The functions of individual actin-binding proteins in particular developmental processes are being revealed through detailed phenotypic analysis of mutant animals. However, these genetic studies have shed little light on the function of these zygotically lethal actin-associated protein loci during very early development. This is due to the large maternal stores of many of these proteins present at fertilization. This maternally contributed protein functions in the early embryo until zygotic transcription is activated at the time of cellularization. Only by depleting maternal stores could a role in early events be revealed. Quite often, this maternally contributed protein persists through embryonic development in sufficient quantities that even the processes that occur later in embryogenesis which require these proteins can proceed in the absence of new synthesis.

In *Drosophila*, genetic mosaic techniques are available that can allow the loss of gene function to be confined to a subset of cells in the adult (Xu and Rubin, 1993; Chou *et al.*, 1993). Using these mosaic techniques, animals carry mutations in the heterozygous state and therefore survive;

patches of homozygous mutant cells are produced by recombination during mitotic cell division during development. The effect of the mutation in the homozygous mutant cells is then revealed through analysis of phenotypic changes. This technique can be used to generate homozygous mutant germ-line clones in an otherwise heterozygous (and therefore phenotypically wild-type) animal. Examination of embryos from such mosaics will, in principle, reveal the function of the mutant protein in the syncytial embryo. Unfortunately, an additional complication is that many of these proteins are likely to be required for oogenesis, so that the loss of function in the germline would prohibit obtaining eggs or fertilized syncytial embryos. In this case, defining the function of the protein of interest in the syncytial blastoderm is difficult. Several different strategies are being pursued to circumvent these difficulties.

One genetic approach which is being used to reveal proteins important in syncytial blastoderm cytoskeletal organization is a search for maternal-effect mutations that specifically affect the syncytial blastoderm organization. This type of phenotypic screen would be expected to identify genes whose products are required only for the specialized processes peculiar to this stage. It also might identify unusual alleles of genes that act both during this stage and at other times. These unusual alleles would only affect the syncytial stage and not other stages. Existing maternal-effect mutant collections have been surveyed for mutants that lay normal-appearing eggs that develop through nuclear migration with no apparent defects, but become aberrant during the syncytial blastoderm divisions (Sullivan *et al.,* 1993). Using nuclear morphology and distribution to define phenotype, examination of 76 existing P element-induced female sterile mutations identified 44 mutants that laid morphologically normal eggs, 32 of which had visible defects in the syncytial stage. Many of these arrested prior to nuclear migration to the cortex (cycle 10) and were not studied further, but 7 specifically affected syncytial blastoderms. These 7 fall into four complementation groups, one of which was previously identified as *staufen. staufen* affects localization of maternally provided informational RNAs (St. Johnston *et al.,* 1991), as well as syncytial blastoderm organization (Sullivan *et al.,* 1993). The three genes identified in the screen not corresponding to known complementation groups were further studied. All have disorganized syncytial blastoderms that are associated with defects in the cytoskeleton. *nuclear fallout (nuf)* and *scrambled (sced)* have specific effects on the actin cytoskeleton. In both these mutants, actin caps are normal, but metaphase furrows are affected. In *sced* mutant embryos, metaphase furrows fail to form. In *nuf* mutant embryos, furrows form but have gaps in them. In both cases the furrow defects lead to subsequent defects in nuclear division due to aberrant microtubule structures forming during mitosis. Since these two mutations affect furrows specifically, it is likely that the encoded proteins are involved in metaphase furrow formation. It will be of great interest to determine where these proteins are located relative to actin and determine their roles in furrow formation. As yet, no information regarding the molecular nature of these proteins has been reported, although cloning of several of the affected genes is in progress (W. Sullivan and W. Theurkauf, personal communication). Since these mutations were induced by P element (transposon)-mediated insertional mutagenesis, cloning should be relatively straightforward.

The third mutation, *grapes* (*grp*), arrests development at metaphase of nuclear cycle 13. The first phenotype visible is a defect in formation of spindle midbodies (Fogarty *et al.*, 1994). The absence of the midbody allows the two daughter nuclei to come back together after division, generating the clumped nuclei characteristic of this mutation (Sullivan *et al.*, 1993). Molecular studies indicate that the *grp* locus encodes a cell cycle checkpoint kinase (Fogarty and Sullivan, manuscript in preparation). Localization suggests that this protein may function to regulate both actin and microtubule structures during the ana/telo/interphase period (W. Sullivan, personal communication).

This screen was rather limited in extent, and other mutations that specifically affect the syncytial blastoderm are likely to exist. Further screening will likely yield some additional mutants with interesting phenotypes. An additional screen is being performed for zygotic lethal loci that have a maternal effect on syncytial blastoderm development (W. Sullivan, personal communication). Temperature-sensitive mutations greatly facilitate experiments in determining whether a lethal locus has a maternal effect, so the screen is performed at 29°C (the restrictive temperature) and any lethal mutant is tested at 22°C for viability. Mutations that are temperature sensitive (a significant function of EMS-induced mutations are temperature sensitive) can then be tested for effects on syncytial blastoderms by keeping adult females at 22°C, but shifting embryos to 29°C. A phenotypic screen of the embryos will reveal if the mutation affects syncytial blastoderm organization. A collection of mutants of this sort should, in principle, identify any protein with an effect on syncytial blastoderm organization.

The maternal-effect mutation, *dal,* has its primary effects on microtubules, as described previously. Some centrosomes fail to separate (Sullivan *et al.*, 1990) and actin-based metaphase furrows fail to form at these sites (Sullivan *et al.*, 1993). The mechanism by which the *dal* gene product has this effect is unknown and no molecular information about *dal* is available.

sponge is another maternal-effect mutation that affects syncytial blastoderm organization. In this mutant, neither actin caps nor furrows form (Postner *et al.*, 1992). The embryos become disorganized through the mechanisms described previously. To date, no molecular information about the *sponge* gene product has been obtained, but further studies of *sponge* mutant embryos have yielded some additional insight into blastoderm organization (Postner *et al.,* 1992; see below).

Three mutations (*nullo, serendipity-α,* and *bottleneck*) have been identified that specifically affect cellularization (Merrill *et al.,* 1988; Simpson and Wieschaus, 1990; Schweiguth *et al.,* 1990; Schejter and Wieschaus, 1993a). These are zygotic lethal loci that do not affect syncytial divisions. They are expressed just prior to cellularization and all three appear to associate with the actin cytoskeleton during cellularization. These proteins appear to work in concert with maternally provided actin-binding proteins to cellularize the embryo.

B. Biochemical and Cell Biological Approaches

While genetic analysis of known actin-binding protein homologs in *Drosophila* has not yet revealed the functions in the syncytial blastoderm (see above), some of these proteins have been localized during syncytial blastoderm divisions. Several spectrin family members are present and associate with the cortex early in syncytial development. An unusually large β-spectrin-related protein, β_{heavy}-spectrin, is present in metaphase furrows during the syncytial blastoderm stage (Thomas and Kiehart, 1994). Its distribution during interphase is not reported. α-Spectrin is also associated with the plasma membrane during the syncytial stage, where it is concentrated in caps that colocalize with actin during cycle 10. However, later during the syncytial blastoderm period, α-spectrin appears to be present uniformly on the plasma membrane during interphase, but is apparently not present in the metaphase furrows (Pesacreta *et al.,* 1989). Spectrin has been suggested to be a stabilizer of actin–plasma membrane interactions. Its localization during syncytial blastoderm divisions is consistent with such a role. Cytoplasmic myosin has been localized during this time (Young *et al.,* 1991; Field and Alberts, 1995; see below), and it is reported to be present in metaphase furrows (Young *et al.,* 1991). Its presence in furrows might reflect a participation in a contractile process important for furrow invagination. Cytoplasmic myosin-specific antibody injections into the early embryo lead to inhibition of cellularization, indicating that this myosin plays a role in this cytokinetic process (Kiehart *et al.,* 1990). This result is consistent with data from other systems. However, effects in the syncytial stages have been somewhat harder to interpret. Antibody injections cause blastoderm disorganization and loss of nuclei from the cortex. These effects are consistent with a disruption of metaphase furrow formation, but data relating to actin and microtubule distribution in antibody disrupted embryos are not available. Further evidence of the participation of these proteins in actin cytoskeletal function during the syncytial stage will eventually be obtained through the use of conditional alleles or other genetic techniques (see above) and perhaps the additional use of antibodies to disrupt function.

Using another approach for the analysis of actin cytoskeletal function in the syncytial blastoderm, a large number of proteins that interact with actin in the early embryo (Miller *et al.*, 1989) have been identified. In this study proteins were selected solely by their ability to bind to actin, and no assumptions were made concerning what type of activity they might have. The proteins thus identified were localized in the early embryo relative to actin. They have a variety of distributions, some localizing in both interphase actin caps and metaphase furrows, while others colocalize with actin only in a subset of structures (Fig. 4). Despite the fact that each protein's distribution had some unique features, all the patterns could be assigned to one of three classes. Those that colocalized in both caps and furrows were termed actin-like; those present only in interphase caps but were not colocalized with actin during mitosis were termed interphase-like; and those present in furrow but not colocalized during interphase were termed metaphase-like. It was hypothesized that interphase-like proteins might be key organizers of appropriate actin structure for cap function, while the metaphase-like proteins would conversely be required to generate actin structures that mediate furrow formation. Proteins present in both types of structures would function more generally in actin stabilization, cross-linking, etc. that would be required in both of these structures. More extensive studies of several of the proteins identified in this screen have provided some insights into their function and support for their roles in helping to generate appropriate actin structures (see below; Postner *et al.*, 1992; Mermall *et al.*, 1994; Mermall and Miller, 1995; Field and Alberts, 1995). Unfortunately, it has been difficult to fit the localizations of actin-binding protein homologues (like the spectrins, whose localization was described previously) into this scheme, because the localizations have not been directly compared under

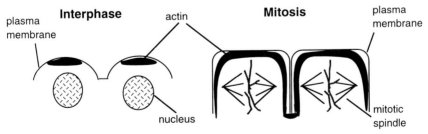

Fig. 4 Schematic representation of the distribution of actin during the syncytial blastoderm mitoses. A small region of the cortex of a syncytial blastoderm embryos is depicted in cross section. Actin is shown in solid black shading.

controlled conditions and, in the published accounts, data have often not been obtained for each protein at comparable stages and cell cycle times.

Two novel proteins that are in the metaphase furrow have been studied by cloning the genes that encode them using antibodies to screen an expression library (Kellerman and Miller, 1992; Field and Alberts, 1995). One of these is a protein that has been named anillin, which localizes to metaphase furrows during syncytial blastoderm mitoses (Miller *et al.,* 1989; Field and Alberts, 1995). It remains in the cortex during interphase, occupying the region outside of the cap where furrows will form during mitosis. Cytoplasmic myosin colocalizes with anillin during interphase, concentrated in the regions between the caps, but it also has a cytoplasmic component not apparent with anillin (Field and Alberts, 1995). During conventional cytokinesis and cellularization of the blastoderm, anillin is also present in the cleavage furrow, where it colocalizes with conventional myosin (Field and Alberts, 1995). However, during interphase in dividing cells, it is not in the cortex—instead it is present in the nucleus. In cells that lack the potential to divide, the protein is not present in the nucleus. This protein's sequence is not particularly revealing—it has no significant homology to other proteins in the database. It has a proline-rich region similar to several to those other cytoskeletal proteins, but the functional significance of this region is not clear. It has several nuclear localization signals, consistent with its nuclear localization during interphase. To date, no evidence for function during cleavage has been presented, but it is postulated to bind to actin in the cleavage furrow to help stabilize it to withstand the forces generated during contraction. Its presence in both metaphase furrows and bonafide cleavage furrows, along with cytoplasmic myosin, suggests that there may be similar processes at work in these two cases. Why it remains in the cortex in interphase in the syncytial blastoderm is as yet unknown, but Field and Alberts (1995) speculate that it is required for a cortical myosin-based contractile process which is important for syncytial embryos during interphase. In support of this, anillin and cytoplasmic myosin appear to be colocalized during inter- and prophase.

Another novel metaphase furrow protein identified in the screen for actin-binding proteins in the syncytial blastoderm, the 95F unconventional myosin, has been molecularly characterized (Kellerman and Miller, 1992) and its function investigated using antibody injection (Mermall *et al.,* 1994; Mermall and Miller, 1995). Localization studies during syncytial blastoderm divisions revealed that this novel myosin is located in particles throughout the cytoplasmic region that surrounds each syncytial blastoderm nucleus during interphase. During mitosis, it is found in the metaphase furrow. The punctate staining during interphase is suggestive of an association with a cytoplasmic structure, perhaps a vesicle (although this remains to be demonstrated). Since these cytoplasmic particles rearrange, it was hypothesized that they might be actively transported. This was intriguing because other unconventional myosins have been suggested as motors for transport of cytoplasmic components such as vesicles along actin filaments. To test this idea, a monoclonal antibody specific for the 95F myosin was labeled with rhodamine and injected into the syncytial embryo at low concentration. The antibody did not interfere with normal development at this low concentration, but it labeled the same structures as those seen in fixed preparations. This allowed visualization of the distribution of 95F myosin in live embryos. Changes in 95F myosin distribution *in vivo* were followed using time-lapse optical sectioning, so that a 3-D volume of cytoplasm could be imaged over time. Particle motion was analyzed by determining the position of individual particles in successive frames of the movie. These coordinates were plotted to generate a representation of the path of each tracked particle through the cytoplasm. Many particles (29% of the total) moved along linear paths for extended time periods (Fig. 5). This linear motion was dependent on actin filaments and ATP, as expected for a myosin-based translocation along actin. Injections of a polyclonal anti-95F myosin antibody that inhibits 95F myosin's binding to actin also stop linear translocation, indicating that 95F myosin catalyzes the movement. These experiments were the first direct measurement of unconventional myosin-mediated cytoplasmic transport. They confirm that unconventional myosins can indeed catalyze transport, as had been postulated. This information is also the first functional information on myosins in this class. In fact, for most members of this ever-growing family, functional information is lacking.

Further experiments have demonstrated that inhibition of 95F myosin activity causes defects in syncytial blastoderm organization (Mermall and

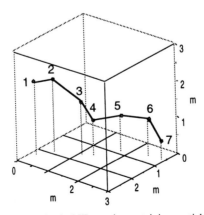

Fig. 5 One example of a linear path of a 95F myosin-containing particle through the cytoplasm is depicted. The three-dimensional coordinates of a particle in successive frames of a time-lapse movie are plotted. Approximately 2 min of time is represented.

Miller, 1995). In these experiments, high concentrations of anti-95F myosin polyclonal antibodies were injected into the embryo just prior to or during the syncytial blastoderm stage. Injected embryos were incubated for 30 min to allow several syncytial divisions to occur and then fixed and stained to reveal different cellular components and the injected antibody. In antibody-disrupted embryos, defects in the distribution of syncytial blastoderm nuclei were observed that were correlated in position to the regions where high concentrations of injected anti-95F myosin antibody were present. Nuclear spacing was irregular and nuclear morphology aberrant. Particularly striking in antibody-disrupted regions are patches of cortex containing some nuclei larger than normal and others that appear multilobed. This type of defect can result from failures in division and/or anaphase fusions. In addition, patches of cortex devoid of nuclei were observed. These regions can be generated when nuclei sink into the embryo's interior as a result of aberrant nuclear divisions. Since defects in cytoskeletal function are the likely cause of nuclear disorganizations such as this, the actin and microtubule cytoskeletons were examined in antibody-disrupted embryos. Both the organization of actin and the microtubules were affected in regions where antibody concentration was high, while few defects were observed in regions where antibody had not diffused or in embryos injected with control antibodies. Since the organization of actin and microtubules is interdependent, it was important to determine the sequence of events that led to the gross abnormalities in organization. Embryos that had been injected with anti-95F myosin antibody but which showed no gross abnormalities were examined. Defects would be in their earliest stages in these embryos; thus, the temporal order of events could be determined. In particular, embryos fixed during mitosis were chosen to visualize metaphase furrow positioning and structure relative to microtubule structures. It seemed likely that if 95F myosin affected blastoderm organization through effects on actin, those effects would primarily be on actin furrows, since 95F myosin colocalizes with actin in these structures. In the antibody-injected embryos, actin accumulates in the regions where furrows normally form. However, despite actin accumulation, metaphase furrows often do not invaginate or they are shallow and microtubules can cross over these defective furrows. This leads to division failures and anaphase collisions of neighboring nuclei (Fig. 6). The role of 95F myosin in the furrow might be direct, by having a role in furrow structure, or indirect, through delivery of other required components. These experiments demonstrate that the 95F unconventional myosin plays an important role in syncytial blastoderm organization. Screens are now being performed to isolate mutations in the gene that encodes this myosin heavy chain (B. Kozel and K. G. Miller, unpublished data). Such mutations will allow the study of the function of this fascinating molecule throughout development.

1. inhibition of 95F myosin   **2. aberrant furrow and cross over of microtubules**

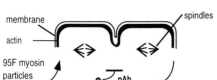

3. spindle collision   **4. nuclear defect**

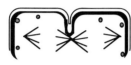

Fig. 6 Schematic diagram depicting how defects are generated in embryos injected with anti-95F myosin antibodies. (1) 95F myosin function is inhibited, so no particles can move to sites of furrow formation. (2) Furrows cannot invaginate if 95F myosin-transported particles are absent. Microtubules can cross from one spindle to adjacent ones. (3). This leads to mitotic aberrations, such as spindle collisions. (4) Misshapen, enlarged, and multipolar nuclei are formed as a result of mitotic abnormalities.

More detailed studies of a cap-specific protein have also been reported. The 13D2 protein colocalizes with actin in caps during interphase but does not rearrange into the transient membrane invaginations during mitosis (Miller *et al.*, 1989; Postner *et al.*, 1992). Instead, it remains associated with the surface membrane (Fig. 7). It was sugggested, based on its distribution, that this protein might play a role in generating or stabilizing the actin cap. Support for a role of this protein in generating the cap structure was

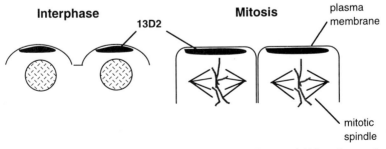

Fig. 7 The distribution of 13D2 protein during the cell cycle in syncytial blastoderm embryos is depicted. While it colocalizes with actin in interphase caps, this protein does not rearrange into the metaphase furrows.

obtained when the distribution of actin and ABPs was examined in embryos whose mothers were homozygous for the maternal-effect mutation, *sponge*. In *sponge* mutant syncytial blastoderm embryos, actin is not normally organized. Neither caps nor furrows are present (Postner *et al.*, 1992). This leads to defects in blastoderm organization similar to those described previously. The effects on actin-binding protein distribution were examined. Most actin-binding proteins were not properly organized, as might be expected. However, the 13D2 protein is organized into cap structures similar to those normally observed (Postner *et al.*, 1992). This was surprising and suggested that 13D2 protein might precede actin in the formation of caps, and not rely on actin itself, but on some other cue for its distribution. Actin would normally be recruited into the cap subsequent to 13D2 protein's localization there. In the mutant, the process of actin recruitment appears to be disrupted. These results suggested that the organization of the 13D2 protein into cap structures might be required for the subsequent recruitment of actin. In addition, 13D2 protein might be detecting the postulated "signal" from the centrosomes/microtubules that positions the cortical actin caps during interphase. Further studies on the nature of the 13D2 protein and its interactions could provide significant information on the coordination between actin and microtubule cytoskeletons.

III. Conclusions and Future Prospects

The studies on the actin cytoskeleton in the syncytial blastoderm are yielding insights into the specific processes peculiar to this cell type. It is clear that some aspects of actin structure are peculiar to this unusual cell; however, it is also clear that many of the mechanisms used to generate the observed cytoskeletal order are shared with other cell types.

During the syncytial stage, the actin and microtubule cytoskeletons cooperate to generate the observed order. As summarized previously, it has been demonstrated that actin and some actin-associated proteins play important roles during this period. In addition, a number of genes important for syncytial development have been identified, some of which specifically affect the actin cytoskeleton. Further detailed studies on all of the components of the actin cytoskeleton present during the syncytial stage through the approaches outlined above will undoubtedly reveal important information about the mechanisms of cytoskeletal organization and function.

From the studies already performed, it is clear that the early *Drosophila* embryo is a useful model system for studies of cytoskeletal function during development. The tools for such studies have been developed, but to date information generated using these tools is limited. Exploitation of this system is at an early stage. The potential to learn about control and coordination of cytoskeletal structure through the use of genetic and cell biological techniques in *Drosophila* is tremendous.

Appendix: Methods for *Drosophila* Embryo Fixation for Immunofluorescence

In this appendix several different methods of fixation that have been used for preserving antigens in early *Drosophila* embryos are summarized. Where appropriate, information is included on the specialized uses for which each method has been designed.

A. Standard (General Purpose) Fixation

This basic procedure has been used for a number of different antigens, some cytoskeletal and others nuclear. It is based on the procedure developed by Mitchison and Sedat (1983) and modified by Karr and Alberts (1986) and others (Miller *et al.,* 1989).

1. Embryos of appropriate ages are collected by timed egg lays on grape juice agar plates. Embryos are harvested by washing from grape juice agar plates into a mesh sieve (120 μM Nitex mesh) with 0.4% NaCl and 0.03% Triton X-100 (Triton–Salt), and dechorionated by incubating in 50% bleach (freshly diluted) for 1.5–2 min. After rinsing thoroughly with Triton–Salt solution in the sieve, the embryos (in the sieve) are blotted dry on a paper towel.

2. The dechorionated embryos (~200–300 μl) are transferred using a spatula to a glass (13 × 100 mm) or plastic (15 ml) screw-top tube containing 5 ml 7.5% formaldehyde in PEM (0.1 M PIPES, 2 mM EGTA, 1 mM MgSO$_4$, pH 6.9), shaken briefly to disperse the embryos, and 5 ml heptane (fixative concentration can be varied over a range from 4 to ~33% formaldehyde; see below) is added. The embryo/fixative mixture is vigorously shaken for 10–20 min (see below). Do not put too large a volume of embryos into the fix; poor, uneven fixation will be the result. The amount of fixative and the tube size can be scaled up if a large quantity of embryos is needed. After incubation, the phases are allowed to separate. The embryos should be at the interface between the layers. The fixative layer (lower) is drawn off using a Pasteur pipette.

3. The fixative is replaced with 90% MeOH and 50 mM EGTA, pH 8. The mixture is shaken vigorously for 2–5 min. The devitellinized embryos will sink to the bottom of the MeOH layer, where they can be collected using a Pasteur pipette.

4. The embryos are transferred to a fresh tube and rinsed three times with 90% MeOH and 50 mM EGTA. They are then rehydrated by rinsing three times with PBS or any other suitable buffer. Some workers have found a stepwise rehydration through 80% MeOH/PBS, 60% MeOH/PBS, 40% MeOH/PBS, and finally into PBS (with 0.02% NaN$_3$) preferable. In the hands of many users of the protocol, no difference is observed between stepwise and one-step rehydration.

Notes: The exact time required for fixation and the concentration of formaldehyde will vary with conditions and antigens, so these parameters are best determined empirically, using the guidelines above for a starting point. If the embryos are too soft (fragment upon subsequent manipulation) or degenerate quickly during storage, increase fixation time or concentration. If the yield of devitellinized embryos is low, then the embryos may be overfixed. Try decreasing time or fixative concentration. Some antigenic determinants may be masked or destroyed by too long of a fixation time. Conversely, some localizations may not be preserved with too short of a fixation time or at low fixative concentration.

If actin's distribution is to be visualized using fluorescent phalloidin, MeOH cannot be used, since dehydration destroys actin's ability to interact with phalloidin. In addition, some antigens may be sensitive to MeOH treatment or may be extracted with MeOH. Also, embryos that have been injected will not devitellinize in MeOH. Manual devitellinization is necessary in these circumstances. If embryos are to be hand devitellinized, at step 3 the fix is replaced with PEM or PBS (no MeOH treatment). The manual devitellinization can be performed immediately or the embryos can be stored for up to 24 hr if refrigerated.

To manually devitellinize, the embryos are transferred from the interface of the heptane/PBS (or other buffer) to a piece of Scotch tape on a glass slide (a few at a time if many embryos are to be devitellinized), taking as little of the heptane along as possible. The heptane is allowed to evaporate and the embryos are allowed to settle. As much of the buffer as possible is removed. A drop of Triton–Salt is added to cover the embryos. Using a Pasteur pipette, the embryos are transferred in a small amount of Triton–Salt to a piece of double-stick tape adjacent on the slide. Again, as much liquid as possible is removed. The embryos should adhere to the tape. A drop of water is added to cover all the transferred embryos. Under a dissecting microscope, each embryo is gently touched with a forceps or blunt tungsten needle to release it from its vitelline membrane. They should pop out easily with minimal pressure, with the vitelline membrane adhering firmly to the double-stick tape. The devitellinized embryos are transferred to a microfuge tube containing PBS and 0.05% Triton X-100. If, during hand devitellinization, the embryos seem too soft, they can be postfixed in 4–8% formaldehyde and PBS for 15 min or the initial fixation time or fixative concentration can be increased. If manual devitellinization is difficult (embryos do not easily pop out of a brittle vitelline membrane), then they may be overfixed.

A less labor-intensive method to manually devitellinize has been successfully used (Theurkauf, 1992). After replacement of the fix with PBS or PEM, fixed embryos are transferred from the aqueous:heptane interface to a frosted slide and covered with a coverslip. The embryos are rolled between the slide and coverslip with pressure exerted. This removes the vitelline membrane from a large proportion of the embryos. The density of embryos, the amount of pressure applied, and the total time of rolling are all important variables to obtain efficient devitellinization and minimize embryo breakage. The embryos are washed from the slide into a tube in PBS or PEM and heptane is added. After shaking, the devitelinized embryos sink to the bottom. The devitellinized embryos are transferred to a new tube and rinsed several times with PBS and 0.05% Triton X-100.

The buffers used for fixation and for incubation/extraction after fixation can affect the quality of the staining and the retention and preservation of antigens of interest. In the above protocol, the buffers used for fixation have no detergent, while those used after fixation and during staining (see below) often have low concentrations (0.05%) of detergent to minimize nonspecific interactions. Some workers have used buffers with high (1%) concentrations of detergent (Triton X-100 or Tween 20) for 1 or 2 hr. to extract embryos after fixation. This procedure may allow better penetration of antibodies and/or better visualization of structures of interest. However, it may result in the loss of important features.

B. Modifications for Fixation of Microtubules

Microtubules are not generally well fixed by the formaldehyde procedure previously described. If taxol (1–2.5 μM) is added to the (4–8%) formaldehyde fixation, spindles and other microtubule structures are preserved (Karr and Alberts, 1986). However, some artifactual polymerization is induced, particularly during interphase (Kellogg *et al.,* 1988). Alternatively, if very high concentrations of formaldehyde (33% formaldehyde, 50 mM EGTA, pH 6.6, or 33% formaldehyde in 1× PBS, i.e., 37% formaldehyde with 1/10 vol of 10× PBS) are used, both actin and microtubules can be preserved in the same specimen (Theurkauf, 1992). However, these conditions may not be suitable for the preservation of other antigens of interest.

To preserve microtubules in as close to *in vivo* distribution as possible [as reflected by *in vivo* observations of embryos injected with fluorescent tubulin (Kellogg *et al.*, 1988)], embryos are fixed by shaking the presence of equal parts of heptane:97% MeOH and 15 mM EGTA for 10 min. Embryos are fixed and devitellinized in this one step and will sink to the bottom of the tube where they can be recovered using a Pasteur pipette and transferred to a new tube. After rinsing three times with MeOH/EGTA, the embryos are incubated for several hours (up to overnight) on ice. The embryos are rehydrated gradually as in step 4 previously described. This method does not preserve actin.

C. Modification for Denatured Antigens

Sometimes antibodies made using denatured antigens do not work well for immunolocalization. A modification of the fixation method that uses heat to fix (and presumably denature proteins) has worked well for some such antibodies. Embryos fixed in this manner should be used immediately, since they degenerate upon storage.

1. Embryos are harvested and dechorionated as described previously.
2. Triton–Salt (10 ml) solution is put into a large glass tube, loosely covered, and heated in a boiling water bath until equilibrated. The tube is removed from the bath and immediately a small scoop of embryos is added to the hot liquid while vortexing. As soon as the embryos are well dispersed, 20 ml of ice-cold Triton–Salt is added while vortexing. The tube is then immersed in a water–ice bath.
3. The embryos are allowed to settle (a few minutes) and the Triton–Salt solution is removed. Ten milliliters of 95% MeOH, 25 mM EGTA, and 10 ml heptane is added and the tube is shaken vigorously for 30 sec.
4. The MeOH/heptane mixture is removed and the embryos are rinsed 3× with 95% MeOH and 25 mM EGTA. The embryos are allowed to incubate on ice for up to several hours in the MeOH/EGTA solution on ice.
5. The embryos are rehydrated through several steps of decreasing MeOH concentration as described previously.

D. Staining of Embryos

Embryos fixed by these methods can be stained using fluorescently labeled or enzyme-linked secondary antibodies, with buffers suitable for the antibodies of interest. A general procedure follows, but this basic protocol can be varied to suit individual needs. The instructions are intentionally vague, since the exact conditions will vary depending on the antigens/antibodies of interest.

1. Embryos are incubated for 1–15 hr in 1–4% BSA in PBS, 0.05% NP-40, or Tween 20 to block nonspecific sites. Other blocking solutions (5% powdered milk or 5% serum) can also be used.
2. An appropriate volume of embryos is transferred to a 0.5-ml microfuge tube (see note). Antibodies are added at an appropriate dilution (empirically determined) in 0.5 ml of the blocking solution and embryos are incubated while being continuously mixed on a rotating wheel. Incubation times range from 1 hr (room temperature) to overnight (4°C), depending on the antigen and antibody of interest. Conditions that provide the best signal and lowest background are often quite different for different antibodies/ antigens.
3. After incubation, embryos are rinsed three times with blocking solution and incubated for 1 hr. After another rinse, secondary antibody is added at an appropriate dilution. Again, incubation times will depend on the experimental particulars, but are in the same range (1–15 hr) as that for primary antibody incubations.
4. After incubation, the embryos are rinsed and incubated as in step 3. The embryos can then be stained with DAPI if desired. Embryos are incubated in 1 μg/ml DAPI in PBS (no detergent) for 2 min and then rinsed three times with PBS. Incubation in PBS for greater than 1 hr with one or two additional rinses is recommended if fluorescence is being used for detection, since any remaining detergent will autofluorescence and increase background.

Note: A 96-well microtiter dish can be used for incubations when many different samples are being assayed. In this case, an amount of embryos sufficient to just cover the bottom of the well is used. No more than 200 μl of solution is used per well. Staining works best if the plates are incubated on a microtiter plate mixer.

Acknowledgments

I wish to thank the members of my lab for their patience while I worked on this article, instead of other things that I should have been spending time on and for reading and making useful suggestions on the manuscript. I thank Bill Sullivan for discussions and permission to use unpublished information and Dan Kiehart for permission to report unpublished experiments. I thank Chris Field for discussion about anillin and its role in the syncytial blastoderm. I especially want to thank Dave Capco for not screaming when this was late, as usual.

References

Baker, J., Theurkauf, W. E., and Schubiger, G. (1993). Dynamic changes in microtubule configuration correlate with nuclear migration in the preblastoderm *Drosophila* embryo. *J. Cell Biol.* **122**, 113–121.

Cant, K., Knowles, B. A., Mooseker, M. S., and Cooley, L. (1994). *Drosophila singed,* a fascin homolog, is required for actin bundle formation during oogenesis and bristle formation. *J. Cell Biol.* **125,** 369–380.

Chou, T., Noll, E., and Perrimon, N. (1993). Autosomal P[*ovo D1*] dominant female-sterile insertions in *Drosophila* and their use in generating germ-line chimeras. *Development* **119,** 1359–1369.

Cooley, L., Verheyen, E., and Ayers, K. (1992). *chickadee* encodes a profilin required for intercellular cytoplasm transport during *Drosophila* oogenesis. *Cell* **69,** 173–184.

Deng H., Lee, J. K., Goldstein, L. S., and Branton, D. (1995). *Drosophila* development requires spectrin network formation. *J. Cell Biol.* **128,** 71–79.

Edgar, B. A., Odell, G. M., and Schubiger, G. (1987). Cytoarchitecture and the patterning of *fushi tarazu* expression in the *Drosophila* blastoderm. *Genes Dev.* **1,** 1126–1132.

Fehon, R. G., Dawson, I. A., and Artavanis-Tsakonas, S. (1994). A *Drosophila* homologue of membrane-skeleton protein 4.1 is associated with septate junctions and is encoded by the *coracle* gene. *Development* **120,** 545–557.

Field, C. M., and Alberts, B. A. (1995). Anillin, a contractile ring protein that cycles from the nucleus to the cell cortex. *J. Cell Biol.,* **131,** in press.

Foe, V. E., and Alberts, B. M. (1983). Studies of the nuclear and cytoplasmic behavior during the five mitotic cycles that precede gastrulation in *Drosophila* embryogenesis. *J. Cell Sci.* **61,** 31–70.

Foe, V. E., Odell, and Edgar, B. A. (1993). Mitosis and morphogenesis in the *Drosophila* embryo: Point and counterpoint. *In* "The Development of *Drosophila melanogaster*" (M. Bate and A. Martinez-Arias, eds.), Vol. I, pp. 149–300. Cold Spring Harbor Laboratory Press, Cold Spring Harbor, New York.

Fogarty, P., Kalpin, R. F., and Sullivan, W. (1994). The *Drosophila* maternal effect mutation *grapes* causes a metaphase arrest at nuclear cycle 13. *Development* **120,** 2131–2142.

Fyrberg, E. (1995). Genetic dissection of *Drosophila* cytoskeletal functions. *Adv. Mol. Cell. Biol.* **10,** in press.

Fyrberg, E., Kelly, M., Ball, E., Fyrberg, C., and M. C. Reedy (1990). Molecular genetics of *Drosophila* Alpha-actinin: Mutant alleles disrupt Z disc integrity and muscle insertions. *J. Cell Biol.* **110,** 1999–2011.

Goldschmidt-Clermont, P. J., Kim, J. W., Machesky, L. M., Rhee, S. G., and Pollard, T. D. (1991). Regulation of phospholipase C-γ1 by profilin and tyrosine phosphorylation. *Science* **251,** 1231–1233.

Goldschmidt-Clermont, P. J., Machesky, L. M., Baldassare, J. J., and Pollard, T. D. (1990). The actin-binding protein profilin binds to PIP2 and inhibits its hydrolysis by phospholipase C. *Science* **247,** 1575–1578.

Hafen, E., Kuroiwa, A., and Gehring, W. J. (1984). Spatial distribution of transcripts from the segmentation gene *fushi tarazu* during *Drosophila* embryonic development. *Cell* **37,** 833–841.

Hatanaka, K., and Okada, M. (1991). Retarded nuclear migration in *Drosophila* embryos with aberrant F-actin reorganization caused by maternal mutations and by cytochalasin. *Development* **111,** 909–920.

Karess, R. E., Chang, X-J., Edwards, K. A., Kulkarni, S., Aguilera, I., and Kiehart, D. P. (1991). The regulatory light chain of nonmuscle myosin is encoded by *spaghetti-squash,* a gene required for cytokinesis in *Drosophila. Cell* **65,** 1177–1189.

Karr, T. L., and Alberts, B. M. (1986). Organization of the cytoskeleton in early *Drosophila* embryos. *J. Cell Biol.* **102,** 1494–1509.

Kellerman, K. A., and Miller, K. G. (1992). An unconventional myosin heavy chain gene from *Drosophila melanogaster. J. Cell Biol.* **119,** 823–834.

Kellogg, D. R., Field, C. M., and Alberts, B. A. (1989). Identification of microtubule associated proteins in the centrosome, spindle, and kinetochore of the early *Drosophila* embryo. *J. Cell Biol.* **109**, 2977–2991.

Kellogg, D. R., Mitchison, T. L., and Alberts, B. M. (1988). Behavior of microtubules and actin filaments in living *Drosophila* embryos. *Development* **103**, 675–686.

Kiehart, D. P., and Feghali, R. (1986). Cytoplasmic myosin from *Drosophila melanogaster. J. Cell Biol.* **103**, 1517–1525.

Kiehart, D. P., Ketchum, A., Young, P., Lutz, D., Alfenito, M. R., Cang, X.-J., Awobuluyi, M., Pesacreta, T. C., Inoue, S., Stewart, C. T., and Chen, T.-L. (1990). Contractile proteins in *Drosophila* development. *Ann. N.Y. Acad. Sci.* **582**, 233–251.

Lee, J. K., Coyne, R. S., Dubreuil, R. R., Goldstein, L. S. B., and Branton, D. (1993). Cell shape and interaction defects in α-spectrin mutants of *Drosophila melanogaster. J. Cell Biol.* **123**, 1797–1810.

Mahajan-Miklos, S., and Cooley, L. (1994). The villin-like protein encoded by the *Drosophila quail* gene is required for actin bundle assembly during oogenesis. *Cell* **78**, 291–301.

Mermall, V., and Miller, K. G. (1995). The 95F unconventional myosin is required for organization of the *Drosophila* syncytial blastoderm. *J. Cell Biol.* **129**, 1575–1588.

Mermall, V., McNally, J. G., and Miller, K. G. (1994). Transport of cytoplasmic particles catalysed by an unconventional myosin in living *Drosophila* embryos. *Nature* **369**, 560–562.

Merrill, P. T., Sweenton, D., and Wieschaus, E. (1988). Requirements for autosomal gene activity during precellular stages of *Drosophila melanogaster. Development* **104**, 495–510.

Miller, K. G., Field, C. M., and Alberts, B. M. (1989). Actin-binding proteins from *Drosophila* embryos: A complex network of interacting proteins detected by F-actin affinity chromatography. *J. Cell Biol.* **109**, 2963–2975.

Mitchison, T. J., and Sedat, J. (1983). Localization of antigenic determinants in whole *Drosophila* embryos. *Dev. Biol.* **99**, 261–264.

Okada, M. (1982). Loss of the ability to form pole cells in *Drosophila* embryos with artificially delayed nuclear arrival at the posterior pole. *In* "Embryonic Development, Part A: Genetic Aspects" (M. M. Burger and G. Weber, eds.), pp. 363–375. A. R. Liss, New York.

Pantaloni, D., and Carlier, M. F. (1993). How profilin promotes actin filament assembly in the presence of thymosin beta 4. *Cell* **75**, 1007–1014.

Pesacreta, T. C., Byers, T. J., Dubreuil, R., Kiehart, D. P., and Branton, D. (1989). *Drosophila* spectrin: The membrane skeleton during embryogenesis. *J. Cell Biol.* **108**, 1697–1709.

Pollard, T. D., and Cooper, J. A. (1986). Actin and actin-binding proteins. A critical evaluation of mechanisms and functions. *Annu. Rev. Biochem.* **55**, 987–1035.

Postner, M. A., Miller, K. G., and Wieschaus, E. F. (1992). Maternal effect mutations of the *sponge* locus affect actin cytoskeletal rearrangements in *Drosophila melanogaster* embryos. *J. Cell Biol.* **119**, 1205–1218.

Raff, J. W., and Glover, D. M. (1989). Centrosomes, and not nuclei, initiate pole cell formation in *Drosophila* embryos. *Cell* **57**, 611–619.

Roulier, E. M., Fyrberg, C., and Fyrberg, E. (1992). Perturbations of Drosophila α-actinin cause muscle paralysis, weakness and atrophy but do not confer obvious nonmuscle phenotypes. *J. Cell Biol.* **116**, 911–922.

Schejter, E. D., and Wieschaus, E. (1993a). *bottleneck* acts as a regulator of the microfilament network governing cellularization of the *Drosophila* embryo. *Cell* **75**, 373–385.

Schejter, E. D., and Wieschaus, E. (1993b). Functional elements of the cytoskeleton in the early *Drosophila* embryo. *Annu. Rev. Cell Biol.* **9**, 67–99.

Simcox, A. A., and Sang, J. H. (1983). When does determination occur in *Drosophila* embryos? *Dev. Biol.* **97**, 212–215.

Simpson, L., and Wieschaus, E. (1990). Zygotic activity of the *nullo* locus is required to stabilize the actin–myosin network during cellularization in *Drosophila. Development* **110**, 851–863.

St. Johnston, D., Beuchle, D., and Nusslein-Volhard, C. (1991). *staufen*, a gene required to localize maternal RNAs in the *Drosophila* egg. *Cell* **66**, 51–63.

Sullivan, W., Fogarty, P., and Theurkauf, W. (1993). Mutations affecting the cytoskeletal organization of syncytial *Drosophila* embryos. *Development* **118**, 1245–1254.

Sullivan, W., Minden, J. S., and Alberts, B. M. (1990). *daughterless-abo-like*, a *Drosophila* maternal-effect mutation that exhibits abnormal centrosome separation during the late blastoderm divisions. *Development* **110**, 311–323.

Schweiguth, F., Lepesant, J. A., and Vincent, A. (1990). The *serendipity alpha* gene encodes a membrane associated protein required for the cellularisation of the *Drosophila* embryo. *Genes Dev.* **4**, 922–931.

Theurkauf, W. E. (1992). Behavior of structurally divergent alpha-tubulin isotypes during *Drosophila* embryogenesis: Evidence for post-translational regulation of isoform abundance. *Dev. Biol.* **154**, 204–217.

Thomas, G. H., and Kiehart, D. P. (1994). β_{heavy}-Spectrin has a restricted tissue and subcellular distribution during *Drosophila* embryogenesis. *Development* **120**, 2039–2050.

Verheyen, E. M., and Cooley, L. (1994). Profilin mutations disrupt multiple actin-dependent processes during *Drosophila* development. *Development* **120**, 717–728.

von Dassow, G., and Schubiger, G. (1994). How an actin network might cause fountain streaming and nuclear migration in the syncytial *Drosophila* embryo. *J. Cell Biol.* **127**, 1637–1653.

Warn, R. M., Flegg, L., and Warn, A. (1987). An investigation of microtubule organization and functions in living *Drosophila* embryos by injection of a fluorescently labeled antibody against tyrosinated α-tubulin. *J. Cell Biol.* **105**, 1721–1730.

Warn, R. M., Magrath, R., and Webb, S. (1984). Distribution of F-actin during cleavage of the *Drosophila* syncytial blastoderm. *J. Cell Biol.* **98**, 156–162.

Warn, R. M., and Warn, A. (1986). Microtubule arrays present during the syncytial and cellular blastoderm stages of the early *Drosophila* embryo. *Exp. Cell Res.* **163**, 201–210.

Whitfield, W. G., Millar, S. E., Saumweber, H., Frasch, M., and Glover, D. M. (1988). Cloning of a gene encoding an antigen associated with the centrosome in *Drosophila. J. Cell Sci.* **89**, 467–480.

Xu, T., and Rubin, G. M. (1993). Analysis of genetic mosaics in developing and adult *Drosophila* tissues. *Development* **117**, 1223–1237.

Yasuda, G. K., Baker, J., and Schubiger, G. (1991). Independent roles of centrosomes and DNA in organizing the *Drosophila* cytoskeleton. *Development* **111**, 379–391.

Young, P. E., Pesacreta, T. C., and Kiehart, D. P. (1991). Dynamic changes in the distribution of cytoplasmic myosin during *Drosophila* embryogenesis. *Development* **111**, 1–14.

Young, P. E., Richman, A. M., Ketchum, A. S., and Kiehart, D. P. (1993). Morphogenesis in *Drosophila* requires nonmuscle myosin heavy chain function. *Genes Dev.* **7**, 29–41.

Zalokar, M., and Erk, I. (1976). Division and migration of nuclei during early embryogenesis of *Drosophila melanogaster. J. Microbiol. Cell.* **25**, 97–106.

7

Role of the Cytoskeleton in the Generation of Spatial Patterns in *Tubifex* Eggs

Takashi Shimizu
Division of Biological Sciences
Graduate School of Science
Hokkaido University
Sapporo 060, Japan

I. Introduction

One of the central issues in developmental biology is how the spatial pattern, which comprises the animal body, is established during embryogenesis. This process is accomplished through a complex series of developmentally coordinated events, which include cytoplasmic reorganization,

Current Topics in Developmental Biology, Vol. 31

cleavages, cell interactions, and morphogenetic movements. These events, if not always, give rise to generation of asymmetry in the egg's or embryo's organization, which is crucial in specifying embryonic axes (Wall, 1990; Gurdon, 1992). Spatial aspects of these events are believed to be defined by positional cues which lie in the single egg cell, even if visible egg organization does not foreshadow eventual embryonic axes. Therefore, one of the challenges in developmental studies is to elucidate how positional information residing in the egg determines the spatial organization of the embryo.

The objective of this chapter is to examine the role of the cytoskeleton in generating spatial patterns in early embryos of the freshwater oligochaete *Tubifex*. I will first focus on the cortical actin cytoskeleton which generates bipolarity in cytoplasmic organization as well as in the cortex. Then I will focus on the role of subcortical actin cytoskeleton in relation to specification of the dorsoventral polarity. Finally, I will discuss factors determining inequality of cleavage divisions with special reference to the anteroposterior polarity in the *Tubifex* embryo.

II. Early Development in *Tubifex*

A. Maturation Divisions and Cleavages

A brief review of early development in *Tubifex* is presented as background for the observations described below (for details, see Shimizu, 1982a). *Tubifex* eggs are oviposited at metaphase of the first meiosis and are believed to be fertilized during cocoon deposition by spermatozoa which are released from spermatozeugmata (i.e., sperm packet) stored in the spermatheca. Following sperm penetration (or activation), the vitelline membrane is separated from the egg surface. Activated *Tubifex* eggs extrude polar bodies twice, and then enter the first cell cycle. During this period of time, eggs undergo three episodes of surface contractile activity. The first two are observed during polar body formation and give rise to deformation movement (Figs. 1c and 1e); the last activity brings about ooplasmic segregation of pole plasms which takes place following the second polar body formation (Fig. 1g; Shimizu, 1982b, 1984, 1986).

The early development of *Tubifex* consists of a stereotyped sequence of cell divisions (Penners, 1992; Shimizu, 1982a). The first cleavage of the *Tubifex* egg is unequal and meridional, and produces a smaller AB cell and a larger CD cell (Fig. 1h). The second cleavage is also meridional and yields cells A, B, C, and D: the CD cell divides into a smaller C cell and a larger D cell, while the AB cell separates into cells A and

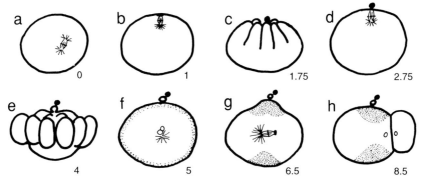

Fig. 1 Schematic representations of *Tubifex* eggs from oviposition through two-cell stage. Numerals in the lower right of each figure indicate time (hr) after oviposition. Except for c and e, meridional sections through the egg axis are presented. The animal pole is up. Filled and open circles at the animal pole are the first and second polar body, respectively. (a) Shortly after oviposition. The egg rounds up on activation; however, it becomes flaccid within 5 min. The metaphase I spindle is located apart from the egg surface. (b) The second rounding up in preparation for the first polar body formation. At this stage, the spindle is located at the animal pole. (c) The first deformation movement accompanying the first polar body formation. (d) Metaphase of the second meiosis. The metaphase II spindle is formed at the animal pole and is localized there. (e) The second deformation accompanying the second polar body formation. (f) Thirty minutes after the second meiosis. The interphase nucleus with an early aster is located at the egg's center. (g) Late metaphase of the first cleavage. Pole plasms (dotted areas) are bipolarly localized at both poles of the egg. (h) Two-cell stage. The first cleavage produces a smaller cell AB (right) and a larger cell CD (left). The nuclei (open circles) are located near the cleavage membrane which is parallel to the egg axis. The pole plasms are inherited by the CD cell.

B of various sizes. From the third cleavage on, the quadrants A–D repeat unequal divisions three or four times, producing micromeres at the animal side (which corresponds to the future dorsal side) and macromeres at the vegetal side (corresponding to the future ventral side of the embryo). The quadrants A–C then divide equally at the sixth cleavage, followed by the D quadrant at the seventh cleavage; the resulting yolky macromeres are endodermal cells, and these repeat equal divisions thereafter. During cleavages, the pole plasms, which have accumulated at both poles of the egg shortly before the first cleavage, are inherited by the D lineage cells (see Fig. 17). They are finally partitioned into the second (2d) and the fourth (4d) micromeres, both of which exclusively possess the ability to generate germ bands of the embryo. The pole plasms appear to be involved not only in the developmental fate specification of blastomeres but also in controlling cleavage cycles (Shimizu, 1993, 1994, 1995b).

B. Polarity

From oviposition through the first polar body formation, *Tubifex* eggs are ellipsoidal in shape with the minor axis in register with the egg axis (see Fig. 3c). During metaphase of the second meiosis, however, they gradually assume a shape of oblate spheroid with the minor axis in register with the egg axis (see Fig. 6a). Thus, *Tubifex* eggs from the second meiosis through the beginning of the first cleavage are radially symmetric about the egg axis. This radial symmetry is broken when first cleavage furrows emerge on the equatorial surface (see Section V,D).

Two-cell embryos are bilaterally symmetric if viewed either from the animal (or vegetal) pole or the side of the embryo. At this stage, the anteroposterior polarity becomes recognizable, since the AB cell is located at the future anterior side, and the CD cell at the future posterior side of the embryo. The anteroposterior polarity is further specified as the D lineage micromeres (cells 1d–4d) are formed at the anterior side of the D quadrant (see Figs. 17c and 17d). On the other hand, the dorsoventral polarity is specified along the egg axis when pole plasms are unified at the animal side of the D quadrant (see Fig. 17c). This polarity is further specified as the D lineage micromeres are produced at the future dorsal side of the embryo (see Section VII,B).

III. Reorganization of Cortical F-Actin upon Activation

The surface of mature eggs of *Tubifex* exhibits numerous microvilli. Mature eggs stained with rhodamine–phalloidin are found to be covered with a dense sheet of fluorescent tiny dots which probably correspond to individual microvilli. As Figure 2b shows, fluorescent tiny dots are distributed uniformly over the egg except for a circular region (20 μm in diameter) in which fluorescent dots are lacking. Such a "dark" circular zone is always located at the point where the minor and major axes of an ellipse cross (Fig. 2a).

Within 30 sec following oviposition, eggs round up and assume the shape of ellipsoid. This rounding up is accompanied by elongation of microvilli and the appearance of reticular organization of cortical actin (Fig. 2a). Probably, elongated microvilli are arranged in such a reticular pattern. As Figure 2c shows, eggs which have rounded up exhibit a relatively darker circular zone at the cortex. The rounding up of eggs lasts for about 5 min, then eggs relax and become flaccid again. At this time, a large part of microvilli is found to have separated from the egg surface, though some short microvilli are retained on the egg surface (Shimizu, 1976a). The reticular organization of cortical actin is no longer detected. Instead, egg

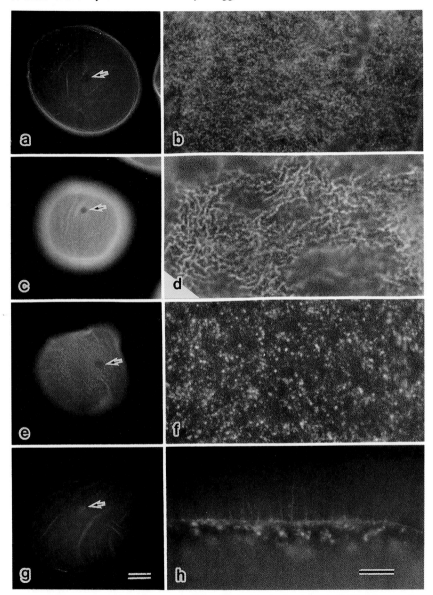

Fig. 2 Fluorescence micrographs of eggs stained with rhodamine–phalloidin. Arrows in a and c point to circular regions deficient in F-actin. The arrow in g points to the animal pole to which the metaphase I spindle has attached. a and b, ovisac eggs; c and d, eggs shortly after fertilization; e and f, eggs at 10 min; g and h, eggs at 30 min. Scale bars: g = 100 μm for a, c, e, and g; h = 10 μm for b, d, f, and h.

cortex contains small aggregates of F-actin (Fig. 2f). During the following 15 min, these aggregates become smaller and decrease in number. Concurrently with these changes in cortical F-actin, a circular region which has been darker than the remaining cortex also becomes obscure (Fig. 2e). After another 15 min, however, a fluorescent spot which is brighter than any other regions appears at the surface (Fig. 2g). This spot corresponds to the first meiotic spindle, which has attached to the cortex (see Fig. 1b). At this time, small aggregates of F-actin are still observed near the surface (Fig. 2h). Similar aggregates of F-actin are also seen during the first polar body formation.

IV. Establishment of Bipolar Organization of Cortical F-Actin

At about 60 min after oviposition (or rounding up) eggs round up again in preparation for the first polar body formation. In eggs stained with rhodamine–phalloidin, cortical fluorescence becomes stronger than before, suggesting an increase of cortical F-actin. At the animal pole, however, a relatively darker, circular area is recognizable (Figs. 3a and 3b), so that eggs at this stage are similar to those shortly after activation. Except for the animal pole, there is no difference in cortical F-actin organization between the animal and vegetal hemisphere.

In eggs undergoing the first polar body formation, F-actin is concentrated at the neck of the polar body (i.e., contractile ring) and grooves which run radially from the animal pole toward the equator (Figs. 3a and 3c). Although the radial streaks of F-actin disappear as deformation movement ceases, the remnant of the contractile ring at the animal pole persists into the second meiosis (Fig. 3b). At the same time, aggregates of F-actin also disappear from the cortex throughout the egg except in the ring encircling the animal pole. As Figures 3e and 3f show, cortices isolated shortly after the first polar body formation exhibit short actin filaments which are scattered uniformly throughout the egg. Thus, at the beginning of the second meiosis, there is no polarity in cortical F-actin organization except that a large aggregate of F-actin is localized at the animal pole (Fig. 3e). During the following 30 min, however, cortical F-actin increases in amount and becomes organized bipolarly. In both hemispheres, actin filaments are distributed in a gradient increasing from the equator to the polar region of the egg (Figs. 3g and 3h). Similar bipolar organization of cortical F-actin is also generated in eggs whose cytoplasm is stratified into four layers by centrifugation (1100g), suggesting that cortical F-actin is organized independently of inner cytoplasm.

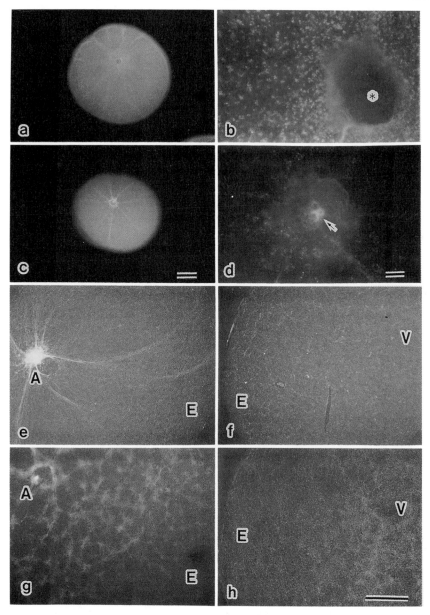

Fig. 3 (a–d) Fluorescence micrographs of eggs before (a and b) and after (c and d) the first polar body formation. Animal pole view. b and d are close ups of the animal pole. The asterisk in b indicates the site to which the first meiotic spindle attaches. The arrow in d points to a contractile ring at the base of the first polar body. (e–h) Distribution of F-actin in isolated cortices. Cortices were prepared shortly after (e and f) and 60 min (g and h) after the first polar body formation and labeled with rhodamine–phalloidin. A, animal pole; E, equator; V, vegetal pole. Scale bars: c = 100 μm for a and c; d = 10 μm for b and d; h = 100 μm for e–h.

As described below, this bipolar organization of cortical F-actin (see Fig. 8b) plays a key role in polar body formation, egg shape change, ooplasmic segregation, and furrow formation in the first cleavage.

V. Developmental Role of Cortical Actin Cytoskeleton

A. Polar Body Formation

About 60 min after the establishment of bipolar organization of cortical F-actin, *Tubifex* eggs begin the second polar body formation. This is a process by which a tiny cell (polar body) of 20 μm in diameter is extruded from a 500-μm egg. This extremely unequal division is brought about via localization of the second meiotic apparatus at the animal pole (see Figs. 1b and 1d). The meiotic spindle orients itself perpendicular to the animal pole surface, and its peripheral pole is firmly attached to the cortex (see Fig. 4a; Shimizu, 1981b). This attachment is brought about by the microtubule-organizing activity of the animal pole cortex, which is composed of F-actin and γ-tubulin (Shimizu, 1981b, 1983, 1990, 1995c). Apparently the attachment of the meiotic apparatus to the surface guarantees the localization of the apparatus at the animal pole.

Polar body formation consists of two steps (Longo, 1972; Shimizu, 1981c): (1) formation of a cytoplasmic bulge for the polar body (during anaphase), and (2) the subsequent development of a cleavage furrow at the base of this bulge (during telophase). These two steps are distinct from each other in their susceptibility to cytoskeleton inhibitors. In cytochalasin-treated eggs, a cytoplasmic bulge does form at the animal pole, although it is much lower yet wider in diameter at the base than the normal bulge (Shimizu, 1981c). However, a cleavage furrow at its base fails to develop in cytochalasin-treated eggs. Similarly, in eggs in which microtubules are depolymerized by colchicine or nocodazole, or the meiotic apparatus is removed from the animal pole, bulge formation occurs normally, but the cleavage furrow fails to form (Shimizu, 1981c, 1990). These results suggest that the operation of bulge formation and furrowing depends on different regulatory mechanisms during polar body formation.

Fig. 4 Fluorescence micrographs showing the distribution of cortical F-actin during the second polar body formation. (a, c, and e) Meridional sections of the animal pole; (b, d, and f) isolated cortices. (a and b) Early anaphase. The arrow points to the thick layer of F-actin at the animal pole surface. (c and d) Mid-anaphase. (e and f) Early telophase. Arrows in c and e point to the top of the cytoplasmic bulge. Scale bar: e = 20 μm for a–f. (From Shimizu, 1990 with permission.)

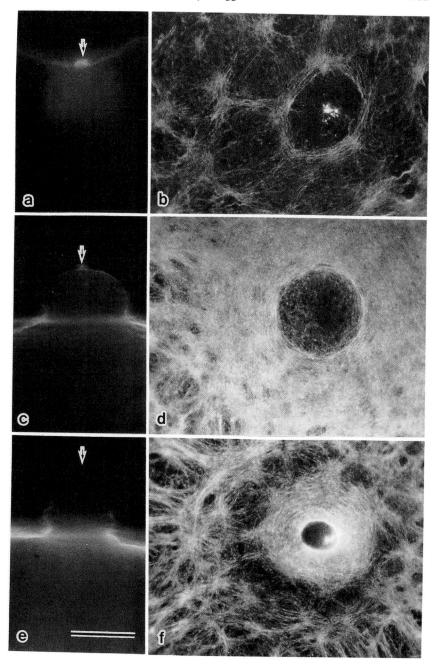

As mentioned previously, the first and second steps of polar body formation in *Tubifex* seem to differ in their susceptibility to cytochalsins. However, an important point regarding the mechanism that underlies polar body formation is that both steps are affected by cytochalasin treatments. This suggests that not only furrow formation, but also bulge formation requires an intact microfilament system if these events are to be accomplished.

1. Reorganization of Cortical F-Actin

To gain an insight into the roles of microfilaments in polar body formation, I examined the actin organization in bisected eggs and isolated cortices stained with rhodamine–phalloidin (Shimizu, 1990). At an early stage of bulge formation, F-actin is distributed differetially in the cortex of the cytoplasmic bulge; it is concentrated at the distal and proximal regions of the bulge (Fig. 4c). A fluorescent spot can be seen at the top of the bulge stained with rhodamine–phalloidin (Figs. 4a and 4c). Cortices isolated at this stage reveal bright bundles of fluorescence that run circumferentially around the polar fluorescent spot (Fig. 4b). This ring of bundles is connected, at five or six points, to the adjacent network of bright bundles of fluorescence. As the cytoplasmic bulge grows, F-actin becomes localized at the base of the bulge. Close examinations of isolated cortices at this stage show that the bright bundles that run circumferentially increase in number and form a circular, dense sheet of F-actin around the base of the bulge (Fig. 4d). At the same time, the fluorescent spot becomes undetected at the bulge top. Instead, fine fluorescent lines appear in the dark zone (which corresponds to the bulge) that is surrounded by the circumferential sheet of bright bundles. Furthermore, these fine lines are distributed as a gradient increasing from the distal to the proximal region of the bulge (Fig. 4d). At the conclusion of bulge formation during the early stage of furrow formation, the fine fluorescent lines in the bulge are very few. However, the circular region that corresponds to the base of the bulge exhibits intense fluorescence (Figs. 4e and 4f). At a later stage of furrow formation (late telophase), a bright ring of fluorescent bundles is located at the base of the bulge and is surrounded by another ring of bundles from which a number of further bundles radiate toward the equator of the egg.

These localization patterns of F-actin during bulge formation are consistent with electron microscopic observations of eggs treated with heavy meromyosin (Shimizu, 1983). The distribution and organization of cortical actin during bulge formation are summarized diagrammatically in Fig. 5. The results suggest that three major changes occur in the organization of cortical actin during bulge formation: (1) the formation of a dense sheet of actin filaments around the base of the bulge, (2) the disappearance of

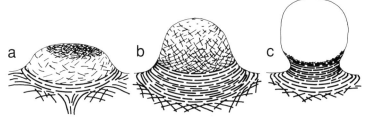

Fig. 5 Diagrammatic summary of the organization of actin bundles in the cortex at the animal pole during formation of the cytoplasmic bulge for the polar body. a, Early anaphase; b, mid-anaphase; c, early telophase. (From Shimizu, 1990 with permission.)

the thick layer of actin filaments from the distal region of the bulge, and (3) the disappearance of small bundles of actin filaments from the sides of the bulge around the time of the onset of furrowing. I propose that these changes result from reorganization of actin filaments directed toward the base of the bulge. The dense sheet may be generated through circumferential alignment of actin filaments around the preexisting ring of actin bundles. The disappearance of small bundles also seems to result from their redistribution in the cortex which gives rise to their accumulation at the base of the bulge. When the polarized distribution of small bundles is taken into account, it is tempting to assume that the actin bundles may move toward the bulge base where the density of cortical actin is higher than that in any other region of the bulge. On the basis of the observation that small bundles increase in number at the bulge wall as the cortical actin in the bulge top decreases, it is conceivable that these actin bundles originate from the thick cortical layer which previously overlaid the metaphase II spindle.

2. Possible Mechanisms

Polar body formation is regarded as a process in which an animal pole bulge protrudes and grows in height but not in width. Apparently this process is accompanied by uplift of the animal pole surface and increase in the bulge surface area. I propose that surface membrane of the bulge is supplied from the egg proper during reorganization of cortical F-actin. This reorganization may possibly be the result of a contraction of the cortex toward the animal pole; this cortical contraction may then cause movement of the surface in the same direction. If so, during bulge formation, the surface of the egg proper at least near the growing bulge moves toward and "accumulates" around the animal pole. Conceivably, a segment of this surface slides over the ring of actin bundles, moves into the area circumscribed by the ring, and contributes to the surface of the bulge (see Hamaguchi and Hiramoto, 1978). Furthermore, during bulge growth, actin filaments are concentrated around the bulge base; in contrast, the cortex of the bulge wall contains much fewer actin filaments. Such a differential distribution pattern of cortical actin can be observed as early as the initiation of bulge formation. It is conceivable that the cortex around the bulge is stiff, whereas the cortex of the bulge is relatively pliant. If the internal pressure of the egg increases (Hamaguchi and Hiramoto, 1978; Shimizu, 1979), a surface deficient in actin filaments may be protruded outward, while the surrounding region underlaid by a dense sheet of actin filaments will remain flat. Since the inner diameter of the ring of actin bundles does not change significantly during bulge formation, if the surface area of the bulge increases from such supplies of surface membrane as those already discussed, the bulge protrudes further and grows in height but not in width.

Reorganization of cortical actin may also be involved in formation of the contractile ring. As mentioned previously, during the later half of bulge formation, small bundles of actin filaments in the bulge wall reorganize toward the bulge base. These actin filaments may accumulate in the bulge base and form a contractile ring together with the preexisting actin bundles of the dense sheet. I therefore suggest that the contractile ring responsible for polar body formation forms as a result of the reorganization of preexisting cortical actin filaments. Possibly, the meiotic apparatus may mediate or trigger this redistribution of actin bundles, since small bundles of actin filaments remain in the bulge wall in eggs where the meiotic apparatus has been removed from the animal pole (Shimizu, 1990).

B. Egg Shape Change

Formation of the second polar body is accompanied by deformation of the eggs. This surface deformation begins with the formation of 12–15 meridionally running grooves on the equatorial surface (Fig. 6b). As grooves deepen, the equatorial surface between two adjacent grooves protrudes (Fig. 6c). Equatorial protrusions thus formed are then constricted at their proximal region since meridional grooves are linked to each other by "latitude" grooves (Fig. 6d). However, these protrusions are never separated from the egg proper; they are resorbed into the egg concurrently with the completion of the second meiosis (Figs. 6e and 6f). At the beginning of telophase, protrusions are also formed at the animal hemisphere, i.e., between the animal pole and the equator of the egg (Fig. 6g). Usually animal protrusions are fewer than equatorial protrusions.

This groove formation in both the animal hemisphere and the equator is a microfilament-dependent process: it is inhibited by cytochalasins (Shimizu, 1978b). In fact, it has been demonstrated using an electron microscope that microfilaments (6 nm in diameter) are localized at the cortex of the

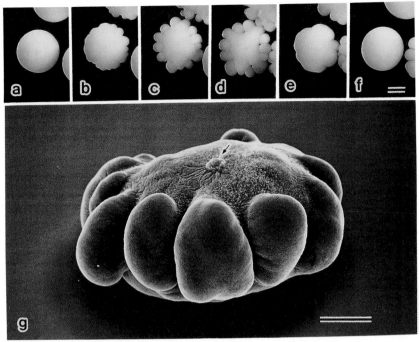

Fig. 6 The second deformation movement in the *Tubifex* egg. (a–f) Animal pole view of a living egg. Photographs were taken at 15-min intervals. (g) Scanning electron micrograph showing the side view of an egg undergoing the deformation movement. The arrow points to the second polar body located at the animal pole. Scale bars: f = 200 μm for a–f; g = 100 μm.

groove bottom (Shimizu, 1975). Furthermore, these cortical filaments are decorated with heavy meromyosin from rabbit skeletal muscle, suggesting that they are actin in nature.

Groove formation is also affected differently by microtubule inhibitors between the animal hemisphere and equator. In eggs treated with colchicine, meiotic division (hence, polar body formation) and the animal hemisphere grooving are both inhibited, but not the equatorial grooving (Shimizu, 1979). The animal hemisphere grooving appears to be closely related to meiotic nuclear events (Shimizu, 1979). It is conceivable that this grooving is triggered by a mechanism which is similar to that involved in the induction of cleavage furrow for polar body formation and mitotic divisions. On the other hand, the equatorial grooving is not dependent on meiotic nuclear divisions. This possibility is supported by the fact that nonnucleate egg fragments produced at metaphase of the second meiosis are able to deform forming grooves on the surface when control intact eggs are undergoing the second deformation movement (Shimizu, 1981a).

How is the pattern of equatorial grooving generated independently of the function of the meiotic apparatus? Views of actin organization revealed in isolated cortices allow us to envision the mode of the process by which cortical F-actin gives rise to formation of meridionally running grooves. Cortices isolated upon groove formation reveal meridional actin bundles crossing the equator of the egg (Fig. 7a). These actin bundles link the actin lattice at both poles of the egg (Fig. 7b). A careful examination suggests that these actin bundles are located at the bottom of grooves just initiated. The discovery of equatorial, meridionally running actin bundles, which are structurally continuous to the polar actin lattice, makes it easy to envision a simple model for equatorial grooving in the *Tubifex* egg. The egg cortex, upon grooving, is regarded as a hull of actin filaments with a dozen of windows around the equator (Fig. 8c). This hull apparently binds to the oolemma. If it contracts, equatorial actin bundles also "shrink" to pull the

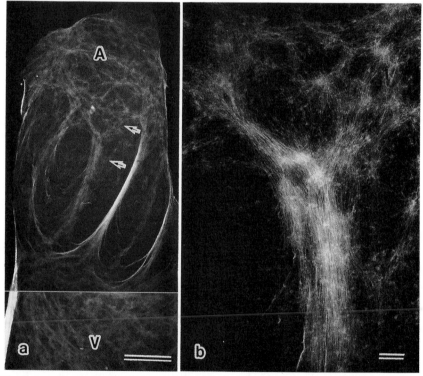

Fig. 7 Fluorescence micrographs showing the organization of actin bundles in the cortex isolated from an egg at the beginning of the second deformation movement. (b) Higher magnification of the actin bundle indicated by a pair of arrows in a. A, animal pole; V, vegetal pole. Scale bars: a = 100 μm; b = 10 μm.

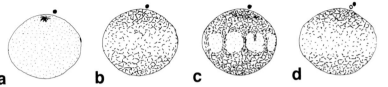

Fig. 8 Diagrammatic summary of the organization of cortical F-actin before (a and b), at the beginning (c), and at the end (d) of the second deformation movement. The animal pole is up. Open and solid circles located at the top of each figure are polar bodies. Polarized organization of cortical F-actin is undetected at the beginning of the second meiosis (a), but clearly seen at 60 min (b).

overlying surface inward, specifically toward the egg axis, since the center of the polar actin lattice corresponds to the poles of the egg. Along with this ingress, the equatorial surface overlying the windows of cortical hull protrudes outward. As grooves deepen, equatorial actin bundles, which have been convex belts of filaments, become straight; at this time two bundles located on the opposite side of a given equatorial protrusion are converted to a closed ring through recruitment of actin filaments which are located at the periphery of the polar actin lattice. As a result of this ring formation, the protrusion might be constricted circumferentially at its base. The nucleus-independent property of the equatorial grooving mentioned previously suggests that cortical contraction occurs independently of the nucleus. In fact, it has been demonstrated that the cortical contraction occurs in nonnucleate egg fragments coincidently with the deformation movement in control eggs (Shimizu, 1981a).

As to the origin of equatorial actin bundles, I suggest that they form through reorganization of preexisting actin filaments. This is based on the observation that equatorial actin filaments which are found outside of the actin bundles are very few compared with those seen in metaphase II eggs. Equatorial groove formation and appearance of actin bundles are preceded by contraction of polar actin lattice. It is conceivable that surface tension generated by this cortical contraction may trigger reorganization of equatorial actin filaments into bundles. As will be described later, similar contraction of cortical actin layers which are organized bipolarly (Fig. 8d) also occurs during ooplasmic segregation (Shimizu, 1984); however, equatorial grooves are never formed at this stage. Therefore, cortical organization in eggs during the second meiosis appears to be distinct from that in later eggs.

C. Ooplasmic Segregation

Ooplasmic segregation in the *Tubifex* egg occurs following the second meiosis (possibly beginning during the second polar body formation albeit very slightly; Shimizu, 1976b). It is a process of accumulation of yolk-free cytoplasm called pole plasm to the animal and vegetal poles. This process consists of two steps (Fig. 9; Shimizu, 1982b). First, mitochondria and endoplasmic reticulum, which are major membranous organelles of the pole plasm, migrate from the inner endoplasmic region outward and form a subcortical layer devoid of yolk granules and lipid droplets. The translocation of this subcortical cytoplasm along the surface is the second step, and it results in the localization of pole plasms at both poles of the egg.

Besides the difference in direction of organelle movement, the two steps of ooplasmic segregation are different from each other in two more respects. In contrast to the increase during the first step, the volume of the subcortical cytoplasm decreases during the second step, which is ascribable to a compact arrangement of organelles rather than their dropping out of the subcortical domain. Although both steps are impaired by cytochalasin, the first step is less sensitive to this inhibitor than the second step. Thus, the first and the second steps of ooplasmic segregation in the *Tubifex* egg appear to be operated by distinct mechanisms.

1. Differentiation of Cytoskeletal Domains

Figure 10 shows a diagrammatic summary of actin organization and its relation to organelles before and during the first step of ooplasmic segregation. Before the second polar body formation begins, a distinct actin network is present in the endoplasmic region (Fig. 10a). This network is disorganized during the polar body formation; however, short actin filaments are caused to aggregate with membranous organelles (Fig. 10b). Following the second meiosis, similar actin filaments become localized in the subcortical layer but not in the underlying yolky region (Fig. 10d).

At 50–60 min after the second meiosis, an elaborate actin network is established in the subcortical layer (Fig. 10d). This subcortical actin network possibly forms *de novo* following the accumulation of "short" actin filaments.

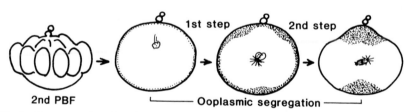

Fig. 9 Two steps of ooplasmic segregation in the *Tubifex* egg. Except for one undergoing the second polar body formation (PBF), meridional sections through the egg axis are presented. Ooplasmic domains to be segregated are indicated by dots. Open circles on the top of eggs represent polar bodies. (From Shimizu, 1984 with permission.)

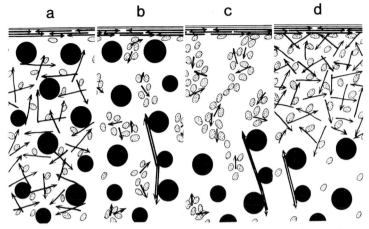

Fig. 10 Diagrammatic summary of the actin organization in the endoplasm and subcortical region. Arrows represent actin filaments; cortical actin filaments are also depicted under the surface. (From Shimizu, 1984 with permission.)

These filaments are derived from the disruption of the endoplasmic actin network and transported along with the centrifugal movement of organelles giving rise to their localization in the subcortical layer (Fig. 10c); thereafter, individual filaments become longer and increase in number to form a network. At this stage the underlying yolky region exhibits filament bundles associated with yolk granules but no actin network (Fig. 10d). In view of the fact that networks and bundles of actin filaments are produced by different kinds of actin cross-linking proteins (Corwin and Hartwig, 1983; Mabuchi, 1986), this may suggest that the mechanism for actin assembly in the endoplasm is different from that in the subcortical layer during ooplasmic segregation. Temporally, disruption of the endoplasmic actin network and elaboration of the subcortical network correspond, respectively, to the beginning of the first and second step of ooplasmic segregation. As discussed under Section V,C,2, these events are prerequisite for respective steps to proceed.

2. Mechanism for the First Step

The first step of ooplasmic segregation in the *Tubifex* egg is the centrifugal movement of membranous organelles or their aggregates from the yolky endoplasm. A key triggering event for such segregation to occur is the disruption of the endoplasmic actin networks; thereby membranous organelles become "free" from yolk granules and can move independently of the latter. During the movements, organelles are organized in aggregates which are often strung up from the subcortical layer inward (Fig. 10c). The domain of organelle aggregates as a whole is expected to contract giving rise to their centrifugal movements because this domain contains actin filaments and anchors to the egg surface. If the nets of yolk granules linked by filament bundles also contract, they would move toward the egg center and, thereby membranous organelles would be segregated efficiently.

3. Mechanism for the Second Step

The second step of ooplasmic segregation is the polarized movement of subcortical cytoplasm along the surface. The presence of the elaborate actin network in this cytoplasm, but not in the inner yolky regions, suggests that the subcortical layer is viscoelastic, while the yolky endoplasm displays a lower viscosity (Lehmann, 1938). Apparently the subcortical cytoplasm moves en mass, and would encounter little resistance from the underlying "fluid" endoplasm. In fact, it has been demonstrated that mitochondria move en mass along the surface in living eggs (Fig. 11; Shimizu, 1986).

As is the first step, the second step is also a microfilament-dependent process. The movement of subcortical cytoplasm could be brought about by the subcortical actin network, the cortical actin lattice, or both. Several lines of evidence suggest that the cortical actin lattice plays a key role in driving the underlying cytoplasm (Shimizu, 1986). First, the egg surface moves together with the underlying cytoplasm in both the animal and the vegetal hemisphere of the egg. This suggests that the cortex contracts

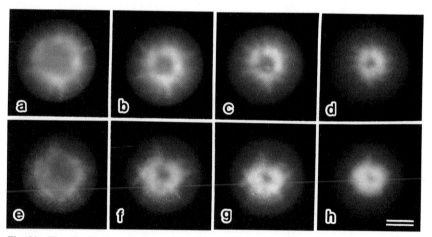

Fig. 11 The distribution of mitochondria in living eggs stained with rhodamine 123 (which is known to be a fluorescent vital dye specific to mitochondria). a–d, animal pole view; e–h, vegetal pole view. Micrographs were taken shortly after (a and e), 40 (b and f), 80 (c and g), and 100 min (d and h) after the second meiosis. Scale bar: h = 200 μm for a–h. (From Shimizu, 1986 with permission.)

in the same direction as the ooplasmic movement. Second, cortical actin filaments reorganize and move toward the pole in both hemispheres of the egg (Fig. 12). This is unambiguous morphological evidence for poleward contraction of the cortical actin lattice. Third, the cortex can contract toward the pole independently of underlying cytoplasm which is movable and can be stratified by centrifugal force (e.g., 1700g). This suggests that the force and directionality of the cortical contraction derive from the cortex itself and not from the inner cytoplasm. Lastly, the subcortical cytoplasm is physically connected to the cortex; this connection is resistant to a centrifugal force of up to 650g.

This strongly suggests that the cortex in the *Tubifex* egg generates not only motive force for movement of subcortical cytoplasm, but also determines its direction. Involvement of an actomyosin force-generating mechanism in this process is suggested by the fact that contractile activities of isolated cortices, which are readily induced by addition of ATP, are completely inhibited by their preincubation with *N*-ethylmaleimide-modified heavy meromyosin (Shimizu, 1985). On the other hand, the polarized organization of cortical F-actin may be a key element of the mechanism for determination of direction. As has been demonstrated in an *in vitro* experiment, if local differences in the actin lattice rigidity are present, contraction of the lattice would cause its directed movement (Stendahl and Stossel, 1980). By analogy with this *in vitro* experiment, contraction of the cortical actin lattice in the *Tubifex* egg is stronger in the polar region than in the equatorial region. The bipolar cortex of the *Tubifex* egg thus forms two focal points for ooplasmic segregation.

At present, it is unclear to what extent the subcortical actin networks contribute to generate force for movement of subcortical cytoplasm. As mentioned previously, membranous organelles become "condensed" in the subcortical layer during the second step of ooplasmic segregation. It is conceivable that contraction of subcortical actin networks gives rise to compact arrangement of organelles. On the other hand, as described under Section VII,A, these actin networks play a role in generating a precise localization pattern of ooplasm during embryogenesis.

In summary, I suggest that ooplasmic segregation in the *Tubifex* egg is a two-step process which is exclusively dependent on the dynamics of actin microfilaments. Key events are the disruption of endoplasmic actin networks, *de novo* formation of subcortical actin networks, and polarized reorganization of cortical actin filaments.

D. Cleavage Furrow Formation

The last event in which cortical actin cytoskeleton plays a key role during the one-cell stage is furrow formation of the first cleavage. As mentioned

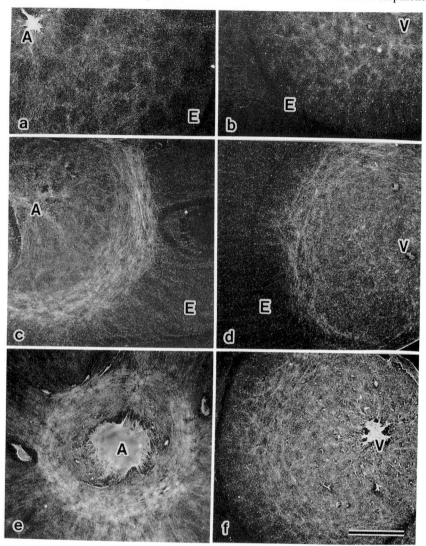

Fig. 12 The distribution of F-actin in isolated cortices of eggs shortly after (a and b), 80 (c and d), and 120 min (e and f) after the second meiosis. Cortex preparations were stained with rhodamine–phalloidin. a, c, and e, Animal hemisphere; b, d, and f, vegetal hemisphere. Cortices of both hemispheres at each stage were prepared from the same egg. A, animal pole; E, equator; V, vegegal pole. Scale bar; f = 100 μm for a–f. (From Shimizu, 1986 with permission.)

previously (see Section II), *Tubifex* eggs divide unequally and meridionally. Compared with other animals, the process of furrow formation in this animal is unique and somewhat complicated. Therefore, before discussing cortical actin organization, I describe this process in some detail.

The first cleavage in *Tubifex* begins with the formation of meridionally running furrows at the two points of the egg's equator which are about 90° apart from each other. During the first 10 min, these furrows deepen approximately toward the center of the egg, i.e., the egg axis (Fig. 13b). At this time, individual leading edges of both equatorial furrows are sepa-

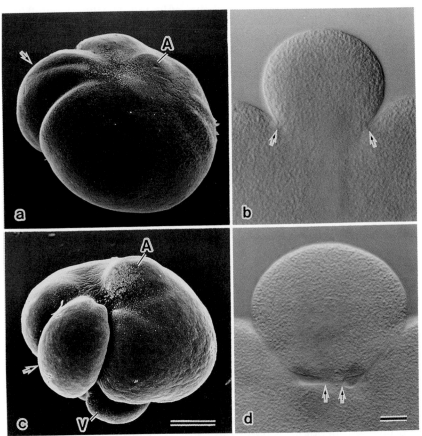

Fig. 13 Scanning electron micrographs (a and c) and differential interference contrast micrographs (b and d) showing furrow formation in the *Tubifex* egg. Eggs were fixed at 10 (a and b) and 20 min (c and d) after the onset of furrowing. Eggs shown in b and d were cleared with Murray's clear (see Appendix) and viewed from the animal pole. Arrows in a and c point to the presumptive AB cells. Arrows in b and d indicate furrow bottoms. A, animal pole; V, vegetal pole. Scale bars: c = 100 μm for a and c; d = 20 μm for b and d.

rated by a flat surface at both the animal and the vegetal sides (Fig. 13a). Two equatorial furrows then deviate from this direction so as to make their bottoms approach each other at a point away from the egg axis (Fig. 13d). Temporally, this shift of furrowing direction corresponds to the time when a surface groove which links the leading edges of the furrows emerges at both the animal and the vegetal sides. Two equatorial furrows are now continuous to each other and comprise a closed ring of cleavage furrow. Constriction of this ring results in the production of two cells of unequal size (Fig. 13c). The ratio of the resulting smaller and larger sister cells is about $1:2$ along the division axis. The cleavage plane is nearly parallel to the egg axis. As in other animals, furrow formation in *Tubifex* is impaired not only by cytochalasin D but also by nocodazole.

To gain an insight into the mechanism by which the pattern of furrow formation described above is generated, organization of cortical F-actin was analyzed in isolated cortices and egg fragments including cleavage furrows stained with rhodamine–phalloidin (Ishii and Shimizu, 1995). Cortices isolated upon furrow formation at the equator show two meridional bundles of F-actin crossing the equator (Fig. 14a). These bundles are structurally continuous to the polar actin lattice at the animal and vegetal poles. A careful examination suggests that these actin bundles are located at the bottom of the cleavage furrow. The equatorial cortex outside these actin bundles exhibits tiny dots stained with rhodamine–phalloidin.

To examine actin organization at furrow bottoms which are located deep inside, furrow regions which had been cut out from eggs were stained with rhodamine–phalloidin. During the first 10 min following the initiation of furrow formation, F-actin layers lining equatorial cleavage furrows were found to be continuous to the polar actin lattice (Fig. 14b). Thereafter, a ring of F-actin (i.e., contractile ring) which contoured the furrow bottom was separated from the polar lattice; this ring, when viewed from the egg's center, was ellipse in shape with a meridional major axis (Fig. 14c). The contractile ring became smaller during the period of 10 min, though its overall shape remained unchanged. As Figure 14d shows, however, its final shape appeared to be a circle. During "shrinkage" of the contractile ring, F-actin was organized in a belt circumscribing the constriction.

On the basis of these observations, I suggest that furrow formation during the first cleavage in *Tubifex* consists of two steps: ingress of the equatorial surface toward the egg axis and an ensuing circular constriction. The first step involves meridional contractile "arcs" of actin filaments which are structurally continuous to the polar actin lattice. Contraction of these arcs together with the polar lattice may bring about an ingress of the equatorial surface to form a meridionally running groove. Apparently this process is similar to that of equatorial groove formation accompanying the second polar body formation (see Section V,B). Furthermore, furrow formation

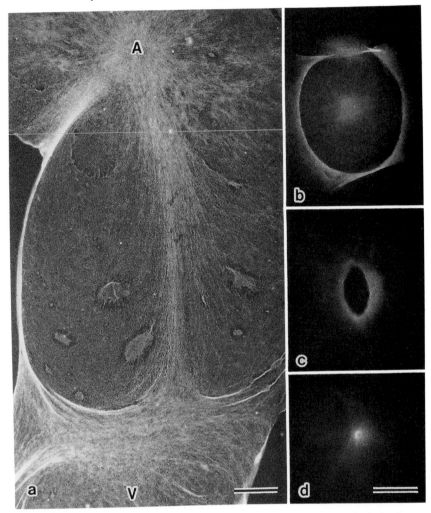

Fig. 14 Fluorescence micrographs showing the organization of cortical F-actin during furrow formation at the first cleavage in the *Tubifex* egg. An isolated cortex (a) and egg fragments (b–d) were stained with rhodamine–phalloidin. (a) The cortex isolated upon furrowing exhibits a meridionally running actin bundle which crosses the equator. A part of the other equatorial actin bundle is seen in the upper left. A, animal pole; V, vegetal pole. (b–d) Contractile rings at 10 (b), 20 (c), and 30 min (d) after the onset of furrow formation. All viewed from the presumptive AB cells. The animal pole is up. Scale bars: a = 50 μm; d = 50 μm for b–d.

during the first step is reminiscent of initial stages of unilateral cleavages which are seen in eggs of coelenterates, crinoids, and amphibians (Schroeder, 1968; Bluemink, 1970; Sawai, 1972; Holland, 1978), although unlike the equatorial location of early furrows in *Tubifex*, furrows in these animals emerge at the animal pole. The second step begins as a contractile ring which is composed of previous contractile arcs separates from the polar actin lattice. Although the early contractile ring assumes the shape of ellipse, it may contract in a similar way to that seen in eggs of other animals such as sea urchins (Mabuchi, 1986).

E. Triggering Mechanisms for Cortical Reorganization

As described previously, polar body formation, deformation movement, ooplasmic segregation, and cleavage furrow formation are all microfilament-dependent processes. Furthermore, these events involve reorganization of cortical F-actin. Despite such similarities, however, they are regulated by different mechanisms. First, as mentioned previously, polar body formation and cleavage furrow formation depend on the meiotic/mitotic apparatus, but the other events do not. Second, 1-(5-isoquinolinesulfonyl)-2-methylpipera-zine dihydrochloride (H-7), a protein kinase inhibitor, inhibits equatorial grooving at the second meiosis (Fig. 15b) and the second step of ooplasmic segregation, but does not inhibit polar body formation or the first cleavage (Fig. 15c). As Figure 15d shows, reorganization of cortical F-actin fails to occur at the equator of an H-7-treated egg, though the animal pole of the same egg exhibits a rather normal arrangement of cortical F-actin. This may suggest that cortical reorganization which leads to generation of equatorial actin bundles is triggered by biochemical pathways different from that in-volved in nucleus-dependent events. Apparently, this difference depends on whether the meiotic/mitotic apparatus is involved. Given that cortical reorga-nization for furrow formation is triggered by "cleavage stimuli" derived from the meiotic/mitotic apparatus (Rappaport, 1986), the triggering mechanism for nucleus-independent contractile activities may include a certain kind of pro-tein phosphorylation which can be bypassed in nucleus-dependent processes.

It has been reported that H-7 blocks cAMP-dependent and cGMP-depen-dent protein kinases as well as protein kinase C (Hidaka *et al.*, 1984). Since another inhibitor HA1004, which blocks the first two kinases much more heavily than protein kinase C, has no inhibitory effect on either nucleus-dependent or -independent contractile activities (Fig. 15a), then inhibition of equatorial grooving and ooplasmic segregation by H-7 may result from blockage of protein kinase C activity. This is supported by the finding that equatorial grooving is induced by calcium ionophore A23187 (Shimizu, 1978a) because it would be anticipated that the calcium-dependent enzyme protein kinase C would be activated by the elevation of intracellular calcium. In addition, activation of protein kinase C has been shown to induce forma-tion of contractile rings (Bement and Capco, 1991) and to induce cortical reorganization (Capco *et al.*, 1992; Ryabova *et al.*, 1994) in amphibian eggs.

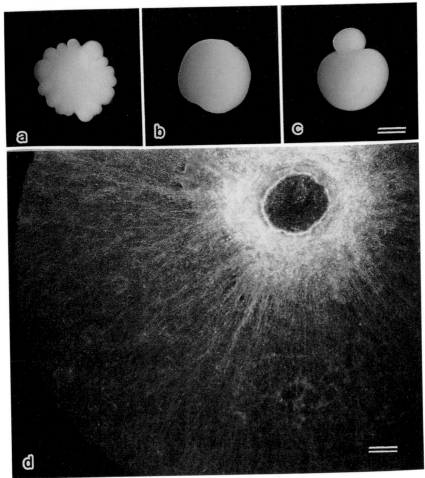

Fig. 15 Effects of protein kinase inhibitors on the deformation movement and the first cleavage. (a–c) Living eggs treated with 2 mM HA1004 (a) and 0.5 mM H-7 (b and c). All viewed from the animal pole. (d) Fluorescence micrograph of a rhodamine–phalloidin-labeled cortex isolated from an egg which had been treated with H-7 at the time of the second deformation movement. A thick sheet of actin bundles circumscribes the animal pole. Scale bars: c = 200 μm for a–c; d = 20 μm.

VI. Polar Cortical Actin Lattice during Early Cleavage

The polar cortical actin lattice which has played a role in cleavage furrow formation persists into cleavage stage. It is inherited by the CD cell of the two-cell embryo and then by the D cell of the four-cell embryo. Figure 16 shows cortices isolated from four-cell embryos. As evident from the figure,

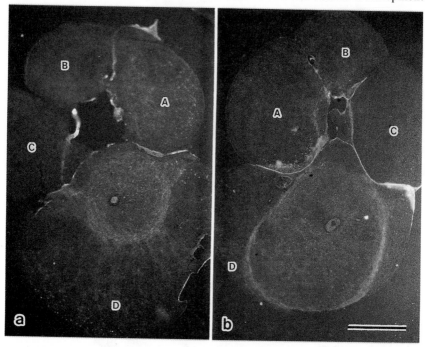

Fig. 16 The organization of cortical F-actin at the four-cell stage. Cortices isolated from the animal (a) and vegetal (b) sides of embryos were stained with rhodamine–phalloidin. Cortical actin lattices are seen in the D cells, but not in other cells (A, B, and C). Scale bar: b = 100 μm for a and b.

cortical F-actin in the D cell is distributed in a bipolar fashion. The actin lattice is composed of a central meshwork and a circumferential ring of actin bundles. The animal actin lattice is smaller and organized more densely than the vegetal one. In other cells (A–C), there is no trace of polarized distribution of cortical F-actin.

Developmental significance of the polar actin lattice in the D lineage cells remains to be explored. They may contribute to generating spatial organization in early embryos. First, the bipolar localization of pole plasms during the first two cleavages (see Fig. 17b) would be retained by the cortical actin lattice. As described under Section VII,A, pole plasms at these stages contain networks of F-actin and are physically connected to the polar cortex. Given that cleavage pattern is strictly regulated in *Tubifex* (see Section VIII), precise segregation of pole plasms to the CD and D cells depends on their precise localization at cleavage. "Anchorage" of pole plasms to the polar cortex would meet this requirement. Second, the polar actin lattice would be involved in

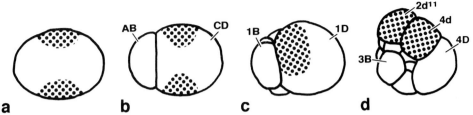

Fig. 17 Diagrammatic summary of the pole plasm localization during early development of *Tubifex*. The pole plasm domains are represented by dotted areas. The anterior end of the embryo is to the left, and the animal pole is up. a, 1-cell stage; b, 2-cell stage; c, 8-cell stage; d, 24-cell stage. The pole plasms are inherited by the CD cell and then by the D cell during the first two cleavages; at this time, they are bipolarly localized (a and b). During the third cleavage, the vegetal pole plasm redistributes toward the animal pole and is unified with the animal pole plasm there in the 1D cell (c). The pole plasms are finally partitioned into precursor cells ($2d^{11}$ and 4d) of teloblasts (d).

translocation of the mitotic apparatus toward the animal pole in the D cell. The mechanism for this translocation resides in the cortex of the D cell and is not affected by the centrifugal force (up to 1100g) which is sufficient to stratify cytoplasm into four layers (Shimizu, 1989). I have recently found that the cortical actin lattice in the D cell is not affected by this centrifugal force either. The cortical movement directed toward the animal pole which is caused by the polar actin lattice might contribute to the translocation of the mitotic apparatus in the D cell. Lastly, the cortical actin lattice might provide a scaffolding for molecules with specific functions (Jeffery, 1989). For example, if a dynactin complex is localized at the animal pole, the poleward translocation of the mitotic apparatus would be facilitated (see Waddle *et al.*, 1994). In view of the recent implication of ectoderm determinants in the animal pole cortex of the leech embryo (Nelson and Weisblat, 1993), it is conceivable that the animal pole cortex of the *Tubifex* embryo also contains similar factors (also see Yisraeli *et al.*, 1990; Elinson *et al.*, 1993).

VII. Developmental Role of Subcortical Actin Cytoskeleton

A. Asymmetric Segregation of Pole Plasms

It has long been known that pole plasms are inherited by teloblasts which play a pivotal role in *Tubifex* embryogenesis (for review, see Shimizu, 1982a). The process of pole plasm localization during cleavage stages is composed of three steps. The first step is asymmetric segregation which results in bipolar localization of pole plasm masses in the CD cell of the two-cell embryos and the D cell of the four-cell embryo (Fig. 17b). The second step is the redistribution of the vegetal pole plasm toward the animal pole and it unification with the animal pole plasm (Fig. 17c). The last step of the localization process is partitioning of unified pole plasm into the precursors (cells 2d and 4d) of teloblasts (Fig. 17d).

Subcortical F-actin which has been localized in polar yolk-free domains before the first cleavage exhibits exactly the same distribution pattern as the pole plasms up to the eight-cell stage (Shimizu, 1988). Its bipolar localization in the CD and D cells is followed by redistribution and unification during the third cleavage. At the eight-cell stage, however, F-actin becomes more concentrated in the nuclear regions than in the surrounding pole plasm. Judging from the fluorescence intensity of rhodamine–phalloidin which binds to pole plasms, the amount of F-actin associated with pole plasms appears to decrease at this time. Similar association of F-actin with nuclear regions (i.e., mitotic apparatus at mitosis and perinuclear cytoplasm at interphase) is also seen in teloblast precursors and macromere 4D (see Fig. 17d).

The changes in distribution of subcortical F-actin at the transition from the four- to the eight-cell stage may suggest that it contributes to maintaining spatial organization of pole plasm to different extents between the eight-cell stage and earlier ones. In fact, cytochalasin treatments cause changes in the pole plasm organization at the two- or four-cell stages, but not at the eight-cell stage. Conversely, the pole plasm organization is sensitive to microtubule inhibitors at the eight-cell stage, but not at the two-cell stage (Shimizu, 1989). These findings suggest that constituents of the pole plasm such as mitochondria are associated with cytoskeleton and that cytoskeletal elements involved are developmental stage specific. It is conceivable that pole plasm is integrated by actin filaments up to the four-cell stage and by microtubules from the eight-cell stage onward. This shift from the involvement of actin filaments to microtubules in pole plasm distribution occurs during the third cleavage. This transition includes not only the association of constituents of pole plasm with microtubules but also changes in the actin organization in pole plasm. Association of constituents of pole plasm with microtubules is seen during the first two cleavages as well, however. Nevertheless, pole plasm does not accumulate around the nucleus at these stages. This may be simply because when astral microtubules disassemble after mitosis, pole plasm domains can revert to their original "shape" through the gelatinous property of pole plasm. Thus, a key element of the cytoskeletal transition at the third cleavage is the decrease of F-actin in the pole plasm. At present, neither the mode nor the mechanism of this decrease are known.

B. Specification of the Dorsoventral Polarity

A significant consequence of the cytoplasmic redistribution is the generation of unipolar cytoplasmic organization in the D cell (Fig. 17c). In *Tubifex,* the animal pole corresponds to the future dorsal side, and the vegetal pole to the ventral side of the embryo (see Section II). Before the third cleavage, however, the two domains of pole plasm in the cells CD and D are located bipolarly at the regions corresponding to the animal and vegetal poles of the precleavage egg, and the overall organization of cytoplasm therefore does not appear to foreshadow the dorsoventral polarity. In this respect, the unipolar cytoplasmic organization generated in the 1D cell at the eight-cell stage is one of the earliest manifestations of the dorsoventral polarity of the embryo. Therefore, the process of the cytoplasmic redistribution leading to localization of pole plasm at the animal side (i.e., the future dorsal side) of the D cell appears to be one of the events specifying the dorsoventral axis of the *Tubifex* embryo.

This cytoplasmic redistribution is driven by the mitotic apparatus, especially by its aster (Shimizu, 1989). That the redistribution is directed to the animal pole but never to the vegetal pole is simply because the mitotic apparatus in the four-cell embryo of *Tubifex* is localized at the animal pole. Given that the mitotic apparatus is normal in every respect, it seems that the cytoplasmic redistribution directed to the animal pole would be accomplished only when the following three requirements are met. First, the mitotic apparatus is localized at the animal pole. Second, it orients itself nearly parallel to the egg axis. Third, the vegetal pole plasm takes up a specific position in the D cell. In four-cell embryos in which the cytoplasmic redistribution is impaired by cytochalasin treatments, mitotic apparatus are directed perpendicular to the egg axis, and pole plasms spread as a sheet along the surface (Shimizu, 1988, 1989). Apparently, in these embryos, the last two of the requirements mentioned previously are not met. Probably, orientation of the mitotic apparatus parallel to the egg axis and the polar localization of pole plasms are both required for the mitotic apparatus to associate with the vegetal pole plasm. These processes are sensitive to cytochalasin treatments, suggesting involvement of actin cytoskeleton. Furthermore, as suggested previously (Section VI), localization of the mitotic apparatus at the animal pole may be brought about by cortical actin cytoskeleton. Thus, it is conceivable that the actin cytoskeleton plays an important role in specifying the dorsoventral axis of the *Tubifex* embryo.

VIII. Role of Centrosomes and Cortex in Generating Cleavage Patterns

The precise segregation of cytoplasm to specific blastomeres depends not only on its controlled intracellular localization but also on patterns of ensuing divisions. Asymmetric segregation of pole plasms in early *Tubifex* embryos results from a stereotyped sequence of unequal cell divisions (see Section II,A). As has been suggested (Penners, 1922; Shimizu, 1982a), inequality of divisions correlates with asymmetry of mitotic apparatus organization. Recent studies have shown that the first two cleavages in *Tubifex* involve distinct mechanisms to generate asymmetry in mitotic apparatus.

A. Involvement of One Centrosome Rather Than Two: The First Cleavage

The first cleavage spindle is in sharp contrast to those in other blastomeres in that it is monastral (Fig. 18a). The astral pole possesses a centrosome and a γ-tubulin-containing sphere which includes the centrosome (Fig. 18c); the other pole lacks these two structures, though it is stained with an anti-γ-tubulin antibody more intensely than other parts of the spindle. The astral pole is organized through the activity of the centrosome, whereas the anastral spindle pole might be organized by a microtubule-organizing center other than centrosomes (Shimizu, 1995c). It is plausible that such a monastral mitotic apparatus is produced only when a single centrosome is involved in spindle assembly. In fact, if two centrosomes originating from the first meiotic apparatus are both retained in an egg at the first cell cycle, they form a symmetrically organized spindle with asters at both poles; it should be mentioned that these eggs undergo *equal* divisions.

The centrosome which participates in the first cleavage appears to be maternal in origin (Shimizu, 1995c). Possibly, a maternal centrosome which has been involved in the second meiosis persists into the first cell cycle without duplicating itself at the transition to the first mitosis. This centrosome, however, duplicates at the transition to the second cell cycle; the resulting centrosomes participate in the spindle assembly in the CD cell (see Fig. 18d). Such a centrosomal behavior implies changes in the ooplasmic conditions which control duplication of centrosomes. Apparently, ooplasmic activity which is responsible for centrosome duplication might be low or lacking at the transition to the first cell cycle and emerge before or at the transition to the second cell cycle.

It seems unusual in animal development that, as we see in *Tubifex,* only one centrosome, rather than two, is involved in the assembly of first cleavage spindles (for review, see Schatten, 1994). It is unlikely, however, that this situation in *Tubifex* eggs evolved accidentally. Rather, it appears that the involvement of a single centrosome in the first cleavage is a prerequisite for the normal development of *Tubifex* eggs. Given that spindle assembly occurs at the egg's center and that *Tubifex* eggs during the first cleavage lack such mechanisms as are seen in other invertebrate (e.g., surf clam)

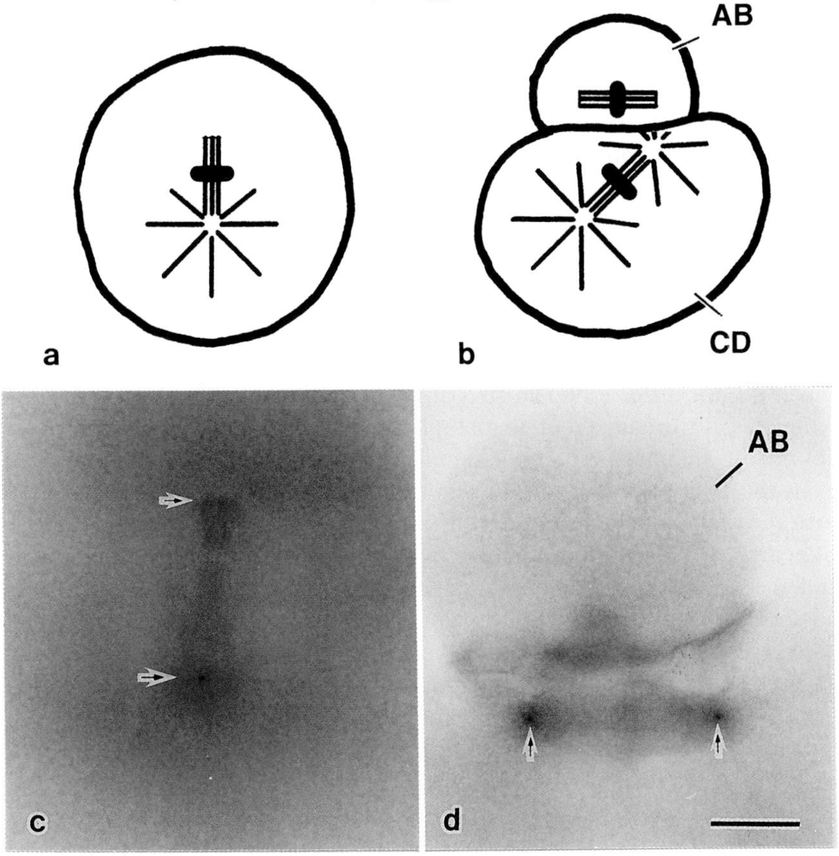

Fig. 18 (a and b) Diagrammatic representations of mitotic apparatus at the first (a) and second (b) cleavage. Animal pole view. A bundle of three lines represent a metaphase spindle; a thick line crossing this bundle indicates chromosomes at the metaphase plate. (c and d) Localization of centrosomes in the first (c) and second (d) cleavage mitotic apparatus. Eggs were labeled with an antibody against γ-tubulin and processed for immunoperoxidase staining. Both c and d show whole-mount preparations viewed from the animal pole. (c) A γ-tubulin-containing sphere (large arrow) is located at the astral spindle pole. A dot-like structure seen at the center of this sphere is a centrosome. At the opposite spindle pole (small arrow), there is no trace of the centrosome. (d) Centrosomes (arrows) are seen at both poles of the spindle in the CD cell. There is no trace of the centrosome in the AB cell. Scale bar: d = 50 μm for c and d.

eggs to translocate the mitotic apparatus eccentrically (Dan and Tanaka, 1990; Ishii and Shimizu, 1995), the formation of symmetric spindles (involving two or no centrosomes) would necessarily lead to equal division of the egg. As mentioned previously, the presence of a pair of centrosomes in an egg gives rise to equal division of the egg. It has long been known that equal first cleavage is disastrous to *Tubifex* development, because it gives rise to the production of embryos with duplicated heads and/or tails (Penners, 1924; Inase, 1960). Thus, it is conceivable that the involvement of a single centrosome in the first cleavage is a developmental strategy that allows *Tubifex* to produce functional organisms.

B. Interactions of Astral Microtubules and Cell Cortex: The Second Cleavage

The mitotic apparatus in the CD cell possesses centrosomes at both poles which do not differ either in size or by γ-tubulin staining (Fig. 18d). It is symmetrically organized up to metaphase, but acquires asymmetric properties during anaphase. The right spindle pole, but not the left (when viewed from the animal pole and CD oriented below AB), becomes associated with the cortex facing the AB cell (i.e., the anterior end of the CD cell). Concurrently, astral microtubules become shortened between the cortex and the (right) spindle pole; as a result, it looks as if this spindle pole possesses a "half" aster (Fig. 18b; Shimizu, 1993). At this stage, this spindle pole is physically connected to the cortex. This association site is localized at or close to the remnant of the midbody of the first cleavage.

In normal embryos, the mitotic apparatus in the CD cell assembles near the cleavage membrane apposed to the AB cell. If the nuclei in CD cells were moved to the side opposite to the AB cells, mitotic apparatus in CD cells were assembled ectopically there. Asters at both poles of the ectopic spindles were not only equal in size, but grew symmetrically. Interestingly, similar symmetric organization of spindle poles was observed in CD cells which had been treated with cytochalasin D. These results suggest that the interaction of the spindle pole with the cell cortex, which may be mediated by F-actin, plays an important role in generating an asymmetric organization of the spindle poles. Since this interaction is confined to the cleavage membrane, especially at the remnant of the midbody, the furrow formation of the first cleavage may contribute to differentiate this membrane for this interaction.

The molecular basis of this spindle pole/cortex interaction remains to be explored. Based on the morphology of the spindle pole which is to be associated with the cortex, it may be envisaged that the cortex of the CD cell facing the AB cell contains at least two kinds of microtubule-associated proteins. First, astral microtubules are shortened between the cortex and the spindle pole, suggesting the presence of molecules which "cap" or depolymerize the plus ends of astral microtubules. Second, there should be molecules which link the microtubules to the cortex. In addition, if such molecules as dynactin are present in the cortex, translocation of the spindle pole toward the cortex would be facilitated (see Waddle *et al.*, 1994).

C. Specification of the Anteroposterior Polarity

The future anteroposterior (AP) axis of the *Tubifex* embryo is in register with the division axis of the first cleavage (see Section II,B). To specify a *single* AP polarity in the *Tubifex* embryo, the following requirements must be met. First cleavage divisions are unequal, and directed perpendicular to the egg axis (see Fig. 17b). As mentioned previously, if *Tubifex* eggs divide equally at the first cleavage, they produce embryos with duplicated AP axes. Furthermore, similar "double" embryos also develop from eggs which have divided along the egg axis during the first cleavage. Therefore, not only inequality but also direction of the first cleavage are important factors for specifying the AP polarity in the *Tubifex* embryo. Recently, we showed that the first cleavage spindle is oriented in any direction in the equatorial plane and that eggs which have divided in any direction around the equator at the first cleavage can establish embyonic axes normally. It seems unlikely that the orientation of the first cleavage spindle around the equator is predetermined in *Tubifex*. On the other hand, how the spindle is oriented perpendicular to the egg axis is currently not known. It would be of interest to examine whether bipolarly localized cytoplasmic domains (i.e., pole plasms) are involved in this control. During the first cleavage, astral microtubules extend into the pole plasm domains. Furthermore, actin networks in the D cell of the four-cell embryo have been demonstrated to be involved in controlling orientation of the mitotic apparatus (Shimizu, 1988).

The AP polarity of the *Tubifex* embryo is further specified by localization of the nucleus to the anterior end of the CD cell (see Fig. 1h). This localization is mediated by the cleavage membrane which results from the first division. As described previously, this membrane is apparently differentiated to "attract" the mitotic aster. Such a property of the cleavage membrane may be specific to the first cleavage, since mitotic asters in the D cell do not attach to the (second) cleavage membrane. Thus, the first cleavage in *Tubifex* comprises a crucial step in specifying the AP axis.

IX. Concluding Remarks

The generation of spatial patterns in the *Tubifex* embryo depends on precise cytoskeletal architecture. Cytoplasmic movements and surface activities involved in this process are directed according to the positional cue of the cortical actin cytoskeleton organization. The bipolar organization of the cortical actin cytoskeleton, which is established at the transition to the second meiosis, persists into early cleavage stage, even though individual actin filaments undergo occasional rearrangement. *Tubifex* eggs may be endowed with not only positional information generating bipolarity in the cortex but also with mechanisms to maintain this cortical organization.

Many questions remain to be answered. How is the polarized organization of cortical actin generated through actin polymerization? What is the mechanism for maintaining this actin organization? What triggers the formation of the equatorial actin bundles without "cleavage stimuli?" What is the mechanism for orienting and localizing the mitotic apparatus in the egg? This chapter suggests some clues to the solution of these problems. However, to answer these questions, we need more information on the organization of plasma membrane, actin dynamics *in vivo,* and the distribution of actin-binding proteins in *Tubifex* eggs.

From an evolutionary point of view, it would be of great interest to compare embryonic cytoskeletal architecture among spiralians. Many developmentally important processes (viz., ooplasmic segregation, specification of D quadrant, establishment of embryonic axes, and unequal segregation of developmental fates to blastomeres) appear to have been preserved in this group of animals (Speksnijder and Dohmen, 1983; Astrow *et al.,* 1989; Holton *et al.,* 1989; Dorresteijn, 1990; Dorresteijn and Kluge, 1990; Freeman and Lundelius, 1992). Interestingly, however, homologous developmental events do not necessarily involve the same cytoskeletal mechanism (Astrow *et al.,* 1989). At present, we have too little information on embryonic cytoskeletal organization in spiralians to discuss the extent to which the cytoskeletal mechanisms have been shifted during evolution. I hope that not only structural but also functional aspects of the cytoskeleton in early embryos will be revealed in a variety of spiralians.

Appendix: Methods

A. Isolated Cortices for Light Microscopy

Egg cortices for light microscopy can be prepared on polylysine-coated coverslips (or glass slides) or on the cover half of a plastic dish. Unless one needs freely floating cortices, I would recommend the use of the latter rather than the former for the following reasons. First, one can use cover halves readily without any pretreatments. Second, eggs (at least those of *Tubifex*) adhere much more firmly to the plastic surface than to the polylysine-coated surface. Third, one can scratch various marks (e.g., the egg's polarity or stream directions) on the plastic surface with a fine needle.

Lastly, there is no need to use newly unpacked dishes. One can use cover halves which have been used once for tissue culture, though one has to wash them before use.

After rinsing several times in an actin stabilization solution (AS solution; 30 mM KCl, 5 mM MgCl$_2$, 5 mM EGTA, 10 mM PIPES, pH 6.9), eggs with vitelline membranes are placed on a gelatin-coated coverslip which has been put in the cover half, and their vitelline membranes are removed with fine forceps. Using a gelatin-coated glass needle, one can move eggs across this coverslip and place them onto the plastic surface orienting them in any directions one desires. By a slow stream of the AS solution, eggs are rolled across the plastic surface and finally subjected to a jet stream of the same solution to wash away the bulk of the cytoplasm. If the animal or vegetal pole is initially placed downward, this procedure isolates cortices of both the animal and the vegetal hemispheres of a single egg. Isolated cortices are fixed with 3.5% formaldehyde in the AS solution for 30 min. (Cold methanol, which has been used in other systems, cannot be used as a fixative for *Tubifex* egg cortices because it causes fragmentation of actin filaments.)

After washed in the AS solution containing 0.1 mg/ml saponin (which is effective to reduce the surface tension of the solution), isolated cortices are labeled with 825 ng/ml rhodamine–phalloidin for 60 min. For immuno-staining, isolated cortices are blocked with 2% bovine serum albumin (BSA) in phosphase-buffered saline (PBS) plus 0.1% sodium azide for 6 hr at room temperature (or overnight at 4°C). Time of subsequent incubations with antibodies depends on their concentrations. For example, isolated cortices are incubated with anti-γ-tubulin antibody (1:1000 in PBS plus 5% BSA and 0.1% sodium azide) for 12 hr at 4°C, followed by rinsing in six changes of PBS containing 0.5% Triton X-100 (each 15 min). The specimens were then incubated with goat anti-rabbit IgG antibody conjugated to FITC (1:50 in PBS plus 5% BSA and 0.1% sodium azide) for 6 hr at room temperature and rinsed again. If necessary, isolated cortices thus labeled with antibodies are next stained with rhodamine–phalloidin as described previously. Specimens labeled with rhodamine–phalloidin, antibodies, or both are sealed with 50% glycerol in PBS containing 2.5% *n*-propyl gallate.

B. Isolated Cortices for Electron Microscopy

Egg cortices can also be isolated on grids coated with collodion and carbon. *Tubifex* eggs firmly adhere to the carbon coat (especially a freshly prepared one) without any additional treatment. The procedure of cortex isolation on the grids is the same as that described previously, except that isolated cortices are fixed with 0.5% glutaraldehyde for 10 min. After rinsing in the AS solution, a drop of 1% aqueous solution of uranyl acetate is put onto grids and immediately absorbed with a small piece of filter paper; grids are then dried in air in a large petri dish. Using this procedure, grids are negatively stained. To identify actin filaments, isolated cortices on grids are incubated with 5 mg/ml heavy meromyosin (prepared from rabbit skeletal muscle) before fixation with glutaraldehyde.

C. Whole-Mount Immunocytochemistry

Tubifex eggs are rather large and opaque. Therefore, if one wishes to examine cytoskeletal organization in whole mounts, it is necessary to devise fixation methods by which antibodies become accessible as far as the egg's center. Furthermore, it is also requisite to make eggs "transparent." The addition of Triton X-100 to a fixative and the use of Murray's clear (a mixture of one part benzyl alcohol : two parts bezyl benzoate) have solved these problems.

Eggs are rinsed briefly in the AS solution and fixed in the AS solution containing 3.5% formaldehyde and 0.9% Triton X-100 for 60 min. After the removal of vitelline membranes in the AS solution, fixed eggs are immersed in methanol/DMSO (4:1) overnight and then in methanol containing 10% H_2O_2 for 24 hr, and stored in methanol at $-20°$ until use.

Before antibody incubation, the eggs stored in methanol are rehydrated in PBS, rinsed in PBS containing 0.5% Triton X-100 three times (each 60 min), and incubated overnight in PBS containing 2% BSA and 0.1% sodium azide. The specimens are next incubated with a mixture of mouse monoclonal antibodies to α- and β-tubulin (1:2000 in PBS plus 5% BSA and 0.1% sodium azide) for 24 hr at 4°C, followed by rinsing in six changes of PBS containing 0.5% Triton X-100 (each more than 60 min). The specimens are then incubated with goat anti-mouse IgG antibody conjugated to horseradish peroxidase (1:2000 in PBS plus 5% BSA) for 24 hr at 4°C and rinsed again. Color development is carried out with 0.25 mg/ml diaminobenzidine and 0.005% H_2O_2. The stained specimens are dehydrated in methanol and cleared with Murray's clear.

References

Astrow, H. S., Holton, B., and Weisblat, D. A. (1989). Teloplasm formation in a leech, *Helobdella triserialis,* is a microtubule-dependent process. *Dev. Biol.* **135,** 306–319.

Bement, W. M., and Capco, D. G. (1991). Analysis of inducible contractile rings suggest a role for protein kinase C in embryonic cytokinesis and wound healing. *Cell Motil. Cytoskeleton* **20,** 145–157.

Bluemink, J. G. (1970). The first cleavage of the amphibian egg. An electron microscope study of the onset of cytokinesis in the egg of *Ambystoma mexicanum*. *J. Ultrastructure Res.* **32**, 142–166.

Capco, D. G., Tutnick, J. M., and Bement, W. M. (1992). The role of protein kinase C in the cortical cytoskeleton during the transition from oocyte to fertilization-competent egg. *J. Exp. Zool.* **264**, 395–405.

Corwin, H. L., and Hartwig, J. H. (1983). Isolation of actin-binding protein and villin from toad oocytes. *Dev. Biol.* **99**, 61–74.

Dan, K., and Tanaka, Y. (1990). Attachment of one spindle pole to the cortex in unequal cleavage. *Ann. N. Y. Acad. Sci.* **582**, 108–119.

Dorresteijn, A. W. C. (1990). Quantitative analysis of cellular differentiation during early embryogenesis of *Platynereis dumerilii*. *Roux's Arch. Dev. Biol.* **199**, 14–30.

Dorresteijn, A. W. C., and Kluge, B. (1990). On the establishment of polarity in polychaete eggs. *In* "Experimental Embryology in Aquatic Plants and Animals" (H.-J. Marthy, ed.), pp. 197–209. Plenum Press, New York.

Elinson, R. P., King, M. L., and Forristall, C. (1993). Isolated vegetal cortex from *Xenopus* oocytes selectively retains localized mRNAs. *Dev. Biol.* **160**, 554–562.

Freeman, G., and Lundelius, J. W. (1992). Evolutionary implications of the mode of D quadrant specification in coelomates with spiral cleavage. *J. Evol. Biol.* **5**, 205–247.

Gurdon, J. B. (1992). The generation of diversity and pattern in animal development. *Cell* **68**, 185–199.

Hamaguchi, M. S., and Hiramoto, Y. (1978). Protoplasmic movement during polar-body formation in starfish oocytes. *Exp. Cell Res.* **112**, 55–62.

Hidaka, H., Inagaki, M., Kawamoto, S., and Sasaki, Y. (1984). Isoquinolinesulfonamides, novel and potent inhibitors of cyclic nucleotide dependent protein kinase and protein kinase C. *Biochemistry* **23**, 5036–5041.

Holland, N. D. (1978). The fine structure of *Comanthus japonica* (Echinodermata: Crinoidea) from zygote through early gastrula. *Tissue Cell* **10**, 93–112.

Holton, B., Astrow, S. H., and Weisblat, D. A. (1989). Animal and vegetal teloplasms mix in the early embryo of the leech, *Helobdella triserialis*. *Dev. Biol.* **131**, 182–188.

Inase, M. (1960). On the double embryo of the aquatic worm *Tubifex hattai*. *Sci. Rep. Tohoku Univ. Ser. IV* **26**, 59–64.

Ishii, R., and Shimizu, T. (1995). The unequal first cleavage in the *Tubifex* egg: Involvement of a monastral mitotic apparatus. *Dev. Growth Differ.* **37**, in press.

Jeffery, W. R. (1989). Localized mRNA and the egg cytoskeleton. *Int. Rev. Cytol.* **119**, 151–195.

Lehmann, F. E. (1938). Zustandständerungen im Ei von *Tubifex* während der Reifungsteilungen. *Arch. Exp. Zellforsch.* **22**, 271–275.

Longo, F. J. (1972). The effects of cytochalasin B on the events of fertilization in the surf clam *Spisula solidissima*. I. Polar body formation. *J. Exp. Zool.* **182**, 321–344.

Mabuchi, I. (1986). Biochemical aspects of cytokinesis. *Int. Rev. Cytol.* **101**, 175–213.

Nelson, B. H., and Weisblat, D. A. (1992). Cytoplasmic and cortical determinants interact to specify ectoderm and mesoderm in the leech embryo. *Development* **115**, 103–115.

Penners, A. (1922). Die Furchung von *Tubifex rivulorum* Lam. *Zool. Jb. Abt. Anat. Ontog.* **43**, 323–367.

Penners, A. (1924). Experimentalle Untersuchungen zum Determinationsproblem an Keim von *Tubifex rivulorum* Lam. I. Die Duplicitas cruciata und Organbildende Keimbezirke. *Arch. Mikrosk. Abt. Entwick. Mechan.* **102**, 51–100.

Rappaport, R. (1986). Establishment of the mechanism of cytokinesis in animal cells. *Int. Rev. Cytol.* **105**, 245–281.

Ryabova, L. V., Vassetzky, S. G., and Capco, D. G. (1994). Development of cortical contractility in the *Xenopus laevis* oocyte mediated by reorganization of the cortical cytoskeleton: A model. *Zygote* **2**, 263–271.

Sawai, T. (1972). Roles of cortical and subcortical components in cleavage furrow formation in amphibia. *J. Cell Sci.* **11,** 543–556.

Schatten, G. (1994). The centrosome and its mode of inheritance: The reduction of the centrosome during gametogenesis and its restoration during fertilization. *Dev. Biol.* **165,** 299–335.

Schroeder, T. E. (1968). Cytokinesis: Filaments in the cleavage furrow. *Exp. Cell Res.* **53,** 272–276.

Shimizu, T. (1975). Occurrence of microfilaments in the *Tubifex* egg undergoing the deformation movement. *J. Fac. Sci. Hokkaido Univ. Ser. VI* **20,** 1–8.

Shimizu, T. (1976a). The fine structure of the *Tubifex* egg before and after fertilization. *J. Fac. Sci. Hokkaido Univ. Ser. VI* **20,** 253–262.

Shimizu, T. (1976b). The staining property of cortical cytoplasm and the appearance of pole plasm in *Tubifex* egg (in Japanese with English abstract). *Zool. Mag.* **85,** 32–39.

Shimizu, T. (1978a). Deformation movement induced by divalent ionophore A23187 in the *Tubifex* egg. *Dev. Growth Differ.* **20,** 27–33.

Shimizu, T. (1978b). Mode of microfilament-arrangement in normal and cytochalasin-treated eggs of *Tubifex* (Annelida, Oligochaeta). *Acta Embryol. Exp.* **1,** 59–74.

Shimizu, T. (1979). Surface contractile activity of the *Tubifex* egg: Its relationship to the meiotic apparatus functions. *J. Exp. Zool.* **208,** 361–378.

Shimizu, T. (1981a). Cyclic changes in shape of a non-nucleate egg fragment of *Tubifex* (Annelida, Oligochaeta). *Dev. Growth Differ.* **23,** 101–109.

Shimizu, T. (1981b). Cortical differentiation of the animal pole during maturation division in fertilized eggs of *Tubifex* (Annelida, Oligochaeta). I. Meiotic apparatus formation. *Dev. Biol.* **85,** 65–76.

Shimizu, T. (1981c). Cortical differentiation of the animal pole during maturation division in fertilized eggs of *Tubifex* (Annelida, Oligochaeta). II. Polar body formation. *Dev. Biol.* **85,** 77–88.

Shimizu, T. (1982a). Development in the freshwater oligochaete *Tubifex*. *In* "Developmental Biology of Freshwater Invertebrates" (F. W. Harrison and R. R. Cowden, eds.), pp. 283–316. A. R. Liss, New York.

Shimizu, T. (1982b). Ooplasmic segregation in the *Tubifex* egg: Mode of pole plasm accumulation and possible involvement of microfilaments. *Roux's Arch. Dev. Biol.* **191,** 246–256.

Shimizu, T. (1983). Organization of actin filaments during polar body formation in eggs of *Tubifex* (Annelida, Oligochaeta). *Eur. J. Cell Biol.* **30,** 74–82.

Shimizu, T. (1984). Dynamics of the actin microfilament system in the *Tubifex* egg during ooplasmic segregation. *Dev. Biol.* **106,** 414–426.

Shimizu, T. (1985). Movements of mitochondria associated with isolated egg cortex. *Dev. Growth Differ.* **27,** 149–154.

Shimizu, T. (1986). Bipolar segregation of mitochondria, actin network and surface in the *Tubifex* egg: Role of cortical polarity. *Dev. Biol.* **116,** 241–251.

Shimizu, T. (1988). Localization of actin networks during early development of *Tubifex* embryos. *Dev. Biol.* **125,** 321–331.

Shimizu, T. (1989). Asymmetric segregation and polarized redistribution of pole plasm during early cleavages in the *Tubifex* embryo: Role of actin networks and mitotic apparatus. *Dev. Growth Differ.* **31,** 283–297.

Shimizu, T. (1990). Polar body formation in *Tubifex* eggs. *Ann. N. Y. Acad. Sci.* **582,** 260–272.

Shimizu, T. (1993). Cleavage asynchrony in the *Tubifex* embryo: Involvement of cytoplasmic and nucleus-associated factors. *Dev. Biol.* **157,** 191–204.

Shimizu, T. (1994). The prevention of smaller blastomeres of early *Tubifex* embryos from entering mitosis by unreplicated DNA. *Dev. Biol.* **161,** 274–284.

Shimizu, T. (1995a). The first two cleavages in *Tubifex* involve distinct mechanisms to generate asymmetry in mitotic apparatus. *Hydrobiologia,* in press.

Shimizu, T. (1995b). Lineage-specific alteration in cell cycle structure in early *Tubifex* embryos. *Dev. Growth Differ.* **37,** 263–272.

Shimizu, T. (1995c). Behavior of centrosomes in early *Tubifex* embryos: Asymmetric segregation and mitotic cycle-dependent duplication. *Roux's Arch. Dev. Biol.,* in press.

Speksnijder, J. E., and Dohmen, M. R. (1983). Local surface modulation correlated with ooplasmic segregation in eggs of *Sabellaria alveolata* (Annelida, Polychaeta). *Roux's Arch. Dev. Biol.* **192,** 248–255.

Stendahl, O. I., and Stossel, T. P. (1980). Actin-binding protein amplifies actomyosin contraction, and gelsolin confers calcium control on the direction of contraction. *Biochem. Biophys. Res. Commun.* **92,** 675–681.

Waddle, J. A., Cooper, J. A., and Waterston, R. H. (1994). Transient localized accumulation of actin in *Caenorhabditis elegans* blastomeres with oriented asymmetric divisions. *Development* **120,** 2317–2328.

Wall, R. (1990). "This Side Up: Spatial Determination in the Early Development of Animals." Cambridge Univ. Press, Cambridge.

Yisraeli, J. K., Sokol, S., and Melton, D. A. (1990). A two-step model for the localization of maternal mRNA in *Xenopus* oocytes: Involvement of microtubules and microfilaments in the translocation and anchoring of Vg1 mRNA. *Development* **108,** 289–298.

Preface to Section II: Chordates

This section of the volume focuses on cytoskeletal mechanisms involved with early development in chordates. The chapters listed in parentheses denote chapters in which comparable cytoskeletal mechanisms also have been reported.

Chapter 8 (by William Jeffery) examines the involvement of the cytoskeleton in one of the most primitive classes of chordates, ascidians. Work performed by this chapter's author using ascidian eggs and embryos first demonstrated the presence of several features of the cytoskeleton in early development; the existence of a cortical cytoskeletal domain, localization of specific mRNA attached to the cytoskeleton, the positioning of morphogenetic determinants associated with the cytoskeleton, and segregation of the determinants into specific locations of the embryo by the cytoskeleton (Chapters 1–7, 12, and 14). In this system the cortical cytoskeletal domain undergoes an extensive remodeling in response to the developmental transition of fertilization. In addition, a cortical rotation, associated with an array of microtubules, occurs comparable to that which occurs in several other classes of organisms (Chapters 2, 5, 12, and 13). The author has recently demonstrated a role for the cytoskeleton in mediating evolutionary change which has been shown by exploiting the difference between tailed and tailless species (a developmental transition at the larval stage).

Two chapters investigate the function of the cytoskeleton in mammalian development. Chapter 9 (by G. Ian Gallicano and David Capco) presents a new feature of cytoskeletal involvement in development, that is, the existence of a unique cytoskeletal structure that appears to be involved with the specialization for placental development in mammals. This unusual cytoskeletal structure forms an extensive network within mammalian eggs. This network does not represent a new cytoskeletal protein, but instead these structures are composed of highly cross-linked arrays of intermediate filaments with precise spacing. This network undergoes extensive remodeling at major developmental transitions and its organization is regulated by calcium and more proximally by the action of protein kinase C and its cytosolic counterpart protein kinase M (Chapters 1, 3–5, 7, 11, 13, and 14). Chapter 10 (by Schatten and co-workers) examines the microtubule and

237

actin filament networks of mammalian eggs. Here the authors provide evidence to demonstrate that microtubules have a role in pronuclear juxtaposition (Chapters 2 and 3). Moreover, evidence is presented which demonstrates an extensive reorganization of the actin filament network at the time of fertilization (Chapters 1, 2, 7, 12, and 13). The authors also examine centrosomal inheritance in the zygote.

Chapter 11 (by Nathan Hart and Richard Fluck) considers the involvement of the cytoskeleton in early development in fish, more specifically teleost eggs and embryos. This system has the advantage of including the genetically manipulable zebrafish, and consequently holds the promise of conducting genetic manipulations of the cytoskeleton and its associated proteins in a system which has a pattern of development different from insects. Comparison of results from such manipulations on *Drosophila* with zebrafish are certain to provide great insight on mechanisms of cytoskeletal involvement in early development. Studies of teleost development reveal evidence for ooplasmic migration mediated by the cytoskeleton as well as a contraction of the ooplasm mediated by cytoskeletal reorganization (Chapters 1, 2, 4, 7, and 13). There is also the establishment of a polarized cortical/subcortical microtubule array (Chapters 2, 5, 7, 12, and 13). Finally, the authors show that cytoplasmic signal transduction events appear to regulate cytoskeletal organization (Chapters 1, 3–5, 7, 9, and 14).

Three chapters evaluate amphibian development, specifically, *Xenopus laevis*. The first of these chapters, Chapter 12 (by David Gard and co-workers) focuses on the microtubule network in amphibian eggs. It demonstrates that an extensive remodeling of the microtubule network accompanies each developmental transition during the process of oogenesis. Data are reviewed on the establishment of the vegetal array of microtubules which form in response to fertilization (Chapters 2, 5, 8, and 11) and are involved with the rotation of the cortical cytoplasm with respect to the internal cytoplasm and appear to mediate axis formation (Chapters 2 and 8). Chapter 12 also reviews the involvement of the cytoskeleton in the localization of specific mRNAs (Chapters 1, 3, 5, 8, and 14) and suggests that localization of mRNA is a two-step process (Chapter 5). Chapter 13 (by Carolyn Larabell) examines the role of the actin cytoskeleton during early development and examines the interaction of the cortical actin and cortical microtubule networks. Chapter 14 (by Michael Klymkowsky) focuses on the intermediate filament network in amphibians and the existence of a cortical cytoskeletal network of intermediate filaments linked to an internal network. The intermediate filament network undergoes extensive remodeling as the oocyte becomes the fertilization-competent egg and the remodeling process is continued as a result of fertilization. This remodeling is under the control of cytoplasmic signal transduction events (Chapters 3, 4, 7, 9, and 11). This chapter also considers a role for localized mRNA in association with the cytoskeleton (Chapters 1, 3, 5, 8, and 12). In addition, this chapter extends consideration of cytoskeletal involvement during later developmental stage and the role of intermediate filaments in cell adhesion.

Section II

Cytoskeletal Mechanisms in Chordate Development

8

Development and Evolution of an Egg Cytoskeletal Domain in Ascidians

William R. Jeffery
Section of Molecular and Cellular Biology
University of California
Davis, California 95616

I. Introduction

The egg is the only metazoan cell that can accumulate yolk, pause in the meiotic cycle, or initiate development of a new individual by fusing with a sperm. Developmental totipotency is also a special quality of the egg. As a totipotent cell, the egg contains instructions for generating the body plan and different cell types of the organism. Considering the unique properties of the egg, it is not surprising that the egg cytoskeleton also performs specialized roles. These roles include maintaining the integrity of a large cell, providing a framework for meiosis and syngamy, and generating a specific cleavage pattern. Perhaps the most unique role of the egg cytoskele-

Current Topics in Developmental Biology, Vol. 31

ton is to sequester and correctly position regulatory factors in the embryo. Maternal mRNAs localized at the poles of the *Drosophila* egg encode transcription factors and other regulatory proteins that determine the anteroposterior axes and germline (reviewed by Nüsslein-Volhard, 1991). These polar mRNAs are maintained in their correct spatial positions by interactions with the cytoskeleton (reviewed by Macdonald, 1992). Maternal mRNAs are also localized in the vegetal hemisphere of *Xenopus* eggs via cytoskeletal elements (Forristall *et al.*, 1995). Other examples of mRNAs that are localized by association with the egg cytoskeleton have been reviewed by Jeffery (1989a).

As a gamete, the egg transmits genotypic information and establishes a phenotype that is influenced by mutation and selection in previous generations. Thus, evolutionary changes are transmitted through the egg and its product, the embryo. Although considerable attention has been paid to the egg cytoskeleton in developmental processes (reviewed by Sardet *et al.*, 1994), its role in mediating evolutionary changes has been virtually ignored. Our lack of knowledge concerning the evolutionary role of the egg cytoskeleton is in part due to a paucity of model systems. A special set of properties is required to define model systems for studying the evolution of development. First, two or more species are required that are closely related but show distinct developmental differences. By studying closely related species, we attempt to minimize differences that have arisen by genetic drift. Moreover, regulatory changes discovered in one species are more likely to be conserved and easier to identify in a closely related species. Second, the evolutionary history of the species should be understood in order to permit changes in development to be viewed as ancestral or derived. Third, the embryology of the species should be well known and they should be amenable to culture and manipulation in the laboratory. Ascidians are organisms that satisfy these criteria (reviewed by Jeffery and Swalla, 1992a).

In this chapter, we review an egg cytoskeletal domain in ascidians and its role in the evolution of development. First, we consider the ascidian life cycle and mode of development. Second, we review the structure and function of the myoplasmic cytoskeletal domain (MCD) and its developmental roles. Third, we describe the ascidian model system consisting of closely related species with different modes of development. Finally, we consider how changes in the MCD may be involved in the evolution of development. The narative is followed by an appendix presenting the methods employed to study the egg cytoskeleton in ascidians.

II. Ascidian Development

Ascidians are urochordates with a life cycle consisting of swimming larval and sessile adult stages. Development from the egg to the tadpole larva has been studied for more than a century (reviewed by Satoh, 1994). Ascidian eggs are fertilized in seawater and complete maturation, cleavage, and development to the larval stage in as little as 12 hr. The tadpole larva contains about 2500 cells and has only six different types of tissue. The larval tissues are epidermis, nervous system (brain and spinal cord), notochord, muscle, mesenchyme, and endoderm. The endoderm, brain, and mesenchyme are located in the head of the tadpole, the spinal cord, notochord, and muscle are located in the tail, and epidermis surrounds the larva. Some of these larval tissues have particularly low cell numbers: the notochord consists of only 40 cells and there are 36–42 muscle cells in the larvae of most ascidian species. This low cell number facilitates cell lineage analysis. Consequently, the fate of each embryonic cell in the ascidian embryo is well defined (reveiwed by Satoh, 1994). After dispersal by swimming, the larva settles on a substrate and undergoes metamorphosis. During metamorphosis the tail is retracted into the head and destroyed, and adult tissues and organs are formed *de novo*.

Embryonic development is shown in Fig. 1. Although several cytoplasmic domains are present in ascidian eggs, this diagram highlights the fate of the myoplasm, a region that is distributed primarily to the muscle cell lineages during cleavage. The myoplasm is localized in the cortex of the unfertilized egg (Fig. 1A). Fertilization triggers the completion of maturation and ooplasmic segregation (Figs. 1A–1C), a translocation of egg cytoplasmic domains that is carried out in two phases (reviewed by Jeffery and Bates, 1989). During the first phase, the myoplasm is concentrated into a cap near the vegetal pole (Fig. 1B), which marks the future dorsal pole of the embryo (Bates and Jeffery, 1987). At the same time, the male pronucleus migrates vegetally and forms an aster in the myoplasm. The microtubules of the sperm aster orchestrate myoplasmic movements during the second phase of ooplasmic segregation (Sawada and Schatten, 1989). During the second phase, most of the myoplasm is extended laterally into a crescent (Fig. 1C)—the famous yellow crescent of Conklin (1905)—which marks the future posterior pole of the embryo. Cleavage is bilateral, synchronous, and equal (Figs. 1D and 1E) until the 16-cell stage, when differences arise in the animal and vegetal blastomeres. The myoplasm is distributed to both blastomeres of the 2-cell embryo, but subsequently to only two cells of 4- and 8-cell embryos, six cells of the 16-cell embryo, eight cells of the 32-cell embryo, and eventually to the progenitors of the tail muscle cells (Figs. 1D–1H). Other cytoplasmic domains are also partitioned to specific blastomeres during cleavage, as described by Conklin (1905).

Following the rapid cleavage phase, the animal pole region contains the prospective ectoderm (epidermis and nervous system), the equatorial zone the prospective mesoderm (notochord, mesenchyme, and muscle), and the vegetal pole region the prospective endoderm. The fate map is conserved

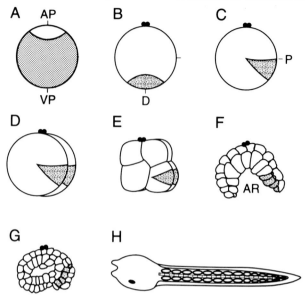

Fig. 1 Ascidian development. (A) An unfertilized egg showing the myoplasm (lightly shaded), the animal pole (AP), and the vegetal pole (VP). (B) A fertilized egg after completion of maturation and the first phase of ooplasmic segregation showing the vegetal cap of myoplasm (shaded), the polar bodies (top), and the future dorsal (vegetal) pole (D). (C) A fertilized egg after completion of the second phase of ooplasmic segregation showing the myoplasmic crescent (shaded) at the posterior pole (P). (D and E) Two- and eight-cell embryos, respectively. (F) An early gastrula showing the archenteron (AR) and muscle cells (shaded). (G) A gastrula. (H) A hatched tadpole larva with an otolith and a tail containing notochord and muscle cells. A–E, whole embryos; F–H, sectioned embryos. Reproduced from Jeffery (1994).

among different ascidian species (Satoh, 1994). During gastrulation, the endoderm cells invaginate into the interior of the embryo near the vegetal pole and are followed by involution of the mesoderm (reviewed by Jeffery, 1992). The position of invagination is near the vegetal pole, the site formerly occupied by the cap of myoplasm at the conclusion of the first phase of ooplasmic segregation (Bates and Jeffery, 1987). The presumptive notochord cells involute over the anterior lip, the prospective mesenchyme cells over the lateral lips, and the presumptive muscle cells over the posterior lip of the blastopore. Gastrulation is completed by epiboly, the spreading of ectoderm cells over the mesoderm and endoderm. A diagram of a late gastrula is shown in Fig. 2E.

As gastrulation is completed, neurulation begins on the ventral (dorsal) surface, the embryo elongates along the anteroposterior axis, and the blastopore closes. An embryo undergoing tail morphogenesis is shown in Fig.

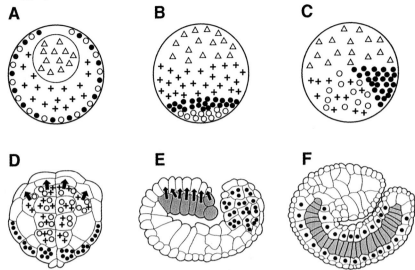

Fig. 2 Control of ascidian embryogenesis. (A) A primary oocyte in section showing ectoplasmic determinants in the germinal vesicle (△), endodermal determinants (+) in the internal cytoplasm, and gastrulation/dorsoventral axis (○) and muscle/anteroposterior (●) axis determinants in the cortical myoplasm. (B) A fertilized zygote after completion of the first phase of ooplasmic segregation showing ectoplasmic determinants in the animal hemisphere cytoplasm, endodermal determinants in the vegetal hemisphere cytoplasm, and muscle/anteroposterior axis and gastrulation/dorsoventral axis determinants layered at the vegetal pole. (C) A fertilized zygote after completion of the second phase of ooplasmic segregation showing ectodermal determinants in the animal hemisphere, endodermal determinants in the vegetal hemisphere, gastrulation/dorsoventral axis determinants ingressed and intermingled with endodermal determinants in the vegetal hemisphere, and muscle/anteroposterior axis determinants in the myoplasmic crescent. (D) A 64-cell embryo viewed from the vegetal pole showing an inductive signal (arrows) transmitted from the endoderm to prospective notochord cells (top). The prospective muscle cells are located at the bottom. (E) A sagittal section of a late gastrula showing an inductive signal (arrows) transmitted from the prospective notochord cells (shaded) to the neural plate cells at the anterior margin of the blastopore. The prospective muscle cells lie at the posterior margin of the blastopore. (F) Sagittal section of an early tailbud embryo showing the extending notochord (shaded) flanked by differentiating muscle cells. A–C, the vegetal (dorsal) pole is located at the bottom; C and D, the posterior pole is located at the right (C) or bottom (D); E and F, the dorsal (vegetal) pole is located at the top and the posterior pole at the right. Reproduced from Jeffery (1994).

2F. The elaboration of the tail involves concerted movements of notochord, muscle, spinal cord, and epidermal cells. These movements are driven by the swelling and posterior extension of the notochord (Miyamoto and Crowther, 1985). The neural tube is later subdivided into the brain and spinal cord, and pigmented sensory organs (the otolith and ocellus) differentiate in the brain. Hatching completes the embryonic phase of the life cycle.

III. Control of Development

Ascidian development has been extensively reviewed (Venuti and Jeffery, 1989; Swalla, 1992; Nishida, 1992a, Satoh, 1994). Therefore, we describe here only those events required to understand how development is changed during evolution.

Regional and cell fate specification in ascidian embryos is mediated by two processes: autonomous and conditional determination. In autonomous determination, the fate of a blastomere is controlled intrinsically by regulatory factors inherited from the egg. The best example of this type of development is the tail muscle cells, which inherit most of the myoplasm. Deletion of the myoplasm from the egg causes defective muscle development, whereas transfer of the myoplasm to nonmuscle blastomeres causes them to adopt a muscle cell fate (Nishida, 1992b). These results show that factors sequestered in the myoplasm can act positively to determine the fate of a muscle cell. The specification of the ectoderm and endoderm is also based on inheritance of factors from the egg. In conditional determination, cell fate is specified by extrinsic factors such as inductive cell interactions. The notochord, nervous system, and a subset of the tail muscle cells are specified by conditional determination (Reverberi et al., 1960; Nishida, 1991; Nakatani and Nishida, 1994). The identity of the factors involved for both types of determination is unknown.

The larval body is established after gastrulation. The nervous system develops at the dorsal pole and the tail at the posterior pole of the embryo. The dorsoventral and anteroposterior axes are determined during ooplasmic segregation, as shown by cytoplasmic deletion experiments. Bates and Jeffery (1987) deleted a portion of the myoplasmic cap from the vegetal (dorsal) pole of the fertilized egg shortly after completion of the first phase of ooplasmic segregation. Embryos lacking vegetal pole cytoplasm failed to gastrulate and produce dorsal structures, although they were able to develop tail muscle cells. Similar results were obtained by ultraviolet (uv) irradiation of the vegetal pole region (Jeffery, 1990a). Nishida (1994) deleted myoplasm from the posterior pole after completion of the second phase of ooplasmic segregation. Embryos lacking the posterior myoplasm gastrulated normally, but developed into radialized larvae lacking an anteroposterior axis and muscle cells. Little is known about the factors involved in axis determination, although the targets of uv irradiation appear to be maternal mRNAs (Jeffery, 1990b).

Ascidian embryogenesis is shown in Fig. 2. Determinants regulate deveopment of (1) the ectoderm, (2) the endoderm, (3) the muscle cells and anteroposterior axis, and (4) gastrulation and the dorsoventral axis (Fig. 2A). The determinants are translocated to different regions of the zygote during ooplasmic segregation and inherited by specific blastomeres (Figs.

2A–2D). The muscle/anteroposterior axis determinants program muscle cell differentiation and the asymmetric cleavages leading to development of the anteroposterior axis. The determinants that enter the endoderm are particularly significant because they initiate an inductive cascade leading to the establishment of the dorsoventral axis. The gastrulation/dorsoventral axis determinants program changes in cell behavior during gastrulation, and the endoderm determinants promote induction of the notochord (Fig. 2D) and possibily the spinal cord by endoderm cells. The next step in the inductive cascade is controversial because some investigators argue that notochord cells (Reverberi *et al.*, 1960), while others argue that spinal cord cells (Nishida, 1992b), induce the brain and neural sensory cells. Figure 2E shows a vertical inductive signal from the notochord to the presumptive neural cells on the dorsal midline of a late gastrula. After performing the presumed inductive function, the prospective notochord cells undergo morphogenetic processes leading to tail formation.

In summary, the myoplasm has three well-defined roles in development: (1) determination of the site of gastrulation, (2) determination of the anterioposterior axis, and (3) determination of the tail muscle cells. According to Sardet *et al.* (1989), a small part of the myoplasm moves anteriorly during the second phase of ooplasmic segregation. The direction of this movement is opposite to that of the major part of this cytoplasmic domain, distributing myoplasm to the anterior region of the zygote. The endoderm, notochord, spinal cord, and neural plate are derived from this part of the fate map. Thus, an additional role of the myoplasm may be to program the inductive cascade leading to the establishment of the dorsoventral axis.

IV. The Myoplasmic Cytoskeletal Domain

The myoplasm was first described by Conklin (1905) as a bright yellow cytoplasm in *Styela partita* eggs. Subsequently, eggs of other ascidian species were described with colored myoplasms. For example, the ascidian *Boltenia villosa* contains an orange myoplasm (Jeffery, 1982a). While all ascidian species do not have colored myoplasms, they exhibit a cytoplasmic domain enriched in organelles characteristic of the myoplasm (Jeffery, 1984a). Ultrastructural studies show that the myoplasm is composed of mitochondria and lipid pigment granules embedded in a reticular matrix (Berg and Humphreys, 1960). The matrix of the myoplasm is a specialized cytoskeletal domain: the MCD. An isolated MCD is shown in Fig. 3.

A. Stucture and Composition

The MCD was discovered in eggs extracted with the nonionic detergent Triton X-100 (Jeffery and Meier, 1983). Nonionic detergents extract lipids and other soluble cellular components, but do not affect cytoskeletal filaments and associated organelles, which are retained in a detergent insoluble residue. Extraction of *Styela* and *Boltenia* eggs with Triton X-100 and scanning electron microscopy of the detergent-insoluble residue revealed the MCD (Fig. 4). The MCD consists of two parts: the plasma membrane lamina (PML) and the deep filamentous lattice (DFL).

The PML is a network of regularly spaced filaments lying immediately below the egg plasma membrane (Fig. 4A). This network is attached to the DFL and also interacts with cell surface components via integral proteins spanning the membrane bilayer. The surface components include lectin-binding sites (Monroy *et al.*, 1973), egg accessory cells (Conklin, 1905), and supernumerary sperm (Sawada and Osanai, 1981). The PML is stained with actin antibodies or phalloidin, interacts with myosin fragments, and is disrupted by treating eggs with DNase I, indicating that it is composed of F-actin (Jeffery and Meier, 1983; Sawada and Osanai, 1985). The PML resembles the plasma membrane skeleton of mammalian erythrocytes and other somatic cells (reviewed by Pumplin and Block, 1993). This analogy is supported by the discovery of spectrin, ankryin, and the transmembrane protein $Na^+K^+ATPase$ in the PML (Jeffery, 1993a; Jeffery and Swalla, 1993).

Figure 5 shows ascidian eggs stained with antibodies that recognize the membrane–skeletal proteins spectrin, ankryin, and $Na^+K^+ATPase$. These proteins are detected in the myoplasm, where they are localized at the interface between the plasma membrane and the MCD (Fig. 5). After fertilization, the spectrin, ankryin, and $Na^+K^+ATPase$ molecules segregate with the myoplasm to the posterior region of the zygote and are inherited by the presumptive muscle cells. Later during embryogenesis, these proteins also appear in other embryonic cells (Jeffery, 1993a; Jeffery and Swalla, 1993). All three proteins are concentrated in the apical margins of the epidermal cells, and $Na^+K^+ATPase$ is present in the mesenchyme cells, which presumably synthesize these proteins *de novo* during embryogenesis. In contrast, the spectrin, ankryin, and $Na^+K^+ATPase$ molecules localized in the MCD are derived from the egg.

The DFL is a lattice of filaments extending throughout the myoplasm (Fig. 2B). Organelles enriched in the myoplasm, including pigment granules, mitochondria, and sheets of endoplasmic reticulum (Berg and Humphreys, 1960; Sardet *et al.*, 1992; Speksnijder *et al.*, 1993), are embedded in this lattice. Moreover, the structure contains localized maternal RNAs (Jeffery, 1984b; Swalla and Jeffery, 1995). Although its composition has not been completely resolved, the DFL contains a major 58-kDa protein (p58) recognized by the monoclonal antibody NN18 (Swalla *et al.*, 1991). The NN18 antibody was produced against the porcine middle-molecular-weight neurofilament protein and reacts with this protein in a variety of vertebrate

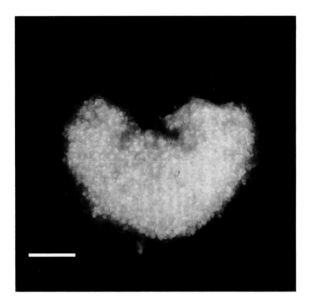

Fig. 3. An MCD isolated from *Styela plicata* eggs. Scale bar, 10 μm. [From Jeffery (1985).]

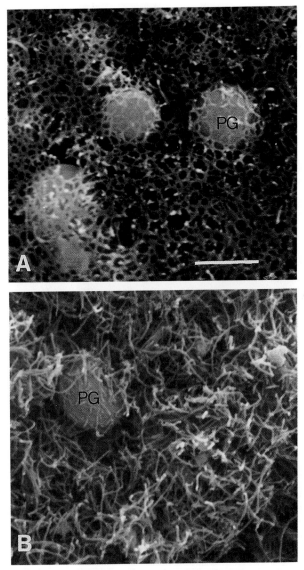

Fig. 4 The MCD in Triton X-100-extracted *Styela plicata* eggs. A scanning electron micrograph of the egg surface showing the PML and pigment granules (PG) below it. (B) A scanning electron micrograph of an egg in which the PML was removed showing the DFL and embedded pigment granules (pg). Scale bar = 2 μm. Reproduced from Jeffery and Meier (1983).

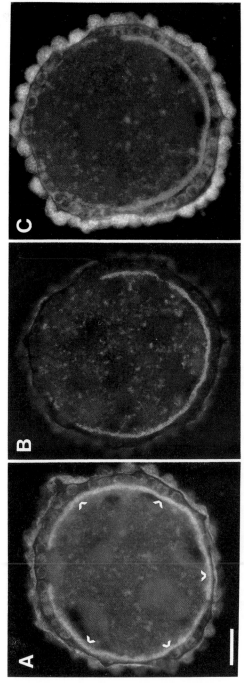

Fig. 5 Localization of Na$^+$K$^+$ATPase (A), ankryin (B), and spectrin (C) in the myoplasm of *Ascidia ceratodes* eggs before and after fertilization. (A) An unfertilized egg showing Na$^+$K$^+$ATPase localized in the myoplasm (arrowheads). (B) A fertilized egg undergoing the first phase of ooplasmic segregation showing ankryin (B) and spectrin (C) localized in the myoplasm. The structure outside the egg plasma membrane is the follicle, which consists of an outer layer of follicle cells (labeled in Fig. 6B), the acellular chorion, and an inner layer of test cells (labeled in Figs. 6A and 6D). In C, the follicle cells and, to a lesser extent, the test cells also contain spectrin. Sectioned eggs stained with Na$^+$K$^+$ATPase, ankryin, and spectrin antibodies are shown with their vegetal poles facing the bottom of each frame. Scale bar = 10μm. Reproduced in part from Jeffery and Swalla (1993).

species (Debus *et al.*, 1983; Shaw *et al.*, 1984). Recognition by NN18 antibody suggests that p58 may share an epitope with the intermediate filament proteins. The p58 protein is distributed to the muscle cell lineage during cleavage and is present in differentiated muscle cells (Fig. 6). Furthermore, p58 is retained in eggs extracted with Triton X-100, suggesting it is tenaciously associated with the MCD (Swalla *et al.*, 1991). Myoplasmin, another protein localized in the DFL, was discovered because it reacts with a mono-

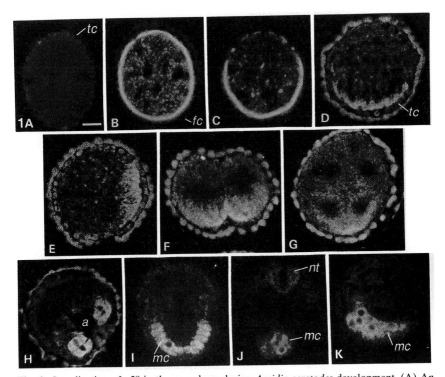

Fig. 6 Localization of p58 in the myoplasm during *Ascidia ceratodes* development. (A) An unfertilized egg which served as a control for antibody staining (tc, test cells). (B) An unfertilized egg showing p58 localization in the cortical myoplasm (fc, follicle cells). (C) A fertilized egg undergoing ooplasmic segregation showing an early stage in p58 translocation to the vegetal pole. (D and E) Fertilized eggs after completion of the first (D) or second (E) phase of ooplasmic segregation showing p58 localized at the vegetal and posterior poles, respectively. (F) A two-cell embryo showing p58 in the myoplasm of both blastomeres. (G) A 4-cell embryo showing p58 localized in the myoplasm of the posterior blastomeres. (H) A gastrula showing p58 localized in the muscle cell precursors (a, archenteron). (I) An early tailbud embryo showing p58 localized in the tail muscle cells (mc). (J and K) Later tailbud embryos in cross (J) and sagittal (K) section showing p58 localized in the muscle cells (mc) and to a lesser extent in the neural tube (nt). Sections were stained with NN18 antibody. In A–F, the vegetal pole is shown at the bottom. In G–K, the posterior pole is shown at the bottom (G, I, and J) or bottom right (H). Scale bar = 20 μm. Reproduced from Swalla *et al.* (1991).

clonal antibody produced from isolated myoplasms (Nishikata *et al.*, 1987a). According to Nishikata *et al.* (1987b), injection of myoplasmin antibody into eggs blocks the appearance of the muscle enzyme acetylcholinesterase, suggesting a role in muscle cell differentiation. It is possible that p58 and myoplasmin interact in the DFL.

Mictrotubules are a third component of the DFL. Although rarely observed in the myoplasm at earlier stages, astral microtubules penetrate the DFL during the second phase of ooplasmic segregation (Sawada and Schatten, 1988; Speksnijder *et al.*, 1993). These microtubules and associated motor proteins may be responsible for translocating myoplasmic organelles to the anterior and posterior poles of the zygote (Sardet *et al.*, 1994).

B. Assembly, Localization, and Segregation

The MCD is assembled and localized in the egg cortex during oogenesis (Swalla *et al.*, 1991; Jeffery, 1993a; Jeffery and Swalla, 1993). In previtellogenic oocytes, actin and spectrin are already associated with the plasma membrane. Before the beginning of vitellogenesis, cytoplasmic vesicles containing ankryin are translocated to the cortex and incorporated into the PML. Based on the distribution of p58, maternal RNA, and myoplasmic organelles, the DFL and PML are assembled into the MCD during vitellogenesis. p58 is first observed in the perinuclear region of previtellogenic oocytes, then gradually moves toward the plasma membrane during vitellogenesis, and is finally localized in the oocyte cortex (see Figs. 12E–12H and 16B and 16D). Maternal RNA molecules are colocalized with p58 during its translocation and incorporation into the MCD (Swalla and Jeffery, 1995). Mitochondria and pigment granules are initially present in a spherical structure located in the cytoplasm on the vegetal side of the germinal vesicle (Hsu, 1963; Jeffery and Bates, 1989). This structure, which resembles the mitochondrial mass of amphibian oocytes (Billet, 1979), disintegrates during vitellogenesis, and its contents are translocated to the cortex. The MCD appears to be assembled from elements of the PML and DFL originating in different parts of the oocyte.

Although the PML and DFL are translocated in concert during both phases of ooplasmic segregation, cell surface components associated with the MCD behave differently. During the first phase, they segregate to the vegetal pole with the MCD, but remain stationary during the second phase (Monroy *et al.*, 1973; Sawada and Osanai, 1981; Bates and Jeffery, 1987) and are incorporated into endoderm cells at the vegetal pole (Bates and Jeffery, 1987). Severing the transmembrane linkage between the MCD and the egg surface components may permit the myoplasm to shift during the second phase of ooplasmic segregation. This shift, which resembles cortical rotation in *Xenopus* eggs (Vincent *et al.,* 1986), is responsible for segregating the gastrulation/dorsoventral axis and muscle/anteroposterior axis determinants to different parts of the zygote (see Figs. 2B and 2C).

Ooplasmic segregation of the MCD in *Styela* eggs is shown in Fig. 7. The model is based on light microscopy of living eggs and scanning electron microscopy of Triton X-100 extracted eggs (Jeffery and Meier, 1983). A similar model has been developed to explain ooplasmic segregation in the ascidian *Ciona* (Sawada, 1983). Both models propose that ooplasmic segregation is mediated by distinct cytoskeletal elements. Several lines of evidence suggest that ooplasmic segregation is initiated by contraction of actin filaments in the PML. First, translocation of the MCD to the vegetal pole is sensitive to cytochalasin B but not to colchicine (Sawada and Osanai, 1981). Second, the PML tightens and is reduced in area as the myoplasm is translocated to the vegetal pole (Jeffery and Meier, 1983). Third, when unfertilized eggs are centrifuged at forces that displace the DFL, the PML still contracts into a cap after fertilization (Jeffery and Meier, 1984). Therefore, contraction of the PML appears to be the motive force for the first phase of ooplasmic segregation. The DFL and associated egg surface components are dragged to the vegetal pole because of their association with the PML.

The second phase of ooplasmic segregation is sensitive to colchicine but not cytochalasin B, implying that it is directed by microtubules (Zalokar, 1974; Sawada and Schatten, 1989). The sperm aster forms in the vegetal hemisphere after completion of the first phase of ooplasmic segregation (Conklin, 1905; Sawada and Schatten, 1988; Sardet *et al.,* 1989). As astral rays extend toward the egg periphery, they meet and displace the MCD into a crescent at the posterior pole. As described previously, a small part of the MCD is also translocated to the anterior pole of the egg (Sardet *et al.,* 1989). Motor proteins associated with the astral microtubules may promote bidirectional MCD segregation.

Although the MCD behaves as a unit during ooplasmic segregation and the first few cleavages, the DFL separates from the PML and returns to its original perinuclear location in the presumptive muscle cells (Jeffery, 1989b). The association of the DFL with the nucleus may be related to the initiation of a specific program of gene expression in the muscle cells.

C. Polarization

The MCD segregates to the vegetal pole during the first phase of ooplasmic segregation. Conklin (1905) proposed that segregation is oriented toward the vegetal pole because it is the site of sperm entry. Conklin's hypothesis was based on his observations of the male pronucleus in the myoplasm

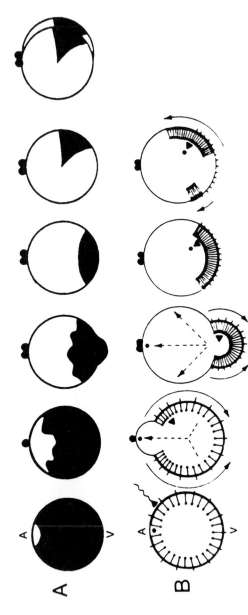

Fig. 7 Ooplasmic segregation of the myoplasm and MCD in ascidian eggs. (A) The translocation of myoplasm (shaded areas) is depicted in whole eggs from fertilization to first cleavage. From left to right the stages are an unfertilized egg, a fertilized egg after completion of the first maturation division, a fertilized egg after completion of second maturation, a fertilized egg after completion of the second maturation division and the first phase of ooplasmic segregation, a fertilized egg after completion of the second phase of ooplasmic segregation, and a two-cell embryo. The filled spheres outside the egg represent polar bodies. A, animal pole; V, vegetal pole. (B) A schematic model for MCD segregation. Egg cytoskeletons are depicted in sections corresponding to the stages shown in A. The thick outer boundaries represent the parts of the egg plasma membrane associated with the underlying PML and the thin boundaries represent regions not associated with PML. The thin lines drawn perpendicular to the outer surface of the plasma membrane represent surface components translocated with the MCD during the first phase of ooplasmic segregation and deposited in the vegetal hemisphere. The lines drawn perpendicular to the inner surface of the plasma membrane and topped with filled spheres represent the DFL and associated myoplasmic organelles. The filled sphere inside the egg represents the female pronucleus and the filled triangle the sperm head (outside the egg) or male pronucleus (inside the egg). The solid arrows represent the direction of MCD translocation during ooplasmic segregation and broken arrows represent the direction of movement of the internal cytoplasmic domains. Reproduced in part from Jeffery (1992).

near the vegetal pole shortly after fertilization. Bates and Jeffery (1988) investigated the possilibity of a restricted sperm entry site by removing the vegetal pole region of unfertilized eggs. They found that eggs lacking the vegetal pole were fertilized and able to develop normally. Speksnijder *et al.* (1990a) investigated the site of sperm entry by fertilizing eggs that were preincubated in a DNA-specific fluorescent dye. They observed that labeled male pronuclei were present in both the animal and the vegetal hemispheres, but concluded that the animal hemisphere is the preferred site of sperm entry. Both groups of investigators rejected the hypothesis that the sperm entry site is restricted to the vegetal pole. Instead, the sperm is likely to enter the egg in the animal hemisphere and to be translocated to the vegetal pole with the advancing myoplasm.

If the site of sperm entry does not focus the MCD, what is the polarizing cue? Another possibility is that polarity is already built into the animal–vegetal axis of the unfertilized egg. According to this idea, sperm penetration would trigger segregation of the MCD, but the direction of segregation would be predetermined in the egg. This hypothesis is supported by investigations of myoplasmic segregation in animal and vegetal egg fragments (Bates and Jeffery, 1988). Consistent with the possibility of a vegetal-to-animal gradient of polarizing factors, the myoplasm segregates to the vegetal pole in fertilized vegetal fragments and the vegetal-most region in fertilized animal fragments. Other studies have shown that the myoplasm can segregate toward an exogenous source of calcium ionophore applied to the surface of an oocyte or unfertilized egg (Jeffery, 1982a; Bates and Jeffery, 1988), suggesting that the polarization gradient may be related to a calcium flux in the vegetal hemisphere (Speksnijder *et al.*, 1990b). Endogenous calcium required for polarizing the MCD appears to be sequestered in the endoplasmic reticulum (Speksnijder *et al.*, 1993).

The factors reponsible for directing the MCD to the posterior pole during the second phase of ooplasmic segregation have not been investigated. However, a specific cue may not be required for the second movement. Instead, the second phase may be fixed relative to the direction of MCD translocation during the first phase of ooplasmic segregation.

V. Evolutionary Changes in Development

The morphology of the tadpole larva is changed during ascidian evolution. Some ascidian species have evolved large tadpoles with robust tails containing hundreds of muscle cells, presumably an adaptation for enhanced larval dispersal (reviewed Jeffery and Swalla, 1992a). Other species have lost the tailed larva, possibly to conserve energy and/or limit dispersal in restricted habitats (reviewed by Jeffery and Swalla, 1990). The species lacking a tailed larva are called anural (or tailless) developers, while those with a tadpole larva are known as urodele (or tailed) developers.

Anural developers have been described in only 2 of the 14 ascidian families: the Styelidae and Molgulidae. Despite this restriction, molecular phylogenetic analysis supports a polyphyletic origin of anural developers from urodele ancestors (Hadfield *et al.*, 1995). The phylogenetic analysis also suggests that anural developers evolved independently, once in the Styelidae and at least four times in the Molgulidae (Fig. 8). In addition, the phylogeny suggests that closely related species can exhibit different modes of development: *Molgula oculata* is a urodele developer and its sister species *Molgula occulta* is an anural developer (Figs. 8 and 9, top). These sister species are being used as a model system to study the mechanisms underlying evolution of anural development (reviewed by Jeffery and Swalla, 1992a).

Embryological studies have revealed the developmental programs that are modified in *M. occulta* embryos (Berrill, 1931; Swalla and Jeffery, 1990; Jeffery and Swalla, 1991). There are no striking differences in egg size, rate of development, cleavage pattern, gastrulation, or neurulation between

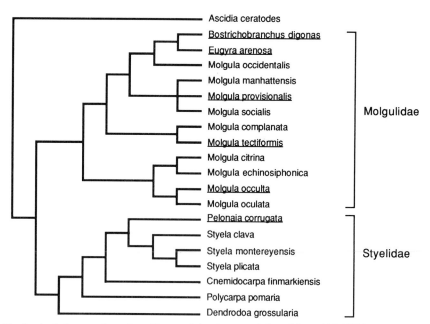

Fig. 8 A phylogeny of species with anural development inferred from 18S rDNA sequences. The anural developers are underlined. All other species are urodele developers. The conventional classification is shown on the left. A strict consensus tree is shown. See Hadfield *et al.* (1995) for further details.

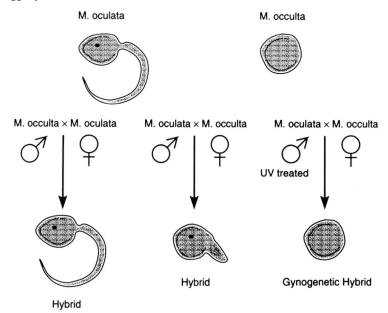

Fig. 9 The morphology of *M. oculata, M. occulta,* and interspecific hybrid larvae. The larval types of the parental species are indicated in the top row: *M. oculata* has a tailed (urodele) larva and *M. occulta* a tailless (anural) larva. Various interspecific crosses (Swalla and Jeffery, 1990) and the resulting hybrid larvae are indicated in the bottom row. (Left) fertilization of *M. oculata* eggs with *M. occulta* sperm produces hybrids with urodele larvae identical to those of *M. oculata.* (Center) fertilization of *M. occulta* eggs with *M. oculata* sperm produces hybrid larvae with restored urodele features, including an otolith and a tail. (Right) fertilization of *M. occulta* eggs with uv-irradiated *M. oculata* sperm produces gynogenetic hybrids with anural larvae identical to those of *M. occulta.* Reproduced from Jeffery (1994).

M. oculata and *M. occulta.* Prospective notochord and muscle cells are generated in *M. occulta,* but they are reduced in number, fail to differentiate, and do not undergo morphogenetic movements. The lack of a tail in *M. occulta* can be explained by the absence of a differentiated notochord, which drives tail formation during urodele development. The neural tube and brain are formed in *M. occulta* embryos but the latter does not differentiate an otolith, either because the notochord cells have lost their inductive properties or the responding tissue (neural plate) has lost the ability to interpret the inductive signal. Thus, the developmental programs responsible for muscle and notochord/neural sensory organ differentiation are modified in the anural developer.

The evolution of anural development in *M. occulta* can be explained by changing two determinant systems (Figs. 2 and 10). First, modification of muscle/anterioposterior axis prevents muscle cell differentiation (compare

A   **B**   **C**

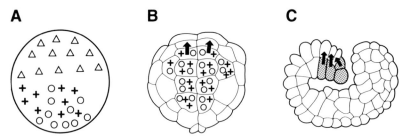

Fig. 10 Changes in developmental programs in an anural developer. (A) A zygote after completion of the second phase of ooplasmic segregation showing the absence of muscle/anteroposterior axis determinants. (B) A 64-cell embryo viewed from the vegetal pole showing a weakened inductive signal emanating from the endoderm cells resulting in minimal notochord induction. (C) A late gastrula stage embryo showing a weakened inductive signal emanating from the prospective notochord cells, which results in no otolith induction. Other details are similar to those shown in Fig. 2. Reproduced from Jeffery (1994).

Fig. 10A and Figs. 2A–2C). Second, changes in the endoderm/dorsoventral axis determinants alter the inductive cascade arising from the endoderm and culminating in notochord/otolith differentiation (compare Figs. 10B and 10C to Figs. 2D and 2E). The muscle/anteroposterior and endodermal/dorsoventral axis determinants originate in the myoplasm (Fig. 2A), suggesting that the MCD may be involved in anural development.

VI. Evolutionary Changes in the Myoplasmic Cytoskeletal Domain

The organization of the myoplasm is also changed in anural developers. The presence of mitochondria was investigated in the anural developer *Molgula arenata* using the Krebs cycle enzyme succinic dehydrogenase (Whittaker, 1979). The results showed that enzyme activity is evenly distributed in the egg cytoplasm, implying that mitochondria are not concentrated in the myoplasm. Similar results have been obtained by electron microscopy and staining for myoplasmic pigment granules in *M. occulta* eggs (W. Jeffery, unpublished results). Thus, the organelles charactersitic of the myoplasm are missing in anural developers, suggesting that this cytoplasmic domain has undergone extensive reorganization.

The composition of the MCD has been studied by comparing the distribution of p58 during oogenesis in *M. oculata* and *M. occulta* (Swalla *et al.*, 1991; Jeffery and Swalla, 1992). Western blots show that p58 is present in *M. oculata* and *M. occulta* oocytes. Although immunofluoresence microscopy indicates that p58 is localized in the conventional manner in *M. oculata*

(Figs. 11A, 11E, 11I, and 12E–12H), the situation is quite different in *M. occulta*. Despite its presence in oocytes, p58 cannot be detected in most clutches of *M. occulta* eggs (Figs. 11D, 11H, and 11L). It should be recalled that p58 appears in the perinuclear region, is translocated to the cortex, and subsequently localized in the MCD. This sequence of events is initiated in *M. occulta* oocytes, but p58 does not become localized in the MCD and is degraded (Figs. 12A–12D). Relatively low levels of p58 persist in the cytoplasm of some clutches of *M. occulta* eggs (Fig. 11B). The significance of this residual p58 will be discussed under Section VII.

Localization of p58 may be mediated by binding sites in the PML. The binding sites could contain actin, spectrin, ankryin, $Na^+K^+ATPase$, or other PML proteins. Immunofluoresence microscopy suggests that actin is present

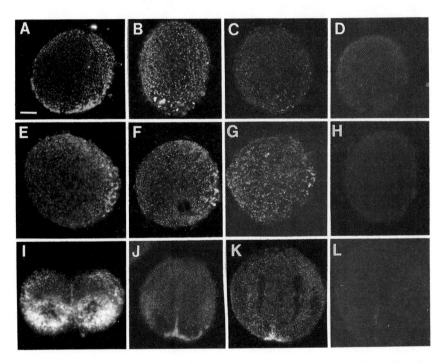

Fig. 11 Distribution and localization of p58 in *M. oculata* zygotes and *M. occulta* × *M. oculata* hybrid zygotes. (A–H) Fertilized eggs after completion of the first (A–D) or second (E–H) phase of ooplasmic segregation. (I–L) Two-cell embryos. A, E, and I, *M. oculata* zygotes; B–D, F–H, and J–L, *M. occulta* × *M. oculata* hybrid zygotes produced from different egg clutches; B, F, and J, strong urodele hybrid-forming clutch; C, G, and K, weak urodele hybrid-forming clutch; D, H, and L, anural hybrid-forming clutch. Staining of sectioned eggs was done with NN18 antibody. Specimens are shown with their vegetal poles facing the bottom and posterior poles facing the right in each frame. Scale bar = 20 μm. Reproduced from Jeffery and Swalla (1992).

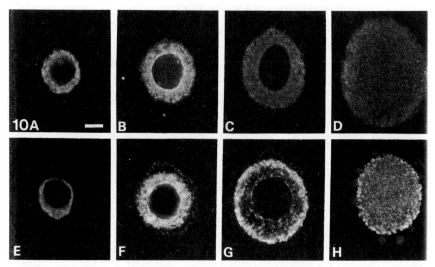

Fig. 12 Distribution and localization of p58 during oogenesis in *M. occulta* (A–D) and *M. oculata* (E–H). A and E, Previtellogenic oocytes; B and F, early vitellogenic oocytes; C and G, late vitellogenic oocytes; D and H, postvitellogenic oocytes after germinal vesicle breakdown. The germinal vesicle is the unstained sphere within the oocyte cytoplasm (A–C and E–G). Staining of sectioned eggs was done with NN18 antibody. Scale bar = 20 μm. Reproduced from Swalla *et al.* (1991).

in the PML of *M. occulta* eggs (Swalla *et al.*, 1991). In contrast, ankryin and spectrin cannot be detected in *M. occulta* eggs (Figs. 13A and 13B) (Jeffery and Swalla, 1993; Jeffery, 1993a). Ankryin is present in young oocytes, but it is not incorporated into the PML and cannot be detected after vitellogenesis. The absence of ankryin is significant because this protein links the internal cytoskeleton to the membrane skeleton of somatic cells (reviewed by Pumplin and Block, 1993). Although Na$^+$K$^+$ATPase is present in *M. occulta* eggs, it is distributed randomly in the cytoplasm (Fig. 13C), rather than localized in the PML. The distribution of Na$^+$K$^+$ATPase resembles p58 in some clutches of *M. occulta* eggs. However, ankryin and spectrin are not present in the *M. occulta* clutches containing residual p58. The loss of PML proteins in *M. occulta* oocytes may prevent p58 localization and promote its degradation.

Despite changes in PML and DFL composition, MCD segregation and polarization is not altered in the anural developer. The small amount of p58 that is evenly distributed in some clutches of *M. occulta* eggs is translocated to the vegetal and posterior poles of the zygote during ooplasmic segregation (Figs. 11B, 11C, 11F and 11G) and enters the (vestigial) muscle lineage. Conventional MCD segregation in *M. occulta*

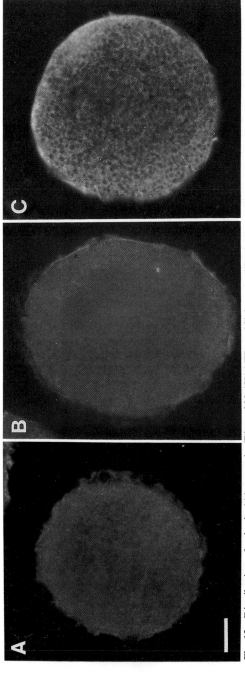

Fig. 13 Distribution of ankryin (A), spectrin (B), and Na⁺K⁺ATPase (C) in unfertilized *M. occulta* eggs. Staining of sectioned eggs was done with specific antibodies. Scale bar = 10 μm.

may be necessary for developmental programs that are conserved in both modes of development.

VII. Role of the Myoplasmic Cytoskeletal Domain in Anural Development

The changes in DFL and MPL proteins described previously suggest that the MCD has an important role in anural development. An ideal means to determine how these proteins actually function would be to delete or modifiy them in an urodele developer and/or introduce them to an anural developer. The means to carry out these experiments are currently being developed. In the meantime, two lines of evidence are presented below that strongly support a fundamental role of the MCD in the evolution of anural development.

A. Restoration of the Urodele Phenotype

The urodele features that are lost in *M. occulta* can be restored by interspecific hybridization (Swalla and Jeffery, 1990; Jeffery and Swalla, 1991, 1992). The results of interspecific crosses between *M. oculata* and *M. occulta* are summarized in Fig. 9. When eggs of the urodele developer are fertilized with sperm of the anural developer they develop into larvae identical to *M. oculata* (Fig. 9, bottom left). In contrast, when eggs of the anural developer are fertilized with sperm of the urodele developer, a short tail with a notochord and an otolith are restored in the hybrid larvae (Fig. 9, bottom center). The restoration of urodele features in hybrid embryos suggests that anural development evolved by the accumulation of hypomorphic mutations in zygotic genes. Consistent with this interpretation, restoration is prevented when eggs of the anural developer are fertilized with uv-irradiated sperm of the urodele developer. The resulting gynogenetic hybrids exhibit the anural phenotype (Fig. 9, bottom right). The *M. oculata* x *M. occulta* hybrids (Fig. 9, bottom center) lack differentiated muscle cells, implying that maternal factors are also involved in the evolution of anural development. Thus, anural development has evolved by changing both maternal and zygotic factors.

An additional aspect of the restoration phenomenon must be considered to relate hybrid phenotypes to changes in the MCD: different clutches of *M. occulta* eggs produce hybrids with varying degrees of urodele features (Jeffery and Swalla, 1992). Some *M. occulta* clutches are strong urodele-hybrid formers, producing hybrids with otoliths and tails (Fig. 14A), some clutches are weak urodele-hybrid formers, producing hybrids with either an otolith or a tail (Figs. 14B and 14C), and other clutches are anural hybrid formers, producing hybrids without urodele structures (Fig. 14D). It is important to note that the three classes of hybrids (strong urodele, weak urodele, and anural) are produced by fertilizing *different* clutches of *M. occulta* eggs with sperm from the *same M. oculata* individual, indicating that the hybrid phenotypes are based on maternal factors. The potential of *M. occulta* egg clutches to restore the urodele phenotype is positively correlated with the quantity of residual p58: *M. occulta* egg clutches lacking detectable p58 (Figs. 11B, 11H, and 11L) produce anural hybrids (Fig. 14C), clutches with low levels of p58 (Fig. 11C, G, K) produce weak urodele hybrids (Figs. 14B and 14C), and clutches with higher levels of p58 (Fig. 11B, 11F, and 11J) produce strong urodele hybrids (Fig. 14A). The restriction of urodele restoration to egg clutches containing p58 suggests that this protein is importart in the evolution of anural development.

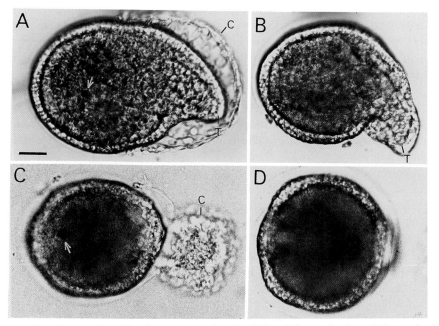

Fig. 14 Classes of hybrid embryos obtained by fertilizing different clutches of *M. occulta* eggs with *M. oculata* sperm. (A) A strong urodele hybrid with a pigmented otolith (arrow) and a short tail (T). The larva is in the process of hatching from the chorion (C). (B and C) Weak urodele hybrids with only a tail (B) or only a pigmented otolith (C). (D) An anural hybrid. Scale bar = 20 μm. Reproduced from Jeffery and Swalla (1992).

B. Convergent Evolution of Anural Development

The styelid *Pelonaia corrugata* and five species in four different molgulid clades have evolved anural development independently (Hadfield *et al.*, 1995). The first molgulid clade contains the anural developers *Eugyra arenosa* and *Bostrichobranchus digonas*, the second clade the anural developer *Molgula provisonalis*, the third clade the anural developer *Molgula tectiformis*, and the fourth clade *M. occulta* (Fig. 8). The styelid and molgulid anural developers have lost the same suite of larval features: the tail, notochord, and muscle cells (Berrill, 1931; Swalla and Jeffery, 1990, 1992; Bates and Mallet, 1991; reviewed by Jeffery and Swalla, 1990). The otolith is also lost in the anural developers, although a vestigial neural pigment cell is present in *B. digonas* (Swalla and Jeffery, 1992). Has the MCD has been modified in other anural developers and are the changes similar or different from those in *M. occulta?*

The DFL and PML has been investigated by comparing p58 distribution during *P. corrugata* and molgulid oogenesis (Swalla *et al.*, 1991; Jeffery and Swalla, manuscript in preparation). *Pelonaia corrugata, B. digonas, E. arenosa, M. provisionalis,* and *M. tectiformis* eggs lack p58, but this protein is present and localized in the myoplasm of closely related urodele developers. In contrast, previtellogenic oocytes of *P. corrugata* and the molgulids exhibit p58, although it is not localized in the cortex and disappears in postvitellogenic oocytes (see Fig. 15). It is remarkable that p58 behaves precisely the same in these widely diverged anural developers as it does in *M. occulta.* This convergence suggests that the alteration in p58 localization mechanisms is a significant factor in the evolution of anural development.

Actin and ankryin have also been investigated in different clades of anural developers (Swalla *et al.*, 1991; Jeffery and Swalla, 1993). Although actin is present in the PML of *M. occulta* and *B. digonas* eggs, it is absent in *M. provisionalis* eggs. Variable actin retention may be a consequence of genetic drift. In contrast, ankryin is missing in *P. corrugata* eggs and in the molgulid anural developers. Similar to p58, ankryin is present in the cytoplasm of previtellogenic oocytes but it is not translocated to the PML and dissappears later in oogenesis. These results suggest that ankryin may be involved in changing p58 localization. However, other PML proteins must also be involved because ankryin is also lacking in eggs of molgulid urodele developers (see Figs. 16A and 16B). Spectrin, which is missing in *M. occulta* but has not been examined in other molgulids, is an obvious candidate for the critical protein. The absence of ankryin in urodele and anural molgulids may be a preadaptation for the evolution of anural development, explaining why this developmental mode evolved almost exclusively in the Molgulidae.

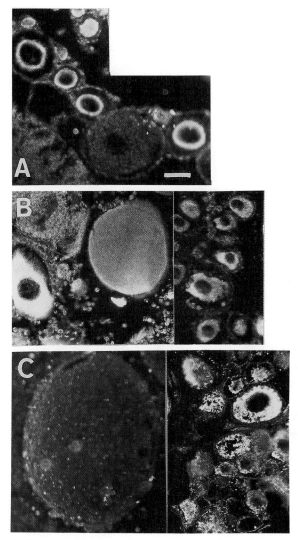

Fig. 15 The distribution of p58 in oocytes and eggs of anural developers. A, *Eugyra arenosa;* B, *Molgula provisionalis;* C, *Pelonaia corrugata.* In each frame, the small but not the large oocytes or eggs contain p58. Sections of gonads were stained with NN18 antibody. Scale bar = 50 μm; magnification is the same in each frame.

VIII. Conclusions and Prospectus

The studies described above provide a scenario for the role of the MCD in the evolution of anural development. It is proposed that alterations in

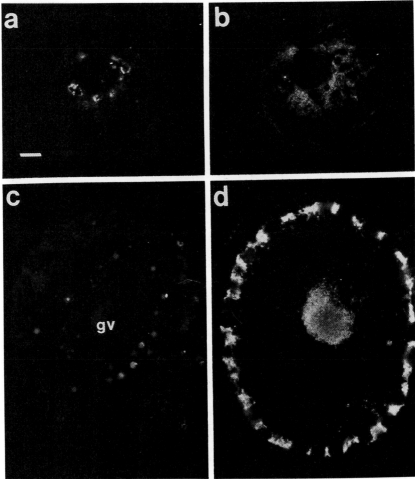

Fig. 16 Distribution of ankryin (A and C) and p58 (B and D) during oogenesis in *Molgula citrina*. (A and B) Previtellogenic oocytes showing ankryin in cytoplasmic vesicles (A) and p58 (B) localized in the perinuclear region. (C and D) Vitellogenic oocytes showing ankryin in the perinuclear region (C) and p58 localized in the cortex (D). Sections of gonads were double stained with ankryin and NN18 antibodies. Scale bar = 20 μm; magnification is the same in each frame. Reproduced from Jeffery and Swalla (1993).

the MCD occurred in at least three steps (Fig. 17). The first step was a modification in the ability of oocytes to localize ankryin in the PML. This modification occurred independently in the styelid lineage in which *P. corrugata* evolved and in the ancestor of the molgulid urodele and anural developers. The ankryin modification did not directly result in anural devel-

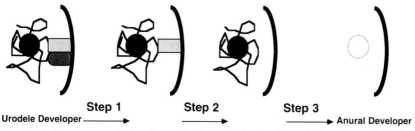

Fig. 17 A three-step model for changes in the MCD during the evolution of anural develop-ment. Thick curved lines, PML; thin latticework, DFL; filled spheres, localized mRNA; open sphere, modified mRNA in anural developer; ellipse, ankryin; rectangles; another PML protein. See text for description of the model.

opment (because molgulid urodele developers also lack ankryin), but served as a predaptation for the next step in the scenario. The second step was a modification in an unknown component that prevented the stable interac-tion between p58 and the PML required for assembly of the MCD during oogenesis. Spectrin, which has not been investigated in the divergent clades of anural developers, is a leading candidate for this component. Lack of p58 localization resulted in the disappearance of the DFL and dispersal of its embedded organelles throughout the cytoplasm. As described previously, maternal mRNAs localized via the egg cytoskeleton control the initial stages of development in other organisms. Maternal RNAs localized in the MCD have also been implicated in ascidian development (Jeffery, 1990a,b; Swalla and Jeffery, 1995). Thus, the third step leading to anural development is proposed to be a modification of maternal mRNAs and/or their binding sites in the MCD. The dispersal and/or degradation of maternal mRNAs could be responsible for changing the activity of muscle/anteroposterior and gastrulation/dorsoventral axis determinants leading to anural development (see Fig. 10). Three genes have been identified by subtractive procedures that are expressed in *M. oculata* but not in *M. occulta* eggs (Swalla *et al.*, 1993). These *urodele* (or *uro*) genes may encode the mRNA determinants that are modified during the evolution of anural development.

The studies described in this chapter open further questions and new avenues for research on the role of the MCD in the evolution of anural development. What is the mechanism of ankryin localization during oogen-esis and how is it changed during evolution? What is the identity of the PML component(s) necessary for DFL localization in the MCD? What is the mechanism of MCD assembly in the cortex and how does its modifica-tion lead to lack of p58 localization during oogenesis in anural developers? Do *uro* mRNAs linked to the MCD control urodele development and how are they altered in anural developers? Finally, what changes in the MCD have occurred in ascidian species that have evolved a large tadpole larva with a robust tail? Does the MCD expand to accommodate the development of hundreds of muscle cells in the larger tail or are other mechanisms generated *de novo* to regulate this evolutionary change? The resolution of these questions should lead to a better understanding of how egg cytoskeletal domains mediate evolutionary changes in development.

Appendix: Methods for Studying the Ascidian Egg Cytoskeleton

A. Ascidians

Ascidians are available from commerial supply companies, including Marinus, Inc. (Long Beach, CA), Gulf Specimens, Inc. (Panacea, FL), and the Marine Biological Laboratory (Woods Hole, MA). Many ascidian species can be easily collected from docks, piers, and floats, but deep water species, such as *M. oculata* and *M. occulta,* must be obtained by dredging.

Adult ascidians can be maintained for short periods in artificial seawater or longer in natural-running seawater. Some species have restricted spawning seasons, whereas others are gravid throughout the year. Spawning in the laboratory can be induced by light–dark regimes in some species (West and Lambert, 1976). Gametes can also be obtained by dissection from the gonads. Fertilized eggs washed free of sperm can be cultured to the tadpole stage and beyond in natural or artificial seawater. Specific details on fertilization and culture of ascidian embryos are described in Costello and Hendley (1971), Tomlinson *et al.* (1987), and Swalla and Jeffery (1990).

B. Removal of the Follicle

The presence of a complex follicle, which presists throughout embryogenesis, is a major obstacle to studying the ascidian egg cytoskeleton. The follicle consists of three parts: (1) an outer layer of follicle cells, (2) an acellular intermediate layer called the chorion, and (3) an inner layer of test cells and associated extracellular matrix (see Figs. 6A, 6B, and 6D). The test and follicle cells are active in RNA and protein synthesis during embryogenesis (Jeffery, 1980) and have their own cytoskeleton with associated mRNAs (Jeffery, 1982b). Removal of the egg envelope can be accomplished manually using tungsten needles but this procedure cannot be used for large-scale preparation of cytoskeletal components.

The following procedure has been developed for removing the follicle in mass (Jeffery, 1993b). The procedure involves digestion with protease

E (type XXV; Sigma Chemical Co, St. Louis, MO). Because the thickness of the follicle is variable, digestion conditions must be developed empirically for each species and clutch of eggs. Crystallin protease is added to about 0.1 ml packed volume of eggs/embryos suspended in 10 ml of artificial seawater, bringing the final protease concentration to 0.1–1 mg/ml. The mixture is placed in a Syracuse dish and repeatedly drawn into and expelled from a Pasteur pipet to promote digestion and physical removal of the follicle. The digestion and removal of the follicle can be followed by placing the dish under a dissection microscope. The process takes about 10 min for unfertilized *Styela clava* eggs and up to 30 min for zygotes at room temperature. After most of the follicles are removed (the remainder are loosened and removed during the following steps), the eggs/embryos are swirled to the center of the dish, as much of the protease solution as possible is removed with a Pasteur pipet, and the eggs/embryos are washed into artificial seawater containg 500 μg/ml glycine. This step is repeated quickly five times, and then the egg/embryos are washed 10 times in artificial sea water. The naked eggs/embryos are very sticky and must be transferred to dishes coated with agarose if they are to be maintained as a culture.

C. Preparation of Cytoskeletal Residues by Detergent Extraction

Methods for preparing egg cytoskeletal residues by detergent extraction were developed by Jeffery and Meier (1983). Prior to detergent extraction, the eggs are washed several times in cytoskeletal extraction buffer (CEB), consisting of 10 mM piperazine-N,N-bis [2-ethanesulfonic acid], 300 mM sucrose, 100 mM KCl, 5 mM magnesium acetate, 1 mM EGTA, and 10 μg/ml leupeptin at pH 8.0. CEB is stored frozen and solid leupeptin is added immediately prior to use for detergent extraction. Eggs are extracted in CEB containing 0.5% Triton X-100 on ice for about 60 min. When cytoskeletal residues are used to obtain RNA for *in situ* hybridization, the CEB must be supplemented with 10 mM vanadyl–ribonucleoside complex to inhibit RNA degradation (Jeffery, 1984b). After extraction, the supernatant is decanted, the detergent-insoluble pellet is washed several times in CEB lacking Triton X-100, and processed.

D. Scanning Electron Microscopy

Detergent-extracted eggs or embryos are fixed for 30 min at 20°C with 2% glutaraldehyde in 0.1 M phosphate buffer (pH 7.2), washed three times in 0.1 M phosphate buffer, postfixed in 1% OsO_4 for 5 min at 20°C, and washed three times in 0.1 M phosphate buffer. The specimens float in the buffer and are mounted by placing a grid beneath them with a fine forceps and slowly lifting the grid to the miniscus. Specimens attached to grids are critical-point dried using CO_2 as the exchange fluid and sputtercoated with 70 nm gold–palladium alloy.

E. Protein Extraction and Western Blotting

Eggs and embryos or detergent-extracted cytoskeletal residues are collected by low-speed centrifugation. The pellets are washed several times in ice-cold homogenization buffer (HB), consisting of 0.01 M Tris–HCl, 0.01 M NaCl, and 0.001 M $MgCl_2$ at pH 7.2. After washing, the pellets are resuspended in HB containing 10 μg/ml leupeptin and homogenized. The homogenates are stored at $-70°C$.

Samples are subjected to one (Laemmli, 1970)- or two (O' Farrell, 1975)-dimensional gel electrophoresis and proteins are transferred to nitrocellulose filters (0.45 μm pore size) by electrolytic elution as described by Towbin *et al.* (1979). The transfer is carrried out at 0.2 A for 3 or 4 hr at 4°C. After transfer, nonspecific binding sites on the filters are blocked by incubation in 0.14 M NaCl, 0.05 M Tris–HCl, 1% bovine serum albumen, and 0.1% Tween 20 at pH 7.2 for 30 min at room temperature. The filters are washed in 0.14 M NaCl, 0.05 M Tris–HCl, 0.1% Tween 20 at pH 7.2 (TBST) and incubated for 30 min in primary antibody diluted with TBST. The filters are then treated with alkline phosphatase conjugated secondary antibody, and the alkaline phosphatase reaction product is developed.

F. Immuofluorescence Microscopy

Eggs and embryos or cytoskeletal residues are collected by low-speed centrifugation and fixed in 100% methanol ($-70°C$) for 20 min, then in 100% ethanol ($-70°C$) for 20 min, and embedded in polyester wax (Steedman, 1957) according to the protocol of Norenburg and Barrett (1987). Sections are cut at 8 μm, dewaxed in 100% ethanol, hydrated through an ethanol series to water, and stained with primary antibody diluted in PBS for 1 hr at room temperature. After staining, the specimens are washed in PBS and treated with rhodamine-labeled secondary antibody for 1 hr at room temperature, washed in PBS, and mounted in 80% glycerol for microscopy. Rhodamine labeling is preferred for secondary antibodies because there is excessive autofluorescence at wavelengths used to detect fluoroscein in ascidian eggs and embryos.

Acknowledgments

This chapter is dedicated to the students and colleagues in my laboratory who have contributed to the studies on the MCD of ascidian eggs, epecially W. R. Bates, S. Meier, and B. J. Swalla. My research on the ascidian MCD has been supported by grants from the National Institutes of Health, the National Science Foundation, and the Muscular Dystrophy Association.

References

Bates, W. R., and Jeffery, W. R. (1987). Localization of axial determinants in the vegetal pole region of ascidian eggs. *Dev. Biol.* **124,** 65–76.

Bates, W. R., and Jeffery, W. R. (1988). Polarization of ooplasmic segregation and dorsal-ventral axis determination in ascidian embryos. *Dev. Biol.* **130,** 98–107.

Bates, W. R., and Mallet, J. E. (1991). Ultrastructural and histchemical study of anural development in the ascidian *Molgula pacifica. Roux's Arch. Dev. Biol.* **200,** 193–201.

Berg, W. E., and Humphreys, W. J. (1960). Electron microscopy of four-cell stages of the ascidians *Ciona* and *Styela. Dev. Biol.* **2,** 42–60.

Berrill, N. J. (1931). Studies in tunicate development. II. Abbreviation of development in the Molgulidae. *Phil. Trans. R. Soc. London B* **219,** 281–346.

Billet, F. S. (1979). Oocyte mitochondria. *In* "Maternal Effects in Development" (D. R. Newth and M. Balls, eds.), pp. 147–166. Cambridge Univ. Press, London.

Conklin, E. G. (1905). The organization and cell lineage of the ascidian egg. *J. Acad. Natl. Sci. Philadelphia* **13,** 1–119.

Costello, D. P., and Hendley, C. (1971). "Methods for Obtaining and Handling Marine Eggs and Embryos," 2nd Ed. Marine Biological Laboratory, Woods Hole, Massachusetts.

Debus, E., Weber, K., and Osborn, M. (1983). Monoclonal antibodies specific for glial fibrillary acidic (GFA) protein and for each of the neurofilament triplet polypeptides. *Differentiation* **25,** 195–203.

Forristall, C., Pondel, M., Chen, L., and King, M. L. (1995). Patterns of localization and cytoskeletal association of two vegetally localized RNAs, *Vg1* and *Xcat-2. Development* **121,** 201–208.

Hadfield, K. A., Swalla, B. J., and Jeffery, W. R. (1995). Multiple origins of anural development in ascidians inferred from rDNA sequences. *J. Mol. Evol.* **40,** 413–427.

Hsu, W. S. (1963). The nuclear envelope in the developing oocytes of the tunicate *Boltenia villosa. Z. Zellforsch.* **58,** 17–26.

Jeffery, W. R. (1980). The follicular envelope of ascidian eggs: A site of messenger RNA and protein synthesis during early embryogenesis. *J. Exp. Zool.* **212,** 279–289.

Jeffery, W. R. (1982a). Calcium ionophore polarizes ooplasmic segregation in ascidian eggs. *Science* **216,** 545–547.

Jeffery, W. R. (1982b). Messenger RNA in the cytoskeletal framework: Analysis by *in situ* hybridization. (1982b). *J. Cell Biol.* **95,** 1–7.

Jeffery, W. R. (1984a). Pattern formation by ooplasmic segregation in ascidian eggs. *Biol. Bull.* **166,** 277–298.

Jeffery, W. R. (1984b). Spatial distribution of messenger RNA in the cytoskeletal framework of ascidian eggs. *Dev. Biol.* **103,** 482–492.

Jeffery, W. R. (1985). Identification of proteins and mRNAs in isolated yellow crescents of ascidian eggs. *J. Embryol. Exp. Morphol.* **89,** 275–287.

Jeffery, W. R. (1989a). Localized mRNA and the egg cytoskeleton. *Int. Rev. Cytol.* **119,** 151–195.

Jeffery, W. R. (1989b). Requirement of cell division for muscle actin expression in the primary muscle lineage of ascidian embryos. *Development* **105,** 75–84.

Jeffery, W. R. (1990a). Ultraviolet irradiation during ooplasmic segregation prevents gastrulation, sensory cell induction, and axis determination in the ascidian embryo. *Dev. Biol.* **140,** 388–400.

Jeffery, W. R. (1990b). An ultraviolet-sensitive mRNA encoding a cytoskeletal protein may be involved in axis formation in the ascidian embryo. *Dev. Biol.* **141,** 141–148.

Jeffery, W. R. (1992). A gastrulation center in the ascidian egg. *Development* Supplement 1992, 53–63.

Jeffery, W. R. (1993a). The myoplasm of ascidian eggs: A plasma membrane skeleton which is modified during evolution. *Biol. Res.* **26,** 481–490.

Jeffery, W. R. (1993b). Role of cell interactions in ascidian muscle and pigment cell specification. *Roux's Arch. Dev. Biol.* **202,** 103–111.

Jeffery, W. R. (1994). A model for ascidian development and developmental modifications during evolution. *J. Marine Biol. Assoc. U. K.* **74,** 35–48.

Jeffery, W. R., and Bates, W. R. (1989). Ooplasmic segregation in the ascidian *Styela. In* "The Molecular Biology of Fertilization" (H. Schatten and G. Schatten, eds), pp. 341–367. Academic Press, New York.

Jeffery, W. R., and Meier, S. (1983). A yellow crescent cytoskeletal domain in ascidian eggs and its role in early development. *Dev. Biol.* **96,** 125–143.

Jeffery, W. R., and Meier, S. (1984). Ooplasmic segregation of the myoplasmic actin network in stratified ascidian eggs. *Roux's Arch. Dev. Biol.* **193,** 257–262.

Jeffery, W. R., and Swalla, B. J. (1990). Anural development in ascidians: Evolutionary modification of the tadpole larva. *Semin. Dev. Biol.* **1,** 253–261.

Jeffery, W. R., and Swalla, B. J. (1991). An evolutionary change in the muscle cell lineage of an anural ascidian embryo is restored by interspecific hybridization with a urodele ascidian. *Dev. Biol.* **145,** 328–337.

Jeffery, W. R., and Swalla, B. J. (1992a). Evolution of alternate modes of development in ascidians. *Bioessays* **14,** 219–226.

Jeffery, W. R., and Swalla, B. J. (1992b). Factors necessary for restoring an evolutionary change in an anural ascidian embryo. *Dev. Biol.* **153,** 194–205.

Jeffery, W. R., and Swalla, B. J. (1993). An ankyrin-like protein in ascidian eggs and its role in the evolution of direct development. *Zygote* **1,** 197–208.

Laemmli, U. K. (1970). Cleavage of structural proteins during the assembly of the head of bacteriophage T4. *Nature* **227,** 680–685.

Macdonald, P. M. (1992). The means to the ends: Localization of maternal messenger RNAs. *Semin. Dev. Biol.* **3,** 413–424.

Monroy, A., Ortolani, G., O'Dell, D. S., and Millonig, G. (1973). Binding of concanavalin A to the surface of unfertilized and fertilized ascidian eggs. *Nature* **242,** 409–410.

Miyamoto, D. M., and Crowther, R. J. (1985). Formation of the notochord in living ascidian embryos. *J. Embryol. Exp. Morphol.* **86,** 1–17.

Nakatani, Y., and Nishida, H. (1994). Induction of notochord during ascidian embryogenesis. *Dev. Biol.* **166,** 289–299.

Nishida, H. (1991). Induction of brain and sensory pigment cells in the ascidian embryo analyzed by experiments with isolated blastomeres. *Development* **112,** 389–395.

Nishida, H. (1992a). Determination of developmental fates in blastomeres of ascidian embryos. *Dev. Growth Differ.* **34,** 253–262.

Nishida, H. (1992b). Regionality of egg cytoplasm that promotes muscle differentiation in embryo of the ascidian *Halocynthia roretzi. Development* **112,** 389–395.

Nishida, H. (1994). Localization of determinants for formation of the anterior–posterior axis in eggs of the ascidian *Halocynthia roretzi*. *Development* **120**, 3093–3104.

Nishikata, T., Mita-Miyazawa, I., Deno, T, and Satoh, N. (1987a). Muscle cell differentiation in ascidian embryos analysed with a tissue-specific monoclonal antibody. *Development* **99**, 163–171.

Nishikata, T., Mita-Miyazawa, I., Deno, T., and Satoh, N. (1987b). Monoclonal antibodies against components of the myoplasm of eggs of the ascidian *Ciona intestinalis* partially block the development of muscle-specific acetylcholinesterase. *Development* **100**, 577–586.

Norenburg, J. L., and Barrett, J. M. (1987). Steedman's polyester wax embeddment and de-embeddment for combined light and scanning electron microscopy. *J. Elec. Microc. Tech.* **6**, 35–41.

Nüsslein-Volhard, C. (1991). Determination of the embryonic axis of *Drosophila*. *Development* Supplement 1991, 1–10.

O' Farrell, P. H. (1975). High resolution two-dimensional electrophoresis of proteins. *J. Biol. Chem.* **250**, 4007–4021.

Pumplin, D. W., and Bloch, R. J. (1993). The membrane skeleton. *Trends Cell Biol.* **3**, 113–117.

Reverberi, G., Ortolani, G., and Farinella-Ferruzza, N. (1960). The causal formation of the brain in the ascidian larva. *Acta Embryol. Exp. Morphol.* **3**, 296–336.

Sardet, C., Speksnijder, J. E., Inoué, S., and Jaffe, L. F. (1989). Fertilization and ooplasmic movements in the ascidian egg. *Development* **105**, 237–249.

Sardet, C., Speksnijder, J., Terasaki, M., and Chang, P. (1992). Polarity of the ascidian egg cortex before fertilization. *Development* **115**, 221–237.

Sardet, C., McDougall, A., and Houliston, E. (1994). Cytoplasmic domains in eggs. *Trends Cell Biol.* **4**, 166–172.

Satoh, N. (1994). "Developmental Biology of Ascidians." Cambridge Univ. Press, Cambridge.

Sawada, T. (1983). How ooplasm segregates bipolarly in ascidian eggs. *Bull. Marine Biol. Stat. Asamushi Tohoku Univ.* **17**, 123–140.

Sawada, T., and Osanai, K. (1981). The cortical contraction related to ooplasmic segregation in *Ciona intestinalis* eggs. *Roux's Arch. Dev. Biol.* **190**, 201–214.

Sawada, T., and Osanai, K. (1985). Distribution of actin in fertilized eggs of the ascidian *Ciona intestinalis*. *Dev. Biol.* **111**, 260–265.

Sawada, T., and Schatten, G. (1988). Microtubules in ascidian eggs during meiosis, fertilization, and mitosis. *Cell Motil. Cytoskeleton* **9**, 219–230.

Sawada, T., and Schatten, G. (1989). Effects of cytoskeletal inhibitors on ooplasmic segregation and microtubule organization during fertilization and early development in the ascidian *Molgula occidentalis*. *Dev. Biol.* **132**, 331–342.

Shaw, G., Debus, E., and Weber, K. (1984). The immunological relatedness of neurofilament proteins of higher vertebrates. *Eur. J. Cell Biol.* **34**, 130–136.

Speksnijder, J. E., Jaffe. L. E., and Sardet, C. (1990a). Polarity of sperm entry in the ascidian egg. *Dev. Biol.* **133**, 180–184.

Speksnijder, J. E., Jaffe, L. F., and Sardet, C. (1990b). The activation wave of calcium in the ascidian egg and its role in ooplasmic segregation. *J. Cell Biol.* **198**, 119–128.

Speksnijder, J. E., Terasaki, M., Hage, W. J., Jaffe. L. F., and Sardet, C. (1993). Polarity and reorganization of the endoplasmic reticulum during fertilization and ooplasmic segregation in the ascidian egg. *J. Cell Biol.* **120**, 1337–1346.

Steedman, H. (1957). Polyester wax. A new ribboning embedding medium for histology. *Nature* **179**, 1345.

Swalla, B. J. (1992). The role of maternal factors in ascidian muscle development. *Semin. Dev. Biol.* **3,** 287–295.

Swalla, B. J., and Jeffery, W. R. (1990). Interspecific hybridization between an anural and urodele ascidian: Differential expression of urodele features suggests multiple mechanisms control anural development. *Dev. Biol.* **142,** 319–334.

Swalla, B. J., and Jeffery, W. R. (1992). Vestigial brain melanocyte development during embryogenesis of an anural ascidian. *Dev. Growth Differ.* **34,** 17–36.

Swalla, B. J., and Jeffery, W. R. (1995). A maternal RNA localized in the yellow crescent is segregated to the larval muscle cells during ascidian development. *Dev. Biol.* **170,** 353–364.

Swalla, B. J., Badgett, M. R., and Jeffery, W. R. (1991). Identification of a cytoskeletal protein localized in the myoplasm of ascidian eggs: Localization is modified during anural development. *Development* **111,** 425–436.

Swalla, B. J., Makabe, K. W., Satoh, N., and Jeffery, W. R. (1993). Novel genes expressed differentially in ascidians with alternate modes of development. *Development* **119,** 307–318.

Tomlinson, C. R., Beach, R. L., and Jeffery, W. R. (1987). Differential expression of a muscle actin gene in muscle cell lineages of ascidian embryos. *Development* **101,** 751–765.

Towbin, H., Staehlin, T., and Gordon, J. (1979). Electrophoretic transfer of proteins from polyacrylamide gels to nitrocellulose sheets: Procedures and some applications. *Proc. Natl. Acad. Sci. USA* **76,** 4350–4354.

Venuti, J. M., and Jeffery, W. R. (1989). Cell lineage and determination of cell fate in ascidian embryos. *Int. J. Dev. Biol.* **33,** 197–212.

Vincent, J-P., Oster, G. F., and Gerhart, J. C. (1986). Kinematics of gray crescent formation in *Xenopus* eggs: The displacement of subcortical cytoplasm relative to the egg surface. *Dev. Biol.* **113,** 484–500.

West, A. B., and Lambert, C. C. (1976). Control of spawning in the tunicate *Styela plicata* by variations in a natural light regime. *J. Exp. Zool.* **195,** 265–270.

Whittaker, J. R. (1979). Development of tail muscle acetylcholinesterase in ascidian embryos lacking mitochrondrial localization and segregation. *Biol. Bull.* **157,** 344–355.

Zalokar, M. (1974). Effect of colchicine and cytochalasin B on ooplasmic segregation of ascidian eggs. *Roux's Arch. Dev. Biol.* **175,** 243–248.

9

Remodeling of the Specialized Intermediate Filament Network in Mammalian Eggs and Embryos during Development: Regulation by Protein Kinase C and Protein Kinase M

G. Ian Gallicano and David G. Capco
Molecular and Cellular Biology Program
Arizona State University
Tempe, Arizona 85287

I. Introduction

A. Early Development in Mammals

Penetration of an egg by a sperm induces within that egg numerous, interdependent structural and biochemical changes that mediate the process of fertilization and transform the egg into the single-celled zygote. The zygote then divides numerous times producing a mass of cells which is remodeled by processes, such as cell migration, cell death, and cellular differentiation, to produce an organism.

Development is accompanied by extensive remodeling at the intracellular level and at the intercellular level. These remodeling events occur at predictable time points during development and are responsible for major transitions in development (referred to as developmental transitions). Such major developmental transitions include fertilization, which is a transition where the remodeling occurs solely at the intracellular level, and embryonic compaction, blastocyst formation, formation of primitive ectoderm and primitive endoderm in the egg cylinder, gastrulation, and subsequent events in which the transitions are accompanied by both intracellular and intercellular remodeling.

The diagram in Fig. 1 illustrates the major developmental transitions as they occur in the reproductive tract of a mammal. While mammalian eggs exhibit some developmental transitions in common with other classes of organisms they also exhibit some developmental transitions which are specialized for mammalian development. The common developmental transitions include fertilization, gastrulation, and subsequent developmental stages, whereas the specialized developmental transitions include embryonic compaction, blastocyst formation, and the formation of the primitive ectoderm and primitive endoderm in the forming egg cylinder. In the latter two specialized transitions, distinct, functional epithelia differentiate well in advance of gastrulation—the developmental point after which differentiated epithelia form in most other classes of organisms.

Embryonic compaction represents a major transition in development because significant remodeling events occur at both the intracellular level and the intercellular level. At this time the embryo acquires an inside–outside embryonic polarity which alters the developmental fate of blastomeres at the interior of the embryo compared to the exterior of the embryo (Johnson and Ziomeck, 1981; Sherman, 1981; Fleming et al., 1991). In addition, the blastomeres at the exterior of the embryo acquire distinct cellular polarity (Ziomeck and Johnson, 1981; Johnson and Maro, 1984; Adler and Ziomeck, 1986). Embryonic polarity forms as the rounded blastomeres flatten on one another to form a spherical ball of cells (Richa et al., 1985; Fleming et al., 1991; Sefton et al., 1993). Once the flattening takes place, outside cells see both an outside environment and an inside environment, whereas inside cells only see an inside environment (Sherman,1981; Ziomeck and Johnson, 1981). The fate of the outside blastomeres is to form the trophectoderm, and subsequently the placenta, whereas the inside blastomeres become fated to form the remainder of the embryo. Cellular polarity occurs primarily in outside blastomeres of the embryo. As these outer

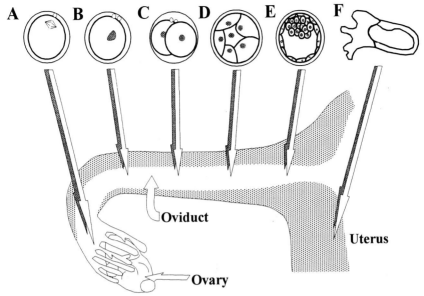

Fig. 1 Fertilization and development of the embryo in the mammalian reproductive tract. Once the egg is released from the ovary (A), and if sperm is present, fertilization will normally occur in the distal region of the oviduct. The fertilized egg will be transformed into the single-celled zygote (B) which will eventually divide into the 2- (C), 4-, and 8-cell stage embryo. At the 8- to 16-cell stage, the rounded blastomeres will flatten on one another forming a morula (D). The hallmark of the morula-stage embryo is the onset of both embryonic polarity and cellular polarity which is seen primarily in outside cells of the embryo as tight junctions, gap junctions, and desmosomes form and a differential insertion of membrane channels and pumps is first detected. As blastomeres divide further and membrane channels and pumps, such as the Na$^+$/K$^+$ ATPase, begin to function, intercellular space will fill with fluid and the blastocyst-stage embryo eventually will form (E). The blastocysts-stage embryo will hatch from its surrounding zona pellucida, form an egg cylinder (F), invade the uterine wall, and then implant into the uterus.

blastomeres flatten the intracellular compartment becomes extensively re-modeled. Intercellular junctional complexes, such as desmosomes and tight junctions, begin to form (Ducibella *et al.*, 1975, 1979; Fleming *et al.*, 1991; Clayton *et al.*, 1993; Riethmacher *et al.*, 1995). This is followed by differential insertion of membrane channels and pumps, such as the Na$^+$/K$^+$ ATPase pump, which is localized to the basal/lateral plasma membranes of these outside blastomeres (Wiley, 1984; Overstrom *et al.*, 1989). In fact, cellular polarization of these blastomeres is actually very similar to the polarization of many types of somatic epithelial cells.

The next developmental transition is formation of the blastocyst which occurs several cell division cycles after embryonic compaction. Here a differentiated epithelia layer forms from the outer cells of the embryo, while the inner cells remain undifferentiated. These outside cells form a primary epithelium known as the trophectoderm, which becomes the placenta after implantation into the uterus. The inside cells, or the inner cell mass cells, will form the embryonic tissues. Once the blastocyst reaches a certain size, usually 12–24 hr after cavitation begins, it will "hatch" from its extracellular matrix known as the zona pellucida. Once hatched, the outer epithelial layer will further differentiate into layers with specialized functions, and these outer epithelial layers will invade the uterine wall pulling the reminder of the embryo behind it.

The third transition specialized for mammalian development is the remodeling that accompanies the conversion of the inner cell mass cells into the primitive ectoderm and primitive endoderm. These two epithelial layers have different developmental fates and exhibit structural characteristics of differentiated epithelia. The primitive ectoderm will form the embryo proper, while the primitive endoderm forms extraembryonic endoderm (Gardner and Rossant, 1979; Gardner, 1984). Both epithelial layers exhibit numerous junctional complexes including tight junctions and desmosomes, and contain bands of tonofilaments (Jackson et al., 1981; Tam et al., 1993; S. Schwarz et al., in press). The primitive endoderm exhibits a brush border on the apical surface of each of its cells and contains numerous vesicles in the apical area (Schwarz et al., 1993; Tam et al., 1993). The primitive ectoderm forms columnar-type cells with a small number of microvilli on the apical surface of its cells (Lawson et al., 1991; Schwarz et al., 1993; Tam et al., 1993).

The functions and challenges presented to this special cell, the egg, during the process of development require specializations in cell structure, specifically the cell's cytoskeleton. Some of the more general challenges to oocytes, eggs, and embryos were outlined in the introduction to this volume. In this chapter we have outlined three unique developmental transitions (i.e., embryonic compaction, blastocyst formation, and formation of the primitive ectoderm and primitive endoderm) and we will show that the cytoskeleton, particularly a highly cross-linked network of intermediate filaments referred to as sheets, has an integral role in each of the developmental transitions.

B. The Three Major Filament Systems

Comparable to most other cells, mammalian eggs and embryos contain intermediate filaments, actin filaments, and microtubules, and a few examples of the function of each of these filament networks will be presented below; however, the focus of this chapter is directed at the intermediate filament network. The suggestion that intermediate filaments are present in mammalian eggs will surprise some readers because there has been a disagreement in the literature concerning the timing of the appearance of intermediate filaments during early development.

The question of whether intemediate filaments exist in the early embryo of the mammal has been controversal. Several investigators have reported that intermediate filaments do not appear until a developmental stage after embryonic compaction (Johnson *et al.*, 1986; Chisholm and Houliston, 1987; Emerson, 1988), whereas others have suggested that they do not appear until after the blastocyst stage (Brulet *et al.*, 1980; Jackson *et al.*, 1980; Paulin *et al.*, 1980). One investigator has attempted to demonstrate that intermediate filaments are not important for early development of mammals by injecting antibodies to cytokeratin 8 into the two-cell mouse embryo to test whether these antibodies impeded development (Emerson, 1988). In contrast, some investigators have detected intermediate filaments in unfertilized mammalian eggs, most notably Lehtonen (1985, 1987).

Intermediate filaments are made during oogenesis and are present in the oocyte, egg, and blastomeres of the embryo. They have not been identified by the investigations noted previously for three reasons. First, and most important, almost all of the investigators noted previously employed the antibodies produced by Oshima and co-workers (Oshima *et al.*, 1983). These antibodies, referred to in the literature as TROMA-1 and ENDO-B, are monoclonal and polyclonal antibodies, respectively, that are monospecific for cytokeratin 8 and 18, respectively. These antibodies do not bind to the major forms of cytokeratin in early mammalian embryos. The major forms of cytokeratin in the early mammalian embryo are at least cytokeratins 5, 6, 16, and a form referred to as "Z" (Gallicano *et al.*, 1994a). Thus, even Emerson's (1988) experiments microinjecting antibodies to cytokeratin 8 into the two-cell embryo to block the biological function of intermediate filaments containing cytokeratin 8 were not testing the major forms of intermediate filaments in the embryo. Second, the intermediate filaments in the egg and early embryo of mammals are in the form of a highly cross-linked network of intermediate filaments (referred to as sheets) and individual 10- or 11-nm filaments do not exist in the early embryo. Thus, ultrastructural analysis would not reveal individual intermediate filaments. Third, the cross-linked network of intermediate filaments is coated with a layer of one or more proteins that prevent the binding of antibodies unless stringent permeabilization techniques are employed (for example, see the methods of Lehtonen, 1985, 1987). This chapter focuses on the role of this cross-linked network of intermediate filaments (i.e., sheets) during development.

Actin filaments form a cortical network in unfertilized eggs organized into two domains. There is a loose network consisting largely of actin filaments, and within this loose network, localized dense networks exist that are composed largely of nonactin filaments. During the time of sperm penetration the loose network of actin filaments associates with the head of the sperm and may be involved with sperm entry (Webster and McGaughey, 1988, 1990). These studies by Webster and McGaughey confirm and extend the earlier investigations of Maro *et al.* (1984) and Longo and Chen (1985) who reported a role for actin filaments during sperm ingression. The actin network undergoes several structural remodeling events during postfertilization development many times in association with developmental transitions (for example, see Ducibella *et al.,* 1977, Lehtonen and Badley, 1980) In addition, there appear to be roles for several actin-binding proteins during development (for example, see Sobel and Alleigro, 1986; Damjanov *et al.,* 1986; Lehtonen *et al.,* 1988), but the involvement of the actin network and its associated proteins during early development will be covered in other chapters.

Microtubules also play an important role in early development, for example, mediating meiotic reduction divisions and mitotic divisions of cleavage (for example, see Wiley and Eglitis, 1980; Schatten *et al.,* 1986; Messinger and Albertini, 1990; Albertini, 1992; Schatten, 1994). Consideration of this particular cytoskeletal network during early development in mammals will be given in Chapter 10 by Schatten and co-workers and several other chapters are present in this volume which focus on the role of microtubules.

II. Specialized Organization of the Intermediate Filament Network (i.e., Cytoskeletal Sheets)

A. Ubiquitous Appearance of Sheets in Mammalian Eggs

As noted previously it is reasonable to assume that cytoskeletal specializations will mediate developmental transitions. Consistent with that premise, Capco and McGaughey (1986) identifed an unusual cytoskeletal network in eggs and embryos of the Syrian hamster. In that study, they detergent-extracted hamster eggs and embryos to obtain the detergent-resistant cy-

---

Fig. 2 Embedment-free sections of a detergent-extracted, ovulated hamster egg demonstrates the detergent-resistant cytoskeleton. (A) Low-magnification view showing the egg surface and interior. The arrowheads show remnants of the detergent-resistant plasma membrane, the plasma lamina. Arrows indicate the cytoskeletal sheets which are in a whorl-like conformation and surrounded by a thin filament network. Notice that the sheets are excluded from the egg periphery. Z, the zona pellucida, C, the remnants of a cumulus cell. (B) High-magnification view of the egg interior shows sheets (arrows) which exhibit a sold, planar structure. The sheets are enmeshed in a thin filament network.

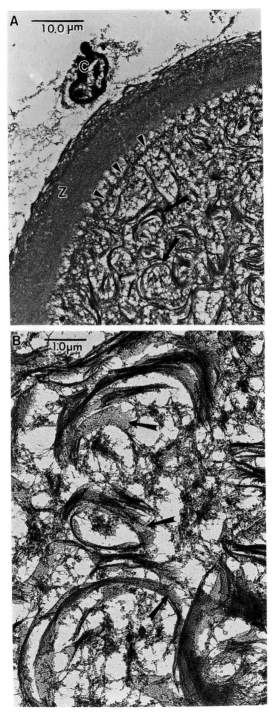

toskeleton which then was examined by embedded-free electron microscopy. This form of microscopy allows examination of relatively thick sections, thus providing more spatial information concerning the cell than conventional thin-section electron microscopy. Low-magnification, embeddment-free, electron microscopic views of detergent-extracted hamster eggs clearly showed numerous, large cytoskeletal structures throughout the interior regions of these eggs (Fig. 2A). When these large cytoskeletal structures were viewed at higher magnifications, they appeared as solid, planar-like structures enmeshed in a thin filament network which was analogous to that found in somatic cells (Fig. 2B). Because of their solid, planar form, Capco and McGaughey (1986) termed these structures cytoskeletal "sheets." The sheets underwent extensive spatial reorganizations at each developmental transition, that is, fertilization, embryonic compaction, and blastocyst formation (Fig. 3).

Before beginning extensive studies on the structure and function of these unusual cytoskeletal elements, the sheets, it was necessary to ask: are the sheets a conserved cytoskeletal component within mammalian eggs and embryos? If so, this would suggest that the sheets may have a common role in the process of development in mammals.

To address this question, eggs from two other species of rodent were examined using embedded-free electron microscopy along with the same detergent-extraction regime used by Capco and McGaughey (1986). In the detergent-extracted mouse egg, the internal region of these eggs contained large electron-dense cytoskeletal structures (Fig. 4A). These large structures also were excluded from the periphery of these eggs much as the sheets were in hamster eggs. At higher magnification, though (Fig. 4B), it was found that these elements were more fibrous looking than those found in the hamster egg. However, because of the abundance of these cytoskeletal components, and the observation that they underwent changes in their spatial organization at developmental transitions comparable to those in the hamster egg, these cytoskeletal components were tentatively considered to be cytoskeletal sheets (Gallicano et al., 1991).

Sheets also existed in eggs and embryos of the rat. Moreover, when viewed at high magnification they appeared as large, planar cytoskeletal components (Fig. 4C) quite comparable to those found in hamster eggs and embryos (Gallicano et al., 1992). Next, several nonrodent species were examined. Analysis of eggs from two species of hoofed animals, porcine and bovine (not shown), showed that they also contained sheets (Gallicano et al., 1992). These sheets were similar to the fibrous type found in mouse eggs and embryos. Human eggs (Fig. 4D), as well as eggs from two breeds of dogs (German Shepherd and Greyhound; G. Gallicano, unpublished results), also contained sheets, each of which was of the fibrous type. Consequently, ultrastructural analysis of eggs from numerous mammalian species

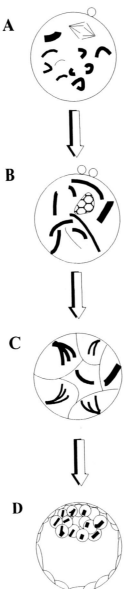

Fig. 3 A diagram depicting the changes in sheet organization at the major developmental transitions, fertilization, embryonic compaction, and blastocyst formation. The sheets (dark lines inside of the egg and blastomeres) are whorled in the unfertilized egg (A) and they are excluded from the periphery. Following fertilization, the sheets linearize and enter into the peripheral cortex region of the zygote (B) and subsequent blastomeres. (C) At embryonic compaction (morula stage), sheets come in close contact with the apical plasma lamina at single focal points. (D) At the blastula stage, sheets are absent in the differentiated trophecto-derm cells, but remain intact in the inner cellmass cells.

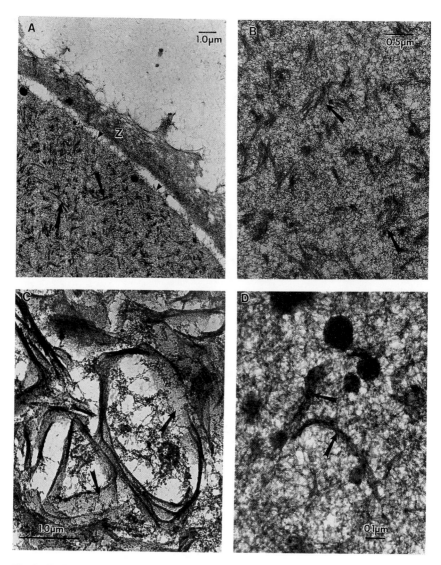

Fig. 4 Embedment-free sections of a detergent-extracted, mammalian egg demonstrates two categories of sheets. (A) Low-magnification view of a portion of a mouse egg shows the egg surface and interior. The arrowheads show remnants of the detergent-resistant plasma membrane, the plasma lamina. Arrows indicate the cytoskeletal sheets which are surrounded by a filament network. Notice that the sheets do not come in close contact with the plasma lamina. Z, the zona pellucida. (B) High-magnification view of the interior of the mouse egg shows the sheets (arrows) which in some areas exhibit a fibrous substructure. The thin filament network encompasses the sheets. (C) High-magnification, embedment-free view from a detergent-extracted rat egg shows that in this species, eggs contain solid, planar sheets (arrows), similar to those in hamster eggs (see Fig. 1B). (D) High-magnification view of a detergent-extracted human egg shows that sheets (arrows) are composed of fibrous elements similar to the sheets found in bovine, porcine, canine (not shown), and mouse eggs. Figures 4A and 4B are reproduced from Gallicano *et al.* (1991) with permission of Wiley, Inc. Figures 4C and 4D are reproduced from Gallicano *et al.* (1992) with permission of Wiley, Inc.

showed that sheets are a conserved component of early mammalian development. Our results also demonstrated that the sheets appeared to exist in two structural configurations: solid, planar sheets which exist in eggs of hamster and rat, and fibrous sheets which exist in eggs of human, mouse, cow, pig, and dog. To determine if the two configurations of sheets represented the same structural component it was necessary to examine their biochemical composition and substructural organization.

B. The Substructural Organization and Biochemical Composition of the Sheets

Eggs and embryos from mouse and hamster were selected for analysis since they represented species which contained sheets of the two structural categories, that is fibrous and planar sheets. In addition, an ample supply of eggs and embryos from these two species could be easily obtained. When mouse eggs were analyzed after detergent extraction in a medium with Tween 20 the sheets had a fibrous appearance; however, detergent extraction in a medium with Triton X-100, which removes more proteins, revealed a distinct substructure to the sheets. They appeared to be composed of 10- or 11-nm filaments lying side by side (Fig. 5A; Gallicano *et al.*, 1991). In contrast, examination of sheets in hamster eggs (solid sheets) after extraction with either Tween 20 or Triton X-100 did not result in a noticeable change in substructure (Fig. 5B). However, when hamster eggs were subjected to detergent extraction in a medium containing a mixed micelle detergent (composed of a nonionic and an ionic detergent) a layer coating the sheets was removed revealing an array of 10- or 11-nm filaments lying side by side (Fig. 5C; Capco *et al.*, 1993).

Intermediate voltage electron microscopy (IVEM) was employed to view the substructural components of both types of sheets at high resolution. This is because IVEM allowed the use of plastic-embedded sections, which were relatively thick, and thus could provide spatial information. Moreover, the plastic used as the embedment prevented the collapse of the sheets onto the surface of the supporting grid as can sometimes happen in embedment-free microscopy; thus, it further aided the analysis of the three-dimensional organization of the sheets. Embedment-free electron microscopy showed that these filaments within sheets were arranged side by side; however, IVEM views showed that these filaments were held in register by distinct crossbridges spaced every 23–25 nm (Fig. 6). Furthermore, sheets also were observed in close contact with membranous components of the cell including vesicles. Cross-sectional IVEM views of sheets revealed electron-dense spots (each spot with a diameter of 10 or 11 nm) linked by crossbridges and arranged in a cylindrical bundle (Fig. 7). This arrangement

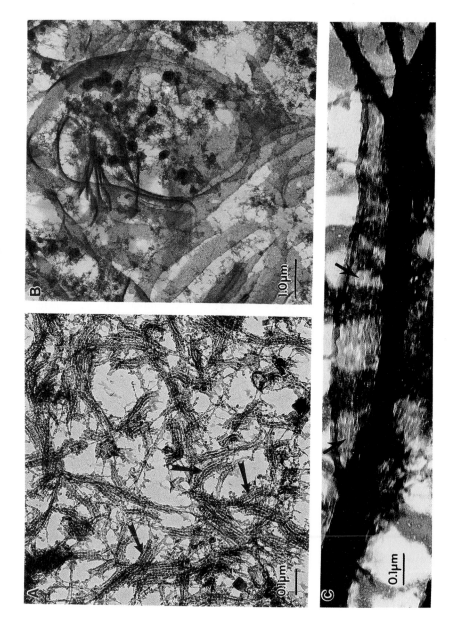

showed that the fibrous sheets were not flat structures in the cytoplasm, but instead folded to form cylinders.

More substructural information about these sheets was obtained by employing another type of electron microscopy known as the quick-freeze, deep-etch (QFDE) electron microscopy. High-magnification views showed sheets arranged in cylindrical bundles, some of which had the top filament of the sheet fractured away allowing filaments deep within the bundle to be observed (Fig. 8). These views confirmed that the 10- or 11-nm filaments were connected to one another by distinct crossbridges spaced every 23–25 nm. Examination of the sheets without detergent extraction simply by freeze fracture revealed that the sheets were not surrounded by membrane (McGaughey and Capco, 1989).

Sheets in mouse eggs were compared to those found in hamster eggs using IVEM. Direct cross-sectional views of sheets in hamster eggs showed that they appeared to be composed of two layers of 10- or 11-nm fibers held in register by distinct crossbridges spaced every 23–25 nm (Fig. 9). However, cross-sectional views that were turned *en face* revealed that each of the two layers of fibers was composed of 10- or 11-nm filaments which were lying side by side. Diffraction analysis of images of these sheets revealed a repeat pattern consistent with a network of two layers of intermediate filaments held in register by both lateral and vertical crossbridges (Fig. 10).

Based on all the structural evidence, models for the two types of sheets were developed (Fig. 11). In hamster eggs (Fig. 11A), a tangential view of a sheet showed that these sheets are composed of two layers of 10- or 11-nm filaments. Within each layer, filaments are connected to each other by lateral crossbridges. A layer of intermediate filaments is connected to another layer by vertical crossbridges. This highly cross-linked network may cause the sheet to be inflexible and thus more planar in appearance. The

---

Fig. 5 Embedment-free electron microscopic views of sheets in a mouse egg and in hamster eggs which were extracted with different detergents. (A) High-magnification view of a mouse egg extracted with Triton X-100 shows that sheets (arrows) are composed of 10- or 11-nm filaments lying side by side. Note, each filament exhibits a periodicity of 20.6 ± 1.6 nm. (B) The sheets in this view of a hamster egg, which is extracted with Triton X-100, do not exhibit a substructure similar to that found in sheets in mouse eggs extracted with Triton X-100. (C) Cytoskeletal sheets in a hamster egg after treatment with a mixed-micelle detergent containing 0.2% deoxycholate and 1.0% NP-40 in the extraction medium prior to fixation shows that these sheets, like those in the mouse egg, are composed of distinct fibers (arrows). Each fiber has a diameter of 10 or 11 nm. Figure 5A is reproduced from Gallicano *et al.* (1991) with permission of Wiley, Inc. Figure 5C is reproduced from Capco *et al.* (1993) with permission of Wiley, Inc.

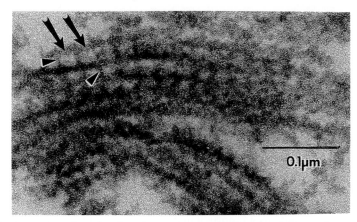

Fig. 6 An *en face* view of a sheet in a mouse egg is viewed at high magnification using IVEM. Distinct 10- or 11-nm filaments (arrows) are held in register with each other by lateral crossbridges (arrowheads) which are spaced every 23.0 ± 3.3 nm apart. Reproduced from Gallicano *et al.* (1994a) with permission of Elsevier Science, Ltd.

sheets in mouse eggs appear to be composed of only one layer of intermediate filaments, which are held in register by lateral crossbridges only (Fig. 11B). Because there are only lateral crossbridges, the sheets may be more flexible and able to fold into cylindrical bundles. For both structural categories of sheets our analysis by QFDE methodology revealed that the 10- or 11-nm filaments were associated with particulate components masking the surface of the individual filaments (Fig. 8; Capco *et al.*, 1993: Gallicano *et al.*, 1994a). We predicted that the presence of this material coating the filaments resulted in the fibrous rather than the filamentous appearance of sheets in mouse eggs, and the solid rather than the filamentous appearance of sheets in hamster eggs. In support of this prediction, we determined that this particulate material can be stripped off the underlying filament network by incubation in a medium containing either elevated ionic strength or a mixed micelle detergent, or in the case of mouse eggs extraction in a medium containing Triton X-100 (Gallicano *et al.*, 1991, 1994). This material stripped from the sheets can be detected by SDS–PAGE and Western analysis which demonstrates that it is not intermediate filament protein (Fig. 12; see below). In the case of sheets in hamster eggs we have used the prominent band from this analysis (M_r 69 kDa) to produce an antibody in rabbits. This antibody binds to the particulate material coating the surface of sheets but not to the intermediate filaments of the sheets (Fig. 13).

Although these two types of sheets appeared to be related at the structural level, it was unknown whether they were comparable at the biochemical level. To examine the biochemical nature of sheets we first employed immunoelectron microscopy using a number of different antibodies directed

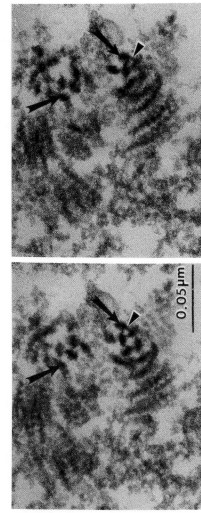

Fig. 7 High-magnification, cross-sectional image of sheets in a mouse egg are viewed as stereoscopic images using IVEM. Note the circular pattern of intermediate filaments (arrows) connected by lateral crossbridges (arrowheads). Reproduced from Gallicano et al. (1994a) with permission of Elsevier Science, Ltd.

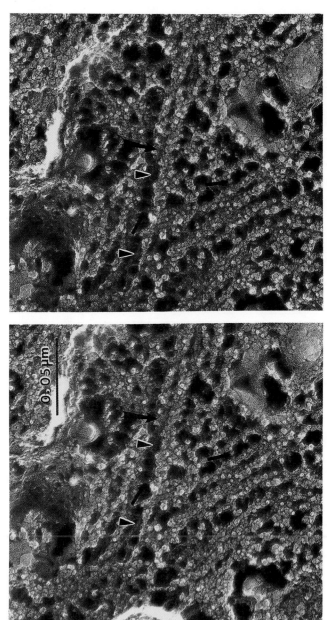

Fig. 8 Quick-freeze, deep-etch, stereoscopic image of sheets viewed *en face*. The fracture plane is cleaved through the middle of a sheet leaving only a portion of the top filament (large arrows) and allowing filaments (arrowheads) deeper within the replica of the sheet to be visible. Crossbridges connecting adjacent filaments (small arrows) are also visible. Reproduced from Gallicano *et al.* (1994a) with permission of Elsevier Science, Ltd.

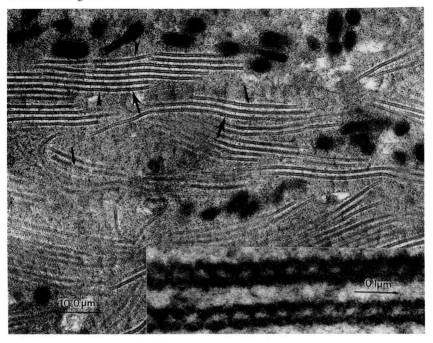

Fig. 9 Low-magnification view of the cytoskeletal sheets in an intact egg viewed using IVEM. Sheets are present in several orientations in this image. Large arrows point to cross-sectional views of the sheets and small arrows point to sheets that are tilted out of cross section. The inset shows a higher-magnification view of the cytoskeletal sheets in cross section. Note that sheets viewed in cross section are surrounded by an area devoid of the granular matrix of the cytoplasm (arrowheads). Reproduced from Capco *et al.* (1993) with permission of Wiley, Inc.

against known cytoskeletal proteins. We predicted that the filaments within these sheets would probably be composed of intermediate filaments. This prediction was based on their resistance to detergents, the periodicity on the filaments, and their diameter (10 or 11 nm). Antibodies to different types of cytokeratin were used to determine if sheets were composed of this protein. The cytokeratin antibodies which gave the best results were the AE1/AE3 set of antibodies. This set of antibodies is known to recognize most of the acidic types of cytokeratins and all of the basic types of cytokeratin (Weiss *et al.*, 1984). By using protein A–gold to localize the first antibody, a high signal was found on sheets in mouse eggs (the fibrous types of sheets). As a control, the zona pellucida from the same eggs that were subjected to antibodies was observed (Fig. 14). These control views showed very little binding of cytokeratin antibodies to the zona demonstrating that the binding of antibody to the sheets was specific. Antibodies directed

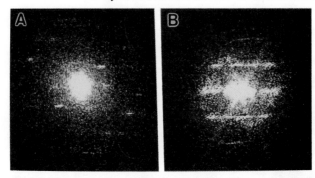

Fig. 10 Optical diffraction patterns from (A) untreated and (B) deoxycholate-extracted sheets. Each area of the micrograph examined includes several sheets seen nearly in cross section, but at various orientations. The figures are oriented with the sheets running vertically, so that the distances in the diffraction patterns measured vertically (from the center horizontal line) correspond to spacings along the sheets. Off-axis spots arise from views such as that in Fig. 9 (small arrows), where lines run across the sheet at an angle. The angle changes with the orientation of the sheet causing horizontal displacement of the spots, leading to the lines seen in these diffraction patterns. Scale bar = 0.05 nm^{-1}. Reproduced from Capco *et al.* (1993) with permission of Wiley, Inc.

against vimentin and tubulin showed virtually no signal, whereas anti-actin antibodies resulted in a very low signal when compared to anti-cytokeratin antibodies. Application of protein A–gold alone resulted in no binding (Fig. 14). These data suggested that the sheets in mouse eggs were composed of cytokeratin(s) (Gallicano *et al.*, 1994a). Sheets in hamster eggs which were similarly analyzed showed virtually identical results. Anti-cytokeratin antibodies followed by protein A–gold resulted in a high signal on sheets, whereas anti-vimentin and anti-tubulin antibodies showed no signal and anti-actin antibodies only a low signal (Capco *et al.*, 1993). The relevance of the low signal of anti-actin antibodies on the sheets in both mouse and hamster eggs is unknown. Thus, it appeared that these sheets were not only structurally similar but also biochemically similar in that they were both composed of some type of cytokeratin. But which type(s) of cytokeratin?

To determine the type(s) of cytokeratin that comprised the sheets, 250 mouse eggs were detergent extracted (in order to obtain detergent-resistant, sheet-enriched fraction), electrophoresed in a two-dimensional polyacrylamide gel, and blotted onto Polyscreen (Fig. 15). Using the AE1/AE3 antibodies followed by chemiluminescence detection system, it was found that the sheets were composed of cytokeratins type 5, 6, 16, and type Z (Fig. 15A). Type Z had been found by Jackson and co-workers (1981) to exist in mouse blastocysts. along with finding these keratins, another important finding from this two-dimensional Western was the detection of K8. This polypeptide had only been found in mouse four–eight cell embryos

A

HAMSTER SHEET

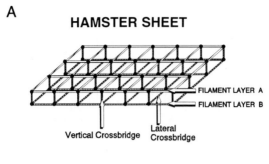

B

MOUSE SHEET

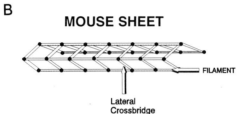

Fig. 11 A model comparing the substructures of sheets in mouse and hamster eggs. (A) Cytoskeletal sheets in hamster eggs are composed of two layers of intermediate filaments. Within each layer, intermediate filaments are held in register by lateral crossbridges. Layers of intermediate filaments are linked to one another by vertical crossbridges which are not present in mouse sheets. (B) Sheets in mouse eggs are composed of a single layer of intermediate filaments which are arranged into a cylinder. Each filament is held in register by lateral crossbridges which are spaced every 23–25 nm. Reproduced from Gallicano *et al.* (1994a) with permission of Elsevier Science, Ltd.

and later developmental stages, but not in eggs (Oshima *et al.*, 1983; Houliston *et al.*, 1987). To determine if this signal was K8, 250 mouse eggs, two-cell embryos, and early blastocysts were analyzed by Western blot using TROMA-1, an antibody specific for K8 in mouse embryos (Oshima *et al.*, 1983). K8 was found in both the unfertilized egg as well as the two-cell embryo, although in low abundance when compared to the early mouse blastocyst (Fig. 15B), demonstrating that the 55-kDa protein, detected in the two-dimensional Western, was indeed K8. The paucity of detectable K8 compared to the other types of keratins, though, made it unlikely that K8 was part of the sheets. One-dimensional Western analysis with the AE1/AE3 set of antibodies also showed the existence of K5, K6, K16, Z, and, in very low abundance, K8 in these eggs (Fig. 15C).

C. Studies by Other Investigators

We were not the first to identify the unusual structures we refer to as sheets. These structures were described in eggs of a variety of mammalian

SOL   CSK1   CSK2

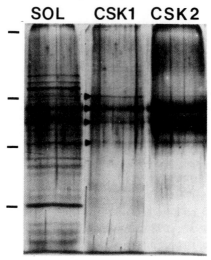

Fig. 12 A representative SDS–polyacrylamide gel of the three fractions of hamster eggs stained with Coomassie blue and silver stain. Approximately 100 eggs were used in this experiment. SOL, the detergent-soluble fraction; CSK1, the proteins solubilized by treatment of the detergent-resistant cytoskeleton with the extraction medium modifies to contain 0.25 M ammonium sulfate; CSK2, proteins remaining insoluble after the detergent and high-salt treatment. Arrowheads indicate the four prominent bands in the CSK-1 fraction. The arrowheads point to proteins with molecular weights of 69, 63, 57, and 45 kDa. The horizontal bars on the left indicate the migration of molecular weight markers (200, 68, 43, and 18.4 kDa from top to bottom of lane). Reproduced from McGaughey and Capco (1989) with permission of Wiley, Inc.

species beginning three decades ago (Enders and Schlafke, 1965; Hadek, 1966; Weakley, 1966, 1967; Schlafke and Enders, 1967; Szolliosi, 1972; Garcia *et al.*, 1979; Green, 1985; Koehler *et al.*, 1985; Lehtonen, 1985). However, these investigators had no clue as to their composition or function, and they referred to them as bilaminar lamellae, paracrystalline arrays, hyaloplasm, yolk, or ribosome storage sites. In defense of the efforts of these earlier investigators it should be pointed out that the studies described in this chapter which identify the composition of the sheets were both technically difficult to conduct because of the small number of eggs obtainable and required technological advancements that have only been available within the past 8 years. It also required a degree of tenacity on the part of the authors of this chapter.

The literature shows that the structures we now recognize as the cross-linked network of intermediate filaments in hamster begin to appear at the very early stage of follicular development, during fetal growth (Weakley, 1969; Garcia *et al.*, 1979; Gallicano *et al.*, 1994b), and our studies have

Fig. 13 Immunoelectron microscopic analysis of sheet components. High-magnification, embedment-free image of a sheet in a hamster egg which was extracted with Triton X-100 and then briefly (30 sec) subjected to 2A culture medium before fixation. A single sheet is seen dissociating (at arrowheads) into its constituative intermediate filaments. Following fixation, the egg was probed with anti-p69 antibodies, followed by gold beads conjugated to protein A. A signal was detected on the intact portion of the sheet, the area of filaments covered by the particulate material. The filaments dissociated from the sheet, the area of naked filaments, are devoid of gold particles.

shown what appears to be the full complement of sheets in unferteilized hamster and mouse eggs in the fully grown primary oocyte. Thus, the sheets appear to be a maternal component produced throughout the process of oogenesis.

III. Functions of the Sheets in Development

A. Storage Sites of Assembled Intermediate Filaments

The sheets serve as a storage site for assembled intermediate filaments of the cytokeratin type. The sheets are present in unfertilized eggs (Capco and McGaughey, 1986; Gallicano et al., 1991, 1992) and may begin to be produced during fetal development (Weakley, 1969; Garcia et al., 1979; Gallicano et al., 1994b). The highly cross-linked network of intermediate filaments which exists in the sheets may serve to organize the mass of filaments in an array that prevents disruption of the process of development.

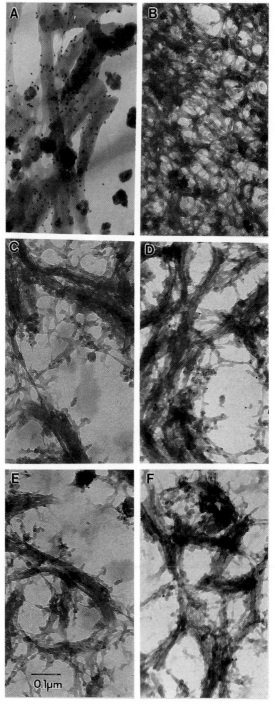

The concept of maternal storage of components in eggs for use in later embryonic development is not unusual, as many cellular components are known to be synthesized by the maternal genome in eggs of a variety of different classes of organisms.

B. Contribution to Forming Junctional Complexes at Embryonic Compaction

Embryonic compaction is a developmental transition that includes a change in shape of the embryo and its blastomeres from a cluster of largely spherical cells to a smooth ball of cells. This shape change is mediated by the establishment of tight junctions and desmosomes that form at the boundaries between blastomeres and progressively flatten the outer surface of cells (Ducibella *et al.*, 1975; Ducibella and Anderson, 1979; Fleming *et al.*, 1991; Clayton *et al.*, 1993; Riethmacher *et al.*, 1995). During this developmental transition, some of the sheets begin to splay apart into individual intermediate filaments and these filaments associate with the forming junctional complexes which resemble desmosomes (Gallicano *et al.*, 1991, 1994). The association of intermediate filaments, particularly those of the cytokeratin type, with desmosomes is a typical role of intermediate filaments in somatic cells. They also are believed to maintain the structural integrity of cells by their association with junctional complex (Klymkowsky *et al.*, 1989; Fuchs, 1994). Thus, some of the stored assemblages of intermediate filaments contribute to the rapid formation of junctional complexes at the time of embryonic compaction. In addition, some of the individual intermediate filaments which splay out from the sheets also associate with the nucleus of each blastomere (Gallicano *et al.*, 1994a). This association of intermediate filaments with the nuclear envelope has been reported to occur in somatic cells and is thought to be involved in the positioning of the nucleus within the somatic cells (Capco *et al.*, 1982; Georgatos and Blobel, 1987; Albers and Fuchs, 1989; Robson, 1989; Skalli and Goldman, 1991). The intermediate filaments may also be positioning the nucleus within the blastomeres beginning at the time of embryonic compaction.

Fig. 14 Immunoelectron microscopic analysis of sheet components. Embedment-free sections from detergent-extracted eggs were challenged with antibodies to keratin (A and B), vimentin (C), and actin (D), and tubulin (E), and viewed after incubation with gold beads conjugated to protein A. The zona pellucida from eggs incubated with antibodies to keratin is shown in B. Virtually no gold beads are detected demonstrating that the signal in A is specific. Reproduced from Gallicano *et al.* (1994a) with permission of Elsevier Science, Ltd.

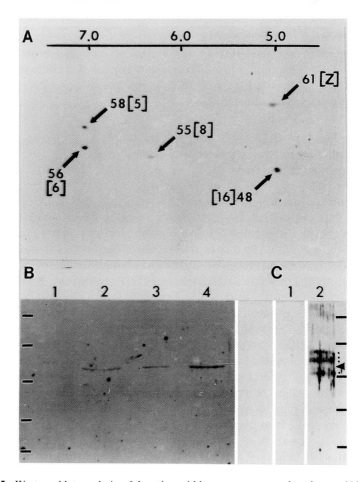

Fig. 15 Western blot analysis of keratins within mouse eggs and embryos. (A) Two-dimensional Western blot of the cytoskeletal fraction from 250 mouse eggs visualized by chemiluminescence indicates polypeptides stained by keratin antibodies, AE1/AE3, at 48, 55, 56, 58, and 61 kDa (arrows). Keratin types based on previous two-dimensional Western blot analyses (Jackson *et al.*, 1980; Moll *et al.*, 1982; Weiss *et al.*, 1984) are in parentheses. (B) One-dimensional Western blot of 250 mouse eggs (lane 2), 250 two-cell embryos (lane 3), and 250 early blastocysts (lane 4) stained with TROMA-1, followed by detection with chemiluminescence. Lane 1 contained 7 μl of the final wash which contained the eggs mixed with 7 μl SDS–sample buffer in order to show no contamination with extraneous cells from the oviduct or finger proteins. (C) One-dimensional Western blot of 250 Triton X-100-extracted eggs stained with AE1/AE3 antibodies (lane 2) reveals five bands (dots). Keratin type 8 is depicted by an arrowhead. Increased background levels were detected because we approached the limit of our chemiluminescence detection system in order to detect all polypeptide bands. Lane 1 contained 7 μl of the final wash which contained the eggs mixed with 7 μl SDS–sample buffer in order to show no contamination with extraneous cells from the oviduct or finger proteins. The horizontal bars on the left indicate the migration of molecular weight markers (97.4, 66, 46, 24, and 18 kDa, from top to bottom. Reproduced from Gallicano *et al.* (1994a) with permission of Elsevier Science, Ltd.

C. Maintenance of Epithelial Integrity during Implantation

There is a correlation between the number of sheets present in the eggs of different mammalian species with the time between fertilization and implantation, that is, the more sheets in the egg the more rapidly implantation occurs (Table I; Gallicano *et al.*, 1992). This suggested to us that the sheets or their derivative (i.e., cytokeratin filaments) may have a role in implantation. Previously, we had shown that the cross-linked network of intermediate filaments completely disassembles into individual intermediate filaments in the trophectoderm cells of the blasto-cyst stage (Capco and McGaughey, 1986; Gallicano *et al.*, 1991). Ongoing work demonstrates that immediately before implantation the sheets also disassemble into individual intermediate filaments in the inner cell mass cells, and that extensive arrays of tonofilaments then appear in all cells of the embryo (Schwarz *et al.*, 1993; Gallicano *et al.*, 1994a; S. Schwarz *et al.*, in press). These tonofilaments associate with desmosomal complexes at the junction between cells and form bands of tonofilaments across individual cells. This arrangement perhaps provides integrity to the embryo as it experiences the mechanical distortion associated with implan-tation into the uterine wall. Our ongoing results indicate that these extensive arrays of tonofilaments are produced from cytokeratins stored in the sheets and few if any are made from cytokeratins 8 and 18, which are synthesized in the embryo at this time (Oshima *et al.*, 1983; Gallicano *et al.*, 1994c; S. Schwarz *et al.*, in press).

The previously discussed results suggest that the maternal supply of assembled cytokeratin filaments in the form of sheets promotes the rapid

Table I Implantation Times and Sheet Densities among the Species

Time to implantation (in days)[a]		Reference	Sheet Density (sheets μm^2)[b]
Early			
Rat	5.0–5.5	Wimsatt (1975), Yochim (1971)	5.94 ± 0.23
Hamster	4.3–5.0	Wimsatt (1975)	4.60 ± 0.24
Mouse	5.0–6.0	Wimsatt (1975), Rumery (1971)	2.65 ± 0.17
Mid			
Human	7.0–9.0	Darnell *et al.* (1986), Wimsatt (1975)	1.40 ± 0.10
Late			
Porcine	11.0–20.0	Crombie (1972), Wimsatt (1975)	0.85 ± 0.04
Bovine	14.0–30.0	Eckstein and Kelly (1977), Wimsatt (1975)	0.45 ± 0.04

[a] Measured from fertilization.
[b] Mean ± standard error of the mean ($n \geq 24$).

assembly of tonofilaments within the embryo that may be essential for implantation into the uterine wall. Eggs with an ample supply of sheets can implant more rapidly, whereas those with less sheets in their eggs may require synthesis of the appropriate cytokeratins from the zygotic genome, and consequently require additional time before implantation.

IV. Regulation of the Sheets by Signal Transduction Mechanisms

Penetration of the egg by the sperm elicits a cascade of cytoplasmic signal transduction events that act to remodel the egg into the zygote. In many species, and certainly in mammalian development, these remodeling events are regulated not by differential gene expression, but rather by responses programmed into the cytoplasm. This is evident not only because the changes occur very rapidly (in a matter of minutes), but also because the chromatin is highly condensed in the form of chromosomes during this time period and not engaged in transcription. Numerous reports demonstrate that an elevation in the level of intracellular free calcium ($[Ca^{2+}]_i$) occurs as a response to sperm penetration in virtually every species studied to date. In fact, the fertilization-competent egg is so preprogrammed to respond to a rise in $[Ca^{2+}]_i$ that if the level of $[Ca^{2+}]_i$ is experimentally induced to elevate in the absence of sperm penetration the egg still undergoes all of the remodeling events that convert it to the structure of a zygote (Cran et al., 1988; Tombes et al., 1992; Kline and Kline, 1992; Gallicano et al., 1993), although death later ensues. The remodeling events that convert the egg into the zygote include cortical granule exocytosis to provide the block to polyspermy, the completion of the reduction division and transit out of M phase of meiosis into interphase, cytokinesis to release the second polar body, reformation of the Golgi apparatus, and resumption of endocytosis and exocytosis (Cran et al., 1988; Kline and Kline; 1992; Gallicano et al., 1993)

We have been investigating the cytoplasmic signaling events that act downstream of the elevation in $[Ca^{2+}]_i$. We have shown that the rise in $[Ca^{2+}]_i$ activates protein kinase C (PKC), a calcium and phospholipid-dependent, serine/threonine kinase which is responsible for many, but not all, of the structural remodeling events that accompany the conversion of the egg into the zygote in both mammals (Gallicano et al., 1993) and amphibians (Bement and Capco, 1989, 1990). We have also shown that experimental activation of PKC under conditions in which the level of $[Ca^{2+}]_i$ is clamped low also results in the remodeling of the egg into the zygote (Gallicano et al., 1993). Our results indicate the PKC is a key signaling agent acting downstream of the rise in $[Ca^{2+}]_i$ to regulate fertilization

(Gallicano *et al.,* 1993; Bement and Capco, 1989, 1990). That conclusion is an important clue to unraveling the regulation of the cytoskeletal sheets. We had noted previously (Capco and McGaughey, 1986; Gallicano *et al.,* 1991) that the developmental transition of fertilization resulted in a dramatic change in the spatial organization of the sheets. Thus, it seemed reasonable that the elevation in $[Ca^{2+}]_i$ or activation of PKC would be the regulator of this structural remodeling.

We initially anticipated that the rise in $[Ca^{2+}]_i$ would be the signaling agent that regulated the fertilization-dependent remodeling of the sheets rather than activation of PKC. This assumption was based on the knowledge that PKC becomes active at membranes of cells, usually the plasma membrane (Ashendel, 1985; Nishizuka, 1988). Since the sheet were not bounded by membrane we thought it unlikely that they would be regulated by PKC. Instead, the rise in $[Ca^{2+}]_i$, which could act on components in the cell interior, was thought to be the likely signaling agent. However, to our surprise the sheets underwent remodeling under conditions in which PKC was experimentally activated and the level of calcium was clamped low. The sheets were not remodeled in the converse condition, that is, where a rise in $[Ca^{2+}]_i$ was induced while clamping the level of activity of PKC low (Gallicano *et al.,* 1995) Thus, it appeared that activation of PKC was responsible for remodeling of the sheets as it was for most of the other egg activation events. Still, the puzzle remained as to how PKC, a membrane-dependent kinase, was acting on the sheets.

The solution to the puzzle resulted from mapping the distribution of the kinase over time and monitoring the molecular weight and cofactor requirements of the kinase (Gallicano *et al.,* 1995). Using the PKC reporter dye, Rim-1, developed by Chen and Poenie (1993), we mapped the distribution of the kinase after egg activation was induced. PKC translocated from a uniform distribution in the cytoplasm in the unfertilized egg to an enrichment at the plasma membrane of the egg by 30 min after the induced rise in $[Ca^{2+}]_i$ that mimicked the sperm-penetration event (Fig. 16). By 50 min after the induced elevation in $[Ca^{2+}]_i$ the kinase was no longer associated with the plasma membrane of the cell, and instead was associated with the cytoskeletal sheets (Fig. 16). This association of the kinase with the sheets correlated precisely with the time when the sheets in hamster eggs were normally undergoing their structural remodeling. Comparison of the location of the kinase in control eggs with eggs 50 min after the induced elevation of $[Ca^{2+}]_i$ by Western analysis demonstrated that the kinase shifted compartments from the soluble to the sheet-enriched fraction as would be expected if the kinase was associated with the sheets (Fig. 17). Just as important, the kinase shifted in molecular weight from 80 to 50 kDa (Fig. 17). This shift to a lower molecular weight provided us with the clue to how the kinase could be acting on the sheets which are not bounded by membrane.

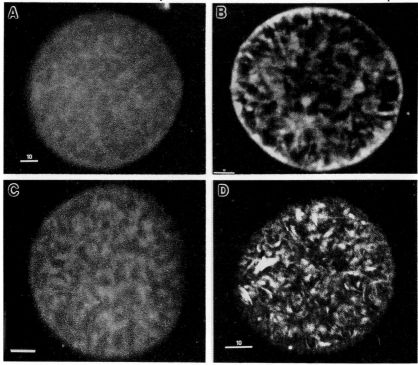

Fig. 16 Confocal images of hamster eggs stained with the PKC reporter dye, Rim-1, before and after activation of eggs with calcium ionophore. (A) An unactivated egg stained with Rim-1 shows a diffuse fluorescence of PKC throughout the interior of the egg. (B) An egg transiently activated with calcium ionophore (see Methods) for 30 min and then stained with Rim-1 demonstrates PKC on the plasma membrane of the egg. (C) An egg transiently activated with calcium ionophore (see Appendix) for 50 min and then stained with Rim-1 shows a decreased signal on the membrane of the egg and an increase in signal on distinct structures in the interior region of the egg. (D) Detergent-extracted, sheet-enriched fraction of eggs after transient, 50-min activation with calcium ionophore shows sheets stained with Rim-1. Reproduced from Gallicano *et al.* (1995) with permission of Academic Press.

One mechanism of downregulating PKC is to cleave it between its membrane-binding domain and the catalytic subunit of the kinase (Kishimoto *et al.*, 1985; Tapley and Murray, 1985; Huang and Huang, 1986; Murray *et al.*, 1987; Jaken, 1990; Hashimoto and Yamamura, 1991). The catalytic subunit, which exhibits a reduced molecular weight, is then free to diffuse away from the membrane, and it becomes both membrane independent and calcium independent. In this form, the active, catalytic subunit of the kinase is referred to as PKM because of early reports in the literature (Inoue *et al.*, 1977; Takai *et al.*, 1977). Our results demonstrated that the kinase associated with the sheets was of the correct molecular weight to

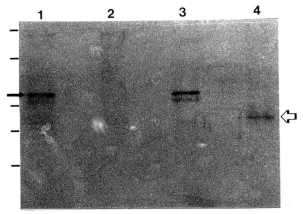

Fig. 17 Western blot analysis of the detergent-soluble and detergent-resistant fractions from unactivated and activated hamster eggs. The monoclonal primary antibody employed was directed against the catalytic subunit of PKC-α. This antibody was detected by a chemiluminescence antibody detection system. Lane 1 shows the detergent-soluble fraction from an unactivated egg. PKC is detected at 80 kDa (solid arrow). Lane 2 is the sheet-enriched fraction showing no antibody staining. Lane 3 is the detergent-soluble fraction from eggs activated with a transient treatment in calcium ionophore and shows a band at 80 kDa representing PKC. Lane 4 is the sheet-enriched fraction from eggs in lane 3 and shows one band at 58 kDa (open arrow) representing the catalytic subunit of the kinase in the detergent-resistant, sheet-enriched fraction. The horizontal bars on the left indicate the migration of molecular weight markers (200, 97.4, 68, 43, and 25 kDa, from top to bottom). Reproduced from Gallicano *et al.* (1995) with permission of Academic Press.

be PKM (no kinase with a molecular weight appropriate for PKC was detectable associated with the sheets), was recognized by an antibody directed against the catalytic subunit of PKC, and was able to phosphorylate an exogenous substrate for PKC/PKM under conditions in which only PKM, and not PKC, would be active. Moreover, inhibitors to protein kinase A, myosin light-chain kinase, and tyrosine kinase, added to the reaction mixture, could not block the phosphorylation of the exogenous substrate; however, when an inhibitor to PKC/PKM was added to the reaction mixture phosphorylation was inhibited. Furthermore, we can show that at the appropriate time PKM phosphorylates two substrates in the sheets, that is cytokeratin 16 and the 69-kDa protein (p69) that coats the cross-linked network of intermediate filaments (Gallicano *et al.,* 1995).

The results described previously demonstrated that PKM is associated with the sheets and phosphorylate components associated with the sheets, however, these results do not demonstrate that the action of PKM is itself responsible for the change in spatial organization of the sheets. To demonstrate this action, we permeabilized hamster eggs and perfused into this system ATP, an ATP-regenerating system, and PKM. Under these conditions, the sheets in the permeabilized system remodeled their organization comparable to what typically occurs as a result of fertilization. When PKM was deleted from the perfusion mixture the sheets did not change their organization (Gallicano *et al.*, 1995). These results strongly suggested that PKM produced as a consequence of fertilization acts to remodel the sheets (Gallicano *et al.*, 1995). This kinase may also act at the subsequent developmental transitions to regulate the structural remodeling. Such an occurrence remains to be determined.

These results also indicated that the two forms of the kinase, that is PKC and PKM, act as a chronometer to regulate the remodeling of the egg at the time of fertilization. This chronometer exhibits both temporal and spatial precision. PKC becomes active first at the cell periphery, triggering among other events cortical granule exocytosis, the event that induces the block to polyspermy. Subsequently, the titer of PKM increases and the kinase gradually becomes active in the cell interior. It seems quite logical to remodel the cell periphery before remodeling the cell interior. Polyspermy must be prevented, otherwise the cell will die. Once the block to polyspermy has been initiated, other internal components of the cell can be remodeled at a more leisurely rate.

V. Summary

The sheets serve as an maternal supply of assembled, cytokeratin, intermediate filaments. They are remodeled at each major developmental transition in mammalian early development, that is fertilization, embryonic compaction, blastocyst formation, and formation of the primitive ectoderm and primitive endoderm during implantation into the uterine wall. Our results indicate that the sheets exist as specialization for placental development as they have a major role in the maintenance of epithelial integrity at the time the embryo is implanting into the uterine wall. They also contribute intermediate filaments to the junctional complexes required for embryonic compaction. Our analyses demonstrate the they are regulated at the time of fertilization by the action of PKC/PKM, a kinase that acts as a cellular chronometer with both temporal and spatial precision that remodels the egg into the zygote.

Appendix: Methods

Investigating the structure and function of the cytoskeleton in mammalian eggs and embryos has been a difficult task because, until a few years ago, the tools and technologies had not been available to exploit the small amount of material obtained from them. To understand the cytoskeleton in greater detail in these specimens, we developed new approaches or modified techniques used in somatic cell research and applied them to study the cytoskeleton of mammalian eggs and embryos.

A. Animal Handling and Acquisition of Eggs/Early Embryos

In the studies discussed in this chapter, we employed the golden Syrian hamster (*Mesocrietus auratus*) and mice of the Swiss albino strain (*Mus musculus*). Hamster embryos are difficult to maintain and do not develop efficiently in culture. Thus, mice were the preferred organism of study for culture experiments. Hamster eggs and embryos, however, contained very large cytoskeletal sheets and thus were easier to use for studies involving these unique cytoskeletal components.

1. Procedure for Obtaining Eggs and Embryos

1. To obtain eggs from mice, female mice are injected with 5–8 IU pregnant mare's serum gonadotropin (PMSG; Sigma Chemical Co., St Louis, MO) around 2:00 PM, followed by injection with 5–8 IU human chorionic gonadotropin (hCG; Sigma) 48 hr later.

2. Eighteen to 20 hr later, the animal is sacrificed by cervical dislocation and the oviducts are removed. Unfertilized eggs are collected in 2A medium (McGaughey, 1977) [20 mM Hepes, pH 7.4, 20 mM sodium lactate, 0.51 mM sodium pyruvate, 86.4 mM NaCl, 4.3 mM KCl, 1.2 mM CaCl$_2$, 1.1 mM KH$_2$PO$_4$, 1.1 mM MgSO$_4$, 25.1 mM NaHCO$_3$, 5.6 mM glucose, 2.0 μM phenol red, 4.5 g/liter dextran T70, penicillin G (1.0×10^5 units/liter), and streptomycin sulfate (3.75×10^4 units/liter)] by teasing the ampulla region of oviducts with Dumont No. 5 forceps. The eggs are denuded of surrounding cumulus cells by adding to the 2A medium 300 units/ml hyaluronidase (Sigma). Once cumulus cells are removed, eggs are transferred to 2A medium without enzyme.

3. To obtain mouse embryos, female mice are injected as described previously, however, 3 hr after hCG injection (approximately 4:00 PM), one female is placed into a cage with one male to allow copulation (fertilization is assumed to occur at midnight). One-, two-, four-, and eight-cell embryos are obtained 10–12, 32, 48, and 60 hr after fertilization, respectively, by teasing the oviducts with Dumont No. 5 forceps.

4. Compacted- and blastocyst-stage embryos are obtained 72 and 84 hr after fertilization, respectively, from the uterus. After removal of the female reproductive tract (oviducts, uterine horns, and

cervix) insert a 26 3/4-gauge needle attached to a 10-cc syringe filled with 2A medium through the cervix. Inject medium so as to balloon the horn outward, then cut the oviduct off and inject more medium. The embryos will drain into the petri dish.

5. To obtain hamster eggs or embryos, females are injected with 25 IU of PMSG on Day 1 of the estrous cycle and 25 IU of hCG on Day 3 of the estrous cycle. Hamsters are sacrificed by CO_2 asphyxiation and unfertilized eggs are obtained 15–18 hr later by teasing the oviducts with Dumont No. 5 forceps. For fertilized eggs a female is placed with one or two males on the evening of Day 3 of the cycle. Subsequent timing and acquisition of obtaining embryos are similar to those of the mouse.

B. Detergent-Extraction and Embedment-Free Electron Microscopy

Cytoskeletal ultrastructure has been studied in great detail in somatic cells using many different types of electron microscopy. The most common type of electron microscopy has been thin-section electron microscopy in which specimens are embedded in a plastic material, sectioned at a thickness of 80–100 nm, and then poststained with heavy metals. This method of electron microscopy has also been used to study the cytoskeleton in mammalian eggs and embryos; however, because of the large size of these eggs and blastomeres in embryos, it is possible that sections only 80–100 μm thick have not provided a representative view of the cytoskeletal components in these specimens.

A different method which provides much more structural information concerning the cytoskeleton during early mammalian development is embedment-free electron microscopy. The power of this technique is recognized when eggs or embryos are first extracted with a nonionic detergent to remove detergent-soluble components leaving behind the detergent-resistant cytoskeleton.

1. Preparation for Detergent-Extraction and Embedment-Free Electron Microscopy

The following materials will be needed: 10% polylysine (100 kDa), No. 1 coverslips, small glass petri dishes, disposable glass pipettes, acetone, 100% ethanol, molten diethylene glycol distearate (DGD) containing 3.0% (v/v) dimethylsulfoxide (DMSO), OsO_4, sodium cacodylate, glutaraldehyde, Dumont No. 5 forceps, razor blade, intracellular buffer, extraction medium, protease inhibitors, and phosphate buffered saline (PBS). The extraction medium used for mammalian eggs and embryos is an intracellular buffer (ICB) designed to resemble the intracellular ionic composition of mammalian cells (Aggeler *et al.*, 1983; Webster and McGaughey, 1990). ICB is composed of 100 mM KCl, 5 mM MgCl$_2$, 3 mM EGTA, 20 mM Hepes, pH 6.8. Extraction buffer is ICB containing 1.0% Tween 20 and 1.2 mM 4-(2-aminoethyl)-benzenesulfonylflouride. Note: when using detergents that extract cells faster than Tween 20 (i.e., Triton X-100 or NP40), 300 mM sucrose can be added to the ICB. The detergents must be added to a final concentration well above their critical micelle concentration.

2. Procedure for Investigating the Cytoskeleton of Eggs and Embryos Using Embedment-Free Electron Microscopy

1. Before sacrificing animals clean and prepare coverslips. Place No. 1 coverslips in a beaker containing acetone and sonicate 3× for 5 min each, followed by sonication in distilled water 3× for 5 min each. Allow coverslips to air dry. Spread 10–20 μl 10% (w/v) polylysine onto coverslips using the pipette tip. If polylysine does not spread evenly or if it retracts from the edges of the coverslip, discard and start with a new coverslip. Allow the polylysine to dry.
2. Obtain eggs or embryos (see above) and wash 3× in PBS at 37°C.
3. Place coverslips in a small glass petri dish and add 1 or 2 ml PBS. Aspirate 1–20 eggs or embryos and slowly add to coverslips. The specimens should adhere immediately. If they do not, reaspirate eggs or embryos onto another prepared coverslip. Using a disposable glass pipette, remove three-fourths volume of PBS, add extraction buffer slowly, and repeat 1×. Allow specimens to extract for 8–10 min at room temperature.
4. Remove three-fourths volume extraction buffer and add extraction buffer containing 2.0% (v/v) glutaraldehyde (2.0% formaldehyde if procedure is used for immunoelectron microscopy). Fix specimens for 1 hr.
5. Remove three-fourths volume fixation medium and wash 3× in 0.1 M sodium cacodylate buffer.
6. Postfix specimens by removing three-fourths volume sodium cacodylate buffer and adding 1.0% (v/v) OsO$_4$ in 0.1 M sodium cacodylate buffer. Specimens should be postfixed for 1 hr, followed by washing 3× in sodium cacodylate buffer without OsO$_4$.
7. Specimens must then be dehydrated through a series of increasing ethanol concentrations. Remove three-fourths volume of sodium cacodylate buffer and add 10% ethanol. Repeat 1× and allow specimens to acclimate in 10% ethanol for 10–15 min. This procedure should be repeated using 30, 50, 85, and 100% ethanol,

respectively. Exchange 100% ethanol 3× for 20 min each with 1:1 100% ethanol:100% n-butyl alcohol. After the third exchange, allow specimens to equilibrate for at least 30 min. Remove three-fourths volume 1:1 100% ethanol:100% n-butyl alcohol and add 100% n-butyl alcohol 3× for 20 min for each exchange. After the third exchange, allow specimens to equilibrate for at least 30 min. Place the petri dish containing the specimens into a prewarmed 70°C oven for 15 min. Pour off the n-butyl alcohol and quickly pour 1:1 100% n-butyl:100% molten DGD into the petri dish containing the specimens. Repeat this 3× for 30 min each. Note: do not allow the DGD to solidify. Pour off 1:1 100% n-butyl alcohol:100% DGD and add 100% molten DGD. Pour off DGD and add fresh DGD 3×. Allow 30 min acclimation before each exchange.

8. Under heat lamps, add molten DGD to each block of a Teflon-coated flat-embedding mold (Ted Pella, Inc., Redding, CA; product 10505). Using a dissecting scope, remove eggs or embryos one at a time from the coverslip by gently nudging the egg or embryo with the tip of a needle. Aspirate the egg in a prewarmed, pulled pipette and quickly place into one mold. Allow the mold to harden at room temperature. More DGD will have to be added to each block due to shrinkage of the wax as it cools and hardens. Remove blocks from mold immediately after they cool to room temperature.

9. Trim and section blocks (Capco *et al.,* 1993).

10. To remove embedding material from sectioned specimens, place electron microscopy Formvar-coated grids containing sections into a petri dish containing 100% n-butyl alcohol for 1 hr at room temperature. After 1 hr, carefully swirl the dish and allow to sit for 15 min. Remove grids and place into a petri dish containing 1:1 100% ethanol:100% n-butyl alcohol for 15 min. Remove grids and place into a petri dish containing 100% ethanol for 1 hr at room temperature. Again, swirl the dish after 1 hr and allow to sit for 15 min. Dry the specimens through the CO_2 critical point and view using any conventional electron microscope.

This procedure can be modified for immunoelectron microscopy (Nickerson *et al.,* 1990; Capco *et al.,* 1993; Gallicano *et al.,* 1994a). It also is likely that by modifying the buffers, eggs and embryos from other species could be analyzed using this technique.

Although this technique is useful for studying cytoskeletal ultrastructure in relatively fine detail, it must be noted that the resolution of embedment-free electron microscopy begins to diminish above magnifications of about

30,000–40,000x. If magnification higher than these is needed, an alternative method of electron microscopy IVEM (Capco *et al.,* 1993; Gallicano *et al.,* 1994a) of plastic-embedded thick sections should be used.

C. Analysis of PKC in Eggs and Embryos

PKC is a ubiquitous serine/threonine kinase composed of at least 10 families, each of which play a pivotal role in specific cellular signal transduction events and cellular growth control (Nishizuka, 1988). Regulation of this enzyme has been shown to be quite unique in that it must translocate to an intracellular membrane (usually the plasma membrane) and interact with certain species of phospholipids and diacylglycerols (and calcium for the major families of PKC) in order to become active. Numerous studies have defined the subcellular distribution of PKC using biochemical and immunocytochemical approaches. A standard procedure for observing the spatial and temporal localization of most cellular components, including PKC, has been immunofluorescence. To date, this technique has been considered highly reliable and reproducible; however, it is not a simple technique and it can be very time consuming and expensive. One procedure that recently was developed to circumvent the drawbacks of immunofluorescence for localizing PKC was the advent of the PKC reporter dyes, Rim-1 and Fim-1 diacetate (Chen and Poenie, 1993). These dyes are composed of a bisindolylmaleimide (Davis *et al.,* 1989), an inhibitor of PKC that functions by competing for the ATP site on the catalitic subunit of the enzyme, conjugated to either rhodamine or FITC (Chen and Poenie, 1993). Both dyes can be used to localize PKC in fixed, permeabilized eggs, embryos, or virtually any other cell type, while Fim-1 diacetate is membrane permeable and can be used to observe the translocation of PKC to cellular membranes in real time.

Rim-1/Fim-1 are available through Teflabs (9503 Capitol View Drive, Austin, TX). Stock solution of either dye should be made with deionized water at 2.0 μM. The working dilution is 200 nM.

1. Procedure for Localizing PKC in Mammalian Eggs or Embryos

1. Eggs or embryos are obtained as explained previously and washed in PBS 3× at 37°C.
2. After experiment (i.e., treating specimens with PMA or calcium ionophore; see below), place specimens into ICB medium (for recipe see above) containing 2.0% formaldehyde (paraformaldehyde) for at least 30 min. It may be necessary to add 0.1% glutaraldehyde for better fixation of cytoskeletal components

(e.g., without 0.1% glutaraldehyde, cytoskeletal sheet proteins are not properly fixed); however, do not use higher concentrations of glutaraldehyde or autofluorescence may occur.

3. After fixation, permeabilize eggs/embryos for 5–10 sec in ICB containing 2.0% formaldehyde and 0.5% Tween 20, followed by washing 3× for 15 min each at room temperature in ICB without fixative or detergent.
4. Place eggs/embryos in ICB containing 200 nM Rim-1 or Fim-1 and place in a darkened room for 30 min.
5. Wash eggs/embryos in ICB 3×, 5 min in the first wash, 15–20 min in the second, and 45 min in the third.
6. While eggs/embryos are washing, thoroughly clean glass slides and coverslips with ethanol. Once slides are cleaned, place four small drops of 1:1 paraffin:vaseline mixture onto the slide arranged in a square pattern. These drops of wax will act as posts onto which the coverslip will be placed.
7. Aspirate eggs/embryos from ICB and place them in the center of the four drops of wax. Place the coverslip on the posts and gently compress the coverslip until it makes contact with the droplet of ICB containing the eggs. Add more ICB under the coverslip (from opposite sides). Using a dissecting scope, compress coverslip further onto the slide until contact is made with an egg/embryo. Do not compress any further! Seal coverslips onto slides with fingernail polish.

When these dyes are compared to immunofluorescence using antibodies directed against PKC, similar results are obtained in eggs and embryos. However, it is advisable that when using a system which has not been tested previously for these dyes, it may be necessary to perform a parallel experiment using standard immunofluorescence techniques to be sure that the dyes provide comparable results.

For optimum results, eggs/embryos should be viewed using a laser-scanning confocal microscope. Because of the large size of eggs and early blastomeres, standard fluorescence microscopy may result in a loss of resolution due to background fluorescence from out of focus portions of the image.

D. Analyzing the Cell Cycle of Eggs and Embryos Using the DNA Dye Hoechst 33342

When analyzing egg activation, it is essential to observe the genetic material in order to determine if the egg has entered interphase or is still arrested in meiotic metaphase II. Hoechst 33342 binds specifically to DNA and requires only minimal preparation.

1. Procedure for Hoechst 33342 Staining of Eggs or Embryos

1. Eggs or embryos should be obtained as previously described. Aspirate eggs/embryos into culture dishes containing PBS (pH 7.4) and 2.0% v/v glutaraldehyde for 45 min.
2. Wash the eggs/embryos in PBS without glutaraldehyde 2× for 15 min each and pipette them into PBS containing 1 μg/ml of Hoechst 33342 dye for >30 min at room temperature. A stock solution of Hoechst should be made at 10 mg/ml in deionized water and frozen at $-20°C$. To minimize photobleaching, the light source on the microscope should be used on a minimum setting and room lights should be dimmed or turned off.
3. Following staining, the specimens are washed in PBS without dye 3× for 5, 20, and 45 min, respectively.
4. During the wash periods prepare and clean slides as described previously. Place eggs/embryos on glass slides and cover with glass coverslips as described previously.

Observations are made using a standard fluorescence microscope equipped with a 50-W mercury arc bulb for epifluorescence and a uv filter block. Fluorescence photographs are taken with T-MAX film (Kodak, ASA 400). If autofluorescence poses a problem increase the wash times twofold after subjecting specimens to the dye.

E. Studying Signal Transduction Pathways in Mammalian Eggs and Embryos

Intracellular signals are a cornerstone for the egg and early embryo to successfully form a new individual. Fertilization-induced alterations of the egg are likely controlled by intracellular signals primarily because both maternal and paternal chromosomes are in a condensed state and transcriptional activity is shut down. In addition, many important events which result after fertilization take only seconds (e.g., the rise in $[Ca^{2+}]_i$) or minutes (e.g., cortical granule exocytosis; second polar body emission) to occur and not the hours or days normally required for transcriptional events to have an effect. Investigating the intracellular signals that control fertilization has been difficult in the past because tools such as agonists and antagonists specific for intracellular signaling agents were not available. Furthermore, the appropriate timing for applying conventional as well as recently developed agonists and antagonists to intracellular signaling agents had not been defined, a probable reason for the conflicting results reported in the literature when the same agonists or antagonsts were applied to eggs and embryos (i.e., Bloom, 1989; Winkel et al., 1990). Consequently, the proper incubation times for specific antagonists and agonists have been found to be essential for successfully studying overall egg activation or individual components of egg activation such as changes in the cytoskeleton.

1. Procedure for Activating or Inhibiting [Ca^{2+}]$_i$ and PKC in Mammalian Eggs and Embryos

1. Eggs can be artificially activated by a transient treatment with the calcium ionophore, A23187 (Sigma). A23187 should be made at a stock solution of 10 mM in DMSO and kept frozen at -20°C until needed. The working dilution is 5 μM made in any medium suitable for culturing mammalian eggs and embryos. Eggs are obtained normally (see above) and placed into medium containing A23187 at 37°C for 5 min. The eggs should be removed and washed 2x in medium free of A23187. Eggs can be assayed for activation by observing emission of the second polar body which takes about 1 hr in mouse eggs and 40 min in hamster eggs.

2. Many egg-activation events can be induced by a brief incubation in PKC agonists such as 250 μM sn-1,2-dioctanoylglycerol (DiC8; Molecular Probes, Eugene, OR) or 100 nM phorbol 12-myristate-13-acetate (PMA; Sigma).

   a). To obtain the working dilution of 250 μM DiC8, 1 mg DiC8 (supplied as a liquid at 1 mg/μl) should be aliquoted into a small Epindorf to which 300 μl of medium is added. This mixture then is sonicated for 15–20 sec. Remove the sonicated DiC8 mixture and add it to 11.3 ml of medium for a final concentration of 250 μM DiC8. Eggs or embryos are obtained normally and then incubated in DiC8 either chronically or transiently for 30 min, followed by washing in medium free of DiC8 (see Gallicano *et al.*, 1993).

   b). A stock solution of PMA should be made at 10 mM in DMSO and frozen at -20°C until used. The working dilution of PMA is 100 nM. Eggs or embryos are obtained normally (see above) and placed into medium containing PMA at 37°C for 5 min. The eggs or embryos are washed 2× in medium free of PMA. If left in medium containing PMA too long (i.e., >20 min), PKC is downregulated resulting in skewed results or even the death of the egg or embryo (Gallicano *et al.*, 1993; Gallicano *et al.*, unpublished results).

   c). As controls for the above agonists, a structurally different phorbol ester, such as 4α-PDD, or a structurally incompatible diacylglycerol, such as sn-1,3-dioctanoylglycerol, both of which do not activate PKC, should be used in parallel.

3. $[Ca^{2+}]_i$ can be clamped low in mammalian eggs and embryos
   without injection, which can be traumatic to eggs or embryos, by
   using 130 μM BAPTA/AM (Sigma). A stock solution of BAPTA/
   AM should be made at 100 mM in DMSO and frozen at $-20°C$.
   Eggs or embryos are incubated in medium containing 130 μM
   BAPTA/AM for 30 min at 37°C. Eggs or embryos then are washed
   in medium free of BAPTA/AM (Gallicano et al., 1993, 1995).
4. Inhibition of PKC activity can be accomplished by a number of
   structurally different inhibitors. We have used 10 μM sphingosine
   (Matreya, Inc., Pleasant Gap, PA), 30 nM staurosporine
   (Calbiochem, La Jolla, CA), 1.0 μM chelerythrine (Calbiochem),
   and 1.0–2.0 μM bisindolylmaleimide (Calbiochem) (please see
   Gallicano et al., 1993, 1995 for justification of concentrations). All
   inhibitors of PKC, except for bisindolylmaleimide, have been used
   on hamster eggs to inhibit PMA-, DiC8-, or A23187-induced
   activation of eggs as measured by emission of a second polar body.
   Bisindolylmaleimide inhibits egg activation only at high
   concentrations (>5 μM) in intact eggs and embryos. This inhibitor
   does not seem to permeate the plasma membrane well; however, it
   is an excellent PKC antagonist when used in homogenates made
   from eggs (Gallicano et al., unpublished data).

Acknowledgment

We express sincere appreciation to the National Institutes of Health for continued support
of our work, currently Grant NIH HD 27151.

References

Adler, R. R., and Ziomeck, C. A. (1986). Cell specific loss of polarity-inducing ability by
    later stage mouse preimplantation. Dev. Biol. **114**, 395–402.
Aggeler, J., Takemura, R., and Werb, Z. (1983). High resolution three-dimensional views
    of membrane-associated clatherin and cytoskeleton in critical-point-dried macrophages.
    J. Cell Biol. **97**, 1452–1458.
Albers, K., and Fuchs, E. (1989). Expression of mutant keratin cDNAs in epithelial cells
    reveals possible mechanisms for initiation and assembly of intermediate filaments.
    J. Cell Biol. **108**, 1477–1493.
Albertini, D. F. (1992). Regulation of meiotic maturation in the mammalian oocyte:
    Interplay between exogenous cues and the microtubule cytoskeleton. Bioessays **14**,
    97–103.
Ashendel, C. (1985). The phorbol ester receptor: A phospholipid-regulated protein kinase.
    Biochim. Biophys. Acta. **822**, 219–242.

Bement, W. M., and Capco, D. G. (1989). Activators of protein kinase C trigger cortical granule exocytosis, cortical contraction, and cleavage furrow formation in *Xenopus laevis* oocytes and eggs. *J. Cell Biol.* **108**, 885–892.

Bement, W. M., and Capco, D. G. (1990). Protein kinase C acts downstream of calcium at entry into first interphase of *Xenopus laevis*. *Cell Regul.* **1**, 315–326.

Bloom, T. L. (1989). The effects of phorbol ester on mouse blastomeres: A role for protein kinase C in compaction? *Development* **106**, 159–171.

Brulet, P., Babinet, C., Kemler, R., and Jacob, F. (1980). Monoclonal antibodies against trophectoderm-specific markers during mouse blastocyst formation. *Proc. Natl. Acad. Sci. USA* **77**, 4113–4117.

Capco, D. G. (1993). Diethylene glycol distearate (DGD): An alternative to PEG as a removable embedment medium. *In* "PEG in Light and Electron Microscopy" (K. X. Gao, ed.). CRC Press, Boca Raton, Florida.

Capco, D. G., Gallicano, G. I., McGaughey, R. W., Downing, K. H., and Larabell, C. A. (1993). Cytoskeletal sheets of mammalian eggs and embryos: A lattice-like network of intermediate filaments. *Cell Motil. Cytoskeleton* **24**, 85–99.

Capco, D. G., Wan, K. M., and Penman, S. (1982). The nuclear matrix: Three dimensional architecture and protein compistion. *Cell* **2**, 847–858.

Capco, D. G., and McGaughey, R. W. (1986). Cytoskeletal reorganization during early mammalian development: Analysis using embedment-free sections. *Dev. Biol.* **115**, 446–450

Chen, C-S., and Poenie, M. (1993). New fluorescent probes for protein kinase C. *J. Biol. Chem.* **268**, 15812–15822.

Chisholm, J. C., and Houliston, E. (1987). Cytokeratin filament assembly in the preimplantation mouse embryo. *Development* **101**, 565–582.

Clayton, L., Stinchcombe, S. V., and Johnson, M. H. (1993). Cell surface localization and stability of uvomorulin during early mouse development. *Zygote* **1**, 333–344.

Cran, D. G., Moor, R. M., and Irvine, R. F. (1988). Initiation of the cortical reaction in hamster and sheep oocytes in response to inositol trisphosphate. *J. Cell Sci.* **91**, 139–144.

Damjanov, I., Damjanov, A., Lehto, V. P., and Virtanen, I. (1986). Spectrin in mouse gametogenesis and embryogenesis. *Dev. Biol.* **114**, 132–140.

Davis, P., Hill, C., Keech, E., Lawton, G., Nixon, J., Sedwick, A., Wadsworth, J., Weatmacott, D., and Wilkinson, S. (1989). Potent selective inhibitors of protein kinase C. *FEBS Lett.* **259**, 61–63.

Ducibella, T., Albertini, D. F., Anderson, E., and Biggers, J. D. (1975). The preimplantaion mammalian embryo: Characterization of intercellular junctions and their appearance during development. *Dev. Biol.* **45**, 231–250.

Ducibella, T., Ukena, T., Karnovsky, M., and Anderson, E. (1977). Changes in cell surface and cortical cytoplasmic organization during early embryogenesis in the preimplantation mouse embryo. *J. Cell Biol.* **74**, 153–167.

Ducibella, T., and Anderson, E. (1979). The effects of calcium deficiency on the formation of the zonula oocludens and blastocoel in the mouse embryo. *Dev. Biol.* **7**, 46–58.

Emerson, J. A. (1988). Disruption of the cytokeratin filament network in the preimplantation mouse embryo. *Development* **104**, 219–234.

Enders, A. C., and Schlafke, S. J. (1965). The fine structure of the blastocyst: Some comparative studies: *In* "Preimplantation Stages of Pregnancy" (G. E. W. Wolstenholme and M. Oconner, eds.). Churchill, London.

Fleming, T. P., Garrod, D. R., and Elsmore, A. J. (1991). Desmosome biogenesis in the mouse preimplantation embryo. *Development* **112**, 527–539.

Fuchs, E. (1994). Intermediate filaments and disease: Mutations that cripple cell strength. *J. Cell Biol.* **125**, 511–516.

Gallicano, G. I., McGaughey, R. W., and Capco, D. G. (1995). Protein kinase M, the cytosolic counterpart of protein kinase C remodels the internal cytoskeleton of the mammalian egg during activation. *Dev. Biol.* **167,** 469–501.

Gallicano, G. I., Larabell, C. A., McGaughey, R. M., and Capco, D. G. (1994a). Novel cytoskeletal elements in the mammalian eggs are composed of a unique arrangement of intermediate filaments. *Mech. Dev.* **45,** 211–226.

Gallicano, G. I., McGaughey, R. W., and Capco, D. G. (1994b). Ontogeny of the cytoskeleton during mammalian oogenesis. *Micro. Res. Tech.* **27,** 134–144.

Gallicano, G. I., Schwarz, S. M., McGaughey, R. W., and Capco, D. G. (1994c). Cytoskeletal sheets in mammalian eggs and early embryos provide as assembled form of intermediate filaments used later in development. *Mol. Biol. Cell.* **5,** 353a.

Gallicano, G. I., Schwarz, S. M., McGaughey, R. W., and Capco, D. G. (1993). Protein kinase C, a pivitol regulator of hamster egg activation, functions after elevation of intracellular free calcium. *Dev. Biol.* **156,** 94–106.

Gallicano, G. I., McGaughey, R. W., and Capco, D. G. (1992). Cytoskeletal sheets appear as universal components of mammalian eggs. *J. Exp. Zool.* **263,** 194–203.

Gallicano, G. I., McGaughey, R. W., and Capco, D. G. (1991). The cytoskeleton of the mouse egg and embryo: Reorganization of planar elements. *Cell Motil. Cytoskeleton* **18,** 143–154.

Garcia, R. B., Pereya-Alfonso, S., and Sotelo, J. R. (1979). Protein synthesizing machinery in the growing oocyte of the cyclic mouse. *Differentiation* **14,** 101–106.

Gardner, R. L. (1984). An in situ cell marker for clonal analysis of development of the extraembryonic endoderm in the mouse. *J. Embryol. Exp. Morphol.* **80,** 251–288.

Gardner, R. L., and Rossant, J. (1979). Investigation of the fate of 4.5 day post coitum mouse inner cell mass cells by blastocyst injection. *J. Embryol. Exp. Morphol.* **52,** 141–152.

Georgatos, S. D., and Blobel, G. (1987). Two distinct attachment sites for vimentin along the plasma membrane and the nuclear envelope in avian erythrocytes: A basis for vectorial assembly of intermediate filaments. *J. Cell Biol.* **15,** 105–115.

Green, D.P. L. (1985). An analysis of the structure of the bilaminar lamellae of the hamster egg. *J. Ultrastructure Res.* **91,** 30–37.

Hadek, R. (1966). Cytoplasmic whorls in the golden hamster oocyte. *J. Cell Sci.* **1,** 281–285.

Hashimoto, E., and Yamamura, H. (1991). Protease-activated protein kinase C in rat liver. *Int. J. Biochem.* **23,** 507–512.

Houliston, E., Pickering, S. J., and Maro, B. (1987). Redistribution of microtubules and pericentriolar material during the development of polarity in mouse blastomeres. *J. Cell Biol.* **104,** 1299–1308.

Huang, K-P., and Huang, F. L. (1986). Immunochemical characterization of rat brain protein kinase C. *J. Biol. Chem.* **261,** 14781–14787.

Inoue, M., Kishimoto, A., Takai, Y., and Nishizuka, Y. (1977). Studies on a cyclic nucleotide-independent protein kinase and its proenzyme in mammalian tissues. *J. Biol. Chem.* **252,** 7610–7616.

Jackson, B. W., Grund, C., Schmid, E., Burki, K., Franke, W. W., and Illmensee, K. (1980). Formation of cytoskeletal elements during mouse embryogenesis. *Differentiation* **17,** 161–179.

Jackson, B. W., Grund, C., Winter, S., Franke, W. W., and Illmensee, K. (1981). Formation of cytoskeletal elements during mouse embryogenesis. II. Epithelial differentiation and intermediate-sized filaments in early postimplantation embryos. *Differentiation* **20,** 203–216.

Jaken, S. (1990). Protein kinase C and tumor promoters. *Curr. Opin. Cell Biol.* **2,** 192–197.

Johnson, M. H., and Maro, B. (1984). The distribution of cytoplasmic actin in mouse 8-cell blastomeres. *J. Embryol. exp. Morphol.* **82**, 97–117.

Johnson, M. H., and Ziomeck, C. A. (1981). The foundation of two distinct cell lineages within the mouse morula. *Cell* **24**, 71–80.

Johnson, M. H., Maro, B., and Takeichi, M. (1986). The role of cell adhesion in the synchronization and orientation of polarization in 8-cell mouse blastomeres. *J. Embryol. Exp. Morphol.* **93**, 239–255.

Kishimoto, A., Nishiyama, K., Nakanishi, H., Uratsuji, Y., Nomura, H., Takeyama, Y., and Nishizuka, Y. (1985). Studies on the phosphorylation of myelin basic protein by protein kinase C and adenosine 3':5'-monophosphate-dependent protein kinase C. *J. Biol. Chem.* **260**, 12492–12499.

Kline, D., and Kline, J. T. (1992). Repetitive calcium transients and the role of calcium in exocytosis and cell cycle activation in the mouse egg. *Dev. Biol.* **149**, 80–89.

Koehler, J. K., Clark, J. M., and Smith, D. (1985). Freeze fracture observations on mammalian oocytes. *Am. J. Anat.* **174**, 317–329.

Klymkowski, M. W., Bachant, J. B., and Domingo, A. (1989). Functions of intermediate filaments. *Cell Motil. Cytoskeleton* **14**, 309–331.

Lawson, K. A., Memese, J. J., and Pederson, R. A. (1991). Clonal analysis of epiblast fate during germ layer formation in the mouse embryo. *Development* **113**, 891–911.

Lehtonen, E. (1985). A monoclonal antibody against mouse oocyte cytoskeleton recognizing cytokeratin-type filaments. *J. Embryol. Exp. Morphol.* **90**, 197–209.

Lehtonen, E. (1987). Cytokeratins in oocytes and preimplantation embryos of the mouse. *Curr. Topics Dev. Biol.* **22**, 153–173.

Lehtonen, E., and Badley, R. A. (1980). Localization of cytoskeletal proteins in preimplantation mouse embryos. *J. Embryol. Exp. Morphol.* **55**, 211–225.

Lehtonen, E., Ordonez, G., and Reima, I. (1988). Cytoskeleton in preimplantation mouse development. *Cell Differ.* **24**, 165–178.

Longo, F. J., and Chen, D. Y. (1985). Development of cortical polarity in mouse eggs: Involvement of the meiotic apparatus. *Dev. Biol.* **17**, 382–394.

Maro, B., Johnson, M. H., Pickering, S. J., and Flach, G. (1984). Changes in actin distrubution during fertilization of the mouse egg. *J. Embryol. Exp. Morphol.* **81**, 211–237.

McGaughey, R. W. (1977). The maturation of porcine oocytes in minimal, defined culture media with varied macromolecular supplements and varied osmolarity. *Exp. Cell Res.* **109**, 25–30.

McGaughey, R. W., and Capco, D. G. (1989). Specialized cytoskeletal elements in mammalian eggs: Structural and biochemical evidence for their composition. *Cell Motil. Cytoskeleton* **13**, 104–111.

Messinger, S. M., and Albertini, D. F. (1990). Centrosome and microtubule dynamics during meiotic progression in the mouse oocyte. *J. Cell Biol.* **100**, 289–298.

Moll, R., Franke, W. W., Schiller, D. L., Geiger, B., and Krepler, R. (1982). The catalog of human cytokeratins: Patterns of expression innormal epithelia, tumors, and cultured cells. *Cell* **31**, 11–24.

Murray, A. W., Fournier, A., and Hardy, S. J. (1987). Proteolytic activation of protein kinase C: A physiological reaction. *TIBS* **12**, 55–56.

Nickerson, J. A., Krockmalinic, G., He, D., and Penman, S. (1990). Immunolocalization in three dimensions: Immunogold staining of cytoskeletal and nuclear matrix proteins in resinless electron microscopy sections. *Proc. Natl. Acad. Sci. USA* **87**, 2259–2263.

Nishizuka, Y. (1988). The molecular heterogeneity of protein kinase C and its implications for cellular regulation. *Nature* **334**, 661–665

Oshima, R. G., Howe, W. E., Klier, F. G., Adamson, E. D., and Shevinsky, L. H. (1983). Intermediate filament protein synthesis in preimplantation murine embryos. *Dev. Biol.* **99**, 447–455.

Overstrom, E. W., Benos, D. J., and Biggers, J. D. (1989). Synthesis of NA$^+$/K$^+$ ATPase by the preimplantation rabbit blastocyst. *J. Reprod. Fertil.* **85**, 283–295.

Paulin, D., Babinet, C., Weber, K., and Osborn, M. (1980). Antibodies as probes of cellular differentaition and cytoskeletal organization in the mouse blastocyst. *Exp. Cell Res.* **130**, 297–304.

Richa, J., Damsky, C. H., Buck, C. A., Knowles, B. B., and Solter, D. (1985). Cell surface glycoproteins mediate compaction, trophoblast attachment, and endoderm formation during early mouse development. *Dev. Biol.* **108**, 513–521.

Riethmacher, D., Brinkmann, V., and Birchmeier, C. (1995). A targeted mutation in the mouse E-cadherin gene results in defective preimplantation development. *Proc. Natl. Acad. Sci. USA* **92**, 855–859.

Robson, R. M. (1989). Intermediate filaments. *Curr. Opin. Cell Biol.* **1**, 36–43.

Schatten, G. (1994). The centrosome and its mode of inheritance: The reduction of the centrosome during gametogenesis and itsrestoration during fertilization. *Dev. Biol.* **165**, 299–335.

Schatten, H., Cheney, R., Balczon, R., Willard, M., Cline, C., Simerly, C., and Schatten, G. (1986). Localization of fodrin during fertilization and early development of sea urchins and mice. *Dev. Biol.* **118**, 457–466.

Schlafke, S., and Enders, A. C. (1967). Cytological changes during cleavage and blastocyst formation in the rat. *J. Anat.* **102**, 13–32.

Schwarz, S. M., McGaughey, R. W., and Capco, D. G. (1993). A role for intermediate filaments in the establishment of the primitive epithelia during mammalian embryogenesis. *Mol. Biol. Cell.* **4**, 250a.

Sefton, M., Johnson, M. H., and Clayton, L. (1993). Synthesis and phosphorylation of uvomorulin during mouse early development. *Development* **115**, 313–318.

Sherman, M. I. (1981). Control of cell fate during early mouse embryogenesis. *In* "Bioregulators of Reproduction." (G. Jagiello and H. J. Vogel, eds.), pp. 559–576. Academic Press, New York.

Skalli, O., and Goldman, R. D. (1991). Recent insights into the assembly, dynamics, and function of intermediate filament networks. *Cell Motil. Cytoskeleton* **17**, 67–79.

Sobel, S. J., and Alleigro, M. A. (1986). Changes in the distribution of a spectrin-like protein during development of the preimplantation mouse embryo. *J. Cell Biol.* **100**, 333–336.

Szolliosi, D. (1972). Changes of some cell organelles during oogenesis in mammals. *In* "Oogenesis" (J. D. Biggers, and A. W. Schuetz, eds.), pp. 47–64. Univ. of Maryland Press, Baltimore.

Takai, Y., Kishimoto, A., Inoue, M., and Nishizuka, Y. (1977). Studies on a cyclic nucleotide-independent protein kinase and its proenzyme in mammalian tissues. *J. Biol. Chem.* **252**, 7603–7609.

Tam, P.P. L., Williams, E. A., and Chan, W. Y. (1993). Gastrulation in the mouse embryos: Ultrastructural and molecular aspects of germ layer morphogenesis. *Microsc. Res. Tech.* **26**, 301–328.

Tapley, P. M., and Murray, A. W. (1985). Evidence that treatment of platelets with phorbol esters causes proteolytic activation of Ca^{2+}-activated, phospholipid-dependent protein kinase. *J. Biochem.* **151**, 419–423.

Tombes, R. M., Simerly, C., Borisy, G. G., and Schatten, G. (1992). Meiosis, egg activation, and nuclear envelope breakdown are differentially reliant on Ca^{2+}, whereas germinal vesical breakdown is Ca^{2+} independent in the mouse oocyte. *J. Cell Biol.* **117**, 799–811.

Weakley, B. S. (1966). Electron microscopy of the oocyte and granulosa cells in the developing ovarian follicles of the golden hamster *Mesocricetus auratus*. *J. Anat.* **100**, 503–534.

Weakley, B. S. (1967). Investigations into the structure and fixation properties of cytoplasmic lamellae in the hamster oocyte. *Z. Zellforsch.* **81**, 91–99.

Webster, S. C., and McGaughey, R. W. (1988). Cytoskeletal interactions between the sperm and the egg at penetration in the Syrian hamster. *J. Cell Biol.* **107**, 178a

Webster, S. C., and McGaughey, R. W. (1990). The cortical cytoskeleton and its role in sperm penetration of the mammalian egg. *Dev. Biol.* **142**, 61–74.

Weiss, R. A., Eichner, R., and Sun, T.-T. (1984). Monoclonal antibody analysis of keratin expression in epidermal disease: A 48- and 56-kdalton keratin as molecular markers for hyperproliferative keratinocytes. *J. Cell Biol.* **98**, 1397–1406.

Wiley, L. M. (1984). Cavitation in the mouse preimplantation embryo: NA/K-ATPase and the origin of nascent blastocoele fluid. *Dev. Biol.* **105**, 330–342.

Wiley, L. M., and Eglitis, M. A. (1980). Effects of colcemid on cavitation during mouse blastocoele formation. *Exp. Cell Res.* **127**, 89–101.

Winkel, G. K., Fergusun, J. E., Takeichi, M., and Nuccitelli, R. (1990). Activation of protein kinase C triggers premature compaction in the 4-cell-stage mouse embryo. *Dev. Biol.* **138**, 1–15.

Ziomeck, C. A., and Johnson, M. H., (1981). Properties of polar and apolar cells from the 16-cell mouse morula. *Roux's Arch. Dev. Biol.* **190**, 287–296.

10

Mammalian Model Systems for Exploring Cytoskeletal Dynamics during Fertilization

Christopher S. Navara, Gwo-Jang Wu,* Calvin Simerly, and Gerald Schatten†
Departments of Zoology and Obstetrics and Gynecology
and the Wisconsin Regional Primate Research Center
University of Wisconsin
Madison, Wisconsin 53706

I. Introduction

Understanding the cellular biology of mammalian fertilization is increasingly more relevant with each passing day. Also, the number of applications of this knowledge is increasing almost exponentially. In the United States an estimated 17–19% of couples are infertile. A large number of these are diagnosed as idiopathic, i.e., of unknown origin. In many countries the exact opposite problem exists, the population is increasing faster than can be supported by the medical or agricultural community. A better understanding of the mechanisms of human fertilization may alleviate both of these problems by either increasing or controlling fertilization.

* Present Address: Department of Ob/Gyn, TSGH, No. 8, 3rd Sec, Ting-chow Rd, Tapei, Taiwan 100 ROC.
† To whom correspondence should be addressed at 1117 West Johnson Street, University of Wisconsin–Madison, Madison, WI 53706. Fax: (608) 262-7319

Current Topics in Developmental Biology, Vol. 31

In addition, a clearer understanding of fertilization in animals of agricultural importance promises to increase production and health of these animals. In those countries where overpopulation is problematic, the food supply is also lacking. The domestic animals which are productive in the United States typically do not do well in harsher climates. Therefore, the aim has been to increase production of native species; however, very little is known regarding fertilization in these species. It would be a herculean task to answer all of the cell biological questions for every species of interest. For that reason, it is crucial that an accurate, tractable, model system exists which can be reliably extrapolated to a large number of other species.

Understanding the role of the cytoskeleton before and during fertilization is an important step to controlling and improving fertilization. The cytoskeleton has numerous, diverse functions in the mammalian unfertilized oocyte and zygote (reviewed by Schatten and Schatten, 1987; Longo, 1989; Schatten, 1994). Actin filaments are involved in the polarization of the oocyte (cortical ring of actin, microvillous-free area, and cortical granule-free zone), the migration of the first meiotic spindle during maturation, and the anchoring of the second meiotic spindle to the plasma membrane. In addition, they assemble at the point of sperm entry in the zygote and are necessary for tail incorporation in mice. They are also required for pronuclear migration in some species. Microtubules are involved in the separation of the chromosomes at meiosis, cytoplasmic organization in the form of cytoplasmic asters in some species (sperm astral microtubules in others), pronuclear migration in all mammalian species studied, and the formation of the first mitotic spindle. In addition, centrioles may or may not participate in the first cell cycle depending on the species studied.

The aim of this chapter is to compare and contrast cytoskeletal features during fertilization among well-studied mammals. Particularly, we compare the data present in the most highly investigated system, the rodents (especially mice), with other mammalian species which have only recently been studied due to advances in *in vitro maturation* and *in vitro fertilization* techniques. We will then speculate on the appropriate animal models for less tractable systems such as exotic species and humans.

II. Cytoskeletal Organization and Dynamics during Rodent Fertilization

A. Actin Filament Organization

Analysis of rodent oocytes using immunocytochemical and ultrastructural techniques localizes actin filaments to the cortical region of oocytes where they participate in the extrusion of polar bodies and cytokinesis. Actin filaments also assemble at the site of sperm–egg fusion and participate in the formation of the incorporation cone after sperm penetration, functions analogous to those described in the invertebrate systems (reviewed by Schatten, 1982). Unlike invertebrate systems, rodent fertilization may not require direct participation of actin filaments for aspects of sperm incorporation. Actin filaments are also required for rotation of the second meiotic spindle during second polar body formation (Maro *et al.*, 1984; Longo and Chen, 1985; Webb *et al.*, 1986).

In the mouse system, there is a complicated interrelationship between the cortical actin filament system, plasma membrane modifications, and meiotic chromosomes (reviewed by Longo, 1989; Schatten and Schatten, 1987). Actin filaments are necessary for the cortical migration of the first meiotic spindle during maturation which restructures the plasma membrane and cortical region of the maturing oocyte. A prominent microvillus-free zone overlying the meiotic spindle forms which is free of cortical granules (Fig. 1B), enriched in submembraneous actin, and lacks an affinity for the lectin concanavalin A (Johnson *et al.*, 1975; Eager *et al.*, 1976; Nicosia *et al.*, 1977; Maro *et al.*, 1984; Longo and Chen, 1985; G. Schatten *et al.*, 1986; Ducibella *et al.*, 1990).

The mature mouse or hamster oocyte arrested at metaphase of second meiosis has a tangentially oriented spindle which must rotate after cell activation in order to form the second polar body (Maro *et al.*, 1984; Okada *et al.*, 1986). Interestingly, preliminary evidence in the mature rat oocyte has shown a uniform distribution of actin filaments in the cortex and no accumulation of F-actin over the site of the second meiotic spindle (Battaglia and Gaddum-Rosse, 1986). The events of meiotic spindle anchoring to the cell cortex, spindle rotation after second meiotic resumption, and second polar body formation are all dependent on actin filament functioning in the mouse oocyte; evidence of similar actin filament behavior in other rodent species is not yet known.

When spindle microtubules are disrupted with depolymerization drugs, chromosomes of the meiotic spindle scatter along the cortex. Actin accumulates and surface microvilli disappear locally over each chromosome mass as depicted by rhodamine–phalloidin and concanavalin A binding (Longo and Chen, 1985; G. Schatten *et al.*, 1986; Maro *et al.*, 1986a). Chromosome scattering is dependent on cortical actin filaments since inclusion of cytochalasin (Maro *et al.*, 1986a) or latrunculin (G. Schatten *et al.*, 1986) along with the microtubule inhibitor will block this process. Recovery from microtubule inhibition results in the formation of miniature spindles at each scattered chromosome site which can rotate and form multiple polar bodies upon oocyte activation (Maro *et al.*, 1986a).

Studies of fertilization in the mouse do not support a required role for actin filaments in sperm head penetration, unlike other nonmammalian animal

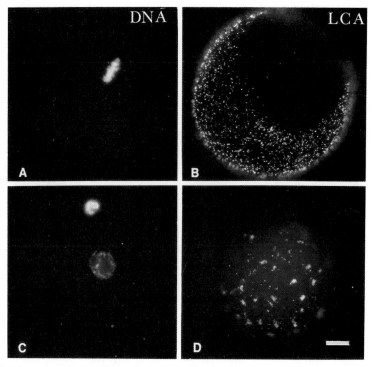

Fig. 1 Detection of cortical granules in unfertilized and artificially activated mouse oocytes. In the mature mouse oocyte arrested at metaphase of second meiosis (A), numerous cortical granules can be detected at the cell cortex except at the site overlying the meiotic spindle region (B). A dramatic reduction in cortical granule numbers (D) occurs in activated oocytes which have entered interphase and formed the female pronucleus (C). All oocytes were double labeled for DAPI DNA stain (A and C) and fluoresceinated-*Lens culinaris* agglutinin (LCA; B and D). Bar = 10 μm. Reprinted from Simerly *et al.*, (1995). Cytoskeletal organization and dynamics in oocytes during maturation and fertilization, *in* "Gametes—The Oocyte" (J. G. Grudzinskas and J. L. Yovich, eds.). Cambridge University Press.

species (maro *et al.*, 1984; H. Schatten *et al.*, 1986; Simerly *et al.*, 1993). Mature mouse spermatozoa do not contain actin at the equatorial site where sperm–egg fusion occurs (Flaherty *et al.*, 1986) and scanning electron microscopy investigations of the surface events during sperm incorporation do not suggest active oocyte microvilli participation in the engulfment of the penetrating sperm head (Shalgi *et al.*, 1978). Living studies of sperm penetration in the presence of actin filament inhibitors, such as latrunculin, reinforce these observations. Actin filament inhibition blocks second polar body formation after sperm incorporation, but does not inhibit completion of second meiosis. Sperm penetration and pronuclear formation occur normally; however, pronuclear centration does not proceed. These observations are in good agreement with the effects noted for cytochalasin inhibition during sperm penetration in the mouse (Maro *et al.*, 1986b; Longo and Chen, 1985).

Actin filaments do appear to be crucial for the incorporation of the sperm tail (Simerly *et al.*, 1993). Time-lapse video microscopy and low-voltage scanning electron microscopy studies of mouse fertilization have shown that tail engulfment into the oocyte is blocked in the presence of actin filament inhibitors. The plasma membrane of the sperm tail fuses at multiple sites with the oocyte's plasma membrane in normal and microtubule-inhibited eggs, but only at microvillus-rich sites on the cell surface. In the presence of cytochalasin the number of microvilli is greatly reduced and the plasma membrane covering the sperm axoneme does not fuse with the oocyte membrane, sperm head penetration is unaffected, and the midpiece of the fertilizing sperm remains at the site of penetration.

Actin filament activity is required for the formation of the incorporation cone (reviewed by Longo, 1989). At the site of gamete fusion, an extension of cytoplasm forms over the penetrating sperm head which is enriched in actin along its plasmalemma surface (Gaddum-Rosse, 1985; Maro *et al.*, 1984). The presence of cytochalasin will prevent cone formation or cause this transient structure to be reabsorbed (Maro *et al.*, 1984; Shalgi *et al.*, 1978).

Pronuclear apposition is affected by actin filament activity (reviewed by Schatten and Schatten, 1987). Actin filaments are detected as a diffuse perinuclear array during mouse pronuclear centration (Maro *et al.*, 1984). This unusual arrangement appears to be required, in addition to the cytoplasmic microtubules, for pronuclear juxtapositioning since cytochalasin (Maro *et al.*, 1984) or latrunculin (G. Schatten *et al.*, 1986) will inhibit pronuclear apposition, but will not inhibit pronuclear development.

B. Microtubule Organization

The dominant microtubule-containing structure in the unfertilized mouse oocyte arrested at metaphase of second meiosis is the meiotic spindle; it is anastral, barrel shaped with broad poles, and anchored to the cell cortex tangentially (Fig. 2A; Schatten *et al.*, 1985; Maro *et al.*, 1986b; and H. Schatten *et al.*, 1986). Approximately a dozen cytoplasmic microtubules (cytasters) are also detected in the mouse oocyte cytoplasm (Fig. 2A). Microtubules in the meiotic spindle are crucial for the proper alignment and separation of the maternal chromosomes during progressive maturation, while the cytoplasmic microtubules are responsible for the motions necessary for pronuclear apposition following sperm insemination (Szöllosi *et al.*, 1972; Wassarman and Fujiwara, 1978; Messinger and Albertini, 1991; H. Schatten *et al.*, 1986).

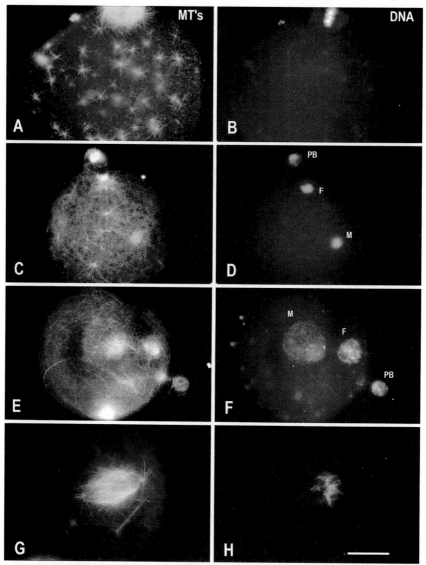

Fig. 2 Microtubules and DNA during mouse fertilization. Microtubules are found as cytoplasmic foci and in the meiotic spindle (A) in unfertilized oocytes which are arrested at metaphase of second meiosis (B). They associate with the developing pronuclei (C–F) as microtubules fill the cytoplasm. At metaphase most of the cytasters have come together to form the poles of the first mitotic spindle (G and H). A, C, E, and G, Anti-tubulin antibody; B, D, G, and H, DNA; M, male pronucleus; F, female pronucleaus. Scale bar = 20 μm. Reprinted in part from G. Schatten, (1994). The centrosome and its mode of inheritance: The reduction of the centrosome during gametogenesis and its restoration during fertilization. *Dev. Biol.* **165,** 299–235.

In most animals, the sperm is thought to introduce the centrosome at fertilization. This organelle creates the sperm aster, the microtubule-based structure responsible for male and female pronuclear movement (Boveri, 1901; reviewed by Schatten, 1994). However, the incorporated mouse sperm does not organize microtubules in the cytoplasm of the oocyte. After fertilization the cytoplasmic asters increase in size (Fig. 2C) filling the egg cytoplasm with a dense matrix of microtubules (Fig. 2E) ultimately responsible for moving the pronuclei into close apposition at the cell center (Fig. 2F). Pronuclear fusion does not occur in the mouse and by the end of first interphase, the dense matrix disassembles, ensheathing the apposed, centered pronuclei in a perinuclear shell of microtubules.

At prophase, the paternal and maternal chromosomes condense separately within the monaster array. As mitosis continues the chromosomes intermix (Fig. 2H) at the equator of the barrel-shaped, anastral metaphase spindle (Fig. 2G). The mouse mitotic spindle is atypical compared to most animal cells and more closely resembles the mitotic spindles of plant cells. Only sparse microtubules appear at the spindle poles in anaphase and few nonspindle microtubules are found in the cytoplasm. By telophase, interzonal microtubules form and aggregate into a midbody structure as a new cytoplasmic array of microtubules develops in the daughter blastomeres. Interestingly, spindle poles in both meiosis and first mitosis are organized without the presence of centrioles (Szöllosi et al., 1972).

III. Cytoskeletal Organization and Dynamics in Other Mammalian Species

A. Actin Filament Organization

Actin filament patterns in those mammalian species that have been studied are distinct from what has been reported for the rodents. A uniform layer of cortical actin has been observed in mature sheep (Le Guen et al., 1989) and bovine oocytes (C. S. Navara, personal observation). No cortical actin has been detected in unfertilized porcine oocytes until after sperm penetration (Albertini, 1987). This actin may be correlated with second meiotic spindle orientation (Le Guen et al., 1989), since in sheep (Le Guen and Crozet, 1989), cow (Long et al., 1993; Navara et al., 1994), rabbit (Yllera-Fernandez et al., 1992), and pigs (Kim et al., 1995) it is oriented radially to the plasma membrane. Investigations in the sheep and cow suggest an active role for actin filaments in cortical spindle attachment since actin filament inhibitors cause the meiotic spindle to migrate away from the plasma membrane and inhibit second polar body formation (Le Guen et al., 1989; C. S. Navara, personal observation).

The mature oocyte in many mammalian species does not appear to be polarized in terms of microvilli or cortical granules. Microvilli are distributed over the entire surface of the cow oocyte (C. S. Navara personal observation), as well as the human oocyte (Johnson et al., 1991; Pickering et al., 1988; Sathananthan et al., 1988). Analysis of the cortical region in many mammalian species [cows (Kruip et al., 1983), humans (Johnson et al., 1991; Pickering et al., 1988; Sathananthan et al., 1988; Van Blerkom, 1990), domestic cats (Byers et al., 1992), rhesus monkeys (G.-J. Wu and G. Schatten, personal observation)] does not demonstrate a cortical granule-free zone as noted in the mouse oocyte but rather a uniform distribution of cortical granules.

Similar to mouse fertilization, sperm head penetration in sheep is not blocked by the presence of actin filament inhibitors; however, sperm tail incorporation is delayed. The incorporation cone also does not form in the presence of cytochalasin (Le Guen et al., 1989). Culturing bovine zygotes in the presence of cytochalasin does not prevent pronuclear apposition, therefore, actin filaments do not appear to be necessary for pronuclear migration during fertilization in cows (C. S. Navara, personal observation).

B. Microtubule Organization

Recent studies in other mammals have indicated that the microtubule patterns observed in the rodent oocyte and zygote are not typical of all mammals and may be unique in the animal kindgom. Microtubule patterns during fertilization in the sheep (Le Guen and Crozet, 1989; Crozet, 1988, 1990), rabbit (Longo, 1976; Yllera-Fernandez et al., 1992), cow (Long et al., 1993; Navara et al., 1994), rhesus monkey (Wu et al., 1995), human (Simerly et al., 1995), and pig (Kim et al., 1995) all differ from those of the mouse and are consistent among one another. Microtubule patterns during bovine fertilization are shown in Fig. 3. The mature bovine oocyte is arrested at metaphase of second meiosis, anastral, and barrel shaped (Fig. 3A). The meiotic spindle is smaller in size than that of the mouse and oriented radially, rather than tangentially, to the cell surface. It is the only detectable microtubule structure in the unfertilized oocyte (Fig. 3A). Shortly after insemination microtubules (Fig. 3C) assemble at the base of the incorporated sperm head (Fig. 3D). As the pronuclei decondense (Fig. 3F), the microtubules elongate (Fig. 3E), eventually filling the entire cytoplasm. The large microtubule array breaks down and splits forming the poles for the first mitotic spindle, which is anastral, may be eccentric in the cytoplasm, and has narrow poles. In addition, the sperm tail remains associated with one of the poles. At anaphase long astral microtubules rapidly return and

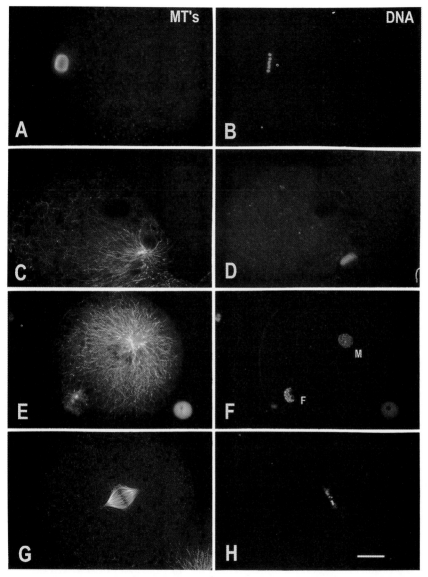

Fig. 3 Microtubules and DNA during bovine fertilization. The only microtubules in the unfertilized oocyte are in the meiotic spindle (A). Shortly after fertilization, microtubules (C) are found in association with the incorporated sperm head (D). These microtubules lengthen to form a sperm aster (C and E) as the sperm DNA decondenses. The only microtubules associated with the female pronucleus are those of the second meiotic spindle (E). At mitosis, microtubules form a bipolar structure and the interphase microtubules have disassembled (G). The metaphase spindle has tightly focused anastral poles (G) and the chromosomes are aligned on the spindle equator (H). A, C, E, and G, anti-tubulin antibody; B, D, F, and H, DNA; M, male pronucleus; F, female pronucleus. Scale bar = 20 (A, B, and E–H) and 30 μm (C and D). Reprinted in part from Navara *et al.* (1994). Microtubule organization in the cow during fertilization, polyspermy, parthenogensis, and nuclear transfer: The role of the sperm aster. *Dev. Biol.* **162.**

by the time of cytokinesis microtubules again fill the cytoplasm (Long *et al.*, 1993; Navara *et al.*, 1994).

The microtubule patterns observed during human fertilization, shown in Fig. 4), are remarkably similar to those seen during bovine fertilization. The meiotic spindle in the unfertilized human oocyte arrested at metaphase of second meiosis is anastral, slightly tapered at the poles, and oriented radially to the cell surface. Only spindle microtubules are detectable in human oocytes at this stage (Fig. 4A; Pickering *et al.*, 1988; Simerly *et al.*, 1995). After fertilization a microtubule aster forms at the base of the incorporated sperm head (Fig. 4C). The microtubules of this aster elongate as the pronuclei decondense (Figs. 4E and 4F). The first mitotic spindle is anastral with relatively narrow poles. A sperm tail can be identified in association with one of the poles which often has two small asters (Figs. 4G and 4H). One of the few differences from bovine fertilization is a small population of cortical microtubules present in the human zygote at mitosis (not shown).

Microtubule patterns in marsupials have also been investigated (Breed *et al.*, 1994). The microtubules of the meiotic spindle are the only microtubules in the oocyte, which is arrested at metaphase of second meiosis. Two different zygotes of the marsupial *Monodelphis domestica* are shown in Fig. 5. After fertilization a microtubule aster is seen to associate with the male pronucleus (Figs. 5F and 5I). The male pronucleus can be identified by its association with the sperm tail (Figs. 5A, 5D, and 5G). Microtubules do not appear to nucleate from the female pronucleus (Fig. 5I).

In several mammalian species centrioles can be found at a single pole of the first mitotic spindle. Ultrastructural studies have shown centrioles at one spindle pole in sheep (Crozet, 1990), cows (H. Sathananthan, personal communication), and humans (Sathananthan *et al.*, 1991). This further strengthens the belief that in these species the sperm contributes components to the centrosome.

The pattern of microtubules after fertilization seems to indicate the paternal inheritance of the centrosome as has been demonstrated in lower animal species from invertebrates through amphibians. This conclusion, however, is blurred by the observation that several mammalian species can be parthenogenetically activated, form bipolar spindles, and divide *in vitro*. Cows (Navara *et al.*, 1994), rabbits (Szöllosi and Ozil, 1991), rhesus monkeys (Wu *et al.*, 1995), and humans (Winston *et al.*, 1991) all develop to the late preimplantation stage after parthenogenetic stimuli. Microtubule patterns after parthenogenesis in the cow are shown in Fig. 6. Four hours after activation a dense meshwork of microtubules can be seen throughout the cytoplasm. Microtubules are also observed nucleating from the remnants of the second meiotic spindle (Fig. 6A). These parthenogenotes have completed meiosis but have been cultured in cytochalasin B for 4 hr to block second polar body formation and form diploid parthenogenotes. The dense

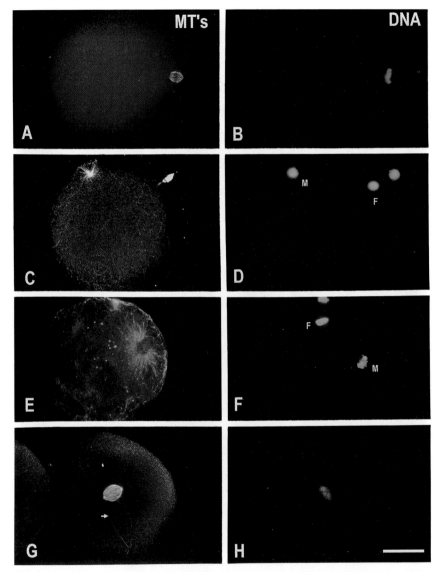

Fig. 4 Detection of microtubules and DNA during human fertilization. The meiotic spindle in mature unfertilized human oocytes is anastral and oriented radially to the cell surface (A). The pole adjacent to the plasma membrane is somewhat narrower (A). No other microtubules are observed in the cytoplasm. Shortly after sperm incorporation microtubules (C) assemble at the base of the sperm head (D). These microtubules elongate forming a radially arrayed sperm aster (E). The first mitotic spindle is anastral and the sperm axoneme remains attached to one pole (G, arrow). A, C, E, and G, anti-tubulin antibody; B, D, F, and H, DNA; M, male pronucleus; F, female pronucleus; arrow, sperm tail. Scale bar = 20 μm. Reprinted from Simerly *et al.* (1995). The paternal inheritance of the centrosome, the cell's microtubule-organizing center in humans, and the implications for infertility. *Nature Med.* **1.**

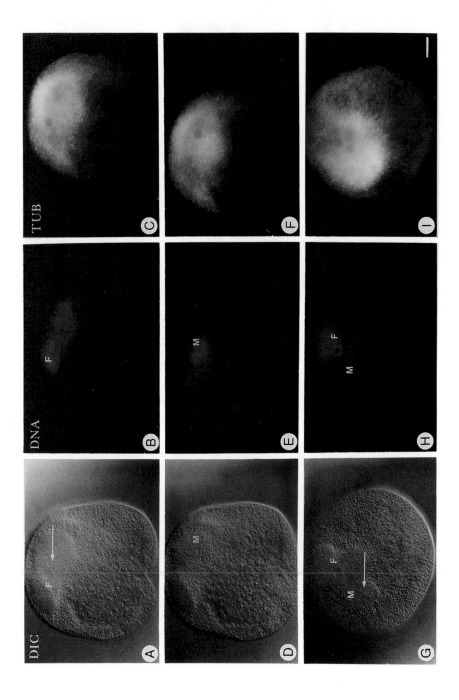

cytoplasmic microtubule array remains (Fig. 6C) as the DNA decondenses (Fig. 6D). Surprisingly, the dense array gives rise to a bipolar spindle which is anastral (Fig. 6E). In this example, some chromosomes are misaligned (Fig. 6F). Anaphase in parthenogenotes is indistinguishable from anaphase during normal fertilization (Fig. 6G).

IV. Summary and Implications

It appears that, in terms of cytoskeletal organization and dynamics during fertilization, rodents are atypical mammals and may perhaps be atypical animals. The differences between rodents and other mammals are summarized in Fig. 7 and Table 1. The unfertilized rodent oocyte [with the exception of actin filament distribution in rat oocytes (Zernika-Goetz et al., 1993)] has a polarized distribution of cortical granules, microvilli, and actin distribution. The nonrodent mammals that have been studied in this manner show no polarization of these three components. The meiotic spindle in rodent oocytes is arranged tangentially to the plasma membrane, while the meiotic spindle in nonrodent mammals is smaller and arranged radially with the plasma membrane. Numerous cytoplasmic asters are found in the unfertilized mouse oocyte. These asters elongate and with the cytoplasmic actin filaments are necessary for pronuclear migration. After fertilization in nonrodent mammals, a radial microtubule aster is formed in association with the incorporated sperm head. These microtubules, but not actin filaments, are necessary for pronuclear migration. Prior to mitosis the numerous cytasters in rodent zygotes come together to form the poles for the mitotic spindle which is centrally located. The sperm tail is not associated with this spindle. The sperm aster in nonrodent mammals splits to form

---

Fig. 5 Microtubules and DNA in zygotes of the marsupial *Monodelphis domestica*. Following sperm penetration, an accumulation of clear cytoplasm distinct from the yolky area surrounds the developing male and female pronuclei (DIC;A, D, and G). The incorporated sperm axoneme could often be seen within this yolk-free cytoplasmic area (arrows in A and G). Microtubules (C and F) were abundant in the cytoplasm surrounding the developing pronuclei (B and E) but did not penetrate the yolky hemisphere at this early stage of development. A large monaster of microtubules (I) surrounds the male and female pronuclei (H) following pronuclear apposition. All oocytes are double labeled with propidium iodide (B, E, and H) and monoclonal β-tubulin antibody (C, F, and I). Differential interference contrast (DIC;A, D, and G). A–C and D–F represent an early pronuclear stage with the oocyte taken at different focal planes to show microtubule patterns around the female and male pronuclei. M, male pronucleus; F, female pronucleus. Scale bar = 20 μm. Reprinted from Breed *et al.* (1991). Distribution of microtubules in eggs and early embryos of the marsupial, *Monodelphis domestica. Dev. Biol.* **164.**

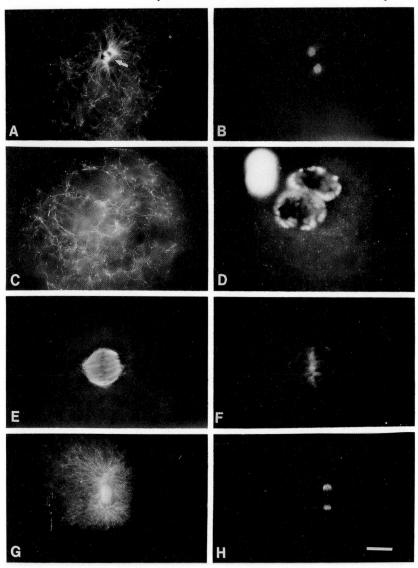

Fig. 6 Epifluorescence microscopy of parthenogenesis in the bovine oocyte. These oocytes were activated using two treatments of ionomycin. Four hours after the first treatment the oocyte has completed meiosis (A and B). This cell has two developing haploid pronuclei due to cytochalasin inhibition of second polar body extrusion (B). The cytoplasm contains a dense meshwork of microtubules (A). Microtubules extend from the remnants of the meiotic spindle (A, arrow). Approximately 10 hr after activation (C and D), the two maternally derived pronuclei have decondensed and come together (D) and a dense network of disarrayed microtubules is apparent (C). At approximately 24 hr after activation, the parthenogenotes are in first mitosis (E and F). The parthenogenotes are able to form an anastral bipolar spindle (E) although sometimes the chromosomes are misaligned on the metaphase plate (F). The

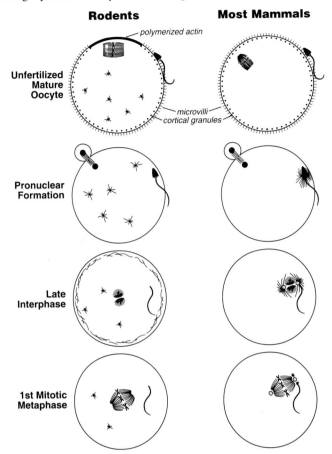

Fig. 7 Schematic comparing microtubules during fertilization in rodents (left) and other mammals (right). The unfertilized rodent oocyte has a polarized distribution of cortical granules, microvilli, and actin distribution. Nonrodent mammals that have been studied in this manner show no polarization of these three components. The meiotic spindle in rodent oocytes is arranged tangentially to the plasma membrane, while the meiotic spindle in nonrodent mammals is smaller and arranged radially. Cytoplasmic asters are found in the unfertilized mouse oocyte. These asters elongate and are necessary for pronuclear migration. After fertilization in nonrodent mammals, a radial microtubule aster is formed in association with the incorporated sperm head. Prior to mitosis the numerous cytasters in rodent zygotes form the poles for the mitotic spindle which is centrally located. The sperm tail is not associated with this spindle. The radial aster in nonrodent mammals splits to form two asters which serve as the poles for the first mitotic spindle. This spindle is often eccentric in the cytoplasm. The sperm tail remains associated with one of the poles.

Fig. 6 (*Continued*) parthenogenotes complete mitosis and during anaphase (G and H) astral microtubules are again present (G). A, C, E, and G, microtubules; B, D, F, H, DNA. Scale bar = 10 μm. Reprinted from Navara *et al.* (1994). Microtubule organization in the cow during fertilization, polyspermy, parthenogenesis, and nuclear transfer: The role of the sperm aster. *Dev. Biol.* **162.**

Table I

					Mammal					
	Mouse	Rat	Hamster	Cow	Human	Marsupial	Pig	Rhesus	Rabbit	Sheep
Met II spindle orientation	Tangential	Tangential	Tangential	Radial	Radial	Radial	Radial	Radial	Radial	Radial
Polarized oocyte	Yes	No	Yes	No	No	?	?	No	No	?
Cytasters	Yes	Yes	Yes	No	No	No	No	No	No	No
Sperm aster	No	No	No	Yes	Yes	Yes	Yes	Yes	Yes	Yes
Sp Tail at mitotic pole	No	No	No	Yes	Yes	Yes	Yes	Yes	?	Yes
Centrioles at 1st mitosis	No	No	No	Yes	Yes	?	?	?	No	Yes
Parthenogenetic cleavages	Yes	Yes	Yes	Yes	Yes	?	Yes	Yes	Yes	?

two asters, which serve as the poles for the first mitotic spindle which may be eccentric within the cytoplasm. The sperm tail remains associated with one of the poles.

Our understanding of the inheritance of the centrosome during fertilization is becoming more complex. The previous thought that centrosomal inheritance is strictly paternal (invertebrates and lower vertebrates) or maternal (rodents) is probably no longer valid (reviewed in Schatten, 1994). Centrosomal inheritance in rodents appears to be primarily maternal. Manipulation of the amount of paternal contribution by polyspermy or parthenogensis does not affect the microtubule patterns during the first cell cycle (Schatten *et al.,* 1991). This does not appear to be the case during fertilization in nonrodent mammals. Polyspermically inseminated oocytes in cows, rhesus monkeys, and humans result in multiple sperm asters and aberrant mitotic figures (Navara *et al.,* 1994; Wu *et al.,* 1995; Simerly *et al.,* 1995). Sperm tails and centrioles are observed at one pole of mitotic spindles in numerous species. These observations are indicative of paternal inheritance of the centrosome. However, several species can be parthenogenetically activated and develop at least to the morula stage which is strong evidence for the maternal inheritance of the centrosome. We believe that centrosomal inheritance in these species is probably biparental as has been proposed for sea urchin fertilization (Holy and Schatten 1991). The sperm brings in a component which recruits maternal centrosomal material to the sperm asters. One of these maternal components is likely to be γ-tubulin. Sperm of the frog *Xenopus laevis* have been shown to recruit γ-tubulin from *Xenopus* egg extracts (Stearns and Kirschener, 1994; Felix *et al.,* 1994).

The cytoskeletal differences between rodents and other mammals raise questions regarding the use and accuracy of using rodent systems as models for other systems in basic biological research. This question becomes much more pressing when applied to the use of rodent oocytes for clinical assays exploring aspects of human infertility. Rodent oocytes have many differences from the human oocyte and may not provide an accurate assessment of human sperm potential. Clearly, another mammalian model system needs to be pursued. The cow is an excellent candidate for such a system. Cytoskeletal characteristics during bovine fertilization are very similar to those of other nonrodent mammals including humans. The IVM and IVF systems for bovine culture are well established and allow for the production of large quantities of zygotes and embryos. In addition, mature bovine oocytes are now commercially available (Bomed Madison, WI). These factors combine to make the bovine system an ideal system for pursuing cell biological questions and applying those answers gained to less tractable systems such as endangered species or humans.

Appendix: Methods

A. Microtubule Visualization

1. Immunocytochemical Labeling of Microtubules in Murine Oocytes and Zygotes

Remove zona pellucidae with a brief incubation in acidified M-2 culture media (pH 2.5). Attach mouse oocytes to polylysine-coated coverslips in Ca^{2+} protein-free M-2. Incubate at 37°C for 2 or 3 min. Remove most of culture medium from plate, being careful not to exposure attached oocytes to liquid–air interface. Carefully pipet buffer M containing 1% Triton X-100, 1 mM 2-mercaptoethanol, and 0.2 mM PMSF (buffer M+ TX, pH 6.8, 37°C) (H. Schatten et al., 1986) into plate and incubate for 10 min at 37°C.

Carefully transfer permeabilized oocytes to dry polylysine-coated coverslip; transfer a minimum amount of fluid with oocytes. Immediately add buffer M without detergent. Do not allow oocytes or zygotes to air dry. Very slowly add −10°C absolute methanol into buffer M until saturation. This will take several exchanges of fluid. Incubate 10 min at room temperature. Remove methanol by rinsing in PBS–TX. Fixed ooctyes can be stored overnight before immunostaining.

Blot excess PBS–TX off coverslip, add 70 μl of 0.1 M PBS, 0.1% Triton X-100, 3 mg/mg^{-1} BSA (to block nonspecific secondary antibody binding sites.) Incubate at 37°C for 30 min in humidified chamber. Again blot coverslip; add 70 μl anti-tubulin antibody (i.e., E7, Chu and Klymkowsky, 1987; Developmental Hybridoma Bank, Ames IA) Incubate at 37°C for 40 min in a humidified chamber. Rinse for 10 min in PBS–TX–BSA at room temperature and repeat twice. Add 70 μl of properly conjugated secondary antibody. Incubate at 37°C for 40 min in humidified chamber. Rinse as above. Add 70 μl of 2.5 μg/ml DAPI DNA stain and incubate for 10 min at room temperature. Rinse one time in dH$_2$O and mount in a solution to prevent photobleaching (Vectashield, Vector Laboratories, Burlingham, CA).

2. Immunocytochemical Labeling of Microtubules in Cow or Rhesus Monkey Oocytes and Zygotes

Remove cumulus cells using 2 mg/ml hyaluronidase and pipetting the oocyte or zygote through a small-bore glass pipette. Zona pellucidae can be removed from oocytes and zygotes by brief treatments with 2% pronase prepared in protein-free TL–Hepes (Navara et al., 1994). Recover zygotes at 39°C for 20 min then attach to polylysine-coated coverslips and fix for

40 min in 2% formaldehyde in 0.1 M PBS without added calcium, magnesium, or protein. Oocytes and embryos are then permeabilized with 0.1 M PBS containing 1% Triton X-100 detergent overnight.

The next morning, reduce the remaining free aldehydes with a 20-min rinse in 0.1 M PBS containing 150 mM glycine. Nonspecific binding of primary and secondary antibodies is blocked with a 20-min incubation in PBS and 3% nonfat dry milk. Antibody and DNA labeling is performed as described previously.

3. Immunocytochemical Labeling of Microtubules in Human Oocytes and Zygotes

The technique for labeling microtubules in mouse oocytes will work for human oocytes except that buffer M and TX should be changed to buffer M, 3% TX-100, and 5% methanol. All other steps remain the same (Simerly *et al.*, 1995).

B. Actin Filament Visualization

1. Fluorescent Labeling of Actin Filaments in Murine, Bovine, Rhesus, or Human Oocytes and Zygotes

The technique for labeling microtubules in bovine oocytes and zygotes works well for actin filament labeling. Substitute 5 μg/ml of phalloidin conjugated to the appropriate fluorochrome for the antibody steps. No additional steps are needed. Phalloidin should be diluted in PBS–TX.

References

Albertini, D. F. (1987). Cytoplasmic reorganization during the resumption of meiosis in cultured pre-ovulatory rat oocytes. *Dev. Biol.* **120,** 121–131.

Battaglia, D. E., and Gaddum-Rosse, P. (1986). The distribution of polymerized actin in the rat egg and its sensitivity to cytochalasin B during fertilization. *J. Exp. Zool.* **237,** 97–105.

Breed, W., Simerly, C., Navara, C. S., Vanderberg, J., and Schatten, G. (1994). Distribution of microtubules in eggs and early embryos of the marsupial, *Monodelphis domestica. Dev. Biol.* **164,** 230–240.

Byers, A. P., Barone, M. A., Donoghue, A. M., and Wildt, D. E. (1992). Mature domestic cat oocyte does not express a cortical granule-free domain. *Biol. Reprod.* **47,** 709–715.

Boveri, T. (1901). "Zellen-studien: ueber die natur der centrosomen," Vol. 4. Fisher, Jena.

Chu, D., and Klymkowsky, M. (1987). Experimental analysis of cytoskeletal function in early *Xenopus laevis* embryos. *First Int. Symp. Cytoskeleton Dev.* **8,** 331–333.

Crozet, N. (1988). Fine structure of sheep fertilization *in vitro. Gamete Res.* **19,** 291–303.

Crozet, N. (1990). Behavior of the sperm centriole during sheep oocyte fertilization. *Eur. J. Cell Biol.* **53,** 326–332.

Ducibella, T., Kurasawa, S., Rangarajan, S., Kopf, G., and Schultz, R. (1990). Precocious loss of cortical granules during mouse oocyte meiotic maturation and correlation with an egg-induced modification of the zona pellucida. *Dev. Biol.* **137,** 46–55.

Eager, D. D., Johnson, M. H., and Thurley, K. W. (1976). Ultrastructural studies on the surface membrane of the mouse egg. *J. Cell Sci.* **22,** 345–353.

Felix, M. A., Antony, C., Wright, M., and Maro, B. (1994). Centrosome assembly *in vitro*. *J. Cell Biol.* **124,** 19–31.

Flaherty, S. P., Winfrey, V. P., and Olson, G. E. (1986). Localization of actin in mammalian spermatozoa: A comparison of eight species. *Anat. Rec.* **216,** 504–515.

Gaddum-Rosse, P. (1985). Mammalian gamete interactions: What can be gained from observation on living eggs? *Am. J. Anat.* **174,** 347–356.

Holy, J., and Schatten, G. (1991). Differential behavior of centrosomes in unequally dividing blastomeres during fourth cleavage of sea urchin embryos. *J. Cell Sci.* **98,** 423–431.

Johnson, M. H., Eager, D., Muggleton-Harris, A., and Grave, H. M. (1975). Mosaicism in the organization of concanavalin A receptors on surface membrane of mouse egg. *Nature* **257,** 321–322.

Kim, N. K., Simerly, C., Funahashi, H., Schatten, G., and Day, B. (1995). Microtubule organization in porcine oocytes during fertilization and parthenogenesis. Submitted for publication.

Kruip, T. A. M., Cran, D. G., Van Beneden, T. H., and Dieleman, S. J. (1983). Structural changes in bovine oocytes during final maturation *in vivo*. *Gamete Res.* **8,** 29–47.

Le Guen, P., and Crozet, N. (1989). Microtubule and centrosome distribution during sheep fertilization. *Eur. J. Cell Biol.* **48,** 239–249.

Le Guen, P., Crozet N., Huneau, D., and Gall, L. (1989). Distribution and role of microfilaments during early events of sheep fertilization. *Gamete Res.* **22,** 411–425.

Long, C. R., Pinto-Correia, C., Duby, R. T., Ponce ce Leon, F. A., Boland M. P., Roche, J. F., and Robl, J. M. (1993). Chromatin and microtubule morphology during the first cell cycle in bovine zygotes. *Mol. Reprod. Dev.* **36,** 23–32.

Longo, F. J. (1976). Sperm aster in rabbit zygotes: Its structure and function. *J. Cell Biol.* **69,** 539–547.

Longo, F. J. (1989). Egg cortical architecture. *In* "The Cell Biology of Fertilization" (H. Schatten and G. Schatten, eds.), pp. 105–158. Academic Press, San Diego.

Longo, F. J., and Chen, D. Y. (1985). Development of cortical polarity in mouse eggs: Involvement of the meiotic apparatus. *Dev. Biol.* **107,** 382–394.

Maro, B., Johnson, M. H., Pickering, S. J., and Flach, G. (1984). Changes in actin distribution during fertilization of the mouse egg. *J. Embryol. Exp. Morphol.* **81,** 211–237.

Maro, B., Johnson, M. H., Webb, M., and Flach, G. (1986a). Mechanism of polar body formation in the mouse oocyte: An interaction between the chromosomes, the cytoskeleton, and the plasma membrane. *J. Embryol. Exp. Morphol.* **92,** 11–32.

Maro, B., Howlett, S. K., and Houliston, E. (1986b). Cytoskeleton dynamics in the mouse egg. *J. Cell Sci. Suppl.* **5,** 343–359.

Messinger, S. M., and Albertini, D. F. (1991). Centrosome and microtubule dynamics during progression in the mouse oocyte. *J. Cell Sci.* **100,** 289–298.

Navara, C. S., First, N. L., and Schatten, G. (1994). Microtubule organization in the cow during fertilization, polyspermy, parthenogenesis and nuclear transfer: The role of the sperm aster. *Dev. Biol.* **162,** 29–40.

Nicosia, S. V., Wolf, D. P., and Inoue, M. (1977). Cortical granule distribution and cell surface characteristics in mouse eggs. *Dev. Biol.* **57,** 56–74.

Okada, A., Yanagimachi, R., and Yanagimachi, H. (1986). Development of a cortical granule-free area of cortex and the perivitelline space in the hamster oocyte during maturation and following ovulation. *J. Submicrosc. Cytol.* **18**, 233–247.

Pickering, S. J., Johnson, M. H., Braude, P. R., and Houliston, E. (1988). Cytoskeletal organization in fresh, aged and spontaneously activated human oocytes. *Hum. Reprod.* **3**, 978–989.

Sathananthan, A. H., Trounson, A., Freeman, L., and Brady, T. (1988). The effect of cooling human oocytes. *Hum. Reprod.* **3**, 968–977.

Sathananthan, A. H., Kola, I., Osborne, I. K. J., Trounson, A., Bongso, S. C., Ng, A., and Ratnam, S. S. (1991). Centrioles in the beginning of human development. *Proc. Natl. Acad. Sci. USA* **88**, 4806–4810.

Schatten, G., Simerly, C., and Schatten, H. (1985). Microtubule configurations during fertilization, mitosis, and early development in the mouse and the requirement for egg microtubule-mediated motility during mammalian fertilization. *Proc. Natl. Acad. Sci. USA* **82**, 4152–4256.

Schatten, G., Schatten, H., Spector, I., Cline, C., Paweletz, N., Simerly, C., and Petzelt, C. (1986). Latrunculin inhibits the microfilament-mediated processes during fertilization, cleavage, and early development in sea urchins and mice. *Exp. Cell Res.* **166**, 191–208.

Schatten, G., Simerly, C., and Schatten, H. (1991). Maternal inheritance of centrosomes in mammals? Studies on parthenogenesis and polyspermy in mice. *Proc. Natl. Acad. Sci. USA* **88**, 6785–6789.

Schatten, G., and Schatten, H. (1987). Cytoskeletal alterations and nuclear architectural changes during mammalian fertilization. *Curr. Topics Dev. Biol.* **23**, 23–54.

Schatten, G. (1982). Motility during fertilization. *Int. Rev. Cytol.* **79**, 59–163.

Schatten, G. (1994). The centrosome and its mode of inheritance: The reduction of the centrosomeduring gametogenesisand its restoration during fertilization. *Dev. Biol.* **165**, 299–335.

Schatten, H., Schatten, G., Mazia, D., Balczon, R., and Simerly, C. (1986). Behavior of centrosomes during fertilization and cell division in mouse oocytes and in sea urchin eggs. *Proc. Natl. Acad. Sci. USA* **83**, 105–109.

Shalgi, R., Phillips, D. M., and Kraicer, P. F. (1978). Observation on the incorporation cone in the rat. *Gamete Res.* **1**, 27–37.

Simerly, C., Hecht, N., Goldberg, E., and Schatten, G. (1993). Tracing the incorporation of the sperm tail in the mouse zygote and early embryo using an anti-testicular–tubulin antibody. *Dev. Biol.* **158**, 536–548.

Simerly, C., Navara, C., Wu, G. J., and Schatten, G. (1995). Cytoskeletal organization and dynamics in mammalian oocytes during maturation and fertilization. *In* "Gametes—The Oocyte" (J. G. Grudzinskas and J. L. Yovich, eds.). Cambridge University Press.

Simerly, C., Wu, G. J., Zoran, S., Ord, T., Rawlins, R., Jones, J., Navara, C., Gerrity, M., Rinehart, J., Binor, Z., Asch, R., and Schatten, G. (1995). The paternal inheritance of the centrosome, the cell's microtubule-organizing center in humans, and the implications for infertility. *Nature Med.* **1**, 47–52.

Stearns, T., and Kirschner, M. (1994). In vitro reconstitution of centrosome assembly and function: The central role of γ-tubulin. *Cell* **76**, 623–637.

Szöllosi, D., Calarco, P., and Donahue, R. P. (1972). Absence of centrioles in the first and second meiotic spindles of mouse oocytes. *J. Cell Sci.* **11**, 521–541.

Szöllosi, D., and Ozil, J. P. (1991). *De novo* formation of centrioles in parthenogenetically activated, diploidized rabbit embryos. *Biol. Cell* **72**, 61–66.

Van Blerkom, J. (1990. Occurrence and developmental consequences of aberrant cellular organization in meiotically mature human oocytes after exogenous ovarian hyperstimulation. *J. Electron Microsc. Tech.* **16**, 324–346.

Wasserman, P. M., and Fujiwara, K. (1978). Immunofluorescent antitubulin staining of
   spindles of mouse oocytes *in vitro. J. Cell Sci.* **29,** 171–188.
Webb, M., Howlett, S. K., and Maro, B. (1986). Parthenogenesis and cytoskeletal
   organization in aging mouse eggs. *J. Embryo. Exp. Morphol.* **95,** 131–145.
Winston, N., Johnson, M., Pickering, S., and Braude, P. (1991). Parthenogenetic activation
   and development of fresh and aged human oocytes. *Fertil. Steril.* **56,** 904–912.
Wu, G. J., Simerly, C., Zoran, S. S., Funte, L. R., Rawlins, R., Binor, Z., and Schatten, G.
   (1995). Microtubule configurations during fertilization, parthenogenesis, polyspermy and
   mitosis in a non-human primate the Rhesus monkey. Submitted for publication.
Yllera-Fernandez, M. D. M., Crozet, N., and Ahmed-Ali, M. (1992). Microtubule
   distribution during fertilization in the rabbit. *Mol. Reprod. Dev.* **32,** 271–276.
Zernika-Goetz, M., Kubiak, J. Z., Antony, C., and Maro, B. (1993). Cytoskeletal
   organization of rat oocytes during metaphase II arrest and following abortive
   activiation: A study by confocal laser scanning microscopy. *Mol. Reprod. Dev.* **35,**
   165–175.

11

Cytoskeleton in Teleost Eggs and Early Embryos: Contributions to Cytoarchitecture and Motile Events

Nathan H. Hart
Department of Biological Sciences
Rutgers University
Piscataway, New Jersey 08855

Richard A. Fluck
Department of Biology
Franklin & Marshall College
Lancaster, Pennsylvania 17604

I. Introduction

Successful interaction between gametes initiates a cascade of pathways that lead to cleavage, cellular differentiation, morphogenesis, and eventual

Current Topics in Developmental Biology, Vol. 31

343

development of an adult organism. Shortly after fertilization of the teleost egg, the egg cortex undergoes a predictable sequence of structural changes in response to intracellular signals. Early transformation of the cortex includes formation of the fertilization cone and second polar body, sperm entry into the egg, cortical granule exocytosis, and surface membrane retrieval by endocytosis. Subsequently, cytoplasm moves toward the animal pole where it accumulates in preparation for cell division and cellular rearrangement. By analogy to other cell systems that show secretion, mitosis, intracellular cytoplasmic flow, and cell migration, one would predict that similar events in eggs are mediated by action of the cytoskeleton. Most of our knowledge about the egg cytoskeleton has been obtained from analyses of sea urchin, amphibian, and mammalian eggs. By comparison, the composition, organization, and function of the egg cytoskeleton in teleost fishes have received considerably less attention.

The eggs of teleost fish have long provided favorable material for the study of vertebrate development (Oppenheimer, 1947; Gilkey, 1981; Powers, 1989; Ho, 1992). Unlike mammalian and amphibian eggs, teleost eggs can be obtained in large numbers, are optically transparent, and are easily accessible for experimental manipulation. Furthermore, teleosts now also offer the possibility of employing genetic and molecular approaches to investigate the role of the cytoskeleton in early development. The zebrafish (*Danio rerio*), for example, is rapidly emerging as an ideal model for studying the genetic mechanisms of vertebrate development. The short 3-month life cycle and the relative ease of making haploid and parthenogenetic diploid embryos is helping in identification and analysis of mutants (Kimmel, 1989). Improved methods for making stable lines of transgenic zebrafish (Stuart *et al.*, 1990) should facilitate the study of genes and their actions in early embryos.

In this chapter, we survey the teleost egg cytoskeleton and evaluate its proposed role(s) in events of fertilization, ooplasmic segregation, and gastrulation. Our discussion will be based principally on contributions made by the zebrafish, medaka (*Oryzias latipes*), and killifish (*Fundulus heteroclitus*).

II. Actin, Myosin, and Spectrin Are Components of the Unfertilized Egg

Unfertilized eggs of teleost fishes are enclosed by an acellular chorion that possesses a single micropyle (Brummett and Dumont, 1979; Kobayashi and Yamamoto, 1981; Hart and Donovan, 1983). The micropyle allows access of the male gamete to the egg surface and is also a convenient marker of egg polarity. The egg is compartmentalized into a central yolk mass and a peripheral, structurally integrated cytoplasmic cortex (Brummett and Dumont, 1979; Kobayashi and Yamamoto, 1981; Hart and Donovan, 1983; Ohta and Iwamatsu, 1983). The cortex of the teleost egg is very pronounced. In the zebrafish, it measures about 15–20 μm in thickness (Hart and Yu, 1980), although it tends to be noticeably thicker at the micropyle. The cortex includes the plasma membrane and a variety of membrane-limited organelles embedded in a matrix of cytoplasm, including an extensive endoplasmic reticulum and cortical granules. The plasma membrane is reflected into either microvilli (Brummett and Dumont, 1981; Kobayashi, 1985) or microplicae (Hart and Donovan, 1983). At the site of sperm entry, the plasma membrane is modified into either a single, short projection (Brummett and Dumont, 1979; Kobayashi and Yamamoto, 1981, 1987) or a circular cluster of microvilli (Hart and Donovan, 1983; Kudo and Sato, 1985). These structures are located at the base of a cone-shaped depression in the egg surface formed by the micropyle.

Similar to sea urchin (Chandler, 1991; Spudich, 1992) and amphibian (Merriam and Clark, 1978; Chow and Elinson, 1993) eggs, actin is a major component of the teleost egg (Ivanenkov et al., 1990; Hart et al., 1992; Becker and Hart, 1995). Extracts of unfertilized loach and zebrafish eggs separated by SDS/PAGE display a single polypeptide having a molecular mass of 43 kDa based on isoelectric focusing (Ivanenkov et al., 1990) and immunoblot (Hart and Becker, 1992) analyses. Zebrafish actin in Triton X-100-treated egg homogenates separates principally to the supernatant fraction upon high-speed centrifugation, which suggests that most of the egg actin is soluble and in nonfilamentous form.

The distribution and organization of actin have been investigated with conventional electron microscopy and fluorescence microscopy methods. Ultrathin sections of zebrafish (Hart et al., 1992) and chum salmon (Kobayashi, 1985) egg cortex reveal a homogeneous-appearing, electron-dense matrix immediately subjacent to the plasma membrane and a deeper, less-dense cytoplasm that extends to the yolk mass. The subplasmalemmal layer of the zebrafish egg is about 200 nm in thickness, except at the site of sperm entry where it measures about 550 nm. Occasionally, 5 to 8-nm filaments are detected in the electron-dense matrix; these are best illustrated in microplicae and microvilli of the sperm entry site where they tend to run parallel to the long axis of the cytoplasmic extension and course down into the electron-dense layer of cytoplasm.

Whole eggs and egg fragments of zebrafish stain with rhodamine–phalloidin (RhPh), a fluorescent probe specifically employed to identify filamentous actin (F-actin). The stain appears as a narrow, continuous rim of fluorescence at the margin of the egg (Hart et al., 1992; Fig. 1a). Through-focus dissection of stained eggs shows that RhPh spatially correlates with the cores of microplicae and the subplasmalemmal layer of cytoplasm. A

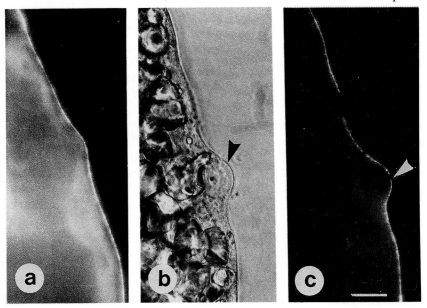

Fig. 1 Rhodamine–phalloidin (RhPh) staining of fixed, whole eggs of the zebrafish. (a) A narrow, continuous rim of fluorescent staining codistributes with the plasma membrane of the unactivated egg. (b) Phase contrast image of an egg activated in dechlorinated tank water to show the position of a cortical granule (arrowhead) just before release of its contents onto the egg surface. (c) Companion image to b showing a distinct interruption in the pattern of RhPh staining (arrowhead) at the site of exocytosis. Scale bar = 25 μm.

similar distribution of F-actin is detected at the site of sperm entry. Additional insight into the organization of F-actin stems from using cortical membrane patches. Using this technique, zebrafish eggs are collected, fixed briefly in 3.7% formaldehyde with an actin-stabilizing buffer, transferred to a poly-L-lysine coated slide, and then mechanically dislodged using tweezers (Hart and Becker, 1992; Becker and Hart, 1995). This technique can yield thin preparations of plasma membrane and associated cortical cytoplasm for analysis. Cortical patches display an elaborate, three-dimensional meshwork of interconnecting RhPh-staining filaments of variable diameter (Hart and Becker, 1992).

The possibility that cortical actin is organized in nonfilamentous form has been studied by indirect immunofluorescent staining with a monoclonal anti-actin antibody (Hart and Becker, 1992; Becker and Hart, 1995). Based on the concept that RhPh stains only F-actin and the anti-actin antibody accesses the total actin pool (Bonder *et al.*, 1989), dual staining has resulted in identification of an actin domain that does not bind RhPh. Semithin, frozen sections, prepared by methods employed for sea urchin eggs (Bonder *et al.,* 1989) and labeled with either RhPh or anti-actin antibody, show the spatial relationships between the cortical F-actin domain, the nonfilamentous actin domain, and various membrane domains (Hart and Becker, 1992; Becker and Hart, 1995). The F-actin network is restricted to the plasma membrane. Nonfilamentous actin alone is localized to the cytoplasm housing the cortical granules and other organelles of the cortex. The actin antibody also detects an apparent enrichment of nonfilamentous actin over the surfaces of cortical granules.

A number of actin-binding proteins that regulate the assembly of actin *in vitro* by linking, severing, or bundling of actin filaments (microfilaments) have been purified and characterized in sea urchin eggs (Spudich, 1992). Little progress has be made in identifying teleost egg proteins with similar properties. Zebrafish egg extracts show a single polypeptide (215–225 kDa) that crossreacts on immunoblots with an antibody against myosin heavy chain (conventional myosin or myosin II) of human platelet cells (Hart and Becker, 1992; Becker and Hart, 1995). The same antibody in immunofluorescence experiments colocalizes with F-actin at the plasma membrane and with the deeper, nonfilamentous actin domain. A Ca^{2+}/Mg^{2+}-dependent ATPase with properties similar to those of actomyosin ATPase has been reported in the egg surface of *Epiplatys* (Jørgensen, 1972). Zebrafish egg extracts also show a polypeptide that comigrates with sea urchin egg spectrin and crossreacts with an anti-sea urchin egg spectrin polyclonal antibody (Becker and Hart, 1994). The spectrin (or spectrin-like) protein colocalizes with F-actin and myosin at the margin of the cortex and with F-actin at the sperm entry site. Spectrin molecules form a meshwork that interconnect short actin filaments at the cell membrane (Luna and Hitt, 1992) and with ankyrin restrict the movement of specific integral membrane proteins (Nelson and Hammerton, 1989).

The peripheral distribution of F-actin, myosin, and spectrin in the teleost egg cortex suggests the existence of an egg plasma membrane cytoskeleton. Results of treating zebrafish eggs with cytochalasins, fungal metabolites that inhibit actin polymerization by either severing or capping the barbed end of actin filaments (Bonder and Mooseker, 1986; Cooper, 1987), support the view that the submembranous cytoskeleton is important in maintaining the structural integrity of the plasma membrane, including stabilizing the depression in the egg surface at the sperm entry site (Wolenski and Hart, 1988a). The actomyosin complex would also appear to mediate the cortical contractions that are observed in fish eggs shortly after fertilization (Gilkey *et al.,* 1978; Abraham *et al.,* 1993). The presence of spatially distinct F-actin and nonfilamentous actin domains is consistent with studies of sea urchin eggs (Spudich *et al.,* 1988; Bonder *et al.,* 1989). These two domains contribute to the overall physicochemical and viscoelastic properties of the egg cortex (Hiramoto, 1970).

III. Fertilization Triggers Changes in the Organization of Actin

A. Formation of the Fertilization Cone and Sperm Entry

The fertilization cone forms at the site of sperm entry as a localized extension of the egg cytoplasm; it is the structural vehicle into which the sperm typically enters the egg (Kobayashi and Yamamoto, 1987; Wolenski and Hart, 1987, 1988a). Although the morphology, timing, and pattern of fertilization cone formation and growth show variation among teleosts, events at the site of sperm entry appear to be fundamentally similar (Hart, 1990).

Correlative electron and fluorescence microscope studies with zebrafish eggs in particular provide insight into the temporal and spatial relationships between the fertilization cone, the fertilizing sperm, and the actin cytoskeleton (Hart et al., 1992). The zebrafish sperm attaches to the microvilli of the sperm entry site within 5–10 sec after insemination (Wolenski and Hart, 1987). Rupture of the fused gamete membranes occurs shortly thereafter and positions the leading edge of the sperm nuclear membrane in direct contact with the subplasmalemmal F-actin meshwork (Hart et al., 1992). The cytoplasm beneath the sperm binding site then quickly elevates, thereby placing the sperm atop the apex of an early developing fertilization cone. Between 30 and 60 sec postinsemination, there is a visible thickening and rearrangement of the actin meshwork associated with the sperm nucleus (Figs. 2a and 2b). RhPh staining reveals a gradual, step-like enclosure of the sperm by the actin meshwork during this time (Fig. 2a). The filaments of the meshwork presumably become linked to the sperm nuclear membrane (Webster and McGaughey, 1990). The fertilization cone itself, now elongated and nipple shaped by 60 sec, shows a two-fold increase in thickness of the cortical actin meshwork. Vesiculation of the sperm nuclear membrane begins at the leading edge of the nucleus and is initiated just before movement of the nucleus into the cone cytoplasm. The nucleus becomes fully incorporated into the cytoplasm of a shortened, flattened cone by 2 min. Movement of the sperm into the egg cytoplasm is more rapid in some species (Brummett et al., 1985). Interestingly, fertilization cones of zebrafish eggs show normal growth when parthenogenetically activated in tank water (Wolenski and Hart, 1988a; Hart et al., 1992). They display intense cortical staining with RhPh (Fig. 2c), indicating that cone actin filaments take origin from the egg cytoskeleton.

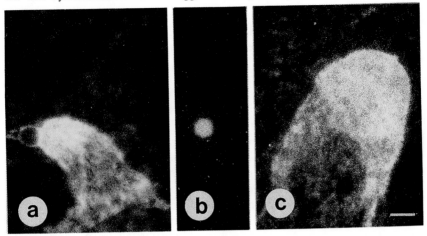

Fig. 2 Rhodamine–phalloidin (RhPh) staining of the zebrafish sperm entry site. (a) *En profile* image of a fertilized egg fixed at 30 sec postinsemination and stained for filamentous actin with RhPh followed by Hoechst 33342 to identify DNA. Note the RhPh staining of the fertilization cone and the rim of fluorescence that encircles the sperm. (b) Companion image to a showing the location of the sperm nucleus using the Hoechst stain. (c) An egg activated in dechlorinated tank water for 15 sec displays a fertilization cone that stains intensely with RhPh. Scale bar = 4 μm. (Reproduced from Hart *et al.*, 1992, copyright Wiley–Liss, Inc.)

Little movement of the egg cytoplasm appears necessary for sperm entry, and the fusing sperm can apparently initiate the alterations of the actin cytoskeleton that accompany sperm incorporation. First, the sperm normally enters the egg of some species before the development of a definitive fertilization cone (Iwamatsu and Ohta, 1981; Brummett *et al.*, 1985). Second, movement of sperm into the egg can take place without either egg activation or fertilization cone formation under experimental conditions (Wolenski and Hart, 1988b). Hence, the reorganization of the F-actin cytoskeleton that spatially and temporally correlates with sperm penetration appears to take place independently of any cytoskeletal alterations required for growth of the fertilization cone.

Exposure of fertilized eggs to cytochalasins has been a standard approach to determine whether actin polymerization is required for sperm incorporation. The sensitivity of sperm movement to these drugs ranges from total inhibition in echinoderm eggs (Schatten and Schatten, 1980) to little effect in mammalian eggs (Maro *et al.*, 1984; Schatten *et al.*, 1986). Treatment of zebrafish eggs with cytochalasins B and D consistently blocks incorporation of the fertilizing sperm in a dose-dependent manner (Wolenski and Hart, 1988a), which strongly suggests that sperm movement into the egg requires the assembly of actin filaments. An increase in actin polymerization at the site of sperm entry has also been reported in fertilized sea urchin (Hamaguchi and Mabuchi, 1988) and amphibian (Chow and Elinson, 1993) eggs. Since zebrafish sperm can enter an unactivated egg, the signal(s) initiating localized actin assembly is either introduced into the egg cytoplasm by the sperm or produced as a consequence of biochemical changes in the plasma membrane (polyphosphoinositide turnover?) upon gamete fusion. If a transient increase in cytosolic calcium is a necessary step in the actin assembly process, it is conceivable that sperm introduces this cation into the egg. The mechanism(s) by which actin polymerization is elevated at the sperm entry site is not known. It could reflect a sudden increase in the nonfilamentous actin pool available for polymerization. Alternatively, existing actin filaments might show an increased affinity for nonfilamentous actin. There are several possibilities by which the assembled cortical actin meshwork could vectorially move the sperm nucleus into the cone cytoplasm. Initially, actin filaments along the leading edge of the nucleus might undergo rearrangement, develop tension, and exert a pulling force. Subsequently, increased tension in the meshwork along the trailing margin coupled with relaxation along the leading margin could effectively "push" the nucleus through the F-actin meshwork and into the cone plasm. Alternatively, the mechanism of movement might be based on localized actomyosin contractility. However, it has not been demonstrated that myosin is localized to the sperm entry site.

The formation and growth of the fertilization cone also appear to depend on reorganization of the actin cytoskeleton. Fertilization cones, for example, form in cytochalasin-treated zebrafish eggs under conditions of both insemination and artificial activation. They tend, however, to develop more rapidly and are often larger than controls (Wolenski and Hart, 1988a), suggesting disturbances in regulation of F-actin rearrangement, assembly, or both. As a polarized extension of the cytoplasm, the fertilization cone bears similarity to the development of a pseudopod (Stossel, 1989; Vasiliev, 1991). Although one or more of several forces may be involved, the dramatic increase in thickness of the cortical actin meshwork (Wolenski and Hart, 1988a) suggests that newly assembled actin may be an important factor in cone growth. Conceptually, following cross-linking to the cytosolic side of the plasma membrane, a rapidly forming isotropic network of actin filaments could create a force that effectively pushes the cone outward. Surprisingly, the assembly of actin appears insensitive to cytochalasin treatment (Wolenski and Hart, 1988a). This may reflect an inability of the drug to completely block barbed-end assembly (Bonder and Mooseker, 1986).

B. Cortical Granule Exocytosis

Regulated secretion is a dramatic, calcium-dependent response of teleost eggs to either binding with sperm (Gilkey *et al.,* 1978) or treatment with an activating agent (Kobayashi, 1985; Donovan and Hart, 1986; Ivanenkov *et al.*, 1990). Exocytosis is stimulated when the intracellular Ca^{2+} concentration is elevated by fertilization (Gilkey *et al.*, 1978), incubation with Ca^{2+} ionophore (Schalkoff and Hart, 1986), or microinjection of either Ca^{2+} or IP_3 (Iwamatsu *et al.*, 1988a,b). During exocytosis, cortical granules (vesicles; alveoli) vectorially move to the oolemma, fuse with it, rupture, and discharge their contents at the egg surface (Brummett and Dumont, 1981; Donovan and Hart, 1986). Granule discharge produces a reorganized egg surface consisting of two distinguishable membrane domains: the original plasma membrane and limiting membranes of cortical granule (exocytotic) profiles (Hart, 1990). Exocytotic profiles gradually disappear as identifiable structures in the egg surface (Brummett and Dumont, 1981). Their transformation, at least in part, is structurally and temporally coupled to the removal of cortical granule membrane by coated vesicle endocytosis (Kobayashi, 1985; Donovan and Hart, 1986; Hart and Collins, 1991).

Subplasmalemmal actin networks have been proposed to act as a barrier to the movement and docking of secretory granules at the plasma membrane (Vitale *et al.*, 1991; Perrin *et al.*, 1992). Cortical granules of zebrafish eggs are immobilized within a cytoskeletal framework and their movement to the plasma membrane would appear to be restricted by the F-actin meshwork. Unstimulated zebrafish eggs treated with cytochalasin D (50 μg/ml) for 5–10 min often spontaneously discharge cortical granules; this is accompanied by loss of the structural integrity of the cortical actin meshwork (Wolenski and Hart, 1988a). When activated, cytochalasin D-treated (50 μg/ml) eggs consistently display degranulation (Wolenski and Hart, 1988a). Conversely, phalloidin microinjected into loach eggs prevents exocytosis; many granules are frequently displaced deeper into the ooplasm because of a thickened actin network (Ivanenkov *et al.*, 1990). Chromaffin cells respond in similar fashion to actin-destabilizing and stabilizing drugs (Lelkes *et al.*, 1986). Also, activated zebrafish eggs labeled with RhPh show a distinct discontinuity in the pattern of fluorescence that can be correlated with the onset of exocytosis of individual granules (Becker and Hart, 1993; Figs. 1b and 1c). Collectively, these studies suggest that (1) migration of granules to the plasma membrane is facilitated by reorganization of the F-actin cytoskeleton, and (2) the remodeling of the F-actin meshwork is probably modulated by proteins that either sever actin filaments or cap the growth ends of actin filaments. Although such proteins have not been identified in teleost eggs, actin-binding proteins with severing and capping properties have been reported in sea urchin (Wang and Bonder, 1991) and amphibian (Ankenbauer *et al.*, 1988) eggs.

Zebrafish eggs show a second, distinct change in F-actin distribution at the plasma membrane. Following fusion between plasma and cortical granule membranes and formation of the cortical granule pore, RhPh-labeled eggs show that F-actin accumulates around the rim of the pore and adjacent walls of exocytotic profiles (Figs. 3a and 3b). This observation is supported by electron microscopy studies of the chum salmon egg (Kobayashi, 1985). Later profiles are flattened and marked in the egg surface by very intense RhPh staining (Fig. 3a). It is tempting to speculate that the F-actin associated with evacuated cortical granule profiles is newly polymerized and assembled from the nonfilamentous actin sequestered at the surfaces of intact cortical granules. Application of techniques of laser scanning confocal microscopy and microinjection of egg actin labeled with specific fluorochromes (Hamaguchi and Mabuchi, 1988) should provide additional insight into the dynamics of actin polymerization around profiles of exocytosis.

The dramatic change in F-actin distribution following exocytosis suggests several possible roles for the cytoskeleton in transforming exocytotic profiles. One possibility is that F-actin stabilizes the limiting membrane of the exocytotic profile either during or after the discharge of granule contents. Alternatively, the actin may be required to drive scission of coated vesicles from exocytotic profiles or produce the invaginations from which they arise (Salisbury et al., 1980; Gottlieb et al., 1993). Experiments with zebrafish eggs and cytochalasins should help to determine whether an actin-based system participates in endocytosis, particularly since it has been recently demonstrated that these drugs act directly on actin molecules during motility events (Ohmori et al., 1992). Also, since molecular motors are thought to be required for endocytosis (Anderson, 1991), we believe that the exocytotic profiles of zebrafish eggs will be useful as a model for studying the molecular processes of endocytosis.

Exocytosis of cortical granules in teleost eggs is dependent on an IP_3-dependent release of cytosolic calcium from internal stores (Iwamatsu et al., 1988a,b; Iwamatsu, 1989). It is reasonable to assume that the calcium signal is somehow transduced to effect changes in the actin cytoskeleton that precede and follow exocytosis. The nature of the events that might signal alterations in cytoskeletal organization remains to be defined. Stimulation of protein kinase C activity induces exocytosis in amphibian (Bement and Capco, 1989) and mammalian (Endo et al., 1987) eggs, suggesting that phosphorylation may be an important step in triggering changes in the actin cytoskeleton. However, since GTP is known to induce exocytosis in medaka eggs (Iwamatsu, 1989), it is also possible that membrane phosphoinositides could interact directly with components of the actin meshwork to initiate cytoskeletal rearrangements (Stossel, 1989).

IV. Ooplasmic Segregation

A. General Features

In the mature medaka egg (diameter ≈ 1200 μm), the ooplasm is present as a thin (15 μm) peripheral layer surrounding a central yolk vacuole much

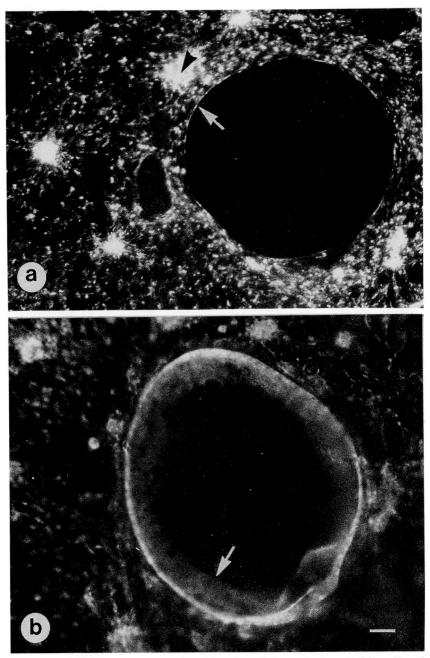

Fig. 3 Rhodamine–phalloidin (RhPh) staining of cortical granule (exocytotic) profiles of activated zebrafish eggs. (a) *En face* image of a cortical granule shortly after discharge of its contents at the egg surface. A narrow rim of staining circumscribes the pore opening (arrow). The intensely stained patches at the plasma membrane (arrowhead) represent sites of exocytotic profiles in terminal stages of transformation. (b) *En face* image of a recently evacuated cortical granule showing RhPh staining around a widened pore and along the adjacent walls (arrow) of the profile. Scale bar: a = 9 μm; b = 5.8 μm.

like the rind covers the flesh of an orange (Abraham *et al.*, 1993). The ooplasm is bounded on one side by the plasma membrane and on the other by the yolk membrane. The peripheral location of ooplasm makes the medaka egg particularly useful for microscopic studies of the movement of the ooplasm and its inclusions. In the zebrafish egg (diameter $\approx$625 μm), ooplasm is distributed throughout the ovum and thus intermingles with yolk (Beams *et al.*, 1985); the same is true of the loach egg (Ivanenkov *et al.*, 1987).

In medaka and zebrafish, as in all teleosts, ooplasmic segregation involves the movement of ooplasm to the animal pole of the zygote and its coalescence into a blastodisc. In medaka, the peripherally placed ooplasm streams toward the animal pole along all meridians, and the growing blastodisc is separated from the yolk by the yolk membrane. In zebrafish, long, opaque streamers of endoplasm form during ooplasmic segregation (Roosen-Runge, 1938; Beams *et al.*, 1985); these streamers are frequently continuous at one end with the forming blastodisc. The movement of ooplasm toward the animal pole can be seen clearly with time-lapse techniques and has been described as bulk flow or streaming (Roosen-Runge, 1938; Gilkey, 1981; Beams *et al.*, 1985; Abraham *et al.*, 1993), with most visible particles in the ooplasm moving simultaneously in the same direction and at the same velocity ($\approx$2 μm/min in medaka; Abraham *et al.*, 1993). Although most ooplasmic inclusions moving toward the animal pole appear to be a part of this bulk flow, some small inclusions ($\approx$1 μm diameter) move saltatorily and more rapidly ($\approx$45 μm/min in medaka; Abraham *et al.*, 1993).

While most ooplasm moves toward the animal pole, some components are either left behind near the vegetal pole or appear to be moved actively toward it. In zebrafish, the former category includes yolk platelets, and the latter the counterstream of ooplasm that moves toward the vegetal pole and accumulates there (Roosen-Runge, 1938). Although there is no evidence for a counterstream of ooplasm toward the vegetal pole in medaka eggs (Gilkey, 1981; Abraham *et al.*, 1993), some inclusions do move toward the vegetal pole in this species. Examples include oil (lipid) droplets (Sakai, 1965; Abraham *et al.*, 1993; Webb *et al.*, 1995) and particles that move saltatorily (Webb *et al.*, 1995).

Ooplasm appears to contract during segregation. In several species, rhythmic contraction waves pass repeatedly over that portion of the egg outside the blastodisc (Yamamoto, 1975). Two types of contractions occur in medaka zygotes. One immediately follows exocytosis of the cortical vesicles and moves as a wave from the animal pole to the vegetal pole (Iwamatsu, 1973). Another contraction occurs immediately after the completion of meiosis. This "second contraction complex" is apparent as the oscillation of ooplasmic inclusions, including oil droplets, along meridians connecting the animal and vegetal poles (Sakai, 1965; Iwamatsu, 1973; Abraham *et al.*,

1993). Inclusions move first toward the vegetal pole then toward the animal pole. This second contraction heralds a radical change in the behavior of the ooplasm in medaka: the beginning of streaming, saltatory movement, and the movement of oil droplets toward the vegetal pole (Abraham *et al.*, 1993).

Although most ooplasm enters the blastodisc before the first cleavage, ooplasm continues to move toward the animal pole long after this stage (Devillers, 1961). In zebrafish, streamers still exist at the first cell division, and endoplasm continues to stream into the blastodisc until after the sixth cleavage (Hisaoka and Firlit, 1960; Beams *et al.*, 1985). Streaming occurs only when the cells are in interphase and stops when prophase begins. In medaka, where streaming toward the animal pole continues until at least the eight-cell stage (V. Abraham and R. Fluck, unpublished observations), the velocity of the streaming particles increases intermittently with the same period as the cell cycle, but a correlation between these pulses and other specific events of the cell cycle has not been made.

Indirect evidence that ooplasm enters the blastodisc after the first cleavage comes from latitudinal microdissection studies in which the developmental capacity of blastoderms was assessed after separating them from the rest of the egg. Such studies found that blastoderms isolated prior to a critical stage (8-cell-stage in the goldfish *Carassius auratus*, 32-cell-stage in *Fundulus*) develop into hyperblastulae, while those isolated later develop into embryos (Rudnick, 1955; Devillers, 1961). Such results suggest that cytoplasmic determinants present in the ooplasm and indispensable for development reach the blastoderm only after the third (or sixth) cleavage. However, there is no direct evidence for qualitative differences in the contents of ooplasm at different latitudes or longitudes of teleost eggs. Ooplasm that enters the blastoderm of zebrafish zygotes after the 16-cell-stage would flow into only the marginal blastomeres, because the four innermost, nonmarginal cells have lost their cytoplasmic connections to the yolk cell by this stage and become cytoplasmically separated from the rest of the embryo (Kimmel and Law, 1985a).

B. Microfilaments (F-Actin)

On the basis of their morphology, microfilaments have been tentatively identified by transmission electron microscopy (TEM) in the subplasmalemmal electron-dense matrix of zebrafish (Katow, 1983) and loach zygotes (Ivanenkov *et al.*, 1987). In zebrafish, this cortical matrix is thicker at the vegetal pole (0.23 ± 0.04 μm) than at the animal pole (0.15 ± 0.02 μm; Katow, 1983). Although Katow (1983) saw no microfilaments (4–8 nm diameter) in streamers examined by TEM, streamers examined by scanning electron microscopy (SEM) do have a filamentous appearance that may be due, at least in part, to the presence of F-actin (Beams *et al.*, 1985). Actin has been identified by staining with rhodamine–phalloidin in the peripheral ooplasm and blastodisc of zebrafish eggs and zygotes (Hart and Becker, 1994), and a molecule that cross-reacts with anti-platelet heavy-chain myosin antibody labels both peripheral ooplasm and the blastodisc of zebrafish zygotes (Hart and Becker, 1994).

Indirect evidence for the presence and action of microfilaments during ooplasmic segregation comes from experiments involving the use of molecules that affect the polymerization and depolymerization of microfilaments. In dechorionated zebrafish zygotes, CCB and CCD cause the meshwork of subplasmalemmal microfilaments to detach from the plasma membrane and prevent formation of the blastodisc and cleavage (Katow, 1983; Hart and Becker, 1994). In medaka, CCD inhibits cytoplasmic streaming and the formation of the blastodisc, but has no apparent effect on oil-droplet movement or saltatory movement (Webb *et al.*, 1995). CCD (up to 10 μg/ml) has no apparent effect on ooplasmic segregation in loach-chorionated zygotes incubated in a medium containing the drug, but it does disrupt segregation when it is microinjected (Ivanenkov *et al.*, 1987). Microinjected DNase I and phalloidin also inhibit ooplasmic segregation, and DNase I disrupts the microfilamentous cortex (Ivanenkov *et al.*, 1987, 1990).

Taken together, these results suggest that (1) microfilaments are necessary for cytoplasmic streaming and formation of the blastodisc in teleost zygotes, and (2) the segregation of some components of the ooplasm (e.g., oil droplets in medaka) occurs independently of microfilaments.

C. Microtubules

Electron microscopic studies have revealed little about the presence of microtubules in teleost zygotes during ooplasmic segregation. No microtubules were seen in either the egg cortex or the streamers of zebrafish zygotes examined by TEM (Katow, 1983), but filamentous structures (25–30 nm diameter) in streamers examined by SEM were tentatively identified as microtubules (Beams *et al.*, 1985). These preliminary findings need to be pursued, perhaps using a more suitable fixative solution that stabilizes microtubules (Gard, 1991).

Indirect immunofluorescence studies have provided a richer picture of the spatiotemporal pattern of microtubules in teleost zygotes. Using a monoclonal antibody against α-tubulin (DM1A), Abraham *et al.* (1995) have shown the development of complex arrays of microtubules between fertilization and the first cleavage in the medaka zygote (Fig. 4). At the animal pole, a radial array of microtubules forms by $T_n \approx 0.3$ (T_n = normalized time, where the time at which the egg is activated is defined as zero and the beginning of the first cleavage as 1.0) and persists until late in the first cell cycle. This array emerges from a microtubule-organizing center (MTOC) near the pronuclei. In interpolar ooplasm, a crisscrossed network of microtubules appears by $T_n \approx 0.08$ and increases in density up to $T_n \approx 0.75$. By $T_n = 1.0$, there is an animal–vegetal gradient of microtubule density in interpolar ooplasm, with few microtubules near the animal pole and many near the vegetal pole. At the vegetal pole, a parallel array of microtubules forms by $T_n \approx 0.24$ and persists until $T_n \approx 0.76$ and then loses its organization. Similar arrays form in prick-activated medaka eggs (Webb and Fluck, 1995). Although less is known about the spatiotemporal pattern of microtubules in zebrafish, it appears to be similar to that in medaka. An array of parallel microtubules is present at the vegetal pole at $T_n \approx 0.67$ (see Fig. 9C in Strähle and Jesuthasan, 1993), and a crisscrossed network is present in interpolar ooplasm near the surface of the zygote (Jesuthasan, personal communication).

Microtubule poisons—colchicine, demecolcine (i.e., colcemid), or noco-dazole–slow the rate of growth of the blastodisc and inhibit oil-droplet movement toward the vegetal pole, saltatory movement toward both poles, and pronuclear movements in medaka zygotes (Abraham et al., 1993; Webb et al., 1995). β-Lumicolchicine, an inactive molecule formed by the photolysis of colchicine, has none of these effects (Abraham et al., 1993). Microtubule poisons also block the appearance of microtubules in medaka zygotes (Abraham et al., 1995; Webb et al., 1995) and appear to solate the ooplasm (Abraham et al., 1993). The latter effect is apparent as an increase in the Brownian motion of small ooplasmic inclusions and in the floating of oil droplets to the top of the zygote. The effect on oil droplets is not apparent until after the second contraction complex, suggesting again that this complex is accompanied by a radical change in the structure and function of ooplasmic cytoskeletal systems. Microtubule poisons, though not inhibiting the streaming of ooplasm toward the animal pole, do cause this movement to become less directed (Abraham et al., 1993). This result suggests that the microfilaments involved in streaming interact with microtubules.

The effects of one microtubule poison (demecolcine) on medaka zygotes can be reversed by irradiation with uv light (360 nm), which photolyzes (and inactivates) demecolcine and colchicine (Aronson and Inoué, 1970). Ultraviolet irradiation of demecolcine-treated medaka zygotes regenerates microtubules, normal oil-droplet movement, and saltatory movement (Webb et al., 1995). The effects of uv light can be restricted to a portion of the zygote—animal pole, vegetal pole, or a patch of interpolar ooplasm—by irradiating with a small beam (diameter $\approx$ 475 μm) of uv light (Webb

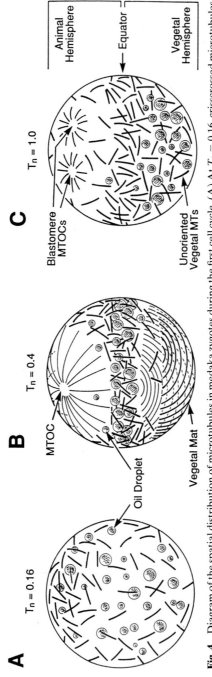

Fig. 4 Diagram of the spatial distribution of microtubules in medaka zygotes during the first cell cycle. (A) At $T_n = 0.16$, crisscrossed microtubules are present throughout the ooplasm. The density of the microtubules is higher near the animal pole. (B) At $T_n = 0.4$, distinct patterns of microtubules are present in three regions of the egg. At the animal pole, a microtubule-organizing center (MTOC) is present; with microtubules radiating from it over a 30° arc in all directions; at the vegetal pole, an array of parallel microtubules extends over an arc of about 30° in all directions. Interpolar ooplasm contains a dense network of crisscrossed microtubules. (C) At $T_n = 1.0$, a MTOC is present in each of the forming blastomeres at the animal pole; the microtubules near the vegetal pole are no longer parallel to each other. Many fewer microtubules are present in interpolar ooplasm near the animal pole than near the vegetal pole.

et al., 1995). The microtubules that form within the irradiated region have the same organization as those in control, untreated zygotes: a radial array at the animal pole, a parallel array at the vegetal pole, and a crisscrossed network in interpolar ooplasm (Webb *et al.,* 1995). These results suggest that the capacity to form microtubules is present throughout the ooplasm of the medaka zygote and that region-specific information about the spatial organization of microtubules is present as well.

By contrast to the pronounced effects of microtubule poisons on medaka zygotes, colchicine has no apparent effect on segregation in zebrafish (Katow, 1983). This difference may be due to the 40-fold lower concentration of the drug used in experiments involving zebrafish zygotes or to a lower permeability of the zebrafish egg to the molecule. It is also possible that ooplasmic segregation in zebrafish does not require the presence of microtubules.

In summary, while it is clear that microtubules are required for ooplasmic segregation in medaka, it is also clear that we still have much to learn about the role of microtubules in segregation in teleosts. For example, is there a trivial explanation for the apparent insensitivity of segregation in zebrafish to microtubule poisons, or is the mechanism of segregation in zebrafish different from that in medaka? Moreover, there is no evidence for microtubule-based motors that could mediate the movement of oil droplets and other inclusions in medaka zygotes. Microinjection of an antibody against kinesin (SUK4; Ingold *et al.,* 1988) has no apparent effect on the movement of oil droplets in medaka eggs (V. Abraham and R. Fluck, unpublished observations). Finally, some of the responses of medaka zygotes to microtubule poisons could be the result of an indirect effect of these poisons on intermediate filament networks (Knapp *et al.,* 1983; Pasdar *et al.,* 1992). At this point, however, such networks are only hypothetical in teleost zygotes.

D. Models of Ooplasmic Segregation

Models of ooplasmic segregation in teleost zygotes have primarily addressed the formation of the blastodisc at the animal pole. The movement of ooplasm and its inclusions toward the vegetal pole has either not been addressed or is considered to be only a consequence of the active movement of ooplasm in the opposite direction. In other words, some inclusions are thought to be left behind when ooplasm moves toward the animal pole. The idea that more than one mechanism might account for ooplasmic segregation in a particular species or that different mechanisms may operate in different species has appeared in the literature only recently.

More than 50 years ago, it was suggested that formation of both the zebrafish blastodisc and the hyaline cap of an amoeba are caused by the same mechanism. Specifically, it was proposed that contraction of the (cortical) gel layer in the vegetal hemisphere of the zebrafish zygote squeezes endoplasm out of the yolk to form the blastodisc (Lewis and Roosen-Runge, 1942, 1943). Although this model has been refined as our understanding of contractile phenomena has deepened, placement of the contractile force in the vegetal hemisphere of zebrafish zygotes remains unchallenged (Hisaoka and Firlit, 1960; Beams and Kessel, 1976; Katow, 1983; Beams *et al.*, 1985). Refinements to this model include the suggestion that cortical tension decreases in the animal hemisphere when it is increasing in the vegetal hemisphere (Beams *et al.*, 1985) and that cortical microfilaments are involved in the changes in cortical tension (Katow, 1983).

The question of the site of the contractile force that forms the blastodisc has been pursued experimentally in loach zygotes by microinjecting molecules that disrupt microfilaments (DNase I and CCD; Ivanenkov *et al.*, 1987). Ooplasmic segregation proceeds normally when these molecules are microinjected into the vegetal hemisphere but not when they are injected into the animal hemisphere. The effects of vegetally injected DNase I are more restricted to the vegetal hemisphere than those of CCD, a result that could be explained by differences in the molecular weights (DNase I, 31,000 Da; CCD, 508 Da) and diffusion constants of these two molecules. These results support a model that places the contractile force in the animal hemisphere, not in the vegetal hemisphere.

A model of ooplasmic segregation in teleosts should address the question of the movement of ooplasm and its inclusions toward the vegetal pole (Roosen-Runge, 1938; Abraham *et al.*, 1993; Webb *et al.*, 1995). That a separate mechanism causes this movement is suggested by experiments in which medaka zygotes were exposed to CO_2–Ringer's solution. Such treatment greatly slows the movement of oil droplets toward the vegetal pole but has little effect on the movement of ooplasm toward the animal pole (Gilkey, 1983). The apparently normal movement of oil droplets toward the vegetal pole of medaka zygotes treated with CCD (Webb *et al.*, 1995) also suggests that movement of oil droplets occurs independently of the movement of ooplasm toward the animal pole and that the molecular basis of oil-droplet movement differs from that of bulk ooplasm. This vegetal-poleward movement may involve microtubules (Abraham *et al.*, 1993; Webb *et al.*, 1995).

A model of ooplasmic segregation in teleosts should also address the fact that ooplasm and its inclusions move along the animal–vegetal axis of the egg (or along meridians that connect the two poles). Although these movements reflect a bipolar differentiation of the egg that arises during oogenesis and is apparent in the ripe ovum (Yamamoto, 1961; Devillers, 1961; Iwamatsu *et al.*, 1994), we know little about the structural and physiological bases of this bipolarity and how it organizes the structure and function of the cytoskeletal systems involved in ooplasmic segregation.

About 20 years ago, Jaffe *et al.* (1974) proposed that cytosolic calcium gradients organize cytoplasmic localization in animal zygotes. Proof of such a theory should have three parts (Gilkey *et al.*, 1978; Fluck *et al.*, 1994): (1) cytosolic calcium gradients should be present at the poles of zygotes during ooplasmic segregation, (2) dissipation of these gradients should disrupt ooplasmic segregation, and (3) artificial induction of a cytosolic calcium gradient should reorganize ooplasmic segregation. Evidence consistent with all three criteria has been found in experiments with medaka zygotes. The first criterion was tested by microinjecting aequorin, a protein that emits light when it binds Ca^{2+} (Miller *et al.*, 1994) into medaka eggs and then observing the eggs with an imaging photon detector (Fluck *et al.*, 1992). Fluorescein–aequorin was coinjected in order to correct for inhomogeneities in the distribution of aequorin in the ooplasm. Zones of elevated cytosolic $[Ca^{2+}]$ were found at both the animal and the vegetal poles of the zygote throughout ooplasmic segregation (Fluck *et al.*, 1992). The second criterion was tested by microinjecting the calcium buffer 5,5'-dibromo-BAPTA (Speksnijder *et al.*, 1989) into medaka zygotes. This buffer was shown to dissipate polar cytosolic calcium gradients (Fluck *et al.*, 1992), inhibit formation of the blastodisc (Fluck *et al.*, 1994), and disrupt oil-droplet movement (Fluck *et al.*, 1994) and the MTOC at the animal pole (Abraham *et al.*, 1995). The third criterion was tested by photolyzing microinjected nitrophenyl-EGTA (NP–EGTA), a photolabile calcium cage (Ellis-Davies and Kaplan, 1994). Such photolysis causes medaka ooplasm to contract and to accumulate in the irradiated region of the zygote (Fluck, 1995).

Other data are consistent with this theory. In both *F. heteroclitus* and medaka eggs, ooplasm accumulates at sites where the egg is pricked (Kao, 1951; Sakai, 1964), an action that would be expected to raise cytosolic $[Ca^{2+}]$. In medaka eggs activated with the calcium ionophore A23187 midway between the animal and vegetal poles, the second contraction is directed toward the site at which the egg was activated instead of toward the animal pole (Stone and Fluck, 1995). Calcium-dependent contractions of cortices isolated from loach zygotes have also been reported (Ivanenkov *et al.*, 1987). Taken together, these results are consistent with a model in which zones of elevated cytosolic $[Ca^{2+}]$ organize the structure and function of the cytoskeletal systems involved in ooplasmic segregation in teleosts.

E. Developmental Significance of the Vegetal Array of Parallel Microtubules

The parallel array of microtubules at the vegetal pole of medaka and zebrafish zygotes is reminiscent of a similar array in amphibians. In amphibians, this array begins to form at $T_n \approx 0.55$, disappears after $T_n \approx 0.8$, and is required for the cortical rotation that forms the gray crescent and establishes the three embryonic axes (Gerhart *et al.*, 1989; Houliston and Elinson, 1991). Treatments that disrupt this array—cold, elevated hydrostatic pressure, uv irradiation, and microtubule poisons—prevent cortical rotation and ventralize the embryos (Gerhart *et al.*, 1989).

Zebrafish embryos with body axis duplications or deficiencies can be produced by irradiating the zygotes with uv light (D. Kane, personal communication; Strähle and Jesuthasan, 1993) or applying hydrostatic pressure to the one-cell stage embryo (G. Streisinger, unpublished results, cited in Kimmel, 1989; D. Kane, personal communication). Irradiation of zygotes with uv light from $T_n \approx 0.22$–0.56 produces a high proportion of embryos that lack any dorsal structures and do not form a well-developed axis (Strähle and Jesuthasan, 1993). It remains to be shown whether a parallel array of microtubules is present in zebrafish zygotes during this crucial period and whether it is affected by uv irradiation.

Efforts to describe cortical rotations in teleosts like those in amphibians have been unsuccessful (Ho, 1992). However, the movement of developmentally significant information along the array of parallel microtubules could occur in the absence of a cortical rotation, and there is evidence from microdissection studies for a "crescent of symmetrization" in teleost zygotes (Devillers, 1961; Clavert, 1962). In primitive (nonteleost) fishes, a visible crescent does form (Ginsburg and Dettlaff, 1991; Bolker, 1993). In the Russian sturgeon (*Acipenser güldenstädti*), for example, a light crescent forms at the margin on one side of the animal region by $T_n \approx 0.55$, giving the egg a bilaterally symmetrical structure. The plane passing through the middle of this crescent and the animal and vegetal poles is the plane of bilateral symmetry of the embryo. It will be important to determine whether a parallel array of microtubules is present in this and related species and whether it has a particular orientation with respect to the crescent and the plane of bilateral symmetry. In zebrafish and medaka, this question could be pursued by microinjecting rhodamine-labeled tubulin (Houliston, 1994) into eggs or zygotes and monitoring the relationship between the orientation of the vegetal array and the plane of bilateral symmetry of the embryo.

V. Epiboly and Gastrulation

A. Cellular Movements

Meroblastic cleavage of the teleost zygote produces a cap of cells positioned atop an uncleaved yolk mass. The general organization of the embryo just before the onset of gastrulation is shown schematically in Fig. 5a for the zebrafish. It is typical of other teleost embryos as well. The blastoderm itself is limited peripherally by a monolayer of cohesive enveloping cells (EVL) beneath which is a population of rounded, more loosely packed deep cells (DC). Immediately subjacent to the blastoderm is a multinucleated yolk syncytial layer (YSL) which arises by the collapse of marginal blastomeres into the yolk cell (Kimmel and Law, 1985b; Trinkaus, 1992). Most of the nuclei are associated with the margin of the YSL or the external yolk syncytial layer (E-YSL). The E-YSL forms a wide belt at the margin of the blastoderm and is continuous with a thin, anuclear yolk cytoplasmic layer (YCL) that encircles the fluid yolk mass. By using lineage tracing molecules (Kimmel and Law, 1985b; Warga and Kimmel, 1990), it has been determined that only deep cells contribute to the germ layers.

The formation of the germ layers and the proper specification of cell types depend on the precise cellular movements of gastrulation. Several coordinated types of movement have been described: epiboly, involution, convergence, and extension (or convergent extension). Although these movements occur together, they are often separated for analytical and experimental purposes.

Epiboly is the dynamic process by which the YSL and the cells comprising the blastoderm spread over the yolk (Trinkaus, 1984). During this time, the marginal cells of the EVL are firmly attached to the YSL by tight and close junctions (Betchaku and Trinkaus, 1978). As the E-YSL advances toward the vegetal pole, the YCL disappears. Experiments with Lucifer yellow show a rapid endocytosis of the surface YCL where it is in contact with the E-YSL (Trinkaus, 1984). This remarkable event of membrane internalization allows advancement of the YSL and is considered part of the mechanism of epiboly. Movements of epiboly continue until the blastoderm completely encloses the yolk sphere.

Movements of deep cells are integrated with epiboly and are a subject of controversy (Wood and Timmermans, 1988; Warga and Kimmel, 1990; Trinkaus, 1994). The conventional view is that teleost embryos gastrulate by involution, a process illustrated in elegant studies of zebrafish in which deep blastomeres were microinjected with fluorescent tracer dyes and viewed by time-lapse video imaging (Warga and Kimmel, 1990). Deep cells move to the margin of the blastoderm, sink inward, and migrate away from the margins using the surfaces of other cells or the yolk cell as a substratum. Movements of involution lead to the formation of a thickened ring (germ ring) at the margin of the blastoderm composed of an inner hypoblast (mesoderm and endoderm) and an overlying, noninvoluted epiblast (ectoderm). Both involuting and noninvoluting cells then converge to the dorsal side of the embryo to form the embryonic shield. As they enter the shield, intercalation between other cells leads to lengthening of the anteroposterior

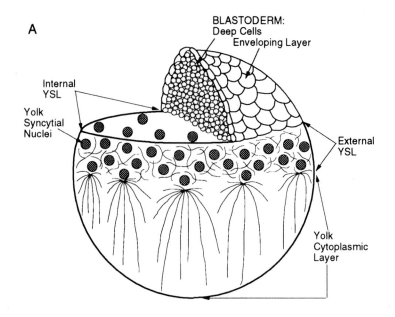

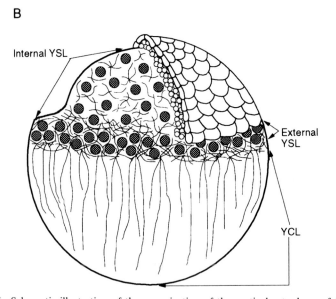

Fig. 5 Schematic illustration of the organization of the cortical cytoplasm of the yolk cell in relationship to other cells of the zebrafish embryo at blastula (A) and 30% epiboly (B) stages. The organization of microtubule networks in the yolk syncytial layer (YSL) and the yolk cytoplasmic layer (YCL) are shown as thin lines. At the blastula stage (A), the blastoderm, consisting of deep cells and a superficial enveloping layer, is positioned atop a flattened YSL. The YSL is populated with nuclei and is composed of internal and external layers. The microtubules of the external YSL form a network. The microtubules of the anuclear YCL radiate from organizing centers of the YSL. At the 30% epiboly stage (B), the external YSL has contracted and shows crowded nuclei and a dense network of microtubules. By 60% epiboly, the YSL nuclei appear stretched along the animal–vegetal axis and lead the epibolic movements toward the vegetal pole. (Reproduced from Solnica-Krezel, L., and Driever, W. (1994). *Development* **120.** The Company of Biologists Ltd.).

axis (Warga and Kimmel, 1990; Trinkaus *et al.*, 1992). Hence, the position of a cell prior to gastrulation not only predicts its future identity, but also indicates the types of movement that cells experience during gastrulation. By contrast, observations by Trinkaus (1994) using killifish embryos are consistent with an ingression of deep cells in the marginal region of the early gastrula rather than involution. Recognizing the diversity among fishes, it is possible that movements of deep cells may be different in large and small teleost eggs. The issue needs further investigation.

B. Epiboly, Microtubules, and Actin Filaments

The absence of visible signs of motility, such as cytoplasmic protrusions, at the advancing edge of the cellular blastoderm clearly indicates that epiboly does not involve crawling movements over the yolk. When the connection between the EVL and the E-YSL is severed, the blastoderm retracts, as if under tension, and stops moving, while the E-YSL continues to move toward the vegetal pole (Betchaku and Trinkaus, 1978). Movement of the YSL continues even if the blastoderm is removed completely. The epibolic movements of the E-YSL would appear to pull the blastoderm over the yolk. To understand the mechanism(s) of epiboly, therefore, one has to know how forces are generated and where they are applied by the YSL to produce vegetal spreading of the blastoderm.

Recent studies with zebrafish (Strähle and Jesuthasan, 1993; Solnica-Krezel and Driever, 1994) point to microtubules as being necessary for normal epiboly. Tubulins have been detected in early epibolizing embryos by Western blot analysis and immunocytochemistry (Solnica-Krezel and Driever, 1994). Monoclonal antibodies against α-tubulin and β-tubulin detect a polypeptide (50 kDa) that comigrates with proteins recognized by the same antibodies in purified fractions of mouse tubulins. Indirect immunofluorescence experiments show that tubulins are organized as arrays of microtubules in both the E-YSL and the YCL (Fig. 6). Microtubules of the E-YSL radiate from organizing centers associated with mitotic and interphase nuclei. Microtubules of the YCL originate from organizing centers of the E-YSL and are aligned along the animal–vegetal axis (Solnica-Krezel and Driever, 1994). It is presumed that microtubules of the YCL exhibit uniform polarity with their plus ends directed toward the vegetal pole. With the onset of epiboly, there is a dramatic constriction and narrowing of the E-YSL, crowding of the yolk nuclei, and an increase in the density of the microtubules of the E-YSL (Solnica-Krezel and Driever, 1994; Fig. 5b).

Studies employing agents known to interfere with the polymerization dynamics of microtubules point to epiboly being driven, at least in part,

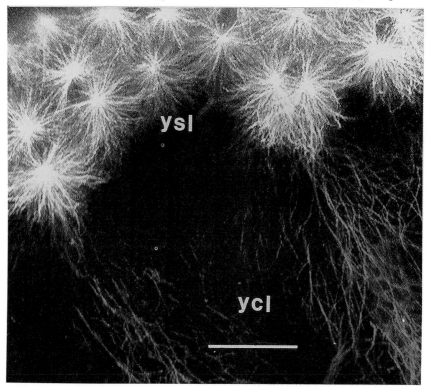

Fig. 6 Organization of the yolk cell microtubules after formation of the yolk syncytial layer (YSL). This fluorescent confocal image of an embryo stained with anti-tubulin antibody shows microtubules of the YSL filling the entire external layer. The yolk cytoplasmic layer (YCL) microtubules originate from organizing centers associated with the vegetal-most yolk syncytial layer nuclei. The animal pole is at the top of the micrograph. Scale bar = 50 μm. (Reproduced from Solnica-Krezel, L., and Driever, W. (1994). *Development* **120.** The Company of Biologists Ltd.).

by microtubules radiating from the YSL into the YCL. Exposure of late-stage zygotes or embryos in various stages of epiboly to uv irradiation retards the movement of the YSL and the blastoderm over the yolk; the extent of the migration is dependent on the uv dose applied (Strähle and Jesuthasan, 1993). The effect of the uv irradiation is noticeable in the YCL, where microtubules are absent, shorter than normal, or not aligned along the animal–vegetal axis. Similar phenotypes are visible in embryos treated with either taxol (Solnica-Krezel and Driever, 1994) or nocodazole (Strähle and Jesuthasan, 1993; Solnica-Krezel and Driever, 1994). Nocodazole completely blocks contraction of the YSL layer and crowding of syncytial nuclei, and disorganizes microtubule arrays of both blastoderm and yolk cell; it is somewhat less effective in arresting epiboly of the DC and the EVL (Solnica-Krezel and Driever, 1994). Embryos incubated in Lucifer yellow and nocodazole display endocytosis of the YCL, suggesting that disruption of microtubules does not alter all epibolic movements equally. In taxol-treated embryos, the YSL contracts, microtubule arrays of the YCL appear denser than those in control embryos, and the vegetal movements of both the YSL and the blastoderm are noticeably retarded (Solnica-Krezel and Driever, 1994). Since microtubules of the YCL become progressively shorter during epiboly (Solnica-Krezel and Driever, 1994), taxol could impair epiboly by either interfering with the normal process of shortening or increasing the stability of microtubules.

Collectively, the drug and uv irradiation studies with zebrafish embryos suggest that microtubules are an important component of the machinery driving teleost epiboly. The extent to which microtubules of the E-YSL and the YCL might contribute to the different movements of epiboly is still unclear. Also, our understanding of how microtubules of the E-YSL and the YCL might mediate blastoderm migration is still unsophisticated and can only be speculated upon. Some insight into the possible role of microtubules during epiboly comes from other studies that have focused on movements of nuclei (Baker et al., 1993) and pronuclei (Manes and Barbieri, 1977; Schatten et al., 1982). One scenario is that the E-YSL advances because plus end-directed microtubule-associated motor proteins move nuclei along tracts formed by networks of microtubules in the YSL and the YCL. Alternatively, microtubule motors positioned between the crisscrossing tracks of microtubules of the YSL could generate a pushing force similar to that in anaphase B of mitosis (Nislow et al., 1992). This might effectively propel the YSL over the surface of the yolk. Distinguishing between these and other possible models of microtubule function during epiboly will require identification and localization of motor proteins that are associated with microtubules. Possible plus end-directed motors include kinesin and kinesin-related proteins such as Eg5, which has recently been implicated in cortical rotation of the Xenopus egg (Houliston et al., 1994). These motors might be anchored at the margin of the yolk syncytial layer (Strähle and Jesuthasan, 1993). If such proteins are present, experiments must then be conducted to determine if they are capable of translocating the blastoderm toward the vegetal pole.

Microtubules are rarely observed in the marginal cells of the EVL, the YSL, and the YCL of Fundulus eggs (Betchaku and Trinkaus, 1978). Indeed, Kessel (1960) has shown that movement of the EVL is relatively unaffected in Fundulus eggs treated with colchicine. Trinkaus and colleagues maintain that epiboly in Fundulus embryos can be accounted for by forces generated by actin filaments. Networks of actin filaments (4 or 5 nm) are detected in the cortical cytoplasm of the YSL and cortices of the YCL and the EVL.

Bundles of actin filaments (10–12 nm) are also evident in the cortex of marginal cells of the EVL and the YSL at sites of contact; these bundles are arranged parallel to each other at the contact site and circumferentially to the whole egg (Betchaku and Trinkaus, 1978). A visible circumferential constriction at the leading edge of the blastoderm is presumably mediated by contraction of actin filaments of the YSL (Trinkaus, 1951). The EVL appears to be a relatively passive participant in blastoderm epiboly, being pulled and stretched across the yolk cell by the YSL. How actin-mediated, circumferential contraction of the YSL directionally propels this region over the surface of the yolk is unclear. Still, failure of uv irradiation and microtubule-disrupting agents to completely inhibit the vegetal migration of the blastoderm in zebrafish eggs (Strähle and Jesuthasan, 1993; Solnica-Krezel and Driever, 1994) suggests that forces generated by cytoskeletal components other than microtubules are probably important in epiboly. Actin filaments do effect the bulk movement of cytoplasm from nurse cells to the *Drosophila* oocyte (Gutzeit, 1986) and translocation of ooplasm during blastodisc formation in zebrafish (Katow, 1983).

The studies with zebrafish and killifish embryos illustrate that regulation of epibolic movements of teleost gastrulation are imperfectly understood. For example, is epiboly of the EVL totally passive? Active rearrangement of cells in this layer has been shown to occur in both *Fundulus* (Keller and Trinkaus, 1987) and the medaka (Kageyama, 1982); they probably occur in zebrafish as well (Warga and Kimmel, 1990). The effect of this cellular rearrangement may be to assist in expanding the EVL. The apparent differences between embryos of these organisms with regard to the role(s) of microtubules and microfilaments require additional experiments to determine how and when these filamentous structures are mobilized and where contractile forces are placed during epiboly. Trinkaus (1993) observed in *Fundulus* that epiboly begins after cessation of nuclear division in the YSL, suggesting that the contractile machinery associated with mitosis could be recruited for circumferential contraction of the E-YSL. Alternatively, attenuation of the mitotic cycle could lead to transcription and expression of genes necessary for the synthesis of contractile proteins. The transition period between mitotic activity and the onset of gastrulation needs serious investigation. Progress in understanding the molecular basis of epiboly may come through mutational analysis of zebrafish. Since uv irradiation of zebrafish zygotes impairs epiboly (Strähle and Jesuthasan, 1993), it is likely that some components of the force-generating system are maternal and stored in the egg. Hence, analysis of maternal-effect mutations could provide insight into specific gene products and the molecular mechanism of their action.

C. Deep Cells, Actin Filaments, and Microtubules

Movements of involution are initiated in many teleost eggs when the blastoderm covers about one-half of the yolk sphere (Warga and Kimmel, 1990; Gevers and Timmermans, 1991). These appear to be regulated by a clock mechanism independent of epiboly (Strähle and Jesuthasan, 1993). The role of the cytoskeleton in moving deep cells is largely inferred from the dramatic changes in their shape that occur during involution and convergent extension. The behavior and alterations in cell shape have been carefully recorded in embryos of zebrafish (Warga and Kimmel, 1990) and *Fundulus* (Trinkaus *et al.*, 1992) with time-lapse video imaging. In zebrafish, individual deep cells that actively migrate toward the margin of the blastoderm display blebbing and filopodia at their surfaces; these same cells show pronounced protrusive activity during involution (Warga and Kimmel, 1990). Once within the germ ring, the cells of *Fundulus* display several modes of motility (Trinkaus *et al.*, 1992). The majority show frequent protrusive, contractile activity along their free surfaces, extending out filopodia, lamellipodia, or a combination of both. A minority of cells move by so-called blebbing locomotion (Fink and Trinkaus, 1988). Other cells are organized as clusters and move by protrusive activity of the cells at their margins. Protrusive activity ceases when individual cells or clusters are in contact with other cells.

An extraordinary feature of teleost gastrulation is the efficient, directional translocation of deep cells toward the dorsal side of the embryo and the embryonic shield. This phenomenon raises questions of how the cells move and what provides the marked, inexorable streaming of cells toward a specific site in the germ ring. The majority of cells move as an organized monolayer consisting of cell clusters and individual cells that form filolamellipodia (Trinkaus *et al.*, 1992). Both individual cells and cell clusters engage in mediolateral intercalation during convergence and extension of the embryonic shield (Warga and Kimmel, 1990; Trinkaus *et al.*, 1992). These observations suggest that intercalation of cells may be an important force in driving the cellular movements of convergent extension and expansion of the blastoderm.

Microtubules contribute to the coordinated movements of both involution and convergent extension. Formation of the germ ring and the convergence of cells to form the embryonic shield are either repressed or inhibited in zebrafish embryos treated with nocodazole (Solnica-Krezel and Driever, 1994). Ultraviolet irradiation slightly retards convergence, but does not block completely the formation of the embryonic shield (Strähle and Jesuthasan, 1993). Movements of deep cells in *Fundulus* are sensitive to colchicine (Trinkaus and Erickson, 1983). How the action of microtubules is translated into the coordinated movements of involution and convergent extension remains to be determined.

The basis for directionality of converging cells also remains an open question. Cells nearer the embryonic shield display accelerated motility and move more rapidly to this site, suggesting that the embryonic shield may produce a factor(s) to which moving cells selectively respond (Trinkaus *et al.*, 1992). Presumably, such a factor or signal might be detected by cytoplasmic extensions of the migrating cells and determine the pathway of movement. Once the signal is detected, information could be transmitted internally and trigger the cytoskeletal remodeling required for moving a cluster, for example, from one capable of moving in several directions to one moving in only one direction. It remains to be determined whether clusters show polarized protrusive activity, either increased activity at the leading edge of a cluster (i.e., nearer to the embryonic shield), decreased activity at the trailing edge of the cluster, or both (Trinkaus *et al.*, 1992).

D. Genetic Analysis of Cellular Movements

Improvement in understanding the mechanics of deep cell motility and epiboly should come from genetic analyses of zebrafish embryos. Two genes that correlate well with our knowledge of cell fate and the cellular movements of gastrulation have been identified in zebrafish: *spadetail* (Kimmel *et al.*, 1989) and *snail* (Hammerschmidt and Nüsslein-Volhard, 1993; Thisse *et al.*, 1993). The *snail* gene, a zygotic gene highly conserved in vertebrates (Smith *et al.*, 1992; Nieto *et al.*, 1992), is expressed in invaginating mesoderm cells during *Drosophila* gastrulation (Alberga *et al.*, 1991). A zebrafish *snail1*-specific RNA probe hybridizes to the dorsal side of the late-stage blastula, the margin of the blastoderm at the beginning of epiboly, cells of the germ ring, and the lateral part of the embryonic shield (Thisse *et al.*, 1993). The labeled cells of the early gastrula can be traced to the paraxial and axial mesoderm (Kimmel and Warga, 1987; Kimmel *et al.*, 1990). *Snail1* expression during convergent movements becomes repressed in cells destined to become axial mesoderm (Thisse *et al.*, 1993).

The zygotic, lethal mutation *spadetail* affects trunk somitic mesoderm whose precursor cells are positioned within the lateral margin of the blastoderm before involution. Application of cell labeling (Kimmel *et al.*, 1989) and cotransplantation (Ho and Kane, 1990) methods show that in wild-type embryos presumptive trunk somite cells always converge dorsally, but the same cells in the mutant move to the tail region and give rise to mesenchyme. The *spadetail* mutation, therefore, appears to selectively affect convergence of the deep cells after involution. Interestingly, cells that express *snail1* in wild-type embryos are misplaced in *spadetail* mutant gastrulas (Thisse *et al.*, 1993). Further experiments are necessary to determine if *snail1*-expressing cells simply fail to converge dorsally in *spadetail* embryos or if *spadetail* somehow disturbs convergence by interfering with the action of *snail1*.

The experiments with *snail1* and *spadetail* genes clearly demonstrate a precise connection between gene expression and control of the cellular movements of gastrulation. The function of these genes in these movements is currently unknown. There are several possible explanations for the altered pathways of DC migration in genetic mutants. The mutations may directly affect the positional identity of the cells before involution and thereby cause them to migrate to a different site. Alternatively, the mutations may disturb the expression of other genes that regulate the cytoskeletal dynamics of moving cells or interfere with the interactions of these cells with the extracellular environment. An initial step in looking at a possible functional connection between these genes and the cytoskeleton would be to compare the behavior of labeled, migrating cells in wild-type and mutant embryos using techniques of cell labeling and video imaging.

VI. Conclusions and Future Directions

We have discussed the cytoskeleton and its proposed role(s) in such diverse events of teleost embryogenesis as fertilization, ooplasmic segregation, and gastrulation. This approach was taken to illustrate that our understanding of the teleost egg cytoskeleton and how it effects changes at these different stages is still very much in its infancy. For example, the remodeling of actin architecture at sites of exocytosis and sperm entry is clearly regulated both spatially and temporally. We need to establish what proteins are present in the egg that modulate the assembly and disassembly of actin, define their spatial distribution, and determine their functional connections to these specific events. This will require focusing on possible signal transduction pathways that link Ca^{2+} to target cytoskeletal proteins. The possibility that protein kinase C may be an important signal transducing protein, as in amphibian eggs (Bement, 1992), needs to be explored. Studies of the mechanisms that produce ooplasmic segregation have only begun. Experiments will be most valuable if they test predictions that follow current models of contractile phenomena, such as the solation–contraction coupling hypothesis (Taylor and Fechheimer, 1982) or the cortical flow hypothesis (Bray and White, 1988). This approach has been useful in the analysis of ooplasmic movements in the fruit fly (von Dassow and Schubiger, 1994). The cellular movements of gastrulation are clearly complex. Analyses of these movements *in vivo* should be enhanced using high-resolution video imaging. We need to know more about what types of cells generate force and how the cytoskeleton is mobilized by specific cells to generate patterns of force that lead to movement. Finally, as the genetics of the zebrafish become increasingly better established, the system should permit the identification and functional analyses of cytoskeletal proteins that control events of early teleost embryogenesis.

Appendix: Methods

A. Collecting and Handling Zebrafish Gametes for Fertilization Studies

An excellent reference manual for maintaining and breeding zebrafish is "The Zebrafish Book" by Monte Westerfield. It is available from the Institute of Neuroscience, University of Oregon.

1. Maintain male and female fish in 10- to 15-gallon aquaria containing aerated, dechlorinated (aged) tap water. Control the day–night cycle with an automatic timer (14 hr light/10 hr dark).
2. Transfer a gravid-appearing (plump) female to a clean, glass fingerbowl (90 or 110 mm in diameter) after rinsing it several times in aged tap water.
3. Collect eggs in aged tap water by immobilizing the female (with fingers or a piece of netting) against the side of the bowl and gently applying pressure to the abdomen. If the female is prepared to lay eggs, little pressure is necessary. Collect only a few eggs at first.
4. Evaluate the quality of eggs under a dissecting microscope. Examine for signs of egg activation such as elevation of the chorion. Zebrafish eggs spontaneously activate in aged tap water.
5. Collect eggs in Ginsburg Fish Ringer's [6.5 g NaCl, 0.25 g KCl, 0.3 g $CaCl_2$ (0.4 g $CaCl_2 \cdot 2H_2O$), and 0.2 g $NaHCO_3$ dissolved in 1 liter of ddH_2O] and transfer in a small volume to a 50-mm watchglass (or 35-mm petri dish). Eggs will remain unactivated in Ringer's for several minutes.
6. Rinse a healthy, male fish and immobilize in a fingerbowl containing Fish Ringer's. Position a glass disposable pipet near the genital pore. While applying firm pressure to the abdomen, aspirate a small volume (100–200 μl) of the milky suspension into the tip of the pipet. Alternatively, the fish can be blotted with a paper towel, placed onto a moistened sponge, and sperm obtained directly from the gential pore.
7. Add the sperm suspension directly to the eggs and gently mix. Add 0.5–1.0 ml of aged water. Flood the gamete suspension with aged water after 1 or 2 min. The success of fertilization can be evaluated

by examination of eggs for cleavage furrow formation between 40 and 50 min postinsemination.

B. Ultracryomicrotomy of Teleost Eggs

Semithin frozen sections have proved useful for localization of actin and actin-associated proteins using conventional fluorescence microscopy. The general protocol outlined below for zebrafish eggs is slightly modified after that described by Bonder *et al.* (1989).

1. Fix eggs in 3.7–5.0% formaldehyde in actin-stabilizing buffer (ASB; after Yonemura and Kinoshita, 1986) for 4–8 hr on ice.
2. Rinse eggs in cold ASB over a period of 60 min. If desired, the chorion can be removed by dissection at this time.
3. Embed eggs with 7.0% molten gelatin (porcine, 300 Bloom) in ASB using shallow rubberized molds. Allow gelatin to harden thoroughly.
4. Cut gelatin into blocks with one egg each using a clean razor blade.
5. Cryoprotect eggs overnight with increasing concentrations of cold sucrose diluted in ASB up to 2.3 *M* sucrose.
6. Mount samples on slotted, stainless-steel pins (12 mm long by 3 mm diameter head) and remove excess sucrose. Trim gelatin block as necessary.
7. Freeze the pin and sample by plunging rapidly into Freon 22 cooled by liquid nitrogen. Store in liquid nitrogen until sectioning.
8. Section samples on an ultracryomicrotome using glass knives. Sectioning parameters (specimen and knife temperature, knife speed) will depend in part on section thickness. Pick up sections using a small stainless-steel or platinum loop containing 2.3 *M* sucrose. Transfer droplet with sections to an acid-cleaned slide.
9. Store sections at −20°C until labeling with antibodies.

C. Indirect Immunofluorescence Staining of Whole Eggs with Anti-actin and Anti-myosin Antibodies

1. Fix eggs in cold 3.7% formaldehyde in ASB for 4–8 hr.
2. Rinse in several changes of ASB and dechorionate with No. 5 Dumont tweezers.
3. Transfer to cold quenching buffer (150 m*M* glycine in ASB) for 30 min.
4. Rinse briefly in ASB.
5. Transfer to cold 0.5% Triton X-100 in ASB for 30 min.

6. Rinse several times in ASB and incubate in cold blocking buffer (1–3% BSA or 1% BSA plus 2% normal goat serum).
7. Incubate in cold primary antibody for about 4 hr (actin) or 12–18 hr (myosin). For actin, dilute anti-chicken monoclonal antibody (Amersham Life Science Inc., Arlington Heights, IL) 1:50 or 1:100 in ASB. For myosin, dilute anti-platelet myosin antibody (Biomedical Technologies Inc., Stoughton, MA) 1:50 in ASB.
8. Rinse thoroughly in several changes of cold ASB.
9. Incubate in FITC-conjugated secondary antibody diluted 1:25 or 1:50 in ASB. Carry out in the cold for 2 hr.
10. Control eggs should be incubated in secondary antibody alone or with primary antibody preadsorbed with antigen followed with secondary antibody.
11. Mount whole eggs on acid-cleaned slides using 2% n-propyl gallate in 50% glycerol. A thin layer of Vaseline can be applied to the edges of the coverglass before mounting.

Acknowledgments

This work was supported in part by NIH Grant HD 17467 to N.H.H. and NSF Grants DCB 9017210 and MCB 9316125 to R.A.F. Figure 4 was prepared by Bob Golder.

References

Abraham, V. C., Gupta, S., and Fluck, R. A. (1993). Ooplasmic segregation in the medaka *Oryzias latipes* egg. *Biol. Bull.* **184**, 115–124.
Abraham, V. C., Miller, A. L., and Fluck, R. A. (1995). Microtuble arrays during ooplasmic segregation in the medaka fish egg *Oryzias laptipes. Biol. Bull.* **188**, 136–145.
Alberga, A., Boulay, J. L., Kempe, E., Dennefeld, C., and Haenlin, M. (1991). The *snail* gene required for mesoderm formation in *Drosophila* is expressed dynamically in derivatives of all three germ layers. *Development* **111**, 983–992.
Anderson, R. G. W. (1991). Molecular motors that shape endocytic membrane. *In* "Intracellular Trafficking of Proteins" (C. J. Steer and J. A. Hanover, eds.), pp. 13–46. Cambridge Univ. Press, Cambridge.
Ankenbauer, T., Kleinschmidt, J. A., Vandekerckhove, J., and Franke, W. W. (1988). Proteins regulating actin assembly in oogenesis and early embryogenesis of *Xenopus laevis:* Gelsolin is the major cytoplasmic actin-binding protein. *J. Cell Biol.* **107**, 1489–1498.
Aronson, J., and Inoué, S. (1970). Reversal by light of the action of *N*-methyl, N-desacetyl colchicine on mitosis. *J. Cell Biol.* **45**, 470–477.
Baker, J., Therukauf, W. E., and Schubiger, G. (1993). Dynamic changes in microtubule configuration correlate with nuclear migration in the preblastoderm *Drosophila* embryo. *J. Cell Biol.* **122**, 113–121.

Beams, H. W., and Kessel, R. G. (1976). Cytokinesis: A comparative study of cytoplasmic division in animal cells. *Am. Sci.* **63,** 279–290.

Beams, H. W., Kessel, R. G., Shih, C. Y., and Tung, H. N. (1985). Scanning electron microscopy studies on blastodisc formation in the zebrafish, *Brachydanio rerio.* *J. Morphol.* **184,** 41–49.

Becker, K. A., and Hart, N. H. (1993). Actin and myosin-II in zebrafish eggs: Reorganization of actin during cortical granule exocytosis. *Mol. Biol. Cell* **4,** 141a.

Becker, K. A., and Hart, N. H. (1994). Spectrin and the actin cytoskeleton in the zebrafish egg. *Mol. Biol. Cell* **5,** 463a.

Becker, K. A., and Hart, N. H. (1995). The cortical actin cytoskeleton of unactivated zebrafish eggs: Spatial organization and distribution of filamentous actin, nonfilamentous actin, and myosin-II. *Mol. Reprod. Dev.,* in press.

Bement, W. M. (1992). Signal transduction by calcium and protein kinase C during egg activation. *J. Exp. Zool.* **263,** 382–397.

Bement, W. M., and Capco, D. G. (1989). Activators of protein kinase C trigger cortical granule exocytosis, cortical contraction, and cleavage furrow formation in *Xenopus laevis* oocytes and eggs. *J. Cell Biol.* **108,** 885–892.

Betchaku, T., and Trinkaus, J. P. (1978). Contact relations, surface activity, and cortical microfilaments of marginal cells of the enveloping layer and of the yolk syncytial and yolk cytoplasmic layers of *Fundulus* before and during epiboly. *J. Exp. Zool.* **206,** 381–426.

Bolker, J. A. (1993). Gastrulation and mesoderm morphogenesis in the white sturgeon. *J. Exp. Zool.* **266,** 116–131.

Bonder, E. M., and Mooseker, M. S. (1986). Cytochalasin B slows but does not prevent monomer addition at the barbed end of the actin filament. *J. Cell Biol.* **102,** 282–288.

Bonder, E. M., Fishkind, D. J., Cotran, N. M., and Begg, D. A. (1989). The cortical actin-membrane cytoskeleton of unfertilized sea urchin eggs: Analysis of the spatial organization and relationship of filamentous actin, nonfilamentous actin, and egg spectrin. *Dev. Biol.* **134,** 327–341.

Bray, D., and White, J. G. (1988). Cortical flow in animal cells. *Science* **239,** 883–888.

Brummett, A. R., and Dumont, J. N. (1979). Initial stages of sperm penetration into the egg of *Fundulus heteroclitus. J. Exp. Zool.* **210,** 417–434.

Brummett, A. R., and Dumont, J. N. (1981). Cortical vesicle breakdown in fertilized eggs of *Fundulus heteroclitus. J. Exp. Zool.* **216,** 63–79.

Brummett, A. R., Dumont, J. N., and Richter, C. S. (1985). Later stages of sperm penetration and second polar body and blastodisc formation in the egg of *Fundulus heteroclitus. J. Exp. Zool.* **234,** 423–439.

Chandler, D. E. (1991). Multiple intracellular signals coordinate structural dynamics in the sea urchin cortex at fertilization. *J. Elect. Micros. Tech.* **17,** 266–293.

Chow, R., and Elinson, R. P. (1993). Local alteration of cortical actin in *Xenopus* eggs by the fertilizing sperm. *Mol. Reprod. Dev.* **35,** 69–75.

Clavert, J. (1962). Symmetrization of the egg of vertebrates. *Adv. Morphogenesis* **2,** 27–60.

Cooper, J. A. (1987). Effects of cytochalasin and phalloidin on actin. *J. Cell Biol.* **105,** 1473–1478.

Devillers, Ch. (1961). Structural and dynamic processes of the development of the teleostean egg. *In* "Advances in Morphogenesis" (M. Abercrombie and J. Brachet, eds.), Vol. 1, pp 379–428. Academic Press, New York.

Donovan, M. J., and Hart, N. H. (1986). Cortical granule exocytosis is coupled with membrane retrieval in eggs of *Brachydanio. J. Exp. Zool.* **237,** 391–405.

Ellis-Davies, C. R., and Kaplan, J. H. (1994). Nitrophenyl–EGTA, a photolabile chelator that selectively binds Ca^{2+} with high affinity and releases it rapidly upon photolysis. *Proc. Natl. Acad. Sci. USA* **91,** 187–191.

Endo, Y., Shultz, R. M., and Kopf, G. S. (1987). Effects of phorbol esters and a diacylglycerol on mouse eggs: Inhibition of fertilization and modification of the zona pellucida. *Dev. Biol.* **119,** 199–209.

Fink, R. D., and Trinkaus, J. P. (1988). *Fundulus* deep cells: Directional migration in response to epithelial wounding. *Dev. Biol.* **129,** 179–190.

Fluck, R. A. (1995). Responses of the medaka fish egg *Oryzias latipes* to the photolysis of microinjected nitrophenyl–EGTA, a photolabile calcium chelator. *Biol. Bull.* **188,** 1–4.

Fluck, R. A., Miller, A. L., and Jaffe, L. F. (1992). High calcium zones at the poles of developing medaka eggs. *Biol. Bull.* **183,** 70–77.

Fluck, R. A., Miller, A. L., Abraham, V. C., and Jaffe, L. F. (1994). Calcium buffer injections inhibit ooplasmic segregation in medaka eggs. *Biol. Bull.* **186,** 254–262.

Gard, D. L. (1991). Organization, nucleation, and acetylation of microtubules in *Xenopus laevis* oocytes: A study by confocal immunofluorescence microscopy. *Dev. Biol.* **143,** 346–362.

Gerhart, J., Danilchik, M., Doniach, T., Roberts, S., Rowning, B., and Stewart, R. (1989). Cortical rotation of the *Xenopus* egg: Consequences for the anteroposterior pattern of embryonic dorsal development. *Development (Suppl. 1989),* 37–51.

Gevers, P., and Timmermans, L. P. M. (1991). Dye-coupling and the formation and fate of the hypoblast in the teleost fish embryo, *Barbus conchonius. Development* **112,** 431–438.

Gilkey, J. C. (1981). Mechanisms of fertilization in fishes. *Am. Zool.* **21,** 359–375.

Gilkey, J. C. (1983). Roles of calcium and pH in activation of eggs of the medaka fish, *Oryzias latipes. J. Cell Biol.* **97,** 669–678.

Gilkey, J. C., Jaffe, L. F., Ridgway, E. B., and Reynolds, G. T. (1978). A free calcium wave traverses the activating egg of the medaka, *Oryzias latipes. J. Cell Biol.* **76,** 448–466.

Ginsburg, A. S., and Dettlaff, T. A. (1991). The Russian sturgeon *Acipenser güldenstädti.* Part 1. Gametes and early development up to time of hatching. *In* "Animal Species for Developmental Studies" (T. A. Dettlaff and S. G. Vassetsky, eds.), Vol. 2, pp. 15–65. Consultants Bureau, New York.

Gottlieb, T. A., Ivanov, I. E., Adesnik, M., and Sabatini, D. D. (1993). Actin microfilaments play a critical role in endocytosis at the apical but not the basolateral surface of polarized epithelial cells. *J. Cell Biol.* **120,** 695–710.

Gutzeit, H. O. (1986). The role of microfilaments in cytoplasmic streaming in *Drosophila* follicles. *J. Cell Sci.* **80,** 159–169.

Hamaguchi, Y., and Mabuchi, I. (1988). Accumulation of fluorescently labeled actin in the cortical layer in sea urchin eggs after fertilization. *Cell Motil. Cytoskeleton* **9,** 153–163.

Hammerschmidt, M., and Nüsslein-Volhard, C. (1993). The expression of a zebrafish gene homologous to *Drosophila snail* suggests a conserved function in invertebrate and vertebrate gastrulation. *Development* **119,** 1107–1118.

Hart, N. H. (1990). Fertilization in teleost fishes: Mechanisms of sperm–egg interactions. *Int. Rev. Cytol.* **121,** 1–66.

Hart, N. H., and Becker, K. A. (1994). Ooplasmic segregation in the zebrafish egg. *Mol. Biol. Cell* **5,** 100a.

Hart, N. H., and Becker, K. A. (1992). Actin and myosin-II in unfertilized zebrafish eggs: Filamentous and nonfilamentous actin domains in the cortex. *Mol. Biol. Cell* **3,** 7a.

Hart, N. H., and G. C. Collins (1991). An electron-microscope and freeze-fracture study of the egg cortex of *Brachydanio rerio. Cell Tissue Res.* **265,** 317–328.

Hart, N. H., and Donovan, M. E. (1983). Fine structure of the chorion and site of sperm entry in the egg of *Brachydanio. J. Exp. Zool.* **227,** 277–296.

Hart, N. H., and Yu, S. (1980). Cortical granule exocytosis and cell surface reorganization in eggs of *Brachydanio. J. Exp. Zool.* **213,** 137–159.

Hart, N. H., Becker, K. A., and Wolenski, J. S. (1992). The sperm entry site during fertilization of the zebrafish egg: Localization of actin. *Mol. Reprod. Dev.* **32**, 217–228.

Hiramoto, Y. (1970). Rheological properties of sea urchin eggs. *Biorheology* **6**, 201–234.

Hisaoka, K. K., and Firlit, C. F. (1960). Further studies on the embryonic development of the zebrafish, *Brachydanio rerio* (Hamilton–Buchanan). *J. Morphol.* **107**, 205–225.

Ho, R. K. (1992). Axis formation in the embryo of the zebrafish, *Brachydanio rerio*. *Semin. Dev. Biol.* **3**, 53–64.

Ho, R. K., and Kane, D. A. (1990). Cell-autonomous action of zebrafish *spadetail–1* mutation in specific mesodermal precursors. *Nature* **348**, 728–730.

Houliston, E. (1994). Microtubule translocation and polymerization during cortical rotation in *Xenopus* eggs. *Development* **120**, 1213–1220.

Houliston, E., and Elinson, R. P. (1991). Patterns of microtubule polymerization relating to cortical rotation in *Xenopus laevis* eggs. *Development* **122**, 107–117,

Houliston, E., Le Guellec, R., Kress, M., Philippe, M., and Le Guellec, K. (1994). The kinesin-related protein Eg5 associates with both interphase and spindle microtubules during *Xenopus* early development. *Dev. Biol.* **164**, 147–159.

Ingold, A. L., Cohn, S. A., and Scholey, J. M. (1988). Inhibition of kinesin-driven microtuble motility by monoclonal antibodies to kinesin heavy chain. *J. Cell Biol.* **107**, 2657–2667.

Ivanenkov, V. V., Minim, A. A., Meshcheryakov, V. N., and Martynova, L. E. (1987). The effect of local microfilament disorganization on ooplasmic segregation in the loach *Misgurnus fossilis* egg. *Cell Differ.* **22**, 19–28.

Ivanenkov, V. V., Minin, A. A., and Ozerova, S. G. (1990). Phalloidin inhibits cortical granule exocytosis and ooplasmic segregation in loach eggs. *Cell Differ. Dev.* **29**, 21–36.

Iwamatsu, T. (1973). On the mechanism of ooplasmic segregation upon fertilization in *Oryzias latipes*. *Jpn. J. Ichthyol.* **20**, 273–278.

Iwamatsu, T. (1989). Exocytosis of cortical alveoli and its initiation time in medaka eggs induced by microinjection of various agents. *Dev. Growth Differ.* **31**, 39–44.

Iwamatsu, T., and Ohta, T. (1981). Scanning electron microscopic observations on sperm penetration in teleostean fish. *J. Exp. Zool.* **218**, 261–277.

Iwamatsu, T., Yoshimoto, Y., and Hiramoto, Y. (1988a). Cytoplasmic Ca^{2+} release induced by microinjection of Ca^{2+} and effects of microinjected divalent cations on Ca^{2+} sequestration and exocytosis of cortical alveoli in the medaka egg. *Dev. Biol.* **125**, 451–457.

Iwamatsu, T., Yoshimoto, Y., and Hiramoto, Y. (1988b). Mechanism of Ca^{2+} release in medaka eggs microinjected with inositol 1,4,5-trisphosphate and Ca^{2+}. *Dev. Biol.* **129**, 191–197.

Iwamatsu, T., Nakashima, S., Onitake, K., Matsushita, A., and Nagahama, Y. (1994). Regional differences in granulosa cells of preovulatory medaka follicles. *Zool. Sci.* **11**, 77–82.

Jaffe, L. F., Robinson, K. R., and Nuccitelli, R. (1974). Local cation entry and self-electrophoresis as an intracellular localization mechanism. *Ann. N.Y. Acad. Sci.* **238**, 372–389.

Jørgensen, N.-C. (1972). Actomyosin-like ATPase at the surface of fish eggs. *Exp. Cell Res.* **71**, 460–464.

Kageyama, T. (1982). Cellular basis of epiboly of the enveloping layer in the embryo of the medaka, *Oryzias latipes*. *J. Exp. Zool.* **219**, 241–256.

Kao, C.-Y. (1951). Micropuncture and the localization of the subsequently-forming blastodisc in the unfertilized *Fundulus* egg. *Biol. Bull.* **101**, 222–223.

Katow, H. (1983). Obstruction of blastodisc formation by cytochalasin B in the zebrafish, *Brachydanio rerio*. *Dev. Growth Differ.* **25**, 477–484.

Keller, R. E., and Trinkaus, J. P. (1987). Rearrangement of enveloping layer cells without disruption of the epithelial permeability barrier as a factor in *Fundulus* epiboly. *Dev. Biol.* **120,** 12–24.

Kessel, R. G. (1960). The role of cell division in gastrulation of *Fundulus heteroclitus. Exp. Cell Res.* **20,** 277–282.

Kimmel, C. B. (1989). Genetics and early development of zebrafish. *Trends Genet.* 5, 283–288.

Kimmel, C. B., and Law, R. D. (1985a). Cell lineage of zebrafish blastomeres. I. Cleavage pattern and cytoplasmic bridges between cells. *Dev. Biol.* **108,** 78–85.

Kimmel, C. B., and Law, R. D. (1985b). Cell lineage of zebrafish blastomeres. II. Formation of the yolk syncytial layer. *Dev. Biol.* **108,** 86–93.

Kimmel, C. B., and Warga, R. M. (1987). Cell lineages generating axial muscle in the zebrafish embryo. *Nature* **327,** 234–237.

Kimmel, C. B., Kane, D. A., Walker, C., Warga, R. .W., and Rothman, M. B. (1989). A mutation that changes cell movement and cell fate in the zebrafish embryo. *Nature* **337,** 358–362.

Kobayashi, W. (1985). Electron microscopic observation of the breakdown of cortical vesicles in the chum salmon egg. *J. Fac. Sci. Hokkaido Univ. Ser. IV Zool.* **24,** 87–102.

Kobayashi, W., and Yamamoto, T. S. (1981). Fine structure of the micropylar apparatus of the chum salmon egg, with a discussion of the mechanism for blocking polyspermy. *J. Exp. Zool.* **217,** 265–275.

Kobayashi, W., and Yamamoto, T. S. (1987). Light and electron microscopic observations of sperm entry in the chum salmon egg. *J. Exp. Zool.* **243,** 311–322.

Knapp, L. W., O'Guin, W. M., and Sawyer, R. H. (1983). Drug-induced alterations of cytokeratin organization in cultured epithelial cells. *Science* **219,** 501–503.

Kudo, S., and Sato, A. (1985). Fertilization cone of carp eggs revealed by scanning electron microscopy. *Dev. Growth Differ.* **27,** 121–128.

Lelkes, P. I., Friedman, J. E., Rosenheck, K., and Oplatka, A. (1986). Destabilization of actin filaments as a requirement for the secretion of catecholamines from permeabilized chromaffin cells. *FEBS Lett.* **208,** 357–362.

Lewis, W. H., and Roosen-Runge, E. C. (1942). The formation of the blastodisc in the egg of the zebra fish, *Brachydanio rerio,* illustrated with motion pictures. *Anat. Rec.* **84,** 463–464.

Lewis, W. H., and Roosen-Ruge, E. C. (1943). The formation of the blastodisc in the egg of the zebra fish, *Brachydanio rerio. Anat. Rec.* **85,** 326.

Luna, E. J., and Hitt, A. L. (1992). Cytoskeleton–plasma membrane interaction. *Science* **258,** 955–964.

Manes, M. E., and Barbieri, F. D. (1977). On the possibility of sperm aster involvement in dorso-ventral polarity and pronuclear migration in the amphibian egg. *J. Embryol. Exp. Morphol.* **40,** 187–197.

Maro, B., Johnson, M. H., Pickering, S. J., and Flach, G. (1984). Changes in actin distribution during fertilization of the mouse egg. *J. Embryol. Exp. Morphol.* **8,** 211–237.

Merriam, R. W., and Clark, T. G. (1978). Actin in *Xenopus* oocytes. II. Intracellular distribution and polymerizability. *J. Cell Biol.* **77,** 439–447.

Miller, A. L., Karplus, E., and Jaffe, L. F. (1994). Imaging $[Ca^{2+}]_i$ with aequorin using a photon imaging detector. *In* "Methods in Cell Biology" (R. Nuccitelli, ed.), Vol. II, pp. 305–338. Academic Press, San Diego.

Nelson, W. J., and Hammerton, R. W. (1989). A membrane–cytoskeletal complex containing Na^+, K^+-ATPase, ankyrin, and fodrin in Madin–Darby canine kidney (MDCK) cells: Implications for the biogenesis of epithelial cell polarity. *J. Cell Biol.* **108,** 893–903.

Nieto, M. A., Bennett, M. F., Sargent, M. G., and Wilkinson, D. G. (1992). Cloning and developmental expression of *Sna,* a murine homologue of the *Drosophila snail* gene. *Development* **116,** 227–237.

Nislow, C., Lombillo, V. A., Kuriyama, R., and McIntosh, J. R. (1992). A plus-end directed motor enzyme that moves antiparallel microtubules *in vitro* localizes to the interzone of mitotic spindles. *Nature* **359,** 543–547.

Ohmori, H., Toyama, S., and Toyama, S. (1992). Direct proof that the primary site of action of cytochalasin on cell motility processes is actin. *J. Cell Biol.* **116,** 933–941.

Ohta, T., and Iwamatsu, T. (1983). Electron microscopic observations on sperm entry into eggs of the rose bitterling, *Rhodeus ocellatus. J. Exp. Zool.* **227,** 109–119.

Oppenheimer, J. M. (1947). Organization of the teleost blastoderm. *Q. Rev. Biol.* **22,** 105–118.

Pasdar, M., Li, Z., and Krzeminski, K. A. (1992). Desmosome assembly in MDCK epithelial cells does not require the presence of functional microtubules. *Cell Motil. Cytoskeleton* **23,** 201–212.

Perrin, D., Möller, K., Hanke, K., and Söling, H.-D. (1992). cAMP and Ca^{2+}-mediated secretion in parotid acinar cells is associated with reversible changes in the organization of the cytoskeleton. *J. Cell Biol.* **116,** 127–134.

Powers, D. A. (1989). Fish as model systems. *Science* **246,** 352–358.

Roosen-Runge, E. C. (1938). On the early development—Bipolar differentiation and cleavage ofthe zebra fish, *Brachydanio rerio. Biol. Bull.* **75,** 119–133.

Rudnick, D. (1955). Teleosts and birds. *In* "Analysis of Development" (B. H. Willier, P. A. Weiss, and V. Hamburger, eds.), pp. 297–314. Saunders, Philadelphia.

Sakai, Y. T. (1964). Studies on the ooplasmic segregation in the egg of the fish *Oryzias latipes.* I. Ooplasmic segregation in egg fragments. *Embryologia* **8,** 129–134.

Sakai, Y. T. (1965). Studies on ooplasmic segregation in the egg of the fish *Oryzias latipes.* III. Analysis of the movement of oil droplets during the process of ooplasmic segregation. *Biol. Bull.* **129,** 189–198.

Salisbury, J. L., Condeelis, J. S., and Satir, P. (1980). Role of coated vesicles, microfilaments, and calmodulin in receptor-mediated endocytosis by cultured B lymphoblastoid cells. *J. Cell Biol.* **87,** 132–141.

Schalkoff, M. E., and Hart, N. H. (1986). Effects of A23187 upon cortical granule exocytosis in eggs of *Brachydanio. Roux's Arch. Dev. Biol.* **195,** 39–48.

Schatten, H., and Schatten, G. (1980). Surface activity at the egg plasma membrane during sperm incorporation and its cytochalasin B sensitivity. *Dev. Biol.* **78,** 435–449.

Schatten, G., Schatten, H., Bestor, T. H., and Balczon, R. (1982). Taxol inhibits the nuclear movements during fertilization and induces asters in unfertilized sea urchin eggs. *J. Cell Biol.* **94,** 455–465.

Schatten, G., Schatten, H., Spector, I., Cline, C., Paweletz, N., and Simerly, C. (1986). Latrunculin inhibits the microfilament-mediated processes during fertilization, cleavage and early development in sea urchin and mice. *Exp. Cell Res.* **166,** 191–208.

Smith, D. E., Franco Del Amo, F., and Gridley, T. (1992). Isolation of *sna,* a mouse gene homologous to the *Drosophila* genes *snail* and *escargot:* Its expression pattern suggests multiple roles during implantation. *Development* **116,** 1033–1039.

Solnica-Krezel, L., and Driever, W. (1994). Microtubule arrays of the zebrafish yolk cell: Organization and function during epiboly. *Development* **120,** 2443–2455.

Speksnijder, J. E., Miller, A. L. Weisenseel, M. H., Chen, T.-H., and Jaffe, L. F. (1989). Calcium buffer injections block fucoid egg development by facilitating calcium diffusion. *Proc. Natl. Acad. Sci. USA* **86,** 6607–6611.

Spudich, A. (1992). Actin organization in the sea urchin egg cortex. *In:* "Current Topics in Developmental Biology" (E. L. Bearer, ed.), Vol. 26, pp. 9–21. Academic Press, San Diego.

Spudich, A., Wrenn, J. T., and Wessells, N. K. (1988). Unfertilized sea urchin eggs contain a discrete cortical shell of actin that is subdivided into two organizational states. *Cell Motil. Cytoskeleton* **9**, 85–96.

Stone, V. J., and Fluck, R. A. (1995). Lateral contraction wave following activation of medaka fish eggs *Oryzias latipes* with calcium ionophore A-23187. *J. Pa. Acad. Sci.* **68**, 194–195.

Stossel, T. P. (1989). From signal to pseudopod: How cells control cytoplasmic actin assembly. *J. Biol. Chem.* **264**, 18261–18264.

Strähle, U., and Jesuthasan, S. (1993). Ultraviolet irradiation impairs epiboly in zebrafish embryos: Evidence for a microtubule-dependent mechanism of epiboly. *Development* **119**, 909–919.

Stuart, G. W., Vielkind, J. R., McMurray, J. V., and Westerfield, M. (1990). Stable lines of transgenic zebrafish exhibit reproducible patterns of transgene expression. *Development* **109**, 577–584.

Taylor, D. L., and Fechheimer, M. (1982). Cytoplasmic structure and contractility: The solation–contraction coupling hypothesis. *Phil. Trans. R. Soc. London. B* **299**, 185–197.

Thisse, C., Thisse, B., Schilling, T. F., and Postlethwait, J. H. (1993). Structure of the zebrafish *snail1* gene and its expression in wild-type, *spadetail* and *no tail* mutant embryos. *Development* **119**, 1203–1215.

Trinkaus, J. P. (1951). A study of the mechanism of epiboly in the egg of *Fundulus heteroclitus*. *J. Exp. Zool.* **118**, 269–320.

Trinkaus, J. P. (1984). Mechanism of *Fundulus* epiboly: A current view. *Am. Zool.* **24**, 673–688.

Trinkaus, J. P. (1992). The midblastula transition, the YSL transition and the onset of gastrulation in *Fundulus*. *Development (Suppl. 1992)*, 75–80.

Trinkaus, J. P. (1993). The yolk syncytial layer of *Fundulus:* Its origin and history and its significance for early embryogenesis. *J. Exp. Zool.* **265**, 258–284.

Trinkaus, J. P. (1994). Ingression of deep cells of the marginal region of the blastoderm during early gastrulation of *Fundulus heteroclitus*. *Zebrafish Sci. Monitor* **3**, issue 3, 3–5.

Trinkaus, J. P., and Erickson, C. A. (1983). Protrusive activity, mode and rate of locomotion, and pattern of adhesion of *Fundulus* deep cells during gastrulation. *J. Exp. Zool.* **228**, 41–70.

Trinkaus, J. P., Trinkaus, M., and Fink, R. D. (1992). On the convergent cell movements of gastrulation in *Fundulus*. *J. Exp. Zool.* **261**, 40–61.

Vasiliev, J. M. (1991). Polarization of pseudopodial activities: Cytoskeletal mechanisms. *J. Cell Sci.* **98**, 1–4.

Vitale, M. L., Rodríguez Del Costillo, A., Tchakarov, L., and Trifarô, J.-M. (1991). Cortical filamentous actin disassembly and scinderin redistribution during chromaffin cell stimulation precede exocytosis, a phenomenon not exhibited by gelsolin. *J. Cell Biol.* **113**, 1057–1067.

von Dassow, G., and Schubiger, G. (1994). How an actin network might cause fountain streaming and nuclear migration in the syncytial *Drosophila* embryo. *J. Cell Biol.* **127**, 1637–1653.

Wang, F.-S., and Bonder, E. M. (1991). Sea urchin egg villin: Identification of villin in a nonepithelial cell from an invertebrate species. *J. Cell Sci.* **100**, 61–71.

Warga, R. M., and Kimmel, C. B. (1990). Cell movements during epiboly and gastrulation in zebrafish. *Development* **108**, 569–580.

Webb, T. A., and Fluck, R. A. (1995). The spatiotemporal pattern of microtubules in parthenogenetically activated medaka fish eggs *Oryzias latipes*. *J. Pa. Acad. Sci.* **68**, 197.

Webb, T. A. Kowalski, W. J., and Fluck, R. A. (1995). Microtubule-based movements during ooplasmic segregation in the medaka fish egg *Oryzias latipes*. *Biol. Bull.* **188**, 146–156.

Webster, S. D., and McGaughey, R. W. (1990). The cortical cytoskeleton and its role in sperm penetration of the mammalian egg. *Dev. Biol.* **142**, 61–74.

Wolenski, J. S., and Hart, N. H. (1987). Scanning electron microscope studies of sperm incorporation into the zebrafish (*Brachydanio*) egg. *J. Exp. Zool.* **243**, 259–273.

Wolenski, J. S., and Hart, N. H. (1988a). Effects of cytochalasins B and D on the fertilization of zebrafish (*Brachydanio*) eggs. *J. Exp. Zool.* **246**, 202–215.

Wolenski, J. S., and Hart, N. H. (1988b). Sperm incorporation independent of fertilization cone formation in the danio egg. *Dev. Growth Differ.* **30**, 619–628.

Wood, A., and Timmermans, L. P. M. (1988). Teleost epiboly: A reassessment of deep cell movement in the germ ring. *Development* **102**, 575–585.

Yamamoto, T.-O. (1961). Physiology of fertilization in fish eggs. *Int. Rev. Cytol.* **12**, 361–405.

Yamamoto, T.-O. (1975). Rhythmical contractile movements. *In* "Medaka (Killifish): Biology and Strains," pp. 59–72. Keigaku, Tokyo.

Yonemura, S., and Kinoshita, S. (1986). Actin filament organization in the sand dollar egg cortex. *Dev. Biol.* **115**, 171–183.

12

Confocal Immunofluorescence Microscopy of Microtubules, Microtubule-Associated Proteins, and Microtubule-Organizing Centers during Amphibian Oogenesis and Early Development

David L. Gard, Byeong Jik Cha, and Marianne M. Schroeder
Department of Biology
University of Utah
Salt Lake City, Utah 84112

I. Introduction

A growing body of evidence suggests that the microtubule (MT) cytoskeleton of amphibian oocytes, eggs, and early embryos participates in the estab-

Current Topics in Developmental Biology, Vol. 31

383

lishment and maintenance of two important developmental axes in amphibians: (1) the animal–vegetal (A–V) axis of oocytes and eggs, and (2) the dorsal–ventral (D–V) axis of developing embryos (for recent reviews, see Bement *et al.*, 1992; Houliston and Elinson, 1992; Gard, 1995). Only a few years ago, however, the existence of MTs in amphibian oocytes was a topic of some controversy. Biochemical evidence indicated that stage VI *Xenopus laevis* oocytes (stages according to Dumont, 1972) and eggs contained large cytoplasmic pools of α- and β-tubulins (TBs), equivalent to 0.1–1.0 μg per oocyte or egg (Elinson, 1985; Gard and Kirschner, 1987a; Heidemann and Kirschner, 1975). Furthermore, fractionation of oocyte cytoplasm revealed that 15–20% of the TBs in *Xenopus* oocytes was in polymeric form (Jessus *et al.*, 1987; D. L. Gard, unpublished observations), suggesting that a single stage VI oocyte contained more than 300 *meters* of MT polymer (Gard, 1991a). However, initial studies using electron microscopy and immunocytochemistry revealed few MTs in stage VI *Xenopus* oocytes (Heidemann and Gallas, 1980; Heidemann *et al.*, 1985; Palecek *et al.*, 1985). Additionally, reports from several laboratories concluded that *Xenopus* oocytes contain factors which inhibit MT assembly *in vitro* (Jessus *et al.*, 1984; Gard and Kirschner, 1987a). Finally, in contrast to the results obtained with activated *Xenopus* eggs, microinjection of centrosomes into stage VI oocytes [which lack functional centrioles and centrosomes (Heidemann and Kirschner, 1975, 1978; Gerhart, 1980, Gard, 1991a; see below)] did not induce the formation of MT asters (Heidemann and Kirschner, 1975, 1978; Karsenti *et al.*, 1984). Together, these observations led us to suggest that, despite their large cytoplasmic pool of tubulin subunits, *Xenopus* oocytes were in some way deficient in MT assembly and lacked substantial numbers of cytoplasmic MTs (Gard and Kirschner, 1987a).

During the past 5 years, our picture of the MT cytoskeleton of amphibian oocytes has changed dramatically, due in large part to refinements in the techniques for preserving and visualizing the oocyte cytoskeleton (for more complete discussions of these methods, see Klymkowsky and Hanken, 1991; Gard, 1993b; Gard and Kropf, 1993). Individual MTs were first identified in the cortex of *Xenopus* oocytes in samples prepared by rapid freezing (Huchon *et al.*, 1988). Subsequently, whole-mount immunocytochemistry and confocal immunofluorescence microscopy has revealed an extensive network of MTs extending throughout the cytoplasm of oocytes from both *X. laevis* and *Rana pipiens* (Dent and Klymkowsky, 1989; Yisraeli *et al.*, 1990; Gard, 1991a, 1995; Wang *et al.*, 1993; D. L. Gard, unpublished observations). In addition, these techniques have revealed that dramatic changes in MT assembly, organization, and dynamics accompany distinct stages in oocyte differentiation, maturation, and early embryonic development (Gard, 1991a,b, 1992; 1993a; Houliston and Elinson, 1991a; Schroeder and Gard, 1992; Gard *et al.*, 1995a).

In the following discussion, we present a view of the assembly and organization of the MT cytoskeleton during oogenesis and early development in the African frog, *X. laevis,* that is based on results obtained in our laboratory as well as in laboratories around the world. Many of the results and images we present have been published previously (Gard, 1991a, 1992, 1993a, 1994; Schroeder and Gard, 1992; Roeder and Gard, 1994; Gard *et al.,* 1995a,b), and we refer interested readers to those reports for more information. Other results have never been published or have been presented only in abstract form (Gard, 1991b) and are included here for the sake of completeness. Results obtained from other amphibian species will be discussed when informative. In addition, we address some of the mechanisms that might serve to regulate MT organization and dynamics during amphibian oogenesis and early development, focusing on the potential contributions of microtubule organizing centers (MTOCs) and microtubule-associated proteins (MAPs).

II. MT Organization during Amphibian Oogenesis and Early Development

The differentiation of *Xenopus* oocytes begins after the last of four oogonial divisions, which gives rise to a cluster or nest of 16 postmitotic oocytes (Al-Mukhtar and Webb, 1971; Coggins, 1973) and ends with the formation of a fully grown, maturation-competent, stage VI oocyte (summarized in Fig. 1) (oocyte stages are according to Dumont, 1972). Although the definitive A–V axis is not apparent until stages IV–VI (Fig. 1E) (Dumont, 1972), light and electron microscopy reveals that early postmitotic *Xenopus* oocytes are highly polarized along an axis established during their final oogonial division (Fig. 1A) (Al-Mukhtar and Webb, 1971; Coggins, 1973; Gard *et al.,* 1995a; Heasman *et al.,* 1984). This "initial" axis is evident in many structural features of the differentiating oocytes. Small postmitotic oocytes exhibit a pear-shaped morphology, with the germinal vesicle (GV, the oocyte nucleus) located toward the broad, distal end of the oocyte. Many of the cytoplasmic organelles, including numerous mitochondria, cluster into a compact perinuclear cap in the narrow, proximal end of the cell. The maternal centrioles are also found in this cytoplasmic cap, as is a dense, compact array of cytoplasmic MTs (see below). Finally, the synaptonemal complexes of the meiotic chromosomes are organized into a characteristic "bouquet" that is directly apposed to the cytoplasmic cap of organelles. This obvious polarization has led to considerable speculation that the initial axis of *Xenopus* oocytes is a functional precursor of the A–V axis that is formed later in oogenesis (Coggins, 1973; Heasman *et al.,* 1984; Gerhart *et al.,* 1983, 1986). However, results from our recent examination of the

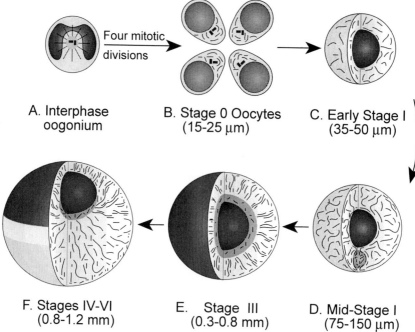

A. Interphase oogonium

Four mitotic divisions

B. Stage 0 Oocytes (15-25 μm)

C. Early Stage I (35-50 μm)

F. Stages IV-VI (0.8-1.2 mm)

E.   Stage III (0.3-0.8 mm)

D. Mid-Stage I (75-150 μm)

Fig. 1 A summary of MT organization during the differentiation of *Xenopus* oocytes. (A) Interphase oogonia contain a sparse array of cytoplasmic MTs radiating from the maternal centrosome, which is commonly located adjacent to the lobate nucleus. (B) Completion of the last of four mitotic divisions results in formation of a nest of 16 stage 0 (postmitotic) oocytes (only 4 of which are shown). The maternal centrosomes and most MTs are located in a cap of cytoplasm at the narrow end of stage 0 oocytes. (C) During early stage I, oocytes contain a dispersed MT array with no obvious MTOC. (D) Larger stage I oocytes contain a complex MT array, with concentrations of MTs in the cortex, perinuclear cytoplasm, and surrounding the mitochondrial mass. (E) MTs radiate from a yolk-free region of cytoplasm surrounding the GV of stage III oocytes. (F) During stages IV–VI, the oocyte MT array becomes polarized along the A–V axis (see text; late stage IV oocyte shown) (Al-Mukhtar and Webb, 1971; Coggins, 1973; Heasman *et al.*, 1984; Gard, 1991; Gard *et al.*, 1995).

organization of MTs and F-actin during early oogenesis have led us to question the relationship between the initial polarity of postmitotic amphibian oocytes and the A–V axis, as well as the roles of the MT and actin cytoskeletons in axis specification (Roeder and Gard, 1994; Gard *et al.*, 1995a; Gard, 1995).

A. The MT Cytoskeleton of *Xenopus* Oocytes Is Reorganized during Early Diplotene

Confocal immunofluorescence microscopy revealed that the first of several changes in MT organization, dynamics, and nucleation that accompany the differentiation and maturation of *Xenopus* oocytes coincides with the final oogonial division (Fig. 2A) (Gard *et al.,* 1995a). The highly polarized morphology of postmitotic oocytes sharply contrasts with the rounded, relatively unpolarized morphology of the oogonia from which they are derived (Al-Mukhtar and Webb, 1971; Coggins, 1973; Gard *et al.,* 1995a). Interphase oogonia, which are readily recognized due to their multilobed nuclei (Al-Mukhtar and Webb, 1971; Coggins, 1973), contain a sparse array of cytoplasmic MTs radiating from a perinuclear centrosome (Fig. 2B) (Gard *et al.,* 1995a). In contrast, most cytoplasmic MTs, including a substantial population containing acetylated α-TB (Fig. 2C), and the maternal centrosome (Fig. 2D) are located in a cap of cytoplasm found at the narrow (proximal) end of pear-shaped, postmitotic oocytes (Al-Mukhtar and Webb, 1971; Heasman *et al.,* 1984; Gard *et al.,* 1995). Acetylation of α-TB is a common characteristic of nondynamic, or stable, MTs (Schulze *et al.,* 1987; Piperno *et al.,* 1987; Webster and Borisy, 1989). The abundance of acetylated MTs throughout the postmitotic differentiation of *Xenopus* oocytes contrasts with the relative lack of acetylated MTs in both interphase and mitotic oogonia (Gard, 1991a; Gard *et al.,* 1995a), suggesting that MT stabilization plays an important role during oocyte differentiation and polarization.

The next dramatic change in oocyte morphology and organization occurs as *Xenopus* oocytes enter the prolonged diplotene stage of growth and differentiation (Al-Mukhtar and Webb, 1971; Coggins, 1973; Gard *et al.,* 1995a). During early diplotene, evidence of the initial axis disappears as oocytes adopt a more symmetrical, rounded morphology (Fig. 1C). The GV moves to a more central location in the cytoplasm, and the bouquet of recombinant chromosomes is replaced by the characteristic lampbrush organization of transcriptionally active chromosomes (Al-Mukhtar and Webb, 1971; Coggins, 1973). The perinuclear cap of mitochondria and other cytoplasmic organelles disperses throughout the cytoplasm (Heasman *et al.,* 1984; Gard *et al.,* 1995a). Finally, the MT and actin-based cytoskeletons both undergo substantial reorganization (Roeder and Gard, 1994; Gard *et al.,* 1995a) (see below). To signify the global changes in cytoplasmic and nuclear organization that accompany entry into the diplotene stage of meiotic prophase, we have recently suggested that an additional stage be added to the nomenclature devised by Dumont (1972) to describe oogenesis in *Xenopus.* This stage, which we have referred to as stage 0 (Gard *et*

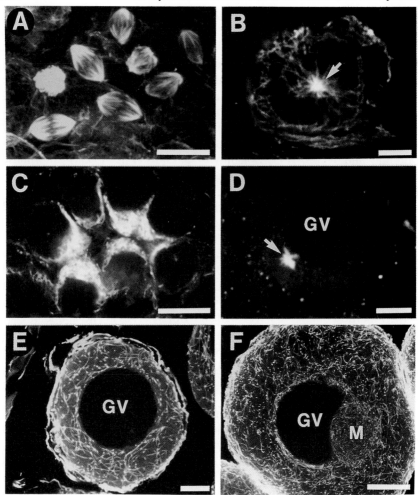

Fig. 2 MT organization in *Xenopus* oogonia and early postmitotic oocytes (stages 0 and I).
(A) A nest of oogonia undergoing their final mitotic division prior to differentiation as
oocytes. Eight spindles are apparent in this projection of 22 optical sections (B) An interphase
oogonium, exhibiting a sparse cytoplasmic MT array radiating from a prominent MTOC
(arrow). (C) A nest of stage 0 oocytes (see text) stained with antibodies specific for acetylated
α-TB. Note the brightly stained network of acetylated MTs in the cytoplasmic cap of each
oocyte. (D) Antibodies specific for *Xenopus* γ-TB reveal the maternal centrosome (arrow)
in this stage 0 oocyte. (E) MTs are dispersed throughout the cytoplasm of this oocyte in early
stage I (stained with anti-α-TB). (F) A single optical section reveals a complex network of
cytoplasmic MTs in this mid-late stage I oocyte. MTs are concentrated in the cortex, surrounding the GV, and in the mitochondrial mass (M). GV in D–F denotes the germinal vesicle
(the oocyte nucleus). Scale bars = 25 μm in A and F, 5 μm in B and D, and 10 μm in C and
E. A, C, and F are reprinted from Gard *et al.* (1995) with permission of Academic Press.

al., 1995a), would include small (12–25 μm), highly polarized, postmitotic oocytes in the preleptotene through zygote stages of meiotic prophase. Stage I would then refer to larger (>35 μm diameter) relatively unpolarized oocytes in the diplotene stage of meiotic prophase.

The transition from stage 0 to stage I of oogenesis is accompanied by the dispersal of the densely packed MT array of stage 0 oocytes. Microtubules, including a subpopulation containing acetylated α-TB, are distributed throughout the cytoplasm of early stage I oocytes with diameters between 35 and 50 μm (Fig. 2E) (Gard *et al.*, 1995a). Although concentrations of MTs are often observed in the cortex and surrounding the GV of early stage I oocytes, there is no evidence of a discrete MTOC or indication that the MT distribution in early stage I oocytes is polarized. In fact, examination of MT regrowth following cold-induced disassembly indicates that the maternal centrosome is inactivated very early in stage I of oogenesis (Gard, 1991a; Gard *et al.*, 1995). While the mechanisms underlying the observed reorganization of cytoplasmic MTs during early stage I remain unknown, the presence of substantial populations of acetylated MTs throughout stages 0 and I (Gard, 1991a; Gard *et al.*, 1995a) suggests that reorganization results, at least in part, from the redistribution of preexisting MTs.

Individual optical sections (Fig. 2F) revealed a multitude of MTs coursing throughout the cytoplasm of oocytes during mid-stage I (75–150 μm in diameter) (Gard, 1991a; Gard *et al.*, 1995a). Concentrations of MTs were apparent in the oocyte cortex and surrounding the GV, and MTs were observed to surround and penetrate the mitochondrial mass characteristic of stage I *Xenopus* oocytes (Gard, 1991a; Gard *et al.*, 1995a; Heasman *et al.*, 1984; Tourte *et al.*, 1991). The density, complexity, and disorder of this MT network can best be appreciated in the stereo pair shown in Fig. 3A, which was constructed from more than 50 serial optical sections obtained by confocal immunofluorescence microscopy. While little order is apparent in the overall distribution of MTs in stage I oocytes, bundles of acetylated MTs are often observed linking the GV, the mitochondrial mass, and the oocyte cortex (shown in stereo in Fig. 3B) (Gard, 1991a; Gard *et al.*, 1995a), suggesting that stable MTs provide a framework for positioning the GV and other organelles within the cytoplasm (see below). This conclusion is supported further by the observation that treatment of stage I oocytes with desacetylcolchicine, which disrupts cytoplasmic MTs, releases the mitochondrial mass and GV to move freely in the cytoplasm (Wylie *et al.*, 1985). Interestingly, neither MT inhibitors nor inhibitors of actin assembly disrupt the structure of the mitochondrial mass itself (Wylie *et al.*, 1985; Gard, 1991a; Gard *et al.*, 1995a; Roeder and Gard, 1994). The mechanism underlying the aggregation and organization of mitochondria into this novel feature of stage I *Xenopus* oocytes thus remains uncertain.

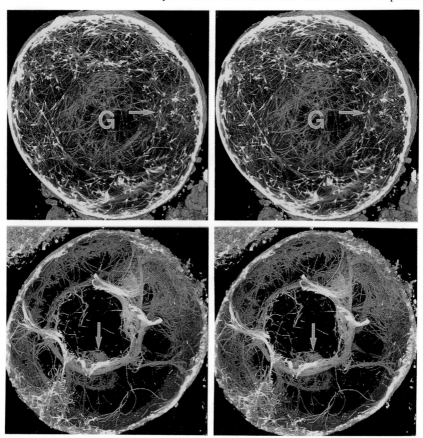

Fig. 3 Stage I oocytes contain a complex array of MTs. (Upper stereo pair) Antibodies to α-TB reveal a dense, three-dimensional array of MTs filling the cytoplasm of this stage I oocyte (85 μm in diameter). The GV is located near the oocyte center. The mitochondrial mass (arrows) is situated between the GV and cortex, toward the right side of this view. (Lower stereo pair) Antibodies to acetylated α-TB reveal a complex network of MT bundles linking the GV to the oocyte cortex of this stage I oocyte (105 μm in diameter). The mitochondrial mass (arrows) is toward the back in this view. Each stereo pair was rendered from more than 50 contiguous optical sections (out of more than 130 per oocyte), using VoxBlast software (VayTek, Fairfield, IA) running on a Pentium (Intel) personal computer.

Despite the observed association of MTs with the mitochondrial mass during mid–late stage I, there is little or no evidence to indicate that the mitochondrial mass plays a unique or singular role in regulating MT assembly or distribution during early amphibian oogenesis. Both MTs and actin cables are observed to surround clusters of mitochondria prior to their aggregation into a compact mitochondrial mass (Gard, 1991a; Gard *et al.*,

1995a; Roeder and Gard, 1994). Some stage I oocytes contain multiple mitochondrial masses, each of which is surrounded by MTs and actin cables (D. L. Gard, unpublished observations, Gard *et al.,* 1995a; Roeder and Gard, 1994). In addition, the mitochondrial mass does not serve as a preferred site for MT nucleation during the recovery of stage I oocytes from cold-induced MT disassembly (Gard, 1991a; Gard *et al.,* 1995a). In fact, the capacity for MT nucleation and regrowth is lost very early in stage I, soon after entry into the diplotene stage of meiotic prophase and prior to formation of the mitochondrial mass. Finally, previtellogenic oocytes from *R. pipiens* and *Ambystoma sp.,* which lack mitochondrial masses, contain MT arrays comparable in all other respects to those of stage I *Xenopus* oocytes (Wang *et al.,* 1993; D. L. Gard, unpublished observations).

The mitochondrial mass characteristic of stage I *Xenopus* oocytes disperses during stage II (Dumont, 1972; Heasman *et al.,* 1984; Hausen and Riebesell, 1990), and components of the mitochondrial mass, including mitochondria, germ plasm, and maternal mRNAs, accumulate in the subcortical cytoplasm of what will become the vegetal hemisphere (Heasman *et al.,* 1984; Hausen and Riebesell, 1990; King *et al.,* 1990; Zhou *et al.,* 1993). The observation of numerous MTs (including stable MTs) surrounding and linking the mitochondrial mass to the cortex of stage I oocytes (Heasman *et al.,* 1984; Gard, 1991a; Tourte *et al.,* 1992; Gard *et al.,* 1995a) makes it tempting to speculate that dispersal of mitochondria and germ plasm to the cortex during stage II of oogenesis is mediated by directed transport along MTs, perhaps mediated by MT-dependent motor proteins (for review, see Skoufias and Scholey, 1993). However, the evidence for MTs playing such an active role is purely circumstantial, and the physiological significance of the association between MTs and the mitochondrial mass of stage I *Xenopus* oocytes remains uncertain.

Stages II and III of oogenesis in *Xenopus* are characterized by the onset of vitellogenesis and the accumulation of pigment in the oocyte cortex (Dumont, 1972), features which render oocytes in all subsequent stages of differentiation opaque. During stage II, *Xenopus* oocytes contain a complex, poorly ordered array of cytoplasmic MTs which is similar in most respects to that seen during stage I (Fig. 4A) (Gard, 1991a). However, the organization of MTs in stage III oocytes differs significantly from the disordered arrays observed during stages I and II (Gard, 1991a). Stage III oocytes contain a loosely organized network of individual MTs and MT bundles radiating from a yolk-free shell of cytoplasm surrounding the centrally located GV (Figs. 1E and 4B). MTs are most commonly observed in yolk-free cytoplasmic channels, or "radii" that extend from the perinuclear region to the oocyte cortex, and are also apparent in the perinuclear shell (Gard, 1991a). The roughly radial organization of MTs emanating from the perinuclear region of stage III oocytes (and all subsequent stages of

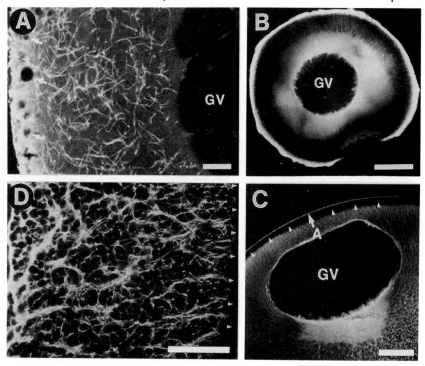

Fig. 4 MT organization during stages II–VI of oogenesis. (A) Stage II oocytes contain a poorly ordered MT array similar to that observed during stage I. (B) During stage III, MTs radiate from a yolk-free shell of cytoplasm surrounding the GV. (C) Loosely organized MT bundles are observed in yolk-free radii that extend from the GV to the animal cortex. Arrowheads indicate the animal surface. (D) Individual MTs and MT bundles extend from the surface of the GV (left edge) to the animal cortex (arrowheads, right edge) of this stage VI oocyte. Scale bars = 10 μm in A, 100 μm in B and C, and 25 μm in D. All figures are reprinted from Gard (1991) with the permission of Academic Press.

oocyte differentiation) suggests that factors associated with the GV, or otherwise localized in the perinuclear cytoplasm, serve as MTOCs during the later stages of oogenesis (Gard, 1991a) (see below).

B. The MT Array of *Xenopus* Oocytes Is Polarized during Formation of the A–V Axis

The MT network of stage III oocytes exhibits little or no evidence of asymmetry or cytoplasmic polarity (Gard, 1991a), consistent with lack of outwardly visible polarity during stage III of oogenesis (Dumont, 1972). The A–V axis of *Xenopus* oocytes first becomes outwardly apparent during stage IV of oogenesis, due to the asymmetric distribution of pigment granules between the animal and vegetal cortex. Many other features of the oocyte become polarized along the A–V axis during stages IV–VI of oogenesis: the GV moves from the center of the oocytes to its final position near the animal pole (Figs. 1F and 4C) (Dumont, 1972), a gradient in the size and distribution of yolk platelets is established through the differential growth and transport of yolk platelets in the vegetal hemisphere (Danilchik and Gerhart, 1987), and specific maternal mRNAs become unequally distributed in the animal and vegetal hemispheres (Capco and Jeffery, 1982; Carpenter and Klein, 1982; King and Barkliss, 1985; Weeks and Melton, 1987; Melton, 1987; Yisraeli et al., 1990; Gururajan et al., 1991; King et al., 1993).

Confocal immunofluorescence microscopy revealed that the MT cytoskeleton of Xenopus oocytes also becomes progressively more polarized during stages IV–VI of oogenesis (Gard, 1991a). In the animal hemisphere, loosely organized bundles of MTs, including a substantial population of acetylated MTs, link the GV to the animal cortex during stages IV–VI (Fig. 4D) (Gard, 1991a). Many of these MTs are concentrated in the yolk-free radii that extend from the GV to the cortex. Interestingly, cables of F-actin and cytokeratin filaments are also concentrated in these yolk-free radii (Roeder and Gard, 1994; Godsave et al., 1984), suggesting that the cytoskeletal networks composed of actin, intermediate filaments, and MTs may be physically linked. Organized bundles of MTs are much less numerous, and radial organization is less apparent, in the vegetal cytoplasm. Individual MTs form a complex network as they course between the large yolk platelets found in the vegetal hemisphere. Acetylated MTs are also less numerous in the vegetal cytoplasm of Xenopus oocytes during stages IV–VI, although a substantial network of acetylated MTs is observed in the vegetal cortex (Gard, 1991a, 1994).

There is little direct evidence to suggest that MTs plays a major role in specification of the A–V axis (for a more extensive discussion, see Gard, 1995). However, substantial evidence indicates that MTs (and F-actin) are required for formation and maintenance of this important developmental axis. Treatment with nocodazole or cytochalasins (which inhibit the assembly of MTs and F-actin, respectively) during stages III–VI slows oocyte growth (Yisraeli et al., 1990), indicating that MTs and F-actin are required for endocytosis and intracellular transport of yolk precursors and yolk platelets. Indeed, the radial organization of MTs observed in stage III oocytes (and in the animal hemisphere during stages IV–VI) is consistent with the hypothesis that MTs participate in the directed translocation of yolk precursors or platelets inward from the oocyte cortex (Danilchik and Gerhart, 1987). However, it is harder to imagine how the large yolk platelets found in the vegetal hemisphere during the later stages of axis formation could be actively transported along the tangled network of vegetal MTs.

The large number of acetylated (presumably stable) MTs linking the GV to the animal cortex throughout stages I–VI of oogenesis suggests that MTs play an important role in the positioning of the oocyte nucleus. This conclusion is further supported by observations that disruption of cytoplasmic MTs results in displacement of the GV from its normal location in the animal hemisphere of stage VI *Xenopus* oocytes (Gard, 1991a, 1993a) or fully grown *Rana* oocytes (Lessman, 1987; D. L. Gard, unpublished observations). Interestingly, extended treatment of stage VI *Xenopus* oocytes with cytochalasin B, which disrupts their actin cytoskeleton (Roeder and Gard, 1994), also disrupts their MT network (B.-J. Cha and D. L. Gard, unpublished observations). Thus, the MT cytoskeleton of *Xenopus* oocytes may be anchored to perinuclear and cortical networks of actin filaments.

Finally, MTs and F-actin are required for the localization of maternal mRNAs in the vegetal cortex of *Xenopus* oocytes. Disassembly of the MT network of *Xenopus* oocytes with nocodazole inhibits localization of Vg1 mRNA to the vegetal cortex, while cytochalasin releases previously localized Vg1 mRNA (Yisraeli *et al.,* 1990), suggesting that localization of maternal mRNAs to the vegetal cortex is a two-step process involving MT-dependent translocation and actin-dependent anchoring. Localization is dependent on the 3' untranslated region of the RNA (Mowry and Melton, 1992) and may be mediated by a specific binding protein (Schwartz *et al.,* 1992). However, the manner by which these signals specifically direct developmentally important RNAs to the vegetal cortex has not been determined.

C. The MT Cytoskeleton Is Remodeled during the Maturation of *Xenopus* Oocytes

One of the earliest indicators of germinal vesicle breakdown (GVBD) during the progesterone-induced maturation of *Xenopus* oocytes is the assembly of a novel MTOC and transient MT array (the MTOC–TMA complex) near the basal (vegetal) surface of the GV (Fig. 5a) (Brachet *et al.,* 1970; Huchon *et al.,* 1981; Jessus *et al.,* 1986; Gard, 1992). This MTOC–TMA complex then transports the condensed meiotic chromosomes to the animal pole (Fig. 5B), where it serves as the immediate precursor of the first meiotic spindle (Gard, 1992). Similar complexes have been observed in maturing oocytes from other amphibian species, including both anurans and urodeles (Beetschen and Gautier, 1989, and references therein), suggesting that this novel MTOC and MT array is a common feature of oocyte maturation in amphibians.

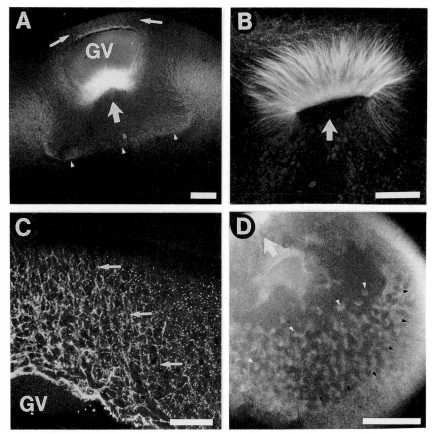

Fig. 5 The maturation of *Xenopus* oocytes is accompanied by the assembly of a novel MTOC and transient MT array. (A) GVBD is accompanied by the assembly of a novel MTOC and transient MT array (large arrow). The MTOC–TMA complex initially forms near the basal surface of the GV (position marked by small arrowheads). Remnants of the oocyte MT array (between arrows) maintain the relative position of the apical surface of the GV and the animal cortex. (B) The MTOC–TMS complex rapidly migrates to the animal pole. (C) Closer examination of the apical surface of the GV showing the remnant of oocyte MTs and the advancing front of MT disassembly (arrows). (D) Numerous small MT aggregates or asters from yolk-free islands in the vegetal hemisphere (white arrowheads), while a front of MT disassembly moves toward the vegetal cortex (black arrowheads). Scale bars = 100 μm in A, 50 μm in B, 25 μm in C, and 250 μm in D. Figure 5B is reprinted from Gard (1992) with the permission of Academic Press.

The mechanisms underlying the directed translocation of the MTOC–TMA complex to the animal pole (covering a distance of 400 μm in 20 min) remain uncertain. Recently, we found that cytochalasin B disrupts the structure of the MTOC–TMA complex, but not its movement toward the cortex (Gard *et al.*, 1995b). From these results, we concluded that F-actin plays an important role in the assembly or organization of the MTOC–TMA, but not it translocation to the cortex (however, see Ryobova *et al.*, 1986). The formation of ectopic meiotic spindles in the cortex or deeper cytoplasm of *Xenopus* oocytes after experimental displacement of their GVs indicates that spindle assembly can occur nearly anywhere in the egg, suggesting that there is no specific signal directing the MTOC–TMA complex to the animal pole (Gard, 1993a). Rather, these observations indicate that the MTOC–TMA complex migrates to the nearest region of cortex at the onset of GVBD. Formation of the meiotic spindles at the animal pole of *Xenopus* eggs is thus a direct consequence of the position of the GV in the animal hemisphere of stage VI oocytes.

Despite the absence of specific signals targeting the MTOC–TMA complex and spindle assembly to the animal pole, the cortex of *Xenopus* oocytes does appear to be functionally polarized along the A–V axis. Based on results from confocal microscopy, we concluded that anchoring and rotation was mediated by interactions between the oocyte cortex and astral MTs emanating from the spindle poles (Gard, 1992). Although they exhibit normal-appearing astral MTs, ectopic spindles assembled near the vegetal cortex often were unable to anchor to the cortex and subsequently failed to undergo spindle rotation (Gard, 1993a), suggesting that the animal and vegetal cortex differ in their ability to support anchoring and rotation of the meiotic spindles. Interestingly, anchoring and rotation of the meiotic spindles was also blocked by cytochalasin B (Gard *et al.*, 1995b), suggesting that spindle rotation is mediated by interactions between astral MTs and cortical F-actin.

As the MTOC–TMA complex migrates toward the animal pole, the extensive MT array of stage VI oocytes is rapidly disassembled (Gard, 1991b). Confocal microscopy of maturing *Xenopus* oocytes indicates that, contrary to popular belief, there is no significant migration of the GV toward the animal pole prior to the onset of GVBD (D. L. Gard, unpublished observations). As the MTOC–TMA complex migrates toward the animal pole, a substantial remnant of the MT array that linked the GV of stage VI oocytes to the animal cortex persists until the final stages of GVBD (Figs. 5A and 5C). This MT cap (visible between the arrows in Fig. 5A) maintains the position of the apical surface of the GV relative to the oocyte surface until GVBD is nearly complete. In the vegetal hemisphere, disassembly of the oocyte MT array proceeds outward from the cell center toward the cortex. In some oocytes, transient bundles of MTs are observed in the peripheral cytoplasm during the early stages of maturation (A. Friend and D. L. Gard, unpublished observations). In most oocytes, numerous small cytasters or MT aggregates appear in the vegetal cytoplasm as the oocyte MT array is disassembled (D. L. Gard, unpublished observations)

(Fig. 5D), forming conspicuous small islands of yolk-free cytoplasm. It is currently unclear whether these MT asters or aggregates represent remnants of the oocyte MT array or transient MT nucleation and *de novo* MT assembly.

During the period of MTOC–TMA migration and spindle assembly, an extensive network of cytoplasmic MTs is assembled in the animal hemisphere of maturing *Xenopus* oocytes (visible in Fig. 5A) (Gard, 1991b; D. L. Gard, unpublished observations). This MT array appears to be organized from a broad region near the center of the oocyte, roughly corresponding to the initial site of MTOC–TMA assembly. MTs extend outward from this region, ending near the animal cortex. The organization of MTs in the animal hemisphere of maturing *Xenopus* oocytes exhibits a remarkable similarity to the zones of yolk rearrangement described by Danilchik and Denegre (1991) (compare Fig. 5A with Fig. 1 in Danilchik and Denegre, 1991), suggesting that MT assembly is responsible for many of the cytoplasmic rearrangements observed during oocyte maturation.

Assembly of both meiotic spindles during the maturation of *Xenopus* oocytes follows a complex pathway (M1 is summarized in Fig. 6, and M2 is shown in Fig. 7) (Gard, 1992) that is characterized by four stages: (1) aggregation and compaction of MTs and chromosomes (Fig. 7A), (2) establishment of a bipolar spindle axis (Fig. 7B), (3) prometaphase elongation in an orientation parallel to the oocyte surface (Fig. 7C), and (4) rotation into alignment with the A–V axis (Fig. 7D). This pathway of spindle assembly thus differs significantly from that observed in early *Xenopus* embryos (Gard et al., 1990) or in mitotic extracts prepared from *Xenopus* eggs *in vitro* (Sawin and Mitchison, 1991). This difference may result from the acentriolar nature of the meiotic spindles in *Xenopus* oocytes (Huchon et al., 1981), which must then depend on other mechanisms for the establishment and organization of the spindle poles (see Rieder et al., 1993, for an interesting comparison of meiotic and mitotic spindle assembly). While the precise mechanisms have yet to be defined, numerous observations suggest that meiotic chromosomes play an important role in establishing the organization of acentriolar spindles during meiosis in *Xenopus* and other species (Karsenti et al., 1984; Church et al., 1986; Steffen et al., 1986; Sawin and Mitchison, 1991; Theurkauf and Hawley, 1992; Gard et al., 1995b) (see below).

Cytoplasmic MTs persist throughout meiosis in *Xenopus,* and an extensive MT array coexists with the metaphase-arrested spindle in unfertilized eggs of both *X. laevis* and *R. pipiens* (Fig. 8) (Jessus et al., 1988; Huchon et al., 1988; Gard, 1991b; D. L. Gard, unpublished observations). Unlike the MT arrays of prophase-arrested oocytes, few cytoplasmic MTs[1] in maturing oo-

---

[1]The second meiotic spindle of unfertilized *Xenopus* eggs stains weakly with antibodies to acetylated α-TB (Chu and Klymkowsky, 1989; D. L. Gard, unpublished observations).

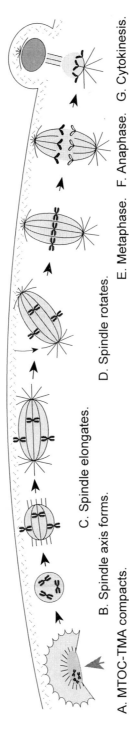

A. MTOC-TMA compacts.

B. Spindle axis forms.

C. Spindle elongates.

D. Spindle rotates.

E. Metaphase. F. Anaphase. G. Cytokinesis.

Fig. 6 A summary of the first meiotic division in *Xenopus* oocytes. Assembly of the first meiotic spindle consists of four stages: (1) Upon reaching the animal pole (A), the MTOC–TMA complex compacts to form a disordered aggregate of MTs and chromosomes; (2) the bipolar axis of the M1 spindle is formed (B); (3) the M1 spindle elongates parallel to the oocyte surface during prometaphase (C); and (4) the M1 spindle anchors to the oocyte cortex and rotates into alignment with the A–V axis (D). Following metaphase (E) and anaphase (F), the first polar body is extruded (G) (Gard, 1992).

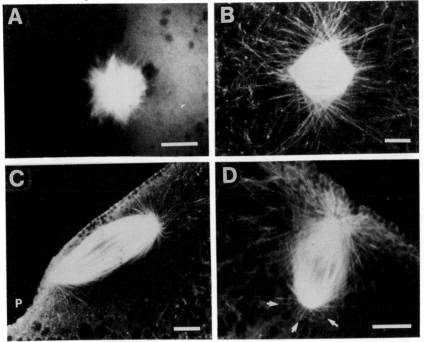

Fig. 7 Confocal microscopy of M2 spindle assembly. (A) Polar view of a microtubule aggregate preceding assembly of the M2 spindle, 80–90 min after formation of the maturation spot (WSF). (B) Polar view of a short M2 spindle oriented parallel to the oocyte surface (120–150 min after WSF). (C) Lateral view of an elongated M2 spindle oriented parallel to the oocyte surface (>180 min after WSF). Note the astral microtubules extending from both spindle poles. (D) Lateral view of an M2 spindle in an unfertilized egg. MT asters from the cortical pole appear to anchor the spindle to the cortex. All scale bars = 10 μm. Reprinted from Gard (1992) with the permission of Academic Press.

cytes or unfertilized eggs are acetylated (D. L. Gard, unpublished observations), indicating that a larger fraction of the individual MTs in maturing oocytes and unfertilized eggs are dynamic, relative to prophase-arrested oocytes. Despite the extensive network of cytoplasmic MTs present throughout maturation (Figs. 8A–8D), assembly of the meiotic spindles in both *Xenopus* and *Rana* oocytes occurs in regions of cytoplasm that are devoid of cytoplasmic MTs (the arrow in Fig. 8A indicates the M2 spindle) (Gard, 1992; D. L. Gard, unpublished observations). This observation underscores the precise spatial regulation of MT nucleation and assembly that is necessary during the maturation of amphibian oocytes.

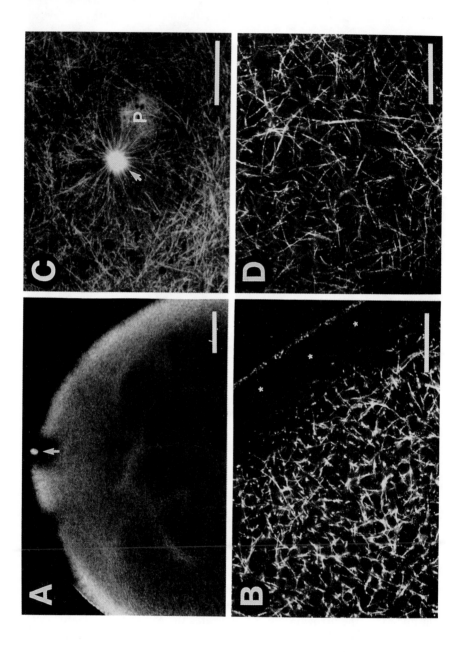

D. Specification of the D–V Axis of Frog Embryos Is Dependent on the Sperm Aster and Subcortical MTs

Fertilization or parthenogenetic activation releases the metaphase arrest of amphibian eggs, allowing completion of second meiotic division and formation of the second polar body, and initiating a significant reorganization of the MT cytoskeleton (summarized in Fig. 9). In *Xenopus* eggs, the fraction of TB in polymeric form (i.e., MTs) falls from 10–15% (in unfertilized eggs) to less than 1% within 10–15 min after fertilization or activation (Elinson, 1985). The MT polymer levels in fertilized eggs recover soon thereafter, as an extensive "sperm aster" is assembled in the animal hemisphere (Figs. 9B, 9C, and 10A) (Manes and Barbieri, 1977; Stewart-Savage and Grey, 1982; Elinson, 1985). This dramatic MT aster is organized by the paternally supplied centrioles and pericentriolar material recruited from a maternal pool of centrosomal proteins (Stearns and Kirschner, 1994; see below). The observation of comparable, though less focused, cytoplasmic asters ("activation asters") in parthenogenetically activated eggs indicates that the maternal pool of centrosomal components is capable of nucleating and organizing a MT aster in the absence of the paternal centrioles (Schroeder and Gard, 1992; Houliston and Elinson, 1991a; see below).

In addition to mediating pronuclear migration (Manes and Barbieri, 1977; Stewart-Savage and Grey, 1982), the sperm aster of fertilized frog eggs participates in specification of the D–V axis of the developing embryo. During the latter half of the first cell cycle, between 0.5 and 0.9 NT (normalized time, where fertilization = 0.0 NT and first cleavage = 1.0 NT), the egg cortex rotates 30° relative to the inner yolk mass (Figs. 9C and 9D) (Vincent *et al.*, 1986). In the absence of other factors, the direction of rotation is established by the location of the sperm entry point (SEP), such that the vegetal cortex moves away from the SEP (see Fig. 9) (Vincent *et al.*, 1986; Gerhart *et al.*, 1986). Inhibition or arrest of the cortical rotation by uv irradiation of the vegetal hemisphere results in embryos with defects in dorsoanterior structures (Malacinski *et al.*, 1975; Manes and Elinson, 1980; Manes *et al.*, 1978; Scharf and Gerhart, 1983), indicating that the

---

Fig. 8 A complex cytoplasmic MT array coexists with the meiotic spindles of unfertilized eggs. (A) A complex, poorly ordered MT array is present in the animal hemisphere of unfertilized *Xenopus* eggs. The second meiotic spindle (arrow) is located at the animal pole, in a small region devoid of cytoplasmic MTs. (B) MTs in the animal cytoplasm of an oocyte during M2 (asterisks denote cortical pigment). (C and D) MTs in the animal (C) and equatorial (D) cortex of an M2-arrested *Xenopus* egg. The arrow in C points to the M2 spindle (P denotes the position of the first polar body in adjacent sections). Scale bars = 100 μm in A, 25 μm in B and D, and 50 μm in C.

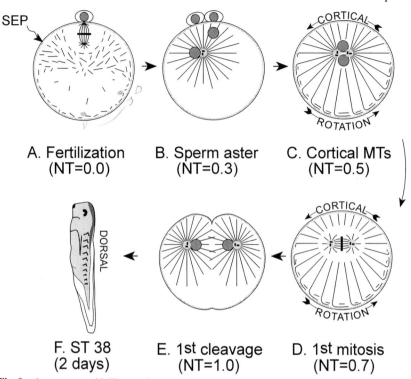

Fig. 9   A summary of MT organization during early development in *Xenopus* (not to scale). (A) Unfertilized eggs are arrested in second meiotic metaphase and contain both spindle and cytoplasmic MTs. (B) Shortly after fertilization and completion of the second meiotic division, an extensive microtubule aster is assembled from the sperm centrosome. (C) Rotation of the egg cortex is dependent on microtubules extending from the sperm aster to form a subcortical network in the vegetal hemisphere. (D) Assembly of the first mitotic spindle is apparent by 0.7 NT, during the peak of cortical rotation. (E) By first cleavage, MTs of the spindle asters nearly fill the egg cytoplasm (replacing the sperm aster), and the cortical MT network has been disassembled. (F) Dorsal embryonic structures are formed from the side of the egg opposite the point of sperm entry (SEP in A).

cortical rotation plays an important role in specification of the D–V axis. In the eggs of some species, displacement of the heavily pigmented animal cortex during cortical rotation results in the exposure of underlying, lightly pigmented cytoplasm and forming the "grey crescent," which is one of the earliest markers of the future dorsal side (see Gerhart, 1980; Gerhart *et al.*, 1986).

The dependence of the cortical rotation, and thus specification of the D–V axis, on MTs was first suggested by the effects of MT inhibitors, cold and hydrostatic pressure on axis formation (Manes and Barbieri, 1977; Manes *et al.*, 1978; Scharf and Gerhart, 1983). Treatment of fertilized eggs

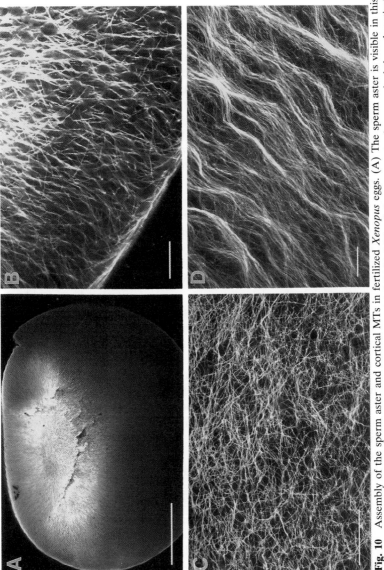

Fig. 10 Assembly of the sperm aster and cortical MTs in fertilized *Xenopus* eggs. (A) The sperm aster is visible in this cross-sectional view of a fertilized egg at 0.3 NT. By 0.35–0.5 NT, MTs of the sperm aster have reached the subcortical cytoplasm of the vegetal hemisphere (B), where they form a dense, disordered network (C). Between 0.5 and 0.8 NT, the network of subcortical MTs becomes aligned along the direction of cortical rotation (D).

with agents which depolymerize MTs during a critical period of the first cell cycle (corresponding to the period of rotation) blocked subsequent differentiation of dorsal–anterior structures, while D_2O, which promotes MT assembly, caused hyperdorsalization (Scharf and Gerhart, 1983). Elinson and Rowning (1988) subsequently described a network of parallel MTs located less than 5 μm beneath the vegetal surface of fertilized and artificially activated frog eggs (from both *X. laevis* and *R. pipiens*), and suggested that these "cortical" MTs, acting as tracks for MT-dependent motor proteins in the cortical or subcortical cytoplasm, might play an important role in cortical rotation and axis specification.

Confocal immunofluorescence microscopy of fertilized or activated eggs revealed that many of the MTs in the subcortical cytoplasm of the vegetal hemisphere were continuous with an underlying network of radial MTs originating from the sperm aster (Fig. 10B) (Schroeder and Gard, 1992). The extensive MT networks observed in the subcortical cytoplasm of parthenogenetically activated eggs (in the absence of a sperm centrosome) can be fully explained by the assembly of an extensive "activation aster," which is organized by maternally supplied centrosomal components (Schroeder and Gard, 1992; see below) (for a different point of view, see Elinson and Palecek, 1993). The continuity of the vegetal MT network with the underlying array of radially organized MTs of the sperm (or activation) aster suggested that the shear zone for cortical rotation lies outside the vegetal MT array (Houlistan and Elinson, 1991a; Schroeder and Gard, 1992), a conclusion that has received additional support from recent observations of the vegetal MT array in living eggs injected with fluorescently labeled TB (Houliston, 1994; Rowning *et al.*, 1994; C. A. Larabell, see Chapter 13 of this volume). In this respect, then, these MTs should more properly be referred to as "subcortical" MTs, since they are located in the cytoplasm immediately underlying the oocyte cortex.

The mechanisms responsible for aligning and bundling the network of subcortical MTs remain largely unknown. MTs of the sperm aster first reach subcortex of the vegetal hemisphere at 0.35–0.4 NT, where they form a disordered array (Fig. 10C) (Schroeder and Gard, 1992). Between 0.5 and 0.7 NT, these subcortical MTs are reorganized to form a dense array of wavy, parallel bundles, which are aligned along the direction of cortical rotation (Fig. 10D) (Elinson and Rowning, 1988; Houliston and Elinson, 1991a; Schroeder and Gard, 1992). The initial disorder exhibited during their assembly (Schroeder and Gard, 1992) and during reassembly following cold-induced disassembly (M. M. Schroeder, R. Worthylake, and D. L. Gard, unpublished observations) suggests that alignment of the subcortical MTs does not result from directed assembly. Rather, cortical rotation and alignment of the subcortical MTs are thought to be interdependent, self-reinforcing processes: MTs are dragged into alignment by the rotation of the overlying cortex, and alignment of the underlying network of MTs reinforces the direction of rotation (Houliston and Elinson, 1991a,b, 1992; Schroeder and Gard, 1992). Recent observations of the alignment of cortical MTs in living eggs injected with fluorescent TB are consistent with this model (Houliston, 1994; Rowning et al., 1994; C. A. Larabell, see Chapter 13 of this volume), which relies on slight asymmetries in the assembly of the sperm aster to provide the initial direction for cortical rotation (Houliston and Elinson, 1992).

While subcortical MTs are present in both animal and vegetal hemispheres (Schroeder and Gard, 1992), only the vegetal hemispheres of hemisected eggs are capable of rotation (Vincent et al., 1987), indicating that the MT-dependent motor(s) that power rotation of the cortex are predominantly restricted to the vegetal hemisphere. Several lines of evidence, including the structural polarity of the cortical MTs, suggest that cortical rotation is powered by a member of the kinesin family of MT-dependent ATPases (Houliston and Elinson, 1991b). Houliston et al. (1994) recently reported that a kinesin-like protein, termed Eg5, is associated with MTs in the vegetal cortex of fertilized eggs, as well as with interphase and spindle MTs during later development. The Eg5 protein is thus a prime candidate for the motor that powers cortical rotation.

Subcortical MTs persist in the animal and vegetal hemispheres throughout the period of rotation, and are disassembled just prior to first cleavage (at about 0.9–0.95 NT), in a process that requires activation of maturation-promoting factor and protein phosphorylation (Schroeder and Gard, 1992). Interestingly, assembly of the first mitotic spindle begins during the peak stages of cortical rotation, at 0.65–0.75 NT. Confocal microscopy reveals that assembly of the mitotic spindle, which initially exhibits few astral MTs (Fig. 11A), occurs in a region of cytoplasm that is relatively devoid of cytoplasmic MTs (B.-J. Cha and D. L. Gard, unpublished observations). The spindle asters enlarge rapidly (Figs. 11B–11D), as disassembly of the sperm aster proceeds outward from the spindle toward the egg cortex. By 0.9–0.95 NT, MTs from the spindle asters have nearly filled the egg cytoplasm and the subcortical MTs of the vegetal hemisphere have been disassembled (Fig. 11D). These observations further underscore the precise spatial regulation of MT assembly and dynamics in amphibian oocytes and eggs.

III. Regulation of MT Assembly and Organization in Oocytes and Eggs: MTOCs and MAPs

The cytoplasmic volume of Xenopus oocytes increases 1000-fold from early stage 0 (12 μm diameter) to mid-stage I (120 μm diameter) and another

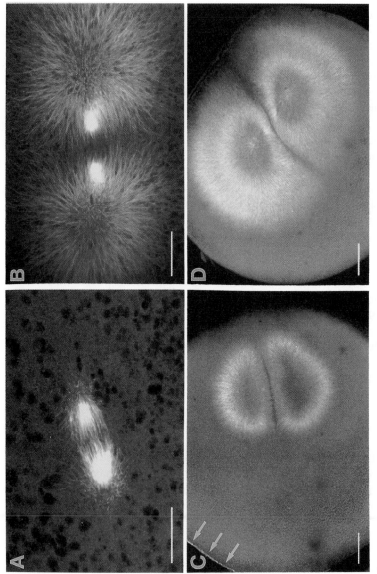

Fig. 11 Assembly of the first mitotic spindle in fertilized *Xenopus* eggs. (A) The first mitotic spindle, shown at 0.7 NT, is assembled in a region devoid of other MTs and initially exhibits few astral MTs. (B) Astral MTs rapidly elongate from the poles of the first mitotic spindle (at 0.8 NT). (C) Images collected at lower magnification reveal the brightly stained spindle asters and the subcortical MTs (arrows) at 0.8 NT. (D) By 0.95 NT, astral MTs fill the animal hemisphere, and the subcortical MTs have been disassembled. Note the appearance of two "waves" of astral MTs in D. Scale bars = 25 μm in A and B and 100 μm in C and D.

500-fold from stage I to stage VI (1.2 mm in diameter) (about 50% of the volume of stage VI oocytes is occupied by yolk). This incredible growth poses unique problems to the regulation of MT assembly and organization during oogenesis. For example, during stages I–VI alone, the cytoplasmic TB concentration of *Xenopus* oocytes increases roughly 10-fold (Pestell *et al.*, 1975; Gard *et al.*, 1995a; B. M. Error and D. L. Gard, unpublished observations), which roughly translates to a 5000-fold increase in the pool of cytoplasmic TB. Based on the fraction of polymer present in stage VI *Xenopus* oocytes,[2] we estimate that the amount of MT polymer increases from about 60 mm in a single stage I oocyte (120 μm in diameter) to more than 300 m in a single stage VI oocyte (1.2 mm in diameter)! How is MT assembly and organization regulated during this phenomenal period of growth?

Microtubule assembly and organization in somatic cells are thought to be regulated by cellular control of two interdependent processes: MT nucleation and MT dynamics (for review, see Kirschner and Mitchison, 1986). In animal cells, most cytoplasmic MTs are nucleated by the centrosome, or other discrete MTOCs (McIntosh, 1983; Brinkley, 1985). The nucleating capacity of centrosomes in somatic cells is regulated during the cell cycle, apparently through the phosphorylation of centrosomal proteins by cyclin-dependent kinases (Vandre and Borisy, 1989; Verde *et al.*, 1991; Bailly *et al.*, 1989; Riabowal *et al.*, 1989; Rattner *et al.*, 1990). While *Xenopus* oocytes lack discrete centrosomes (Huchon *et al.*, 1981; Gard, 1991a; Gard *et al.*, 1995a), mounting evidence suggests that a maternal pool of centrosomal components plays an important role in the nucleation and organization of MTs during oogenesis and early development (Gard *et al.*, 1990; Verde *et al.*, 1991; Stearns *et al.*, 1991) (see below).

In contrast to the static images of MTs obtained by immunofluorescence or electron microscopy, most cellular MTs are highly dynamic polymers that exist in one of two states or phases (for reviews, see Kirschner and Mitchison, 1986; Cassimeris *et al.*, 1987): (1) elongation, in which TB subunits add to MT ends; and (2) rapid shortening, in which TB subunits are lost from MT ends. Interconversion between these states occurs abruptly, through apparently stochastic transitions termed catastrophe (the transition from elongation to shortening) and rescue (the transition from shortening to elongation). This behavior, termed dynamic instability (Mitchison and Kirschner, 1984) has been well documented in MTs grown *in vitro* (Walker

---

[2]Based on published estimates of the TB pool of stage VI oocytes (0.5–1.0 μg/oocyte), the fraction of TB in polymer in stage VI oocytes and eggs (15–20%; Elinson, 1985; Jessus *et al.*, 1987; D. L. Gard, unpublished observations), and the increase in TB concentration (relative to total soluble protein) between stages I and VI (10-fold; B. M. Error and D. L. Gard, unpublished observations).

et al., 1988) as well as in individual MTs in cultured mammalian cells (Cassimeris *et al.,* 1988; Sheldon and Wadsworth, 1993). Regulation of MT assembly is thought to occur, in part, by the regulation of dynamic instability through the binding of MAPs (Drechsel *et al.,* 1992; Pryer *et al.,* 1992; Kowalski and Williams, 1993). While it is likely that MTOCs and MAPs also regulate MT assembly and organization in amphibian oocytes, eggs, and early embryos, the spatial and temporal scale of MT assembly in these cells place a number of unique constraints on the regulation of MT nucleation and dynamics.

A. The Role of Centrosomal Proteins during Oogenesis

The capacity to nucleate and regrow cytoplasmic MTs after cold-induced disassembly decreases dramatically during stage I of oogenesis (Gard, 1991a; Gard *et al.,* 1995a), indicating that the maternal centrosome is inactivated very early in the differentiation of *Xenopus* oocytes. While the mechanism of centrosome inactivation and the fate of the maternal centrioles remain unknown, confocal immunofluorescence microscopy revealed that centrosome inactivation coincides with dispersal of the centrosomal protein γ-TB to multiple cytoplasmic foci, which are incapable of serving as MTOCs (Gard *et al.,* 1995a). Previous results have suggested that centrosome activity is regulated by cell cycle-dependent phosphorylation (Vandre and Borisy, 1989; Bailly *et al.,* 1989; Rattner *et al.,* 1990; Riabowal *et al.,* 1989; Iwao and Elinson, 1990), leading us to suggest that inactivation of the maternal centrosome during the differentiation of *Xenopus* oocytes is a consequence of the arrest of their cell cycle in meiotic prophase (Gard *et al.,* 1995a).

The dramatic episodes of MT reorganization during oogenesis and oocyte maturation thus occur in the absence of a classical centrosome (Gard, 1991a, 1992; Gard *et al.,* 1995a). However, despite lacking classical centrosomes (and presumably centrioles), *Xenopus* oocytes and eggs contain a substantial pool of centrosomal proteins. Early evidence that *Xenopus* embryos (and by extension, oocytes and eggs) contain a maternal pool of centrosome components was provided by our observation that centrosome duplication continued in cycloheximide-arrested *Xenopus* blastulae (Gard *et al.,* 1990). Further support was provided by the nucleation and organization of MT asters in cytoplasmic extracts of *Xenopus* eggs (Verde *et al.,* 1991), and by the assembly of an extensive activation aster in parthenogenetically activated *Xenopus* eggs (Houliston and Elinson, 1991a; Schroeder and Gard, 1992). Finally, Stearns *et al.* (1991) demonstrated that *Xenopus* eggs contained a substantial pool of γ-TB, a component of centrosomes, spindle poles, and other MTOCs in cells from many diverse species (Oakley and Oakley, 1989; Oakley *et al.,* 1990; Zheng *et al.,* 1991; Stearns *et al.,*

1991; Joshi *et al.*, 1992). Undoubtedly, this maternal pool of γ-TB and other centrosomal proteins (Doxsey *et al.*, 1994) provides a store of centrosomal components needed during the rapid division cycles of early cleavage. Recently, maternally supplied γ-TB has been shown to be required for the assembly of functional centrosomes from *Xenopus* sperm centrioles (Stearns and Kirschner, 1994). Recent evidence also suggests that the maternally supplied pool of γ-TB (and perhaps other centrosomal components) plays an important role in regulating the assembly and organization of the MTs during oogenesis and oocyte maturation (Gard, 1994; see below).

As a first step in understanding the mechanisms underlying reorganization of the MT during formation of the A–V axis, we examined the distribution of γ-TB during stages III–VI of oogenesis (Gard, 1994). γ-TB was found to be concentrated in the perinuclear cytoplasm of stage VI oocytes, consistent with previously published results suggesting that the GV functions as an MTOC during late oogenesis. Surprisingly, however, substantial concentrations of γ-TB also were apparent in the cortex of stage VI oocytes (Figs. 12A and 12B). The presence of γ-TB in the cortex of *Xenopus* oocytes raises a number of interesting questions about the structural polarity and organization of MTs in stage VI *Xenopus* oocytes. γ-TB has been observed at centrosomes and spindle poles in cells of many diverse species (Zheng *et al.*, 1991; Stearns *et al.*, 1991; Joshi *et al.*, 1992), leading to suggestions that it associates primarily with the minus ends of MTs (Joshi *et al.*, 1992). Localization of γ-TB in the cortex of *Xenopus* oocytes, then, suggests that a substantial fraction of the MTs in stage VI oocytes have their minus ends anchored in the cortex. This "minus-end out" orientation contrasts with the orientation of MTs in most somatic cells, in which the minus ends are anchored at the centrosome and plus ends extend into the cell periphery (Euteneuer and McIntosh, 1981a,b). Further examination by hook decoration and electron microscopy will be required to directly ascertain the polarity of oocyte MTs, which may have important consequences for the transport or localization of developmentally-important maternal mRNAs (as well as other oocyte components).

Confocal immunofluorescence microscopy revealed that small foci of γ-TB were evenly distributed throughout the cortex of stage III *Xenopus* oocytes (Gard, 1994). However, during stages IV–VI, the distribution of γ-TB became polarized along the developing A–V axis. In the vegetal cortex, brightly stained foci of γ-TB formed a characteristic network of short linear aggregates (Fig. 12B), which were associated with MTs (including acetylated MTs) in the vegetal cortex (Gard, 1994). In the animal hemisphere, less brightly stained foci of γ-TB foci were more evenly distributed throughout the cortex. The distribution of γ-TB in the cortex of stage VI oocytes was not disrupted by complete depolymerization of the oocyte MT array, indicating that the distribution of γ-TB is not dependent on MTs.

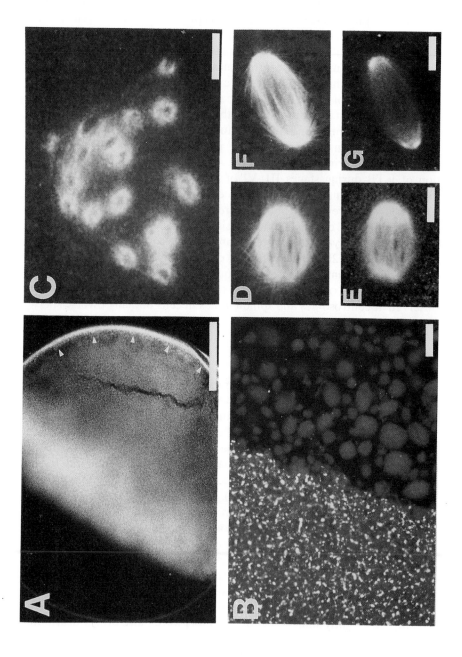

Although additional studies are clearly necessary, the temporal coincidence between the establishment of the polarized distribution of γ-TB in the cortex of *Xenopus* oocytes with the reorganization of the oocyte MTs suggests that cortical γ-TB plays an important role in regulating MT organization during formation of the A–V axis through the differential nucleation or stabilization of oocyte MTs.

γ-TB may also serve to integrate components of the cytoskeleton of *Xenopus* oocytes. Extended exposure of stage VI *Xenopus* oocytes to cytochalasin B, an inhibitor of actin assembly, releases γ-TB from the cortex (Gard, 1994) and disrupts the normal distribution of oocyte MTs (B.-J. Cha and D. L. Gard, unpublished observations). These observations suggest that oocyte MTs are anchored, perhaps through γ-TB, to perinuclear and cortical networks of F-actin .

The role of γ-TB in the dramatic reorganization of MTs that accompanies oocyte maturation remains largely unexplored. γ-TB is lost from the vegetal cortex of *Xenopus* oocytes soon after GVBD (D. L. Gard, unpublished observations). Interestingly, disassembly of the extensive vegetal network of cytokeratin filaments (Klymkowsky *et al.*, 1987, 1991) and release of Vg1 mRNA from the vegetal cortex (Pondel and King, 1988; Klymkowsky *et al.*, 1991; Yisraeli *et al.*, 1990) also coincide with GVBD, suggesting that the oocyte cortex is extensively remodeled during the early stages of oocyte maturation.

Preliminary results indicated that the discoidal MTOC that forms during GVBD in *Xenopus* oocytes stains only weakly with antibodies to γ-TB. During the initial stages of spindle assembly, however, individual meiotic chromosomes are surrounded by brightly stained shells of γ-TB (Fig. 12C) (D. L. Gard, unpublished observations), consistent with published reports that chromatin and chromosomes play an important role in the nucleation and assembly of spindle MTs during meiosis (Bajer, 1972; Church *et al.*, 1986; Steffen *et al.*, 1986; Gard, 1992; Gard *et al.*, 1995b). γ-TB is bound along the length of the spindle MTs during the formation of the bipolar

---

Fig. 12 γ-TB is located in the cortex of stage VI oocytes and associates with meiotic chromosomes and spindles. (A) A prominent rim of fluorescence was apparent in the vegetal cortex of stage VI oocytes stained with antibodies sepcific for *Xenopus* γ-TB (arrowheads). (B) Grazing optical sections of the vegetal cortex reveal a complex pattern of γ-TB foci, often organized into short linear arrays (arrows; the vegetal cortex was partially peeled away in B). (C) γ-TB is concentrated around the chromosomes during the early stages of assembly of the M1 spindle (seen in a top view of the animal pole). (D and E) γ-TB is bound along the length of this early bipolar M1 spindle (D is stained with α-TB and E is stained with γ-TB antibodies). (F and G) γ-TB is concentrated at the poles of this fully elongated M1 spindle (F is stained with α-TB and G is stained with γ-TB antibodies). Scale bars = 250 μm in A and 10 μm in B–G.

spindle axis (Figs. 12D and 12E), and only after prometaphase elongation does γ-TB become heavily concentrated to the spindle poles (Figs. 12F and 12G). Confocal immunofluorescence microscopy suggests that only a fraction of the total pool of γ-TB associates with the meiotic spindles (D. L. Gard, unpublished observations), while the rest apparently remains dispersed in the cytoplasm. Further study will be required to elucidate the role(s) of γ-TB in the assembly and organization of the meiotic spindles and the complex reorganization of the MT cytoskeleton that accompanies oocyte maturation.

B. *Xenopus* Oocytes and Eggs Contain Novel Factors That May Regulate MT Assembly

Amphibian oocytes and eggs face a number of potentially unique problems in the regulation of MT assembly dynamics. For example, the extended period of oocyte differentiation and growth requires the maintenance of a stable framework of MTs over a period of many months, as evident in the extensive populations of acetylated, presumably stable, MTs observed throughout the postmitotic stages of oogenesis in both *X. laevis* and *R. pipiens* (Gard, 1991a; Gard *et al.*, 1995a; Wang *et al.*, 1993; D. L. Gard, unpublished observations). These stable MTs may serve several functions during oogenesis. First, the organization of acetylated MTs throughout differentiation of oocytes in both *Xenopus* and *Rana* suggests that stable MTs provide a cytoskeletal framework for maintaining cytoplasmic organization and polarization during this period of extraordinary oocyte growth. Further support for this hypothesis is provided by our observation that a large fraction of the MTs in fully grown oocytes from *R. pipiens* are cold stable (Wang *et al.*, 1993; D. L. Gard, unpublished observations). The evolution of cold-stable MTs in *R. pipiens* oocytes, in which oocyte growth and differentiation extends through several growing seasons that are interspersed with periods of winter hibernation, further underscores the importance of the oocyte MT array in maintenance of the A–V axis. In addition, the extensive network of stable MTs observed in amphibian oocytes may provide sites for the assembly or elongation of MTs following inactivation of the maternal centrosome, in a manner analogous to that of stable MTs in neuronal axons (Baas and Ahmad, 1992; and references therein).

The large size (up to 2 mm in diameter) and rapid cell cycles exhibited by amphibian oocytes, eggs, and embryos also require rates of MT assembly and disassembly that stretch to the limits our current models of MT dynamics based on the dynamics of MTs *in vitro* (Walker *et al.*, 1988) and in cultured mammalian cells (Cassimeris *et al.*, 1988; Sheldon and Wadsworth, 1993). During maturation, for example, the oocyte MT cytoskeleton, which is estimated to contain more than half a million MTs with an average length of 0.6 mm (600 μm) (Gard, 1991a), is completely remodeled within a span of 30 min (Gard, 1991b, 1992). Furthermore, assembly of the sperm aster in fertilized *Xenopus* eggs requires MT elongation at rates of 20–50 μm min^{-1} (Stewart-Savage and Grey, 1982; Gard and Kirschner, 1987b), which are substantially faster than those observed in most somatic cells (Cassimeris *et al.*, 1988; Sheldon and Wadsworth, 1993) or *in vitro* (Walker *et al.*, 1988; Belmont *et al.*, 1990). Finally, the 20- to 30-min cell cycles during early cleavage require rapid cycling between the extensive cytoplasmic MT arrays of interphase and the characteristic bipolar organization of the mitotic spindles. These episodes of assembly and disassembly during oogenesis and early development may depend on the presence and regulation of two distinct mechanisms: microtubule-severing factors that are activated during M phase to facilitate MT disassembly, and microtubule-associated proteins that promote both MT assembly and dynamics.

C. MT-Severing Proteins May Mediate the Rapid MT Disassembly during Oocyte Maturation and Early Development

Addition and loss of TB subunits is normally restricted to the free plus ends of MTs. However, endwise disassembly, at published rates of 12–20 μm min^{-1} (Walker *et al.*, 1988; Cassimeris *et al.*, 1988; Belmont *et al.*, 1990), may be insufficient to account for the rapid reorganization of MTs during oocyte maturation and early development. The rapid disassembly of MTs during maturation and early development may be mediated or aided by the activation of cell cycle-dependent MT-severing factors, an activity was first observed in M phase-arrested *Xenopus* egg extracts by Vale (1991). A MT-severing factor has subsequently been purified from *Xenopus* eggs (Shiina *et al.*, 1992). However, differences in the ATP dependence of the severing activities of the purified protein and unfractionated egg cytoplasm suggest that other severing factors may yet be identified in *Xenopus* eggs (McNally and Vale, 1993).

The physiologic significance and function(s) of MT severing during oogenesis and early development remain unknown. MT severing is not observed in interphase extracts from *Xenopus* eggs (Vale, 1991; Shiina *et al.*, 1992), indicating that the activity of the severing factor is regulated during the cell cycle, perhaps by MPF-dependent phosphorylation (Shiina *et al.*, 1992). Presumably, severing of MTs creates additional ends for MT depolymerization during M phase, and thus might play an important role during the rapid disassembly of MTs that accompanies oocyte maturation, as well as during disassembly of the sperm aster and cortical MT array of fertilized eggs. Free MT ends generated by the action of M phase-severing proteins might also serve as nucleation sites for MT assembly, analogous to the stimulation of actin nucleation by the gelsolin-mediated severing of actin filaments during platelet activation (Hartwig, 1992). However, severing of MTs *in vitro* by *Xenopus* egg cytoplasm and by the severing factor isolated from *Xenopus* eggs is inhibited by MAPs and monomeric TB (Vale, 1991; Shiina *et al.*, 1992), proteins that are present in great abundance in *Xenopus* oocytes and eggs. Further analysis will be required to conclusively establish the role of these novel MT-severing factors during amphibian oogenesis and early development.

D. *Xenopus* Oocytes and Eggs Contain Functionally Distinct Microtubule-Associated Proteins

Xenopus oocytes and eggs contain at least two biochemically and functionally-distinct, high-molecular-weight microtubule-associated proteins (XMAPs)[3]: XMAP215 (Gard and Kirschner, 1987b) and XMAP230 (Shiina *et al.*, 1992; Andersen *et al.*, 1994; Cha *et al.*, 1994). XMAP215 was first identified and isolated from activated egg cytoplasm based on its ability to promote centrosome- and axoneme-nucleated MT assembly *in vitro* (Gard and Kirschner, 1987b). Further characterization revealed that XMAP215 was a potent promoter of MT assembly that exhibited the (as yet) unique ability to specifically promote elongation of MT plus ends (Gard and Kirschner, 1987b; Vasquez *et al.*, 1994).

Recently, several labs have independently isolated heat-stable high-molecular-weight (M_r 220–230 kDa) MAPs from activated *Xenopus* eggs (Shiina *et al.*, 1992; Faruki and Karsenti, 1994; Anderson *et al.*, 1995; Cha *et al.*, 1994). Similarities in their biochemical properties suggest these proteins, previously referred to as p220 (Shiina *et al.*), XMAP230 (Faruki and Karsenti, 1994; Andersen *et al.*, 1994), and MAP250 (Cha *et al.*, 1994) represent independent isolations of the same protein, which we shall refer to as XMAP230.

The structural relationship(s) between XMAP215 and XMAP230, and between XMAPs and other vertebrate MAPs, remains uncertain. XMAP215 and XMAP230 are immunologically, biochemically, and functionally distinct (see below), suggesting that they are structurally unrelated (Shiina *et al.*, 1992; Andersen *et al.*, 1994; Vasquez *et al.*, 1994; B.-J. Cha and

---

[3]We have adopted the nomenclature suggested by Andersen *et al.* (1994), in which the term "XMAP" collectively refers to MAPs in *Xenopus* oocytes and eggs. Individual XMAPs are then referred to by their apparent molecular weight. In this nomenclature, the protein previously called XMAP (Gard and Kirschner, 1987) is now called XMAP215. XMAP230 refers to proteins previously called p220 (Shiina *et al.*, 1991), XMAP230 (Andersen *et al.*, 1994), and MAP250 (Cha *et al.*, 1994).

D. L. Gard, unpublished observations). Similarly, the lack of immunological cross-reaction and dissimilarities in biochemical and functional characteristics suggest that XMAP215 is unrelated to previously characterized vertebrate MAPs (antibodies to XMAP215 do detect a protein of similar molecular weight in *Rana* oocytes) (Gard and Kirschner, 1987b; D. L. Gard, unpublished observations). In contrast, XMAP230 has been reported to cross-react with antibodies to mammalian MAP1b and MAP4 (Andersen *et al.*, 1994; Faruki and Karsenti; 1994). XMAP230 may also be identical to O-MAP (Fellous *et al.*, 1991), which was identified in *Xenopus* oocytes with antibodies to MAP2. Determination of the evolutionary and structural relationships between XMAPs and other vertebrate MAPs thus awaits molecular analysis of their primary structure.

Both XMAP230 and XMAP215 proteins are predominantly found in oocytes, eggs, and early embryos. The cytoplasmic concentration of XMAP230 increases 10-fold (relative to total soluble proteins) between stages I and VI of oogenesis, with the major increase occurring during stages III and IV (Cha *et al.*, 1994; B.-J. Cha, B. M. Error, and D. L. Gard, unpublished observations). XMAP215 accumulates somewhat later, during stages IV–VI of oogenesis in *Xenopus* (D. L. Gard and M. M. Schroeder, unpublished observations). Interestingly, the greatest increase in the cytoplasmic concentration of tubulin occurs during stages II and III of oogenesis, prior to accumulation of either XMAP (B. M. Error and D. L. Gard, unpublished observations). Levels of both proteins subsequently decline (more than 10-fold) during embryogenesis (postneurulation) (B.-J. Cha, B. M. Error, M. M. Schroeder, T. Mueller, and D. L. Gard, unpublished observations). In adults, both XMAP proteins are found predominantly in ovaries (oocytes), with lesser amounts detected in brain and testes (tissues which are characterized by an abundance of MTs) (Gard and Kirschner, 1987b; B.-J. Cha, B. M. Error, and D. L. Gard, unpublished observations). Western immunoblots fail to detect significant amounts of XMAP215 protein in other adult tissues. The presence of minor amounts of XMAP230 in other somatic tissues of adult frogs remains controversial, with conflicting results being reported by different laboratories (Shiina *et al.*, 1992; Andersen *et al.*, 1994; B.-J. Cha, B. M. Error, and D. L. Gard, unpublished observations). However, the prevalence of these proteins in oocytes, eggs, and early embryos suggests that they play a major role in the regulation of MT assembly during these stages of development.

XMAP215 and XMAP230 are both cell cycle-dependent phosphoproteins, exhibiting increased phosphorylation during both meiosis and mitosis (Gard and Kirschner, 1987b; Shiina *et al.*, 1992; Andersen *et al.*, 1994; D. L. Gard, unpublished observations). Phosphorylation by p34^{cdc2} or MAP kinases is reported to reduce the binding of XMAP230 to MTs (Shiina *et al.*, 1992), and thus reduce MT assembly promotion. Preliminary results suggest that XMAP215 is phosphorylated by p34^{cdc2}, but not MAP kinase (R. J. Vasquez, D. L. Gard, and L. U. Cassimeris, unpublished observations). However, the effects of phosphorylation on the activity of XMAP215 have yet to be studied in detail. The cell cycle-dependent phosphorylation of one or both of these proteins may thus be responsible for the dramatic increase in MT dynamics observed during M phase (Belmont *et al.*, 1990; Shiina *et al.*, 1992; Verde *et al.*, 1992).

In vitro analysis of MT assembly revealed that XMAP215 and XMAP230 have markedly different effects on the dynamic instability of individual MTs. Andersen *et al.* (1994) reported that XMAP230 promotes MT assembly and stabilizes MTs by suppressing dynamic instability *in vitro*, in a manner analogous to the vertebrate brain MAPs Tau and MAP2 (Drechsel *et al.*, 1992; Pryer *et al.*, 1992; Kowalski and Williams, 1993). Addition of XMAP230 to purified TB (1) suppressed catastrophe (F_{Cat}), (2) increased rescue (F_{Resc}); (3) decreased the rate of rapid shortening (V_S), and (4) promoted a modest (two- to four-fold) increase in the rate of elongation (V_E). Like previously characterized brain MAPs, XMAP230 promoted assembly of both the plus and the minus ends of MTs. However, unlike Tau and MAP2, XMAP230 did not promote MT nucleation (Andersen *et al.*, 1994).

In contrast to the effects observed with XMAP230 and brain MAPs, Vasquez *et al.* (1994) found that XMAP215 promoted both the assembly and the dynamics of MT plus ends, while having only minimal effects on minus-end assembly. Promotion of plus-end assembly by XMAP215 resulted from a dramatic (7- to 10-fold) increase in V_E. However, XMAP215 also promoted a 2- or 3-fold increase in V_S and nearly eliminated rescue, while having little effect on F_{Cat}. XMAP215 thus promotes the assembly of long, but extremely dynamic, MTs.

Based on their observed effects on MT assembly *in vitro*, XMAP215 and XMAP230 have been proposed to play important roles in the regulation of MT assembly and dynamics during the cell cycle and early development. However, their precise functions *in vivo* remain to be determined. Their dramatically different effects on MT dynamics *in vitro*, and the potential to independently regulate their activity through phosphorylation, suggest that XMAP215 and XMAP230 have distinct functions during oogenesis and early development. For example, the 7- to 10-fold increase in MT V_E promoted by XMAP215 may play a major role in the assembly of the sperm aster and cortical MTs of fertilized eggs, which require MT assembly at rates estimated to be 20–50 μm min^{-1} (Gard and Kirschner, 1987b). The 2- or 3-fold increase in V_S promoted by XMAP215 (from 19 to 50 μm min^{-1}) (Vasquez *et al.*, 1994), in conjunction with the activation of MT-severing activities during M phase (Vale, 1991; Shiina *et al.*, 1992), may also play an important role in promoting the rapid disassembly of MTs in maturing oocytes, fertilized eggs, and cleaving embryos. In contrast, XMAP230 may serve to stabilize MTs during oogenesis and early development through its dramatic effect on F_{Cat} (Andersen *et al.*, 1994). XMAPs may also act in combination: the promotion of V_E by XMAP215 and suppression of catastrophe by XMAP230 may both be required for the rapid assembly of the sperm aster and cortical MTs in fertilized *Xenopus* eggs.

Results obtained by immunofluorescence microscopy also suggest that XMAP215 and XMAP230 serve to regulate MT assembly *in vivo*. Antibodies to XMAP230 stain interphase MTs (Andersen *et al.*, 1994; Shiina *et al.*, 1992) and the central spindle of *Xenopus* epithelial cells grown in cell culture (Andersen *et al.*, 1994; however, see Shiina *et al.*, 1992). However, spindle staining was not observed in mitotic prophase, nor were the extensive polar asters of metaphase spindles stained with XMAP230 antibody (Andersen *et al.*, 1994). Results from preliminary confocal immunofluorescence microscopy revealed that XMAP230 is associated with interphase MTs (Fig. 13A) and spindle MTs (including astral MTs) (Fig. 13B) throughout the cell cycle of *Xenopus* blastomeres (B.-J. Cha and D. L. Gard, unpublished observations). More interestingly, affinity-purified antiserum to XMAP230 stains the cortical MT network of fertilized *Xenopus* eggs (Fig. 13C) (B.-J. Cha and D. L. Gard, unpublished observations), suggesting that XMAP230 plays an important role in regulating the assembly of this novel, developmentally important MT array. Less is known regarding the intracellular location of XMAP215. However, preliminary confocal microscopy indicates that XMAP215 antibodies stain mitotic spindles and centrosomes in *Xenopus* oogonia and blastulae-stage embryos (see Fig. 13D) (M. M. Schroeder and D. L. Gard, unpublished observations), confirming that XMAP215 also associates with MTs *in vivo*.

IV. Concluding Remarks: Toward an Understanding of MT Assembly during Amphibian Oogenesis and Early Development

The studies described previously have dramatically changed our view of the microtubule cytoskeleton of amphibian oocytes and eggs. In particular, whole-mount immunocytochemistry and confocal immunofluorescence microscopy have provided our first global view of the complex patterns of MT present during oocyte differentiation, polarization, and maturation, and have provided a clearer picture of the rapid changes in MT assembly and organization that accompany fertilization and early development. While the images obtained using these techniques have greatly added to our knowledge of MT organization during oogenesis and early development, many questions remain unanswered: How is the maternal centrosome inacti-

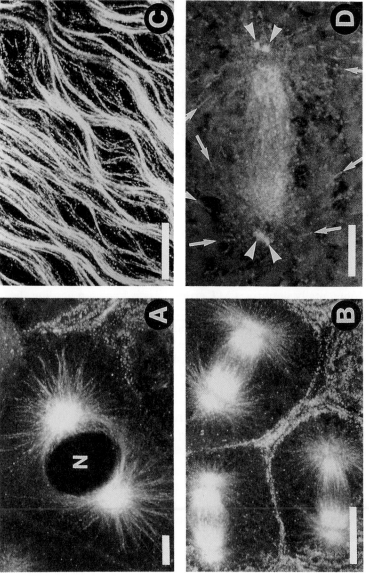

Fig. 13 XMAPs are associated with interphase and spindle MTs during early development. (A and B) Affinity-purified XMAP230 antibodies stain interphase (A) and spindle MTs (B) in *Xenopus* blastomeres. (C) Affinity-purified XMAP230 antibodies stain the network of subcortical MTs in the vegetal hemisphere of fertilized Xenopus eggs. (D) Affinity-purified XMAP215 antibodies stain centrioles (arrowheads) and spindle MTs (arrows denote astral MTs) in *Xenopus* blastomeres. A–C were fixed in FGT, while D was fixed in 100% methanol. Scale bars = 5 μm in A and B and 10 μm in C and D.

vated, and what is the fate of the centrioles inherited from the final oogonial division? How is microtubule nucleation and organization regulated in amphibian oocytes, in the absence of centrosomes or other discrete organizing centers? How are the extreme rates of MT assembly and disassembly required in oocytes, eggs, and early embryos achieved? What role(s) do stable MTs play in the formation of the A–V axis?

Achieving a comprehensive understanding of MT assembly and regulation during oogenesis and early development will depend on the identification and characterization of the protein factors which regulate MT nucleation and assembly. Several such factors, including the centrosomal protein γ-TB, microtubule-severing factors, and XMAPs, have already been isolated, and their roles in regulating MT nucleation and assembly are being characterized *in vitro*. Undoubtedly, other factors remain to be identified and characterized. Some, such as γ-Tb, may be identical to factors found in somatic cells. Others, such as XMAP215, may exhibit unique properties that provide solutions to problems posed by the large size and rapid cell cycles of developing embryos. However, the identification of MT assembly factors and characterization of their activity *in vitro* is but a first step in understanding MT regulation in oocytes and eggs. From there, it will be necessary to examine the function of these proteins *in vivo*, for it is only in intact cells that the complex spatial and temporal patterns of MT assembly are fully apparent. We trust that the combination of *in vitro* and *in vivo* analysis will soon result in a comprehensive understanding of the regulation of MT assembly and organization during amphibian oogenesis and early development, and that this knowledge will provide the necessary foundation for understanding the role of the MT cytoskeleton in these important developmental processes.

Appendix: Confocal Microscopy of MTs in Amphibian Oocytes, Eggs, and Early Embryos

Many of the techniques that we have used for studying the organization of MTs and F-actin during oogenesis and early development in amphibians have been reviewed previously (Gard, 1993b; Gard and Kropf, 1993; Klymkowsky and Hanken, 1991). Additional discussions of confocal microscopy can be found in Gard (1993b), other chapters in Matsumoto (1993), and Stevens *et al.*, (1994). Rather than repeating those lengthy discussions, we consider here some of the more important aspects of confocal immunofluorescence microscopy of microtubules in amphibian oocytes and eggs.

A. Fixation and Sample Preparation

In many regards, the preservation of cytoplasmic MTs provides one of the most stringent tests of any fixation protocol. In addition to being highly dynamic structures, MTs are sensitive to cold, Ca^{2+}, and ionic strength. These characteristics make cytoplasmic MTs difficult to preserve, particularly in large cells or tissues such as amphibian oocytes and eggs. During the past 5 years, we have evaluated more than a dozen different protocols for preserving MTs in ovaries, oocytes, eggs, and early embryos from different amphibian species. These protocols ranged from simple fixation in methanol or acetone (at room temperature or cold) to complex protocols involving preextraction in microtubule-stabilizing buffer followed by fixation with aldehydes or other chemical cross-linking reagents. To summarize the results of these numerous comparisons, the best (and only consistent) preservation of MTs was obtained with combinations of formaldehyde and glutaraldehyde.

Organic solvents, such as methanol or acetone (alone, or in combination with taxol, formaldehyde, TCA, DMSO, or other additives), provided the least consistent preservation of cytoplasmic MTs in oocytes and unfertilized eggs, although fixation of meiotic and mitotic spindles in blastulae was often adequate (Gard, 1991a, 1992; B.-J. Cha and D. L. Gard, unpublished observations). Preservation of the cortical MTs of fertilized *Xenopus* eggs by methanol fixation was also inconsistent, and the underlying array of radially organized MTs was often destroyed by fixation in methanol (Schroeder and Gard, 1992). Formaldehyde alone (commercial preparations or freshly prepared from paraformaldehyde) also did not provide consistent MT preservation. Inclusion of taxol (a MT-stabilizing agent) during formaldehyde fixation and postfixation in methanol improved MT preservation (Gard, 1991a). Preservation of cytoplasmic MTs with glutaraldehyde alone (0.25–3%) was also less than optimal (B.-J. Cha and D. L. Gard, unpublished observations), possibly as a result of the slower penetration of glutaraldehyde into the interior of these large cells and tissues.

The most consistent preservation of MTs in all of the cell types we examined (as assayed by confocal immunofluorescence microscopy) was obtained with FGT fixative [3.7 % formaldehyde, 0.25% glutaraldehyde, and 0.5 μM taxol in 80 mM kPIPES, 1 mM MgCl$_2$, 5 mM EGTA, and 0.2% Triton X-100 (pH 6.8)] (Gard, 1991a, 1992, 1993; Gard and Kropf, 1993; Gard *et al.*, 1995a). Although we include taxol during fixation and follow with postfixation in 100% methanol, neither are absolutely required (B.-J. Cha, A. D. Roeder, M. M. Schroeder, and D. L. Gard, unpublished observations). We routinely use commercial stocks of formaldehyde (37% AR; Mallinkrodt, Paris, KY) and glutaraldehyde (50% electron microscopy grade; Ted Pella, Redding, CA), but have used electron microscopy-grade formaldehyde (18%, methanol free; Ted Pella) and paraformaldehyde with equal success (B.-J. Cha and D. L. Gard, unpublished observations).

After fixation or postfixation in methanol, samples are rehydrated into phosphate- or Tris-buffered saline. Larger oocytes, eggs, and early embryos are often hemisected to facilitate penetration of antibodies into the cell interior. Fixation in FGT, or in other fixatives containing glutaraldehyde, should be followed with reduction of the unreacted aldehydes with sodium borohydride (100 mM in buffered saline) for 4–24 hr. Reduction in borohydride significantly reduces the autofluorescence caused by glutaraldehyde fixation, but is not necessary for samples fixed in formaldehyde or methanol. Borohydride solutions are effervescent and should not contain any detergents. For the same reason, containers or tubes used for borohydride steps should be left uncapped, or be vented, to prevent mishaps due to the buildup of pressure. Sodium borohydride is a strong reducing agent and is potentially explosive. Due caution should be used to avoid contact with the skin or eyes. After reduction, samples can be returned to phosphate- or Tris-buffered saline containing 0.1% NP-40 or Triton X-100.

B. Antibodies

We use a mouse monoclonal specific for α-TB (DM1A; ICN Immunologicals, Lisle, IL) to visualize MTs in FGT-fixed oocytes and egg, and have used a number of other commercially available TB antibodies with excellent results (see Table I). In contrast, the epitopes recognized by many autoanti-

Table I Antibodies for Immunofluorescence Microscopy of Amphibian MTs

Antibody	Species[a]	Fixation[b]	Supplier	Reference
Anti-TB				
DM1A (α)	Mouse; M	MeOH/FGT	ICN immunologicals (Lisle, IL)	Blose et al. (1984)
DM1B (β)	Mouse; M	MeOH/FGT	ICN immunologicals (Lisle, IL)	Blose et al. (1984)
KMX-1 (β)	Mouse; M	MeOH/FGT	Boehringer-Mannheim (Indianapolis, IN)	Birkett et al. (1985)
TUB1A2 (α)[c]	Mouse; M	MeOH/FGT	Sigma (St. Louis, MO)	Kreis (1987)
6-11B-1 (α)[d]	Mouse; M	MeOH/FGT	Sigma (St. Louis, MO)	Piperno et al. (1987)
Anti-centrosome				
XGAM (γ-TB)	Rabbit; P	MeOH/FGT	NC[e]	Stearns et al. (1991)
Pericentrin	Rabbit; P	MeOH	NC	Doxsey et al. (1994)
SJ1	Human; A	MeOH	NC	Gard et al. (1990)
5051	Human; A	MeOH	NC	Calarco-Gillam et al. (1983)
Anti-XMAP				
XMAP215	Rabbit; P	MeOH	NC	Gard and Kirschner (1987b)
XMAP230	Mouse; M	MeOH/FGT	NC	Andersen et al. (1994)
XMAP230	Rabbit; P	MeOH/FGT	NC	Cha et al. (1994)

[a] Antibodies labeled M are monoclonals, P are polyconal, and A are autoantibodies.
[b] Antigens are stable to fixation in methanol (MeOH) or aldehydes (FGT).
[c] Specific for tyrosinated α-TB.
[d] Specific for acetylated α-TB.
[e] Not commercially available.

bodies to centrosomal proteins, and some antibodies to *Xenopus* MAPs (most noticeably, rabbit antibody to XMAP215), are destroyed by aldehyde fixation (M. M. Schroeder and D. L. Gard, unpublished observations). In these instances, we have resorted to fixation in 100% methanol, despite its mediocre preservation of cytoplasmic MTs. In many cases, methanol preservation of spindle MTs, though not optimal, is sufficient to allow localization of antigens by confocal immunofluorescence microscopy. We have not found it necessary to preblock samples when using commercially available anti-TB antibodies. Inclusion of nonionic detergents and BSA in antibody solutions decreases nonspecific binding of primary and secondary antibodies.

We use rhodamine-labeled secondary antibodies [from Orginon-Cappell (Malvern, PA) or Molecular Probes (Eugene, OR)] for most immunofluorescence microscopy. Although the excitation of rhodamines by the argon ion laser on many first-generation confocal microscopes is less efficient than excitation of fluorescein, we find that rhodamine provides images with better contrast and is much less susceptible to photobleaching. New fluorochromes have recently become available, many of which may prove useful (the Molecular Probes catalog is an excellent reference for identifying potentially useful fluorochromes).

C. Clearing and Mounting

The clearing solution developed by Murray and Kirschner (BA:BB: 1 part benzyl alcohol to 2 parts benzylbenzoate) has proven invaluable to the study of cytoskeletal organization in amphibian oocytes and eggs (Dent and Klymkowsky, 1989; Klymkowsky and Hanken, 1991; Gard, 1993b; Gard and Kropf, 1993). Unfortunately, BA:BB clearing solution is immiscible with aqueous buffers, and samples must be completely dehydrated in an organic solvent, such as methanol, ethanol, or acetone, prior to clearing. This prevents the use of BA:BB as a clearing and mounting agent for samples that are not compatible with organic solvents, including those stained with fluorescent phalloidins (Roeder and Gard, 1994). BA:BB should be considered toxic, and due caution should be used to avoid contact with the skin or ingestion. BA:BB is also an excellent solvent for many plastics, and care should be taken to avoid damage to expensive instruments.

The mounting techniques used for amphibian oocytes and eggs vary based on the size and geometry of the individual samples. Intact ovaries from juvenile frogs are typically mounted on standard glass slides, or in a chamber fashioned from spacers (cut from No. 2 coverslips) sandwiched between a slide and coverslip (Gard *et al.,* 1995b). Larger oocytes (stages IV–VI) and eggs are often hemisected prior to processing with antibodies, and are then mounted in commercially available glass well slides (0.5 mm wells). Intact oocytes, eggs, or early embryos are mounted in chamber slides machined from aluminum (chamber thicknesses of 0.7 and 1.0 mm), with coverslips cemented to both faces. Cells are mounted in BA:BB clearing solution, and coverslips are sealed with clear fingernail polish. Fingernail polishes or other cements should be tested with BA:BB, since many will not harden when used with this clearing agent [Sally Hansen's "Hard as Nails" (Farmingdale, NY) works well].

Properly mounted samples can be stored in BA:BB for many months with no apparent loss of fluorescence. Although BA:BB by itself retards photobleaching of rhodamines, addition of 25–50 mg/ml propyl gallate (note that propyl gallate dissolves very slowly in BA:BB) can help retard photobleaching of fluorescein. The slight yellow–brown tint caused by adding propyl gallate to the clearing solution does not appear to interfere with fluorescence. However, samples are not stable in BA:BB plus propyl gallate, and should be examined within a few days of clearing.

D. Confocal Microscopy

The theory and practice of confocal microscopy are covered extensively in Matsumoto (1993) and Stevens *et al.* (1994), and will not be discussed in detail here. Our experience indicates that obtaining the best possible images of cells, such as amphibian oocytes and eggs, is dependent on a number of features. First, optimum imaging of individual MTs requires high-resolution objectives with large numerical apertures (NA 1.4). Unfortunately, these objectives also have limited working distances (<200 μm). In practice, this means that, although single stage I oocytes and small ovaries can be optically sectioned in their entirety, larger oocytes and eggs often must be physically sectioned. We simply hemisect large oocytes and eggs with a fresh scalpel (either laterally or along the equator) before antibody processing, and then use the confocal to optically section below the area of knife damage.

Next, the laser intensity, photomultiplier gain, pinhole aperture, and filter mode (for averaging images) must be optimized for each sample. All of these factors interact to affect the brightness of the collected image, as well as the rate of photobleaching. Optimized images will use as much of the dynamic range of the photomultiplier tubes as possible (usually 8 bits, or 256 levels of gray). While the digital images resulting from confocal microscopy are very amenable to enhancement using the established techniques of digital image processing [discussed in great detail in Russ (1994) and available through commercial and public domain image processing software], it is usually better to collect the best possible image by optimizing the conditions of image collection. We routinely limit postcollection image processing of digital images to minor adjustments of contrast. Finally, laser-scanning confocal microscopes are relatively easy and enjoyable to use, and we find that the quality of the images collected improves dramatically with the experience and enjoyment of the operator!

Acknowledgments

The authors thank Drs. T. Stearns, M. Kirschner, and G. Piperno for providing antibodies used for many of our studies, Ms. Pauline Jenkins for assistance with the figures, and Dr. Ed King for his encouraging and insightful comments throughout the course of these experiments. Our laboratory has been supported by Grant MCB-9117477 from the National Science Foundation and by grants from the University of Utah Research Committee and Research Instrumentation Fund.

References

Al–Mukhtar, K. A. K., and Webb, A. (1971). An ultrastructural study of primordial germ cells, oogonia, and early oocytes in *Xenopus* laevis. *J. Embryol. Exp. Morphol.* **26,** 195–217.

Andersen, S. S. L., Buendia, B., Dominguez, J. E., Sawyer, A., and Karsenti, E. (1994). Effect on microtubule dynamics of XMAP230, a microtubule-associated protein present in *Xenopus* laevis eggs and dividing cells. *J. Cell Biol.* **127,** 1289–1300.

Baas, P. W., and Ahmad, F. J. (1992). The plus-ends of stable microtubules are the exclusive nucleating structures for microtubules in the axon. *J. Cell Biol.* **116,** 1231–1241.

Bailly, E., Doree, M., Nurse, P., and Bornens, M. (1989). p34^{cdc2} is located in both nucleus and cytoplasm: Part is centrosomally associated at G$_2$/M and enters vesicles at anaphase. *EMBO J.* **8,** 3985–3995.

Bajer, A. (1972). Spindle dynamics and chromosome movements. *Int. Rev. Cytol. Suppl.* **3,** 1–271.

Beetschen, J.-C., and Gautier, J. (1989). Oogenesis. *In* "Developmental Biology of the Axolotl" (J. B. Armstrong and G. M. Malacinski, eds.), pp. 25–35. Oxford Univ. Press, New York.

Belmont, L. D., Hyman, A. A., Sawin, K. E., and Mitchison, T. J. (1990). Real time visualization of cell cycle dependent changes in microtubule dynamics in cytoplasmic extracts. *Cell* **62,** 579–589.

Bement, W. M., Gallicano, G. I., and Capco, David G. (1992). Role of the cytoskeleton during early development. *Microsc. Res. Tech.* **22,** 23–48.

Birkett, C. R., *et al.* (1985). *FEBS Lett.* **187,** 211.

Blose, S. H., Meltzer, D. I., and Feramisco, J. R. (1984). 10-nm filaments are induced to collapse in living cells microinjected with monoclonal and polyclonal antibodies against tubulin. *J. Cell Biol.* **98,** 847–858.

Brachet, J., Hanocq, F., and Van Gansen, P. (1970). A cytochemical and ultrastructural analysis of in vitro maturation in amphibian oocytes. *Dev. Biol.* **21,** 157–195.

Brinkley, B. R. (1985). Microtubule organizing centers. *Annu. Rev. Cell Biol.* **1,** 145–172.

Calarco-Gillam, P. Siebert, M., Hubble, R., Mitchison, T., and Kirschner, M. (1983). Centrosome development in early mouse embryos as defined by an autoantibody against pericentriolar material. *Cell* **35**, 621–629.

Capco, D. G., and Jeffery, W. R. (1982). Transient localizations of messenger RNA in *Xenopus laevis*. *Dev. Biol.* **89**, 1–12.

Carpenter, C. D., and Klein, W. (1982). A gradient of Poly (A)+ RNA sequences in *Xenopus laevis* eggs and embryos. *Dev. Biol.* **89**, 1–12.

Cassimeris, L., Pryer, N. K., and Salmon, E. D. (1988). Real-time observations of microtubule dynamic instability in living cells. *J. Cell Biol.* **107**, 2223–2231.

Cassimeris, L. U., Walker, R. A., Pryer, N. K., and Salmon, E. D. (1987). Dynamic instability of microtubules. *Bioessays* **7**, 149–154.

Cha, B.-J., Error, B. M., and Gard, D. L. (1994). The spatial and temporal distribution of a 250KD microtubule-associated protein (MAP250) in *Xenopus* oocytes and embryos. *Mol. Biol. Cell* **5**, 169a.

Chu, D. T. W., and Klymkowsky, M. W. (1989). The appearance of acetylated α-tubulin during early development and cellular differentiation in *Xenopus*. *Dev. Biol.* **136**, 104–117.

Church, K., Niklas, R. B., and Lin, H.-P. P. (1986). Micromanipulated bivalents can trigger mini-spindle formation in *Drosophila melanogaster* spermatocyte cytoplasm. *J. Cell Biol.* **103**, 2765–2773.

Coggins, L. W. (1973). An ultrastructural and radioautographic study of early oogenesis in the toad *Xenopus laevis*. *J. Cell Sci.* **12**, 71–93.

Danilchik, M., and Denegre, J. (1991). Deep cytoplasmic rearrangements during early development in *Xenopus laevis*. *Development* **111**, 845–856.

Danilchik, M., and Gerhart, J. (1987). Differentiation of the animal–vegetal axis in *Xenopus laevis* oocytes. I. Polarized intracellular translocation of platelets established the yolk gradient. *Dev. Biol.* **122**, 101–112.

Dent, J. A., and Klymkowsky, M. W. (1989). Whole-mount analysis of cytoskeletal reorganization and function during oogenesis and early embryogenesis in *Xenopus*. *In* "The Cell Biology of Fertilization" (H. Schatten and G. Schatten, eds.). pp. 63–103. Academic Press, Orlando.

Dent, J. A., Polson, A. G., and Klymkowsky, M. W. (1989). A whole-mount immunocytochemical analysis of the expression of the intermediate filament protein vimentin in *Xenopus*. *Development* **105**, 61–74.

Doxsey, S. J., Stein, P., Evans, L., Calarco, P., and Kirschner, M. (1994). Pericentrin, a highly conserved centrosome protein involved in microtubule organization. *Cell* **76**, 639–650.

Drechsel, D. N., Hyman, A. A., Cobb, M. H., and Kirschner, M. W. (1992). Modulation of dynamic instability of tubulin assembly by the microtubule-associated protein tau. *Mol. Biol. Cell* **3**, 1141–1154.

Dumont, J. (1972). Oogenesis in *Xenopus laevis* (Daudin). I. Stages of oocyte development in laboratory maintained animals. *J. Morphol.* **136**, 153–180.

Elinson, R. P. (1985). Changes in levels of polymeric tubulin associated with activation and dorso-ventral polarization of the frog egg. *Dev. Biol.* **109**, 224–233.

Elinson, R., and Rowning, B. (1988). A transient array of parallel microtubules in frog eggs: Potential tracks for a cytoplasmic rotation that specifies the dorso-ventral axis. *Dev. Biol.* **128**, 185–197.

Elinson, R. P., and Palecek, J. (1993). Independence of two microtubule systems in fertilized frog eggs: The sperm aster and vegetal parallel array. *Roux's Arch. Dev. Biol.* **202**, 224–232.

Euteneuer, U., and McIntosh, J. R. (1981a). Polarity of some motility-related microtubules. *Proc. Natl. Acad. Sci. USA* **78**, 372–376.

Euteneuer, U., and McIntosh, J. R. (1981b). Structural polarity of kinetochore microtubules in PtK1 cells. *J. Cell Biol.* **89,** 338–345.

Faruki, S., and Karsenti, E. (1994). Purification of microtubule proteins from *Xenopus* egg extracts: Identification of a 230K MAP4-like protein. *Cell Motil. Cytoskeleton* **28,** 108–118.

Fellous, A., Huchon, D., Thibier, C., and Jessus, C. (1991). Intracellular location of MAP2-related protein (O-map) in prophase I and metaphase II oocytes of *Xenopus. Mech. Dev.* **33,** 139–146.

Gard, D. L. (1991a). Organization, nucleation, and acetylation of microtubules in *Xenopus laevis* oocytes: A study by confocal immunofluorescence microscopy. *Dev. Biol.* **143,** 346–362.

Gard, D. L. (1991b). MT organization and spindle assembly during meiotic maturation of *Xenopus* oocytes. *J. Cell Biol.* **115,** 46a.

Gard, D. L. (1992). Microtubule organization during maturation of *Xenopus* oocytes: Assembly and rotation of the meiotic spindles. *Dev. Biol.* **151,** 516–530.

Gard, D. L. (1993a). Ectopic spindle assembly during maturation of *Xenopus* oocytes: Evidence for functional polarization of the oocyte cortex. *Dev. Biol.* **159,** 298–310.

Gard, D. L. (1993b). Confocal microscopy of microtubules in amphibian oocytes and eggs. *In* "Cell Biological Applications of Confocal Microscopy" (Methods in Cell Biology, Vol. 38) (B. Matsumoto, ed.), pp. 241–264. Academic Press, San Diego.

Gard, D. L. (1994). γ-Tubulin is asymmetrically distributed in the cortex of *Xenopus* oocytes. *Dev. Biol.* **161,** 131–140.

Gard, D. L. (1995). Axis formation during amphibian oogenesis: Reevaluating the role of the cytoskeleton. *Curr. Topics Dev. Biol.* **30,** 213–250.

Gard, D. L., Affleck, D., and Error, B. (1995a) Microtubule organization, acetylation, and nucleation in *Xenopus laevis* oocytes: II. A developmental transition in microtubule organization during early diplotene. *Dev. Biol.* **163,** 189–201.

Gard, D. L., Cha, B.-J., and Roeder, A. D. (1995b). F-actin is required for spindle anchoring and rotation in *Xenopus* oocytes: A re-examination of the effects of cytochalasin B on oocyte maturation. *Zygote* **3,** 17–26.

Gard, D. L., Hafezi, S., Zhang, T., and Doxsey, S. J. (1990). Centrosome duplication continues in cycloheximide-treated *Xenopus* blastulae in the absence of a detectable cell cycle. *J. Cell Biol.* **110,** 2033–2042,

Gard, D. L., and Kirschner, M. (1987a). Microtubule assembly in cytoplasmic extracts of *Xenopus* oocytes and eggs. *J. Cell Biol.* **105,** 2191–2201.

Gard, D. L., and Kirschner, M. (1987b). A microtubule-associated protein from *Xenopus* eggs that specifically promotes assembly at the plus-end. *J. Cell Biol.* **105,** 2203–2215.

Gard, D. L., and Kropf, D. L. (1993). Confocal immunofluorescence microscopy of plant and animal oocytes, eggs, and embryos. *In* "Antibodies in Cell Biology" (Methods in Cell Biology, Vol. 37) (D. Asai, ed.), pp. 147–169. Academic Press, San Diego.

Gerhart, J. C. (1980). Mechanisms regulating pattern formation in the amphibian egg and early embryo. *In* "Biological Regulation and Development" (R. F. Goldberger, ed.), pp. 133–316. Plenum Press, New York.

Gerhart, J., Black, S., Gimlich, R., and Schart, S. (1983). Control of polarity in the amphibian egg. *In* "Time, Space, and Pattern in Embryonic Development" (W. R. Jeffrey and R. A. Raff, eds), pp. 261–286. A. R. Liss, New York.

Gerhart, J., Danilchik, M., Roberts, J., Rowning, B., and Vincent, J.-P. (1986). Primary and secondary polarity of the Amphibian oocyte and egg. *In* "Gametogenesis and the Early Embryo," pp. 305–319. A. R. Liss, New York.

Godsave, S. F., Wylie, C. C., Lane, E. B., and Anderton, B. H. (1984). Intermediate filaments in the *Xenopus* oocyte: The appearance and distribution of cytokeratin-containing filaments. *J. Embryol. Exp. Morphol.* **83,** 157–167.

Gururajan, R., Perry-O'keefe, H., Melton, D., and Weeks, D. (1991). The *Xenopus* localized messenger RNA An3 may encode an ATP-dependent RNA helicase. *Nature* **349**, 717–719.

Hartwig, J. H. (1992). Mechanisms of actin rearrangements mediating platelet activation. *J. Cell Biol.* **118**, 1421–1442.

Hausen, P., and Riebesell, M. (1990). "The Early Development of Xenopis laevis. An Atlas of Histology." Springer-Verlag, Berlin.

Heasman, J., Quarmby, J., and Wylie, C. C. (1984). The Mitochondrial Cloud of *Xenopus* Oocytes: The source of germinal granule material. *Dev. Biol.* **105**, 458–469.

Heidemann, S. R., Hambborg, M. A., Balasz, J. E., and Lindley, S. (1985). Microtubules in immature oocytes of *Xenopus* laevis. *J. Cell Sci.* **77**, 129–141.

Heidemann, S. R., and Gallas, P. T. (1980). The effect of taxol on living eggs of *Xenopus* laevis. *Dev. Biol.* **80**, 489–494.

Heidemann, S. R., and Kirschner, M. W. (1975). Aster formation in eggs of *Xenopus laevis.* Induction by isolated basal bodies. *J. Cell Biol.* **67**, 105–117.

Heidemann, S. R., and Kirschner, M. W. (1978). Induced formation of asters and cleavage furrows in oocytes of *Xenopus laevis* during in vitro maturation. *J. Exp. Zool.* **206**, 431–444.

Houliston, E. (1994). Microtubule translocation and polymerization during cortical rotation in *Xenopus* eggs. *Development* **120**, 1213–1220.

Houliston, E., and Elinson, R. P. (1991a). Patterns of MT polymerization relating to cortica rotation in *Xenopus laevis* eggs. *Development* **112**, 107–117.

Houliston, E., and Elinson, R. P. (1991b). Evidence for the involvement of microtubules, ER, and kinesin in the cortical rotation of fertilized frog eggs. *J. Cell Biol.* **114**, 1017–1028.

Houliston, E., and Elinson, R. P. (1992). Microtubules and cytoplasmic reorganization in the frog egg. *Curr. Topics Dev. Biol.* **26**, 53–70.

Houliston, E., Le Guellec, R., Kress, M., Phillipe, M., and LeGuellic, K. (1994). The kinesin-related protein Eg5 associates with both interphase and spindle microtubules during *Xenopus* early development. *Dev. Biol.* **164**, 147–159.

Huchon, D., Crozet, N., Cantenot, N., and Ozon, R. (1981). Germinal vesicle breakdown in the *Xenopus laevis* oocyte: Description of a transient microtubular structure. *Reprod. Nutr. Dev.* **21**, 135–148.

Huchon, D., Jessus, C., Thibier, C., and Ozon, R. (1988). Presence of microtubules in isolated cortices of prophase I and metaphase II oocytes in *Xenopus laevis.* *Cell Tissue Res.* **154**, 415–420.

Iwao, Y., and Elinson, R. P. (1990). Control of sperm nuclear behavior in physiologically polyspermic newt eggs: Possible involvement of MPF. *Dev. Biol.* **142**, 301–312.

Jessus, C., Friederich, E., Francon, J., and Ozon, R. (1984). In vitro inhibition of tubulin assembly by a ribonucleoprotein complex associated with the free ribosomal fraction isolated from *Xenopus laevis* oocytes: Effect at the level of microtubule-associated proteins. *Cell Differ.* **14**, 179–187.

Jessus, C., Huchon, D., and Ozon, R. (1986). Distribution of microtubules during the breakdown of the nuclear envelope of the *Xenopus* oocyte: An immunocytochemical study. *Biol. Cell* **56**, 113–120.

Jessus, C., Thibier, C., and Ozon, R. (1987). Levels of microtubules during meiotic maturation of the *Xenopus* oocyte. *J. Cell Sci.* **87**, 705–712.

Joshi, H. C., Palacios, M. J., McNamera, L., and Cleveland, D. (1992). γ-Tubulin is a centrosomal protein required for cell-cycle-dependent microtubule nucleation. *Nature* **356**, 80–83.

Karsenti, E., Newport, J., Hubble, R., and Kirschner, M. (1984) . Interconversion of metaphase and interphase microtubule arrays, as studied by the injection of centrosomes and nuclei into *Xenopus* eggs. *J. Cell Biol.* **98**, 1730–1745.

King, M. L., and Barklis, E. (1985). Regional distribution of maternal messenger RNA in the amphibian oocyte. *Dev. Biol.* **112,** 203–212.

King, M. L., Forristall, C., and Zhou, Y. (1993). RNA localization to the vegetal cortex of *Xenopus* oocytes. *Mol. Biol. Cell* **4** (Suppl.), 5a.

Kirschner, M., and Mitchison, T. (1986). Beyond self-assembly: From microtubules to morphogenesis. *Cell* **45,** 329–342.

Klymkowsky, M., and Hanken, J. (1991). Whole-mount staining of *Xenopus* and other vertebrates. *Methods Cell Biol.* **36,** 419–441.

Klymkowsky, M. W., Maynell, L. A., and Polson, A. G. (1987). Polar asymmetry in the organization of the cortical cytokeratin system of *Xenopus laevis* oocytes and embryos. *Development* **100,** 543–557.

Klymkowsky, M. W., Maynell, L. A., and Nislow, C. (1991). Cytokeratin phosphorylation,cytokeratin filament severing, and the solubilization of the maternal mRNA Vg1. *J. Cell Biol.* **114,** 787–797.

Kowalski, R., and Williams, R. C. (1993). Microtubule-associated protein 2 alters the dynamic properties of microtubule assembly and disassembly. *J. Biol. Chem.* **268,** 9847–9855.

Kreis, T. E. (1987). Microtubules containing detyrosinated tubulin are less dynamic. *EMBO J.* **6,** 2597–2606.

Lessman, C. A. (1987). Germinal vesicle migration and dissolution in *Rana pipiens* oocytes: Effects of steroids and microtubule poisons. *Cell Differ.* **20,** 238–251.

Malacinski, G. M., Benford, H., and Chung, H. M. (1975). Association of an ultraviolet irradiation senesditive cytoplasmic localization with the future dorsal side of the amphibian egg. *J. Exp. Zool.* **191,** 97–110.

Manes, M. E., and Barbieri, F. D. (1977). On the possibility of sperm aster involvement in dorso-ventral polarization and pronuclear migration in the amphibian egg. *J. Embryol. Exp. Morphol.* **40,** 187–197.

Manes, M. E., and Elinson, R. P. (1980). Ultraviolet light inhibits grey crescent formation on the frog egg. *Roux's Arch. Dev. Biol.* **189,** 73–76.

Manes, M. E., Elinson, R. P., and Barbieri, F. D. (1978). Formation of the amphibian egg grey crescent: Effects of colchicine and cytochalasin B. *Roux's Arch. Dev. Biol.* **185,** 99–104.

Matsumoto, B. (ed.) (1993). Cell biological applications of confocal microscopy. *In* "Methods in Cell Biology" Vol. 38, pp. 380.

McIntosh, J. R. (1983). The centrosome as an organizer of the cytoskeleton. *Mod. Cell Biol.* **2,** 115–142.

McNally, F. J., and Vale, R. D. (1993). Identification of Katanin, and ATPase that severs and disassembles stable microtubules. *Cell* **75,** 419–429.

Melton, D. (1987). Translocation of a localized maternal mRNA to the vegetal pole of *Xenopus* oocytes. *Nature* **328,** 80–82.

Mitchison, T. J., and Kirschner, M. W. (1984). Microtubule assembly nucleated by isolated centrosomes. *Nature (London)* **312,** 232–236.

Mowry, K. L., and Melton, D. A. (1992). Vegetal messenger RNA localization directed by a 340-nt RNA sequence element in *Xenopus* oocytes. *Science* **255,** 991–994.

Oakley, B. R., Oakley, C. E., Yoon, Y., and Jung, M. K. (1990). γ-Tubulin is a component of the spindle pole body that is essential for microtubule function in *Aspergillus nidulans. Cell* **61,** 1289–1301.

Oakley, C. E., and Oakley, B. R. (1989). Identification of γ-Tubulin, a new member of the tubulin superfamily encoded by the mipA gene of *Aspergillus nidulans. Nature (London)* **338,** 662–664.

Palecek, J., Habrova, V., Nedvidek, J., and Romanovsky, A. (1985). Dynamics of tubulin structures in *Xenopus laevis* oogenesis. *J. Embryol. Exp. Morphol.* **87,** 75–86.

Pestell, R. Q. W. (1975). Microtubule protein synthesis during oogenesis and early embryogenesis in *Xenopus laevis*. *Biochem. J.* **145**, 527–534.

Piperno, G., LeDizet, M., and Chang, X.-J. (1987). Microtubules containing acetylated α-tubulin in mammalian cells in culture. *J. Cell Biol.* **104**, 289–302.

Pondel, M., and King, M. L. (1988). Localized maternal mRNA related to transforming growth factor b mRNA is concentrated in a cytokeratin-enriched fraction from *Xenopus* oocytes. *Proc. Natl. Acad. Sci. USA* **85**, 7612–7616.

Pryer, N. K., Walker, R. A., Skeen, V. P., Bourns, B. D., Soboeiro, M. F., and Salmon, E. D. (1992). Microtubule-associated proteins modulate microtubule dynamic instability in vitro: Real-time observations using video microscopy. *J. Cell Sci.* **103**, 965–976.

Rattner, J. B., Lew, J., and Wang, J. H. (1990). p34^{cdc2} kinase is localized to distinct domains within the mitotic apparatus. *Cell Motil. Cytoskeleton* **17**, 227–235.

Riabowal, K., Draetta, G., Bizuela, L., Vandre, D., and Beach, D. (1989). The cdc2 kinase is a nuclear protein that is essential for mitosis in mammalian cells. *Cell* **57**, 393–401.

Rieder, C. L., Ault, J. G., Eichenlaub-Ritter, U., and Sluder, G. (1993). Morphogenesis of the mitotic and meiotic spindle: Conclusions obtained from one system are not necessarily applicable to the other. *In* "Chromosome Segregation and Aneuploidy" (B. K. Vig and A. Kappas, eds.). Springer-Verlag, New York.

Roeder, A. D., and Gard, D. L. Gard (1994). Confocal microscopy of F-actin distribution in *Xenopus* oocytes *Zygote* **2**, 111–124.

Rowning, B. A., Larabell, C. A., Wells, J. C., Wu, M., and Gerhart, J. C. (1994). Inverted confocal microscopy analysis of microtubule arrays during cortical rotation in the first cell cycle of living *Xenopus* eggs. *Mol. Biol. Cell* **5**, 256a.

Russ, J. C. (1994). "The Image Processing Handbook," pp. 674. CRC Press, Boca Raton, Florida.

Ryabova, L. P., Betina, M. A., and Vassetzky, S. G. (1986). Influence of cytochalasin B on oocyte maturation in *Xenopus laevis*. *Cell Differ.* **19**, 89–96.

Sawin, K., and Mitchison, T. (1991). Mitotic assembly by two different pathways *in vitro*. *J. Cell Biol.* **112**, 925–940.

Scharf, S., and Gerhart, J. (1983). Axis determination in eggs of *Xenopus laevis:* A critical period before first cleavage, identified by the common effects of cold, pressure, and ultraviolet irradiation. *Dev. Biol.* **99**, 75–87.

Schroeder, M. M., and Gard, D. L. (1992). Organization and regulation of cortical microtubules during the first cell cycle of *Xenopus* eggs. *Development* **114**, 699–709.

Schulze, E., Asai, D. J., Bulinski, J. C., and Kirschner, M. W. (1987). Post-translational modification and microtubule stability. *J. Cell Biol.* **105**, 2167–2177.

Schwartz, S. P., Aisenthal, L., Elisha, Z., Oberman, F., and Yisraeili, J. K. (1992). A 69-kDa RNA-binding protein from *Xenopus* oocytes recognizes a common motif in two vegetally localized maternal mRNAs. *Proc. Natl. Acad. Sci. USA* **89**, 11895–11899.

Sheldon, E., and Wadsworth, P. (1993). Observation and quantification of individual microtubule behavior in vivo: Microtubule dynamics are cell-type specific. *J. Cell Biol.* **120**, 935–945.

Shiina, N., Moriguchi, T., Ohta, K., Gotoh, Y., and Nishida, E. (1992). Regulation of a major microtubule-associated protein by MPF and MAP kinase. *EMBO J.* **11**, 3977–3984.

Shiina, N., Gotoh, Y., and Nishida, E. (1992). A novel homo-oligomeric protein responsible for an MPF-dependent microtubule-severing activity. *EMBO J.* **11**, 4723–4731.

Skoufias, D. A., and Scholey, J. M. (1993). Cytoplasmic microtubule-base motor proteins. *Curr. Opin. Cell Biol.* **5**, 95–104.

Stearns, T., Evans, L., and Kirschner, M. (1991). γ-Tubulin is a highly conserved component of the centrosome. *Cell* **65**, 825–836.

Stearns, T., and Kirschner, M. (1994). In vitro reconstitution of centrosome assembly and function: The central role of γ-tubulin. *Cell* **76,** 623–637.

Steffen, W., Fuge, H., Dietz, R., Bastmeyer, M., and Muller, G. (1986). Aster-free spindle polesin insect spermatocytes: Evidence for chromosome-induced spindle formation. *J. Cell Biol.* **102,** 1679–1687.

Stevens, J. K., Mills, L. R., and Trogadis, J. E. (eds.) (1994). "Three Dimensional Confocal Microscopy: Volume Investigation of Biological Specimens," pp. 507. Academic Press (New York).

Stewart-Savage, J., and Grey, R. D. (1982). The temporal and spatial relationships between cortical contraction, sperm trail fromation, and pronuclear migration in fertilized *Xenopus* eggs. *Roux's Arch. Dev. Biol.* **191,** 241–245.

Theurkauf, W. E., and Hawley, R. S. (1992). Meiotic spindle assembly in *Drosophila* females: Behavior of non-exchange chromosomes and the effects of mutations in the *Nod* kinesin-like protein. *J. Cell Biol.* **116,** 1167–1180.

Tourte, M., Besse, C., and Mouunolou, J. C. (1991). Cytochemical evidence of an organized microtubule cytoskeleton in *Xenopus laevis* oocytes: Involvement in the segregation of mitochondrial populations. *Mol. Reprod. Dev.* **30,** 353–359.

Vale, R. D. (1991). Severing of stable MTs by a mitotically-activated protein in *Xenopus* egg extracts. *Cell* **64,** 827–839.

Vandre, D. D., and Borisy G. G. (1989). The centrosome cycle in animal cells. *In* "Mitosis: Molecules and Mechanisms" (J. S. Hyams and B. R. Brinkley, eds.), pp. 39–75. Academic Press, New York.

Vasquez, R., Gard, D. L., and Cassimeris, L. (1994). XMAP from *Xenopus* eggs promotes rapid plus end assembly of microtubules and rapid microtubule turnover. *J. Cell Biol.* **127,** 985–994.

Verde, F., Berrez, J.-M., Antony, C., and Karsenti, E. (1991). Taxol-induced microtubule asters in mitotic extracts of *Xenopus* eggs: Requirements for phosphorylated factors and cytoplasmic dynein. *J. Cell Biol.* **112,** 1177–1187.

Verde, F., Dogterom, M., Stelzer, E., Karsenti, E., and Leibler, S. (1992). Control of microtubule dynamics vy cyclin-A and cyclin-B-dependent protein kinases in *Xenopus* egg extracts. *J. Cell Biol.* **118,** 1097–1108.

Vincent, J.-P., Oster, G., and Gerhart, J. (1986). Kinematics of gray crescent formation in *Xenopus* eggs: The displacement of subcortical cytoplasm relative to the egg surface. *Dev. Biol.* **113,** 484–500.

Vincent, J.-P. Scharf, S. R., and Gerhart, J. C. (1987) Subcortical rotation in *Xenopus* eggs: A preliminary study of its mechanochemical basis. *Cell Motil. Cytoskeleton* **8,** 143–154.

Walker, R. A., O'Brien, E. T., Pryer, N. K., Soboeiro, M. F., Voter, W. A., Erickson, H. P., and Salmon, E. D. (1988). Dynamic instability of individual microtubules analyzed by video light microscopy: Rate constants and transition frequencies. *J. Cell Biol.* **107,** 1437–1448.

Wang, T., Lessman, C. A., and Gard, D. L. (1993). Developmental regulation of cold-stable microtubules during oogenesis in Rana pipiens. *Mol. Biol. Cell* **4** (Suppl.), 26a.

Webster, D. R., and Borisy, G. G. (1989). Microtubules are acetylated in domains that turn over slowly. *J. Cell Sci.* **92,** 57–65.

Weeks, D. L., and Melton, D. A. (1987). A maternal mRNA localized to the vegetal hemisphere in *Xenopus* eggs codes for a growth factor related to TGF-β. *Cell* **51,** 861–867.

Wylie, C. C., Brown, D., Godsave, S. F., Quarmby, J., and Heasman, J. (1985). The cytoskeleton of *Xenopus* oocytes and its role in development. *J. Embryol. Exp. Morphol.* **89** (Suppl.), 1–15.

Yisraeli, J. K., Sokol, S., and Melton, D. A. (1990). A two-step model for the localization of maternal mRNA in *Xenopus* oocytes: Involvement of microtubules and microfilaments in the translocation and anchoring of Vg1 mRNA. *Development* **108,** 289–298.

Zheng, Y., Jung, M. K., and Oakley, B. R. (1991). γ-Tubulin is present in *Drosophila melanogaster* and *Homo sapiens* and is associated with the centrosome. *Cell* **65,** 817–824.

Zhou, Y., Forristall, C., and King, M. L. (1993). XCAT-2 RNA localization in *Xenopus* oocytes. *Mol. Biol. Cell* **4** (Suppl.), 25a.

13

Cortical Cytoskeleton of the *Xenopus* Oocyte, Egg, and Early Embryo

Carolyn A. Larabell
Lawrence Berkeley National Laboratory
University of California
Berkeley, California 94720

I. Introduction
II. Cortical Cytoskeleton of the Meiotically Immature Oocyte
    A. Structure and Composition of the Oocyte Cortex
    B. Onset of Contractility: Role of Protein Kinase C
    C. Localizations of RNA
III. Cortical Cytoskeleton of the Unfertilized Egg
    A. Organization of the Unfertilized Egg Cortex
    B. Cortical Actin
    C. Other Cytoskeletal Proteins
IV. Cortical Cytoskeleton during the First Cell Cycle
    A. The Zygote Cortex: A New Structure, a New Function
    B. Cortical Actin
    C. Microtubule Arrays during Cortical Rotation in Live Embryos
V. Conclusion
    References

I. Introduction

The cortex of eggs and embryos is a highly specialized region of the cell that undergoes rapid changes in structure and function, in response to biochemical signals, throughout oogenesis and development. Progesterone triggers resumption of meiosis, which results in alignment of the cortical granules beneath the plasma membrane, formation of the endoplasmic reticulum, and conversion of the unfertilizable oocyte into the fertilizable egg. Biochemical signals at fertilization trigger further modifications of the cortex. For example, activation of protein kinase C triggers cortical granule exocytosis, the process that delivers proteins to the vitelline envelope to convert it to the sperm-blocking fertilization envelope. Another major modification of the egg occurs midway through the first cell cycle. There is a remarkable 30° displacement of the newly remodeled cortex with respect to the inner cytoplasmic mass; this rotation is critical for specification of the dorsal–ventral axis and normal development. All of these cortical

Current Topics in Developmental Biology, Vol. 31
433

changes, and many others that occur throughout development, are accompanied by a reorganization of the cortical cytoskeleton and are required to generate a normal, healthy organism.

While in most cells the cortex is considered to be the outer 1 or 2 μm of the cell surface, I will refer to the cortex of *Xenopus* oocytes, eggs, and unicellular zygotes as the outer 4 or 5 μm of the cell, based on structural and functional distinctions. If one assumes an average somatic cell is approximately 10 μm in diameter, then a cortex 1-μm thick comprises 4.89% of the cell volume. Since *Xenopus* eggs are very large cells, between 1.2 and 1.4 μm in diameter, the cortex of the egg is an even smaller relative portion of the total cell volume. If one describes the cortex as the outer 1 μm of this 1200 μm diameter egg, the cortex comprises only 0.55% of the cell volume, while a 5-μm thick cortex contains 2.51% of the egg volume. Whether one defines the cortex as the outer 1 μm of the cell or the outer 5 μm, the *Xenopus* egg cortex comprises a very small portion of the cell that is responsible for orchestrating major events throughout development. The role of the cytoskeleton in the dynamic reorganizations of the egg cortex throughout oogenesis and development is of great interest.

II. Cortical Cytoskeleton of the Meiotically Immature Oocyte

The structural organization of the cortex of the stage VI, meiotically immature oocyte is quite different from that of the fertilizable egg; this most likely reflects the change in function that accompanies the progression through meiosis. The function of the stage VI oocyte, which is arrested at prophase I of meiosis, is that of storage and synthesis. The cortex contains specialized distributions of cytoskeletal proteins, localizations of mRNAs, and compartmentalized proteins that are required at fertilization or during later development. The cortex of the fertilizable egg, on the other hand, is designed to rapidly propagate the waves of increased intracellular free calcium, cortical granule exocytosis, microvillar elongation, and cortical contraction—rapid reorganizations that are critical for the onset of normal development.

A. Structure and Composition of the Oocyte Cortex

The plasma membrane of the oocyte is highly folded with many long projections, referred to as macrovilli, extending into the perivitelline space and long, thin microvilli covering the entire egg surface (Larabell and Chandler, 1988a). The cortical granules are irregularly distributed throughout the egg cortex rather than positioned directly beneath the plasma membrane as required for exocytosis. Small vesicles, mitochondria, and pigment granules are also seen in the cortex, but the elaborate calcium-sequestering endoplasmic reticulum seen in fertilizable eggs is not yet present.

1. Cortical Actin

The most abundant cytoskeletal protein in the oocyte cortex is actin. Actin has been seen in the microvilli and in a band beneath the plasma membrane using immunofluorescence light microscopy (Colombo *et al.*, 1981), transmission electron microscopy (Franke *et al.*, 1976), and immunogold analyses (Gall *et al.*, 1983; Ryabova, 1982, 1990; Ryabova *et al.*, 1992, 1994). Figure 1 shows the distribution of actin in the cortex isolated from an oocyte as described previously (Larabell, 1993) and labeled with anti-actin antibodies

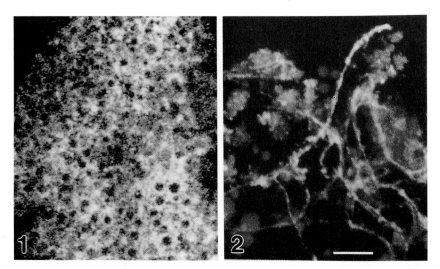

Fig. 1 Micrograph of a cortex isolated from a meiotically immature (stage VI) oocyte that was fixed in 2% paraformaldehyde and 0.1% glutaraldehyde, then incubated in anti-actin antibodies. Micrograph is a projection of three optical sections from the MRC 1000 laser scanning confocal microscope. The highly folded oocyte surface yielded optical sections showing different regions of the cortex with very different information; consequently, a projection of multiple sections was required. A dense mat of actin is seen surrounding the irregularly placed cortical granules (black, circular shadows). In some regions, however, only punctate labeling of microvilli can be seen. Scale bar = 10 μm.

Fig. 2 Isolated oocyte cortex, fixed in 2% paraformaldehyde and 0.1% glutaraldehyde then labeled with rhodamine–phalloidin, showing long actin filaments that are located at least 5 μm beneath the plasma membrane. Scale bar = 10 μm.

then viewed in the confocal microscope. This confocal micrograph is a projection of three optical sections obtained parallel to the cell surface. There is a dense blanket of actin with randomly distributed secretory granules and some punctate staining representing microvilli. Since the cortical granules are situated at varying distances from the plasma membrane, which is also highly folded, there are regions in this view of the cortex with large amounts of actin and no cortical granules. This pattern is very different from that seen in cortices isolated from meiotically mature, unfertilized eggs which display an orderly distribution of cortical granules (see Fig. 7).

2. Interaction of Cortex with Inner Cytoplasm

Actin has also been reported to extend from the cortex into regions of the cytoplasm referred to as cytoplasmic corridors (Ryabova *et al.,* 1992, 1994; Ryabova and Vassetzky, 1993). These corridors are large areas of yolk-free cytoplasm in the animal hemisphere that contain membrane cisternae, mitochondria, and annulate lamellae (Bement and Capco, 1989a). It has been proposed that the cytoplasmic corridors link the cortex with the inner cytoplasm and provide pathways for movement of organelles to and from the nucleus. It is quite possible that organelles and other egg components travel along these actin filaments from the nucleus to their destination in the cortex, or that endocytotic vesicles use them to travel from the cell surface to their cytoplasmic destination.

Cortices isolated from oocytes then fixed and incubated in rhodamine-labeled phalloidin revealed fluorescently labeled actin filaments (Fig. 2) that may represent the actin tracks connecting the cortex with the inner cytoplasm. This possibility is supported by studies performed using isolated cortices that are maintained in ATP-regenerating buffer and viewed using differential interference contrast microscopy (C. Larabell, unpublished data). The region just beneath the plasma membrane demonstrated numerous organelles securely attached to the cell surface and showed no organellar movement whatsoever. These were presumably the cortical granules that are enmeshed in actin. The region slightly deeper than the cortical granules showed a number of pigment granules, as well as other organelles (including mitochondria and endoplasmic reticulum) that demonstrated Brownian movements. Of great interest is that in thicker regions of the isolated cortex, there were long fibers decorated with organelles. These organelles demonstrated unidirectional movements at velocities of between 0.3 and 4.8 μm/sec. Figure 3 shows organelles moving along one of these tracks at 0.7 μm/sec. Directional displacement of organelles continued for as long as 1 hr during visualization, as long as ATP was in the buffer. Cortices isolated from unfertilized eggs, however, demonstrated neither long fibers nor directional organelle movements. The presence of long fibers

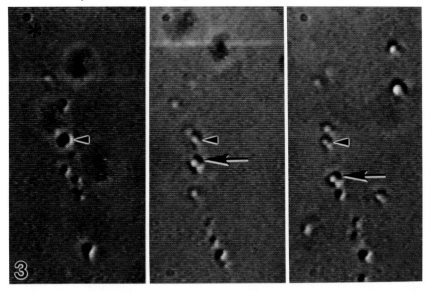

Fig. 3 Isolated oocyte cortex viewed with differential interference contrast optics. Images were collected using a CCD camera and time-lapse video recorder. Organelles can be seen moving along fibers at a velocity of 0.7 μm/sec. The starting point of selected organelles is labeled with an arrowhead, and the new position with an arrow. The asterisk points out a nonmoving marker.

extending from the oocyte cortex and their absence in unfertilized eggs correlates with the disappearance of cytoplasmic corridors during meiotic maturation (Bement and Capco, 1989b), further suggesting that the filaments seen in oocyte cortices represent pathways between the inner cytoplasm and cortex.

3. Other Cytoskeletal Proteins

The actin-binding proteins, talin (Evans *et al.*, 1990) and spectrin (Ryabova *et al.*, 1992, 1994), are also found in the cortex of the meiotically immature oocyte. While talin remains in the cortex during meiotic maturation and is found in the unfertilized egg cortex, spectrin has not been found in the cortex of the fertilizable egg (Ryabova *et al.*, 1992, 1994). Vinculin, on the other hand, has been seen in the inner cytoplasm of the oocyte, but not in the cortex (Evans *et al.*, 1990). After the resumption of meiosis, however, vinculin is found colocalized with talin in the cortex of the fertilizable egg (Evans *et al.*, 1990). The role of these cytoskeleton proteins during development has not been well defined. The reorganization of actin-binding proteins in the cortex may reflect the altered function of the cortex of the fertilizable egg. The function of the oocyte cytoskeleton is to secure components in the cortex for future usage at fertilization, while the cytoskeleton of the fertilizable egg is poised for action. Upon sperm/egg interaction, the cortex undergoes rapid transformations as waves of cortical granule exocytosis, microvillar elongation, and cortical contraction cross the egg.

A unique distribution of intermediate filaments, including cytokeratin and vimentin, is seen in the oocyte cortex, where cytokeratin filaments form a geodesic array (Franz *et al.*, 1983; Gall *et al.*, 1983; Godsave *et al.*, 1984a,b; Klymkowsky *et al.*, 1987, 1991); these filaments virtually disappear during meiotic maturation (Godsave *et al.*, 1984b; Gall and Karsenti, 1987; Klymkowsky *et al.*, 1987). Intermediate filaments have also been seen in the cytoplasmic corridors in the oocyte (Godsave *et al.*, 1984b, Tang *et al.*, 1988; Torpey *et al.*, 1992). For a detailed discussion of intermediate filaments in the *Xenopus* egg, see Chapter 14 in this book by Klymkowsky.

A complex network of microtubules is also seen in the oocyte and egg cortex, as well as throughout the cytoplasm (Huchon *et al.*, 1988; Gard, 1991). An example of an interaction of microtubules with a mitochondrion is demonstrated in stereo electron micrographs obtained from a cortex isolated from the meiotically immature oocyte and viewed as a whole mount in the intermediate voltage electron microscope (Fig. 4). The distribution of microtubules, and their function in *Xenopus* oocytes and eggs, is discussed in detail by Gard *et al.* in Chapter 12 of this book. The oocyte cortex also contains myosin (Ryabova *et al.*, 1992, 1994), but the meiotically immature oocyte is not capable of undergoing an actin/myosin contraction. Contractile ability is acquired during meiotic maturation (Meeusen and Cande, 1979; Christensen *et al.*, 1984; Ezzell *et al.*, 1985), as discussed below.

B. Onset of Contractility: Role of Protein Kinase C

The cortex of the newly fertilized egg undergoes a wave of contraction that is believed to play a role in bringing the male and female pronuclei together (Elinson, 1980). This contraction is believed to be mediated by the interaction of the cortical actin filament network with myosin, since contraction is sensitive to *N*-ethylmelamide-modified heavy meromyosin *in vitro* and *in vivo* (Christensen *et al.*, 1984; Ezzell *et al.*, 1985). Depletion of myosin from eggs sliced in half prevents contraction, and addition of myosin to the bisected eggs restores contractility (Christensen *et al.*, 1984). The meiotically immature oocyte, however, is not yet capable of undergoing a contractile response, even though both actin and myosin are present in the cortex. The cortex acquires the ability to undergo a cortical contraction during resumption of meiosis (Gingell, 1970; Schroeder and Strickland, 1974; Mer-

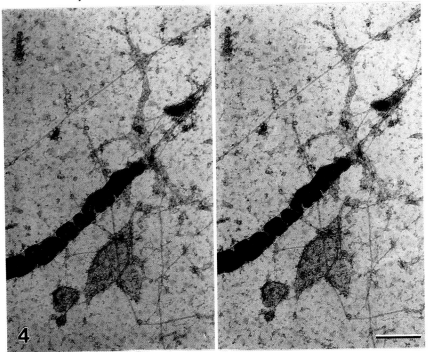

Fig. 4 Microtubules in the isolated oocyte cortex viewed as a whole mount at 400 kV in a JEOL 4000 intermediate voltage electron microscope. A long mitochondrion is in direct contact with the microtubules.

riam and Sauterer, 1983; Ryabova, 1982; Ezzell *et al.*, 1985) within a few hours after addition of progesterone (Capco *et al.*, 1992).

The oocyte cortex acquires a contractile ability immediately after treatment with progesterone, less than 5% of the way through meiotic resumption, which is depicted upon bisection of these oocytes as the ability to form a contractile ring at the bisected surface (Capco *et al.*, 1992). The ability to form contractile rings is gradually lost during the meiotic resumption and is replaced by the ability of the cortex to undergo a complete cortical contraction about 80% of the way through meiosis (Capco *et al.*, 1992). However, contractile rings can be induced to form in hemispheres of bisected oocytes in the absence of progesterone by activation of the enzyme protein kinase C (PKC), suggesting that PKC has a role in the reorganization of the cortical cytoskeleton that is ordinarily triggered by progesterone (Bement and Capco, 1991). It has also been shown that progesterone induces activation of PKC (Olson and Capco, 1992), and that PKC antagonists block the progesterone-induced reorganization of the cortical cytoskeleton

(Capco *et al.*, 1992). These data suggest that activation of PKC is the requisite signal for the onset of contractility. Of interest is the fact that cytochalasin B does not block cortical contraction in ionophore-activated metaphase II eggs (Merriam and Sauterer, 1983) or PKC-activated eggs (Capco *et al.*, 1992), but does block contraction induced by PKC agonists in the meiotically immature oocyte (Capco *et al.*, 1992). This suggests that the organization of actin in the oocyte is different than that in the egg. The specific cytoskeletal reorganization that occurs, however, is not yet known.

The onset of contractility has recently been temporally correlated with the disruption of the cytoplasmic corridors during meiotic maturation, which begins near the equator of the oocyte and progresses to the animal pole (Ryabova *et al.*, 1992, 1994). In addition, the loss of sensitivity of cortical actin to cytochalasin B both spatially and temporally accompanies the disassembly of cytoplasmic corridors (Ryabova *et al.*, 1994). These authors suggest that disruption of the corridors frees the cortex from constraints of the inner cytoplasm, facilitating a reorganization of cortical actin into a configuration capable of manifesting a contractile response (Ryabova *et al.*, 1994).

Contractility of the oocyte cortex can also be brought about by expression of *src* in oocytes. Deregulated pp60$^{c\text{-}src}$ (pp60$^{d\text{-}src}$), which lacks kinase activity due to loss of a carboxy terminal tyrosine residue, results in tyrosine phosphorylation of a 84- and a 100-kDa protein in *Xenopus* oocytes (Unger and Steele, 1992). This triggers a local aggregation of pigment granules and apparent invagination of the cortex in the region around the injection site. The accumulation of pigment granules, which resembles the pigment accumulation seen at fertilization, can be seen beginning approximately 90 min following injection of pp60$^{d\text{-}src}$. Ultrastructural examination demonstrates a thick band of actin beneath the plasma membrane which gets thicker and denser with time following injection. After several hours, the contraction becomes so robust that all cortical organelles (cortical granules, pigment granules, and endoplasmic reticulum) lose contact with the plasma membrane and are displaced from the cortex, sinking deeper into the inner cytoplasm (Larabell *et al.*, 1993). A similar loss of organelle contact with the cell surface and resultant sinking into the inner cytoplasm was seen as a response to treatment of maturing oocytes with cytochalasin B (Ryabova *et al.*, 1994). These responses emphasize the importance of the cortical cytoskeleton in maintaining the structure of the cortex, as well as the entire oocyte.

C. Localizations of RNA

Since the major functions of the oocyte are storage and synthesis, one would expect that there would be storage sites of RNA. The finding that there is a heterogeneous distribution of RNA in the oocyte was particularly exciting. Using *in situ* hybridization, it was shown that there is a localization of poly(A)$^+$RNA in the periphery of the vegetal hemisphere of *Xenopus* oocytes (Capco and Jeffery, 1982; Larabell and Capco, 1988). Localizations of specific mRNAs have also been found in *Xenopus* oocytes. For example, the oocyte periphery contains localizations of actin and tubulin mRNA (Larabell and Capco, 1988; Perry and Capco, 1988). While the localization of tubulin is disrupted during meiotic maturation, and high concentrations of tubulin mRNA can be found in the center of the unfertilized egg, the concentration of actin mRNA in the center of the egg increases only slightly; after fertilization, tubulin and actin mRNA are once again found localized in the periphery of the zygote (Larabell and Capco, 1988; Perry and Capco, 1988). Since there is a dramatic reorganization of cortical actin filaments and microtubules during meiotic maturation and fertilization, it is not too surprising that there are high concentrations of these mRNAs in the periphery of the oocyte and zygote. The presence of these mRNAs in the periphery supports the hypothesis that mRNA localizations can provide a way of spatially controlling macromolecular assembly reactions.

It has also been suggested that localizations of mRNA provide a way to assure the presence of protein gradients for cell fate determination during early development. Several specific RNAs that may be involved in axial patterning of the embryo have been found in the vegetal cortex of the *Xenopus* oocyte to support that hypothesis. These include *Vg1* mRNA, a TGF-β homologue (Weeks and Melton, 1987; Melton, 1987), *Xwnt11* (Ku and Melton, 1993), and *Xcat2,* which encodes a nanos-like molecule (Mosquera *et al.,* 1993). It has been reported that *Xlsirts* (nontranslatable short interspersed repeat sequence transcripts) and *Xcat-3* (Elinson *et al.,* 1993) also localize to the vegetal cortex of oocytes (Kloc *et al.,* 1993). In addition, four RNAs are found localized in the animal hemisphere, including *An1* (Linnen *et al.,* 1993), *An2* (Weeks and Melton, 1987), *An3* (Gururajan *et al.,* 1994) and *xlan4* (Reddy *et al.,* 1992).

The specialized localizations of RNA appear to be dependent on interactions with the cytoskeleton. As much as 80% of actin and tubulin mRNA are associated with the detergent-resistant cytoskeleton in oocytes and eggs (Hauptman *et al.,* 1989), and these mRNAs colocalize in oocyte periphery with the cytokeratin and actin networks (Larabell and Capco, 1988; Perry and Capco, 1988). *Vg1* mRNA is associated with the detergent-resistant cytoskeleton of oocytes, but is associated with soluble components of the egg (Pondel and King, 1988). It has recently been shown that intact microtubules are required to move *Vg1* mRNA to the vegetal pole, but not to keep it in the cortex (Forristall *et al.,* 1995). Precisely how *Vg1* mRNA is anchored to the cortex is not clear; it is partially dependent on microfilaments, since 70% of *Vg1* can be released with cytochalasin B (Yisraeli *et al.,* 1990), but there is no direct link between *Vg1* and cytokeratins (Klymkowsky *et al.,* 1991). *Xcat-2* moves to the cortex with the mitochondrial cloud in a process that is also dependent on intact microtubules (Yisraeli *et al.,* 1990) and is always recovered in a detergent-insoluble fraction (Forristall *et al.,* 1995).

III. Cortical Cytoskeleton of the Unfertilized Egg

A. Organization of the Unfertilized Egg Cortex

A dramatic structural modification of the oocyte cortex accompanies meiotic maturation to prepare the egg for fertilization. The cell surface flattens and the cortical granules become positioned just beneath the plasma membrane, where they await the signal to secrete their contents to the perivitelline space (Balinsky, 1966; Grey *et al.,* 1974; Larabell and Chandler, 1988a). An elaborate endoplasmic reticulum forms which twists and turns throughout the cortex, wrapping around the cortical granules and frequently contacting the plasma membrane (Charbonneau and Grey, 1984; Campanella *et al.,* 1984; Larabell and Chandler, 1988a). This reticular structure is believed to be the calcium-sequestering organelle that releases calcium at fertilization. In addition, there is an accompanying decrease in the number and length of microvilli (Balinsky and Devis, 1963; Campanella *et al.,* 1984; Larabell and Chandler, 1988; Bement and Capco, 1989b).

Figure 5 shows a view of the newly remodeled cortex of the unfertilized egg as seen using quick-freeze, deep-etch, rotary-shadow electron microscopy. An elaborate cytomatrix is seen in the cortex, with numerous filaments contacting the cortical granules and a dense filamentous matrix situated between the plasma membrane and cortical granules. Segments of the endoplasmic reticulum can also be seen wrapping around the cortical granule. Stereo electron micrographs of an isolated egg cortex show the same region of the egg, but from a different perspective (Fig. 6). These images are parallel to the egg surface, looking down at the inside of the plasma membrane, and soluble components have been rinsed away during preparation. A network of filaments, referred to as the submembrane cytoskeleton, consists of numerous short, interconnecting filaments approximately 12 nm in diameter located between the plasma membrane and cortical granule (Larabell, 1993). This filamentous network is in a position to act as a barrier to cortical granule exocytosis and would, presumably, have to be disassembled in order to allow fusion of the cortical granules with the plasma membrane. Other filaments, also about 12 or 13 nm in diameter, can be seen interconnecting the pigment granules, endoplasmic reticulum,

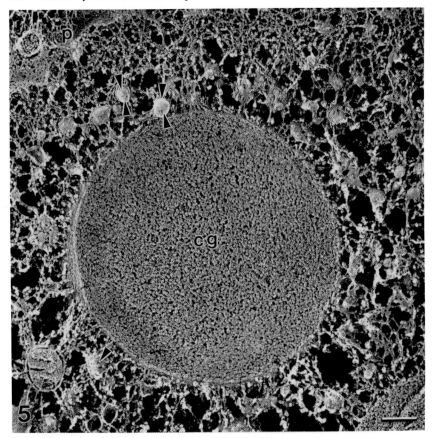

Fig. 5 Platinum replica of the cortex of an unfertilized egg. Specimen was rapidly frozen in liquid helium, freeze-fractured at liquid nitrogen temperatures, allowed to sublime, and then a platinum/carbon replica of the exposed surface was made. The replica was then viewed in a transmission electron microscope. p, plasma membrane; cg, cortical granule; arrowheads, endoplasmic reticulum. Scale bar = 0.1 μm.

and cortical granules, and securing these organelles to the submembrane cytoskeleton.

B. Cortical Actin

Immunocytochemical localization of actin using anti-actin antibodies demonstrates the increased organization of the unfertilized egg cortex compared with that seen in the oocyte. A single optical section of the unfertilized egg, which now has a very even cell surface, provides a view of an orderly

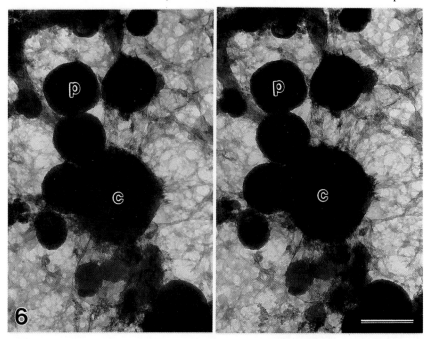

Fig. 6 Stereo electron micrograph of a cortex isolated from an unfertilized egg. A dense meshwork of filaments is seen just beneath the plasma membrane. A large cortical granule (c) rests on top of those filaments. Long filaments are seen interconnecting pigment granules (p) and the tubular cortical endoplasmic reticulum. Scale bar = 0.5 μm.

distribution of microvilli, as represented by the punctate pattern of bright fluorescence between the shadows of cortical granules (Fig. 7a). The next optical section of that same region, 1 μm deeper into the egg, demonstrates the dense distribution of actin surrounding the very orderly arrangement of cortical granules (Fig. 7b). The majority of actin in the unfertilized egg is unpolymerized (Merriam and Clark, 1978), and it has recently been shown that more than 80% of the unpolymerized actin is ATP bound (Rosenblatt *et al.*, 1995). These authors also present data suggesting that this soluble actin is complexed with the sequestering protein thymosin-β4 (Rosenblatt *et al.*, 1995). This provides a large pool of actin that is readily available for polymerization as needed at fertilization. (For a detailed discussion of intermediate filaments and microtubules, see Chapters 14 and 12 in this book by Klymkowsky and Gard *et al.*, respectively.)

C. Other Cytoskeletal Proteins

It has been reported that myosin and talin also remain in the cortex after resumption of meiosis (Franke *et al.*, 1976; Colombo *et al.*, 1981; Gall *et*

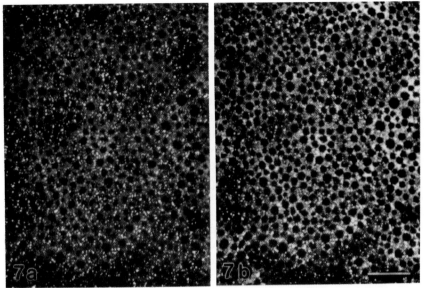

Fig. 7 Isolated cortex from an unfertilized egg prepared as described in the legend of Fig. 1. Figure 7a is a single optical section of the cortex showing punctate spots labeling the actin filaments in the microvilli at the outermost region of the cortex. Figure 7b is an optical section 1 μm deeper in the cortex showing the dense network of actin surrounding the orderly array of cortical granules (black, circular shadows). Scale bar = 10 μm.

al., 1983; Ryabova, 1982, 1990; Ryabova *et al.*, 1992, 1994), but that spectrin is not seen in the cortex of the unfertilized egg (Ryabova *et al.*, 1992, 1994). Vinculin, which was found in the inner cytoplasm of the oocyte, however, is now colocalized with talin in the unfertilized egg cortex (Evans *et al.*, 1990). The remarkable geodesic pattern of cytokeratin previously seen in the oocyte cortex is not found in unfertilized eggs, and only residual elements of cytokeratin can be seen in the cortex of the vegetal hemisphere (Godsave *et al.*, 1984b; Klymkowsky *et al.*, 1987). Vimentin, however, remains uniformly distributed throughout the egg, including the cortex (Godsave *et al.*, 1984a).

IV. Cortical Cytoskeleton during the First Cell Cycle

A. The Zygote Cortex: A New Structure, a New Function

The oocyte and egg are the same cell at different stages of competence for fertilization, and the cortical organization at these stages reflects their impending function: preparation for sperm entry. The zygote, however, is starting on the developmental pathway, and the cortex undergoes a dramatic reorganization to facilitate that purpose. At fertilization, the cortical granules fuse with the plasma membrane and secrete their contents into the perivitelline space, where they convert the sperm-binding vitelline envelope into the sperm-blocking fertilization envelope (Grey *et al.,* 1974; Hara and Tydeman, 1979; Gerton and Hedrick, 1986; Larabell and Chandler, 1988a, 1988b, 1990). Fusion of the cortical granule membranes with the plasma membrane results in the addition of a large amount of membrane to the zygote. However, the cell does not increase in size. One way in which the added membrane is accommodated is by effecting a rapid increase in the size of the microvilli (Takeichi and Kubota, 1984), a process which is actomyosin based (Ezzell *et al.,* 1985). Another way in which the increased amount of membrane is accommodated is by the process of endocytosis, which follows fertilization (Bernardini *et al.,* 1986, 1987). Without the large secretory granules and their surrounding endoplasmic reticulum, which is now situated beneath (and parallel to) the plasma membrane, the cortex of the zygote no longer bears any resemblance to the cortex of the oocyte or egg.

B. Cortical Actin

The cytoskeleton also undergoes a major reorganization after fertilization. The cytoskeletal components that were previously attached to the membranes of the cell surface and cortical granules (Figs. 5 and 6) must be reorganized as the secretory granules fuse with the plasma membrane. During this time, the cell surface is also being remodeled to incorporate the additional membrane contributed to the cortical granule membranes. The exact mechanism by which this membrane remodeling occurs, however, is not known. It is known that the microvilli elongate to accommodate additional membrane via an actomyosin process (Ezzell *et al.,* 1985). During this time, however, the cytoskeleton must retain the capacity to conduct the actomyosin-based cortical contraction that crosses the egg immediately following exocytosis. Midway through the first cell cycle, at the time of cortical rotation, the actin network is once again remodeled. A dense band of actin has been described beneath the plasma membrane in methanol-fixed zygotes (Houliston and Elinson, 1991). We have recently shown that, in embryos fixed with aldehydes, there is an elaborate actin network (in addition to the band of actin beneath the plasma membrane) that extends 3 or 4 μm beneath the plasma membrane and encircles numerous cortical organelles (Fig. 8; Larabell *et al.,* 1995). The identity of these organelles is not yet known; they may represent endosomes or, on the other hand, may be newly assembled secretory granules.

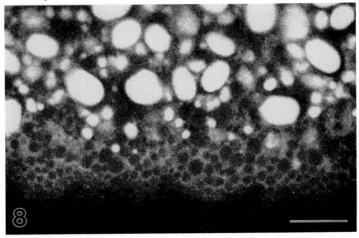

Fig. 8 Cortex of fertilized egg fixed at 0.6 normalized time in 2% paraformaldehyde and 0.1% glutaraldehyde then labeled with anti-actin antibodies and viewed in the confocal microscope. Actin is seen surrounding small vesicles in the cortex; yolk platelet autofluorescence is also seen. Scale bar = 10 μm.

During the first cell cycle, a cytokeratin network reappears in the cortex of the vegetal hemisphere, while only a sparse cytokeratin network is seen in the animal hemisphere cortex, and vimentin remains uniformly distributed throughout the egg (Klymkowsky *et al.,* 1987). In addition, talin and vinculin are reported to colocalize in the cortex of cleavage furrows and blastomeres (Evans *et al.,* 1990). The most pronounced cytoskeletal reorganization seen in the zygote periphery, however, is the formation of the elaborate array of microtubules which effect cortical rotation.

C. Microtubule Arrays during Cortical Rotation in Live Embryos

Cortical rotation, the 30° rotation of the cortex with respect to the inner cytoplasm, is critical for dorsal axis specification (see Gerhart *et al.,* 1989). This microtubule-dependent movement has been reported to begin midway through the first cell cycle and end just before first cleavage. An array of aligned microtubules appears in the vegetal hemisphere at this time and is responsible for generating this movement (Elinson and Rowning, 1988). The array is oriented parallel to the direction of movement (Elinson and Rowning, 1988), with the plus ends of microtubules pointing in the direction of cortical displacement (Houliston and Elinson, 1991).

We recently analyzed the formation of the parallel array of microtubules and organelle movements during cortical rotation using confocal microscopy of living *Xenopus* eggs (Rowning *et al.,* 1994; Larabell *et al.,* 1995). We were able to monitor microtubule polymerization using fluorescently labeled tubulin while monitoring the movement of the Nile red-labeled yolk platelets. Using this approach, we showed that the microtubule array is located in that region of the subcortex between 4 and 10 μm beneath the plasma membrane (Larabell *et al.,* 1995) and that the microtubules move with the inner cytoplasm, as previously proposed (Houliston, 1994). The movement at the vegetal pole occurs as early as 0.3 normalized time (NT; time from fertilization to first cleavage), rather than at 0.45–0.5 NT as previously reported (Elinson and Rowning, 1988; Houliston, 1994). However, we do not detect newly polymerized microtubules until between 0.45 and 0.5 NT. One possible explanation is that movement is generated using preexisting microtubules that we do not visualize, and that we detect eventual elongation of these microtubules. It is also possible, however, that movement is initiated using another mechanism other than microtubules, or that movement is initiated via microtubules in another region of the egg.

Figure 9 shows an optical section of the vegetal hemisphere obtained by confocal microscopy during formation of the parallel array of microtubules in the living egg. The egg had been injected with rhodamine-labeled tubulin, enabling visualization of newly polymerized microtubules. This tangential section provides a view of the immobilized cortex at the lower right of the image and the formation of microtubules in the shear zone. A dense collection of organelles can be seen in the cortex which may contain the motors required to move the egg periphery along the microtubules during rotation. Long microtubules of the newly formed parallel array can be seen in the deeper region of the section, at the upper left corner of the image. Although it is clear that cortical rotation is critical for dorsal/ventral axis specification and normal development, it is not known precisely how rotation determines where the axis will form. We are currently addressing this question using the increased resolution provided by confocal microscopy of living eggs during cortical rotation.

V. Conclusion

The cortical cytoskeleton is an extremely dynamic structure that undergoes repeated reorganizations as the function of the egg changes. The importance of these cytoskeletal changes during meiotic maturation, fertilization, and early development is clear since disrupting these rearrangements results in abnormal growth and development. Although we are beginning to discover a few of the regulatory mechanisms for the cytoskeletal remodeling, there are still many missing pieces of the regulatory puzzle. Likewise, we have a very limited understanding of the way in which the cytoskeleton affects the steps along the developmental pathway. A cell as large as the *Xenopus*

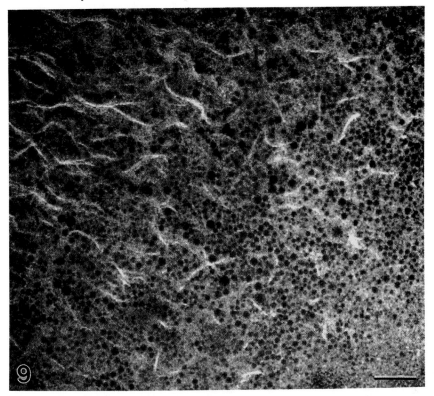

Fig. 9 Cortex of fertilized egg viewed at 0.6 normalized time. Egg was injected with rhoda-mine-labeled tubulin at 0.2 normalized time, immobilized in Ficoll, then viewed in the confocal microscope as microtubules are polymerized during cortical rotation. The optical section provides a tangential view of the shear zone showing the immobilized cortex on the right, progressing into the subcortical region where microtubules begin to appear (center), to the subcortical region where the parallel array of microtubules is forming. Scale bar = 20 μm.

egg requires specialized regions with distinct functions. Delivery of unique components to these regions via the cytoskeleton can resolve the developmental challenges faced by such a large egg. As the tools of molecular biology and confocal microscopy provide increasing amounts of new information about the repositioning of these unique components, the role of the cytoskeleton in early development will become better understood.

References

Balinsky, B. I. (1966). Changes in the ultrastructure of amphibian eggs following fertilization. *Acta. Morphol. Embryol. Exp.* **9,** 132–154.

Balinsky, B. I., and Devis, R. J. (1963). Origin and differentiation of cytoplasmic structures in the oocytes of *Xenopus laevis. Acta. Morphol. Embryol. Exp.* **6,** 55–108.

Bement, W. M., and Capco, D. G. (1989a). Activators of protein kinase C trigger cortical granule exocytosis, cortical contraction, and cleavage furrow formation in *Xenopus laevis* oocytes and eggs. *J. Cell Biol.* **108,** 885–892.

Bement, W. M., and Capco, D. G. (1989b). Intracellular signals trigger ultrastructural events characteristic of meiotic maturation in oocytes of *Xenopus laevis. Cell Tissue Res.* **255,** 183–191.

Bement, W. M., and Capco, D. G. (1991). Analysis of inducible contractile rings suggest a role for protein kinase C in embryonic cytokinesis and wound healing. *Cell Motil. Cytoskeleton* **20,** 145–157.

Bernardini, G., Ferraguti, M., and Peres, A. (1986). The decrease of *Xenopus* egg membrane capacity during activation might be due to endocytosis. *Gamete Res.* **14,** 123–127.

Bernardini, G., Ferraguti, M., and Peres, A. (1987). Fertilization induces endocytosis in *Xenopus* eggs. *Cell Differ.* **21,** 255–260.

Campanella, C., Andreuccetti, P., Taddei, C., and Talevi, R. (1984). The modifications of cortical endoplasmic reticulum during in vitro maturation of *Xenopus laevis* oocytes and its involvement in cortical granule exocytosis. *J. Exp. Zool.* **229,** 283–293.

Capco, D. G., and Jeffery, W. (1982). Transient localizations of messenger RNA in *Xenopus laevis* oocytes. *Dev. Biol.* **89,** 1–12.

Capco, D. G., Tutnick, J. M., and Bement, W. M. (1992). The role of protein kinase C in reorganization of the cortical cytoskeleton during the transition from oocyte to fertilization-competent egg. *J. Exp. Zool.* **264,** 395–405.

Charbonneau, M., and Grey, R. D. (1984). The onset of activation responsiveness during maturation coincides with the formation of the cortical endoplasmic reticulum in oocytes of *Xenopus laevis. Dev. Biol.* **102,** 90–97.

Charbonneau, M., and Picheral, B. (1983). Early events in anuran amphibian fertilization: An ultrastructural study of changes occurring in the course of monospermic fertilization and artificial activation. *Dev. Growth Differ.* **225,** 23–37.

Christensen, K., Sauterer, R., and Merriam, R. W. (1984). Role of soluble myosin in cortical contractions of *Xenopus* eggs. *Nature* **310,** 150–151.

Colombo, R., Benedusi, P., and Valle, G. (1981). Actin in *Xenopus* development: Indirect immunofluorescence study of actin localization. *Differentiation* **20,** 45–51.

Elinson, R. P. (1980). The amphibian egg cortex in fertilization and early development. *In* "The Cell Surface: Mediator of Developmental Processes"(S. Subtelney and N. K. Wessels, eds.), pp. 217–234. Academic Press, New York.

Elinson, R. P., and Rowning, B. (1988). A transient array of parallel microtubules in frog eggs: Potential tracks for a cytoplasmic rotation that specifies the dorso-ventral axis. *Dev. Biol.* **128,** 185–197.

Elinson, R. P., King, M. L., and Forristall, C. (1993). Isolated vegetal cortex from *Xenopus* oocytes selectively retains localized mRNAs. *Dev. Biol.* **160,** 554–562.

Evans, J. P., Page, B. D., and Kay, B. K. (1990). Talin and vinculin in the oocytes, eggs, and early embryos of *Xenopus laevis:* A developmentally regulated change in distribution. *Dev. Biol.* **137,** 403–413.

Ezzell, R. M., Cande, W. Z., and Brothers, A. J. (1985). Ca^{2+}–ionophore-induced microvilli and cortical contractions in *Xenopus* eggs. *Proc. Natl. Acad. Sci. USA* **77,** 462–466.

Forristall, C., Pondel, M., Chen, L., and King, M. L. (1995). Patterns of localization and cytoskeletal association of two vegetally localized RNAs, *Vg1* and *Xcat-2. Development* **121,** 201–208.

Franke, W., Rathke, P., Seib, E., Trendelenburg, M., Osborn, M., and Weber, K. (1976). Distribution and mode of arrangement of microfilamentous structures and actin in the cortex of the amphibian oocyte. *Cytobiology* **14,** 111–130.

Franz, J., Gall, L., Williams, M., Picheral, B., and Franke, W. (1983). Intermediate-size filaments in a germ cell: Expression of cytokeratins in oocytes and eggs of the frog *Xenopus*. *Proc. Natl. Acad. Sci. USA* **80,** 6254–6258.

Gall, L., and Karsenti, E. (1987). Soluble cytokeratins in *Xenopus laevis* oocytes and eggs. *Biol. Cell* **61,** 33–38.

Gall, L., Picheral, B., and Gounon, P. (1983). Cytochemical evidence for the presence of intermediate filaments and microfilaments in the egg of *Xenopus laevis*. *Biol. Cell* **47,** 331–342.

Gard, D. L. (1991). Organization, nucleation, and acetylation of microtubules in *Xenopus laevis* oocytes: A study by confocal immunofluorescence microscopy. *Dev. Biol.* **143,** 346–362.

Gerhart, J. C., Danilchik, M., Doniach, T., Roberts, S., Rowning, B., and Stewart, R. (1989). Cortical rotation of the *Xenopus* egg: Consequences for the anterioposterior pattern of embryonic dorsal development. *Development* **107**(Suppl), 37–51.

Gerton, G. L., and Hedrick, J. L. (1986). The vitelline envelope to fertilization envelope conversion in eggs of *Xenopus laevis*. *Dev. Biol.* **116,** 1–7.

Gingell, D. (1970). Contractile responses at the surface of an amphibian egg. *J. Embryol. Exp. Morphol.* **23,** 583–609.

Godsave, S., Anderton, B., Heasman, J., and Wylie, C. (1984a). Oocytes and early embryos of *Xenopus laevis* contain intermediate filaments which react with anti-mammalian vimentin antibodies. *J. Embryol. Exp. Morphol.* **83,** 169–187.

Godsave, S., Wylie, C., Lane, E., and Anderton, B. (1984b). Intermediate filaments in the *Xenopus* oocyte: The appearance and distribution of cytokeratin-containing filaments. *J. Embryol. Exp. Morphol.* **83,** 157–167.

Grey, R. D., Wolf, D. P., and Hedrick, J. L. (1974). Formation and structure of the fertilization envelope in *Xenopus laevis*. *Dev. Biol.* **36,** 44–61.

Gururajan, R., Mathews, L., Longo, F., and Weeks, D. (1994). An3 mRNA encodes an RNA helicase that colocalizes with nucleoli in *Xenopus* oocytes in a stage-specific manner. *Proc. Natl. Acad. Sci. USA* **91,** 2056–2060.

Hara, K., and Tydeman, P. (1979). Cinematographic observation of an "activation wave" (AW) on the locally inseminated egg of *Xenopus laevis*. *Roux's Arch. Dev. Biol.* **186,** 91–94.

Hauptman, R. J., Perry, B. A., and Capco, D. G. (1989). A freeze-sectioning method for preparation of the detergent-resistant cytoskeleton identifies stage-specific cytoskeletal proteins and associated mRNA in *Xenopus* oocytes and embryos. *Dev. Growth Differ.* **31,** 157–164.

Houliston, E. (1994). Microtubule translocation and polymerization during cortical rotation in *Xenopus* eggs. *Development* **120,** 1213–1220.

Houliston, E., and Elinson, R. P. (1991). Evidence for the involvement of microtubules, ER, and kinesin in the cortical rotation of fertilized frog eggs. *J. Cell Biol.* **114,** 1017–1028.

Huchon, D., Jessus, C., Thibier, C., and Ozon, R. (1988). Presence of microtubules in isolated cortices of prophase I and metaphase II oocytes in *Xenopus laevis*. *Cell Tissue Res.* **254,** 415–420.

Kloc, M., Spohr, G., and Etkin, L. D. (1993). Translocation of repetitive RNA sequences with the germ plasm in *Xenopus* oocytes. *Science* **262,** 1712–1714.

Klymkowsky, M. W., Maynell, L., and Polson, A. (1987). Polar asymmetry in the organization of the cortical cytokeratin system of *Xenopus laevis* oocytes and embryos. *Development* **100,** 543–557.

Klymkowsky, M. W., Maynell, L. A., and Nislow, C. (1991). Cytokeratin phosphorylation, cytokeratin filament severing and the solubilization of the maternal mRNA Vg1. *J. Cell Biol.* **114,** 787–797.

Ku, M., and Melton, D. A. (1993). *Xlwnt-11:* A maternally expressed *Xenopus* wnt gene. *Development* **119**, 1161–1173.

Larabell, C. A. (1993). A new technique for isolation and visualization of the *Xenopus* egg cortex reveals a complex cytoskeleton. *J. Struct. Biol.* **110**, 154–159.

Larabell, C., and Capco, D. (1988). Role of calcium in the localization of maternal poly-(A)⁺RNA and tubulin mRNA in *Xenopus* oocytes. *Roux's Arch. Dev. Biol.* **197**, 175–183.

Larabell, C. A., and Chandler, D. E. (1988a). Freeze-fracture analysis of structural reorganization during meiotic maturation in oocytes of *Xenopus laevis. Cell Tissue Res.* **251**, 129–136.

Larabell, C. A., and Chandler, D. E. (1988b). *In vitro* formation of the "S" layer, a unique component of the fertilization envelope in *Xenopus laevis* eggs. *Dev. Biol.* **130**, 356–364.

Larabell, C. A., and Chandler, D. E. (1990). Stepwise transformation of the vitelline envelope of *Xenopus* eggs at activation: A quick-freeze, deep-etch analysis. *Dev. Biol.* **139**, 263–268.

Larabell, C. A., Unger, T. F., and Steele, R. E. (1993). Deregulated pp60^c-src triggers contraction of the cortical cytoskeleton in *Xenopus* oocytes. *Mol. Biol. Cell* **4**(Suppl.), 170a.

Larabell, C. A., Rowning, B. A., Wells, J. C., Wu, M., and Gerhart, J. C. (1995). Confocal microscopy analysis of living *Xenopus* eggs and the mechanism of cortical rotation. *Development,* submitted for publication.

Linnen, J., Bailey, C., and Weeks, D. (1993). Two related localized mRNAs from *Xenopus laevis* encode ubiquitin-like fusion proteins. *Gene* **128**, 181–188.

Meeusen, R. L., and Cande, W. Z. (1979). N-ethylmaleimide-modified heavy meromyosin: A probe for actomyosin interactions. *J. Cell Biol.* **82**, 57–65.

Melton, D. A. (1987). Translocation of a localized maternal mRNA to the vegetal pole of *Xenopus* oocytes. *Nature* **38**, 80–82.

Merriam, R. W., and Clark, T. G. (1978). Actin in *Xenopus* oocytes. II. Intracellular distribution and polymerizability. *J. Cell Biol.* **77**, 439–447.

Merriam, R. W., and Sauterer, R. A. (1983). Localization of a pigment-containing structure near the surface of *Xenopus* eggs which contracts in response to calcium. *J. Embryol. Exp. Morphol.* **76**, 51–65.

Mosquera, L., Forristall, C., Zhou, Y., and King, M. L. (1993). A mRNA localized to the vegetal cortex of *Xenopus* oocytes encodes a protein with a *nanos*-like zinc finger domain. *Development* **117**, 377–386.

Olson, J. H., and Capco, D. G. (1992). Crosstalk between MPF and protein kinase C in the organization of the cortical cytoskeleton. *Mol. Biol. Cell* **3**, 7a.

Perry, B. A., and Capco, D. G. (1988). Spatial reorganization of actin, tubulin and histone mRNA's during meiotic maturation and fertilization in *Xenopus* oocytes. *Cell Differ. Dev.* **25**, 99–108.

Pondel, M. D., and King, M. L. (1988). Localized maternal mRNA related to transforming growth factor β mRNA is concentrated in a cytokeratin-enriched fraction from *Xenopus* oocytes. *Proc. Natl. Acad. Sci. USA* **85**, 7612–7616.

Reddy, B. A., Kloc, M., and Etkin, L. (1992). The cloning and characterization of a localized maternal transcript in *Xenopus laevis* whose zygotic counterpart is detected in the CNS. *Mech. Dev.* 1–8.

Rosenblatt, J., Peluso, P., and Mitchison, T. J. (1995). The bulk of unpolymerized actin in *Xenopus* egg extracts is ATP-bound. *Mol. Biol. Cell* **6**, 227–236

Rowning, B. A., Larabell, C. A., Wells, J. C., Wu, M., and Gerhart, J. C. (1994). Inverted confocal microscopy analysis of microtubule arrays during cortical rotation in the first cell cycle of living *Xenopus* eggs. *Mol. Biol. Cell* **4**(Suppl.), 256a.

Ryabova, L. V. (1982). Electron microscopic study of the development of oocyte cortical contractibility in the common frog. *Sov. J. Dev. Biol.* **13,** 266–273.

Ryabova, L. V. (1990). Organization of the cortical layer in amphibian eggs. II. Actin-containing structure in the cortex of *Xenopus laevis* oocytes and eggs. *Sov. J. Dev. Biol.* **21,** 369–375.

Ryabova, L. V., Virtanen, I., Wartiovaara, J., and Vassetzky, S. G. (1992). Contractile proteins and non-erythroid spectrin in oogenesis of *Xenopus laevis. Ontogenez* **23,** 487–500.

Ryabova, L. V., Vassetzky, S. G., and Capco, D. G. (1994). Development of cortical contractility in the *Xenopus laevis* oocyte mediated by reorganisation of the cortical cytoskeleton: A model. *Zygote* **2,** 263–271.

Ryabova, L. V., and Vassetzky, S. G. (1993). Involvement of polymerized actin in maintenance of the spatial organization. *Xenopus laevis* oocytes and its visualization in deep layers of the ooplasm. *Ontogenez.* **24,** 122–127.

Schroeder, T. E., and Strickland, D. L. (1974). Ionophore A23187, calcium, and contractility in frog eggs. *Exp. Cell Res.* **83,** 139–142.

Takeichi, T., and Kubota, H. Y. (1984). Structural basis of the activation wave in the egg of *Xenopus laevis. J. Ebryol. Exp. Morphol.* **81,** 1–16.

Tang, P., Sharpe, C. R., Mohun, T. J., and Wylie, C. C. (1988). Vimentin expression in oocytes, eggs and early embryos of *Xenopus laevis. Development* **103,** 279–287.

Torpey, N. P., Heasman, J., and Wylie, C. C. (1992). Distinct distribution of vimentin and cytokeratin in *Xenopus* oocytes and early embryos. *J. Cell Sci.* **101,** 151–160.

Unger, T. F., and Steele, R. E. (1992). Biochemical and cytological changes associated with expression of deregulated pp60src in Xenopus oocytes. *Mol. Cell Biol.* **12,** 5485–5498.

Weeks, D. L., and Melton, D. A. (1987). A maternal mRNA localized to the vegetal hemisphere in *Xenopus* eggs codes for a growth factor related to TGF-β. *Cell* **51,** 861–867.

Yisraeli, J. K., Sokol, S., and Melton, D. A. (1990). A two-step model for the localization of maternal mRNA in *Xenopus* oocytes: Involvement of microtubules and microfilaments in the translocation and anchoring of *Vg1* mRNA. *Development* **108,** 289–298.

14

Intermediate Filament Organization, Reorganization, and Function in the Clawed Frog *Xenopus*

Michael W. Klymkowsky
Department of Molecular, Cellular, and Developmental Biology
University of Colorado
Boulder, Colorado 80309

I. Introduction

Cytoplasmic intermediate filaments (IFs) are arguably the most enigmatic component of the eukaryotic cytoskeleton. IF subunit proteins (IFPs) are a highly conserved common feature of vertebrate cells,[1] where they often form a substantial and largely insoluble network. Yet the disruption of IF organization in cultured cells has little if any apparent effect on cellular morphology or behavior (see Klymkowsky *et al.*, 1989). Even in the context of the intact

---

[1] Even in the metazoans, however, IFs are not ubiquitous; IFs appear to be absent from the arthropods in general and the fruit fly *Drosophila melanogaster* in particular (Bartnik and Weber, 1989).

organism, the absence of normally abundant IF networks has been seen to produce surprisingly subtle effects (Colucci *et al.*, 1994; Gomi *et al.*, 1995; Pekney *et al.*, 1995). On the other hand, it is also clear that IFs play critical roles in the maintenance of epidermal integrity, neuronal axon diameter, and muscle–extracellular matrix attachment sites (see Klymkowsky 1995).

We chose the African clawed frog *Xenopus laevis* for our studies of IF organization and function based on its economy (particularly when compared to mice) and the fact that *Xenopus* has been the focus of many recent studies on the cell biology of axis formation and cellular differentiation during embryogenesis. *Xenopus* offers unique experimental opportunities to study (1) the cellular control of IF organization, (2) the processes of IF assembly/disassembly during the cell cycle, and (3) the role of IFs and associated structures in complex cellular functions.

II. IF Proteins in *Xenopus*

Compared to the invertebrates, vertebrates have undergone an explosive proliferation of IFPs. Based on their ability to copolymerize with one another, these proteins fall into three distinct groups: the vimentin/neuronal IF proteins, the lens-specific IF proteins, and the keratins (Klymkowsky, 1995). Vimentin and other closely related proteins [i.e., desmin, peripherin, and glial fibrillary acidic protein (GFAP), sometimes referred to as type III IFPs] are all capable of forming homopolymeric IFs. They all can copolymerize with one another. The neuronal IFPs [i.e., NFL, NFM, NFH, α-internexin (referred to as the type IV IFPs) and nestin (referred to as a type VI IFP)] are all capable of copolymerizing with vimentin. They also share similarities in genomic organization that indicate their evolutionary relationship (Liem, 1993). The lens IFPs, filensin and phakinin, appear to copolymerize only with one another. They appear to be involved in the formation of "beaded filaments," which are structures unique to the lens. The keratins (or cytokeratins) are by far the largest group of IFPs. In humans, more than 40 different keratins have been described. The keratins are classed as a group based on the fact that keratin filaments are obligate heteropolymers: to form an IF type I (acidic) and the type II (neutral/basic) keratins must combine with a one-to-one stoichiometry (see Fuchs and Weber, 1994).

The nuclear lamins are another set of proteins that are clearly closely related to the IFPs of vertebrates (Franke, 1987). They differ from the vertebrate IFPs in features of primary structure and in the type of filaments they form (which are associated with the inner surface of the nuclear envelope). The nuclear lamins are more closely related to the cytoplasmic IFPs of invertebrates than to those of vertebrates (see Klymkowsky, 1995).

In terms of IF proteins, the main difference between *X. laevis* and the other vertebrates studied to date is that *X. laevis* is effectively tetraploid (see Kobel and Pasquier, 1986). This means that there are often two distinct genes per haploid genome for each of the IF proteins normally present as a single locus in other vertebrates. This feature of *X. laevis* IFPs has been best described for vimentin. Herrmann *et al.* (1989a) originally isolated two distinct *X. laevis* cDNAs. We have also isolated variants of these two cDNA, which we termed vimentin-1 (55 kDa) and vimentin-2 (57 kDa) (Dent, 1992). When the sequences are compared to one another and other vertebrate vimentins (Fig. 1), it is readily apparent that the two vimentins of *X. laevis* are evolving (or drifting) apart from one another rather quickly.

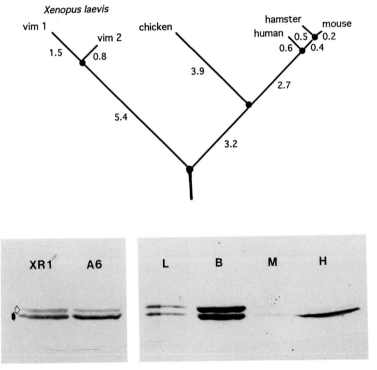

Fig. 1 The divergence of vimentins in *X. laevis*. (Top) the two vimentins of *X. laevis* were compared with other vimentin sequences and a tree diagram was generated using the alignment program of Feng and Dolittle (1990). Branches are drawn to scale with genetic distances. Note that the two *X. laevis* vimentins are more divergent from one another than are mouse and human vimentin. (Bottom) differential accumulation of vimentin-1 (V1) and vimentin-2 (V2) in cultured cells and adult tissues was monitored by SDS–PAGE/Western blot using the anti-vimentin antibody 14h7. XR1 and A6 are cell lines, while L, B, M, and H refer to liver, brain, skeletal muscle, and heart, respectively. Figure modified from Dent (1992).

These two vimentins can be distinguished based on the reactivity with the anti-vimentin antibodies 14h7 and RV202: 14h7 reacts with both polypeptides, whereas RV202 reacts only with vimentin-1. When the 14h7 antibody is used to analyze various *X. laevis* cell lines and tissues, it reveals a clear difference in the accumulation of these two polypeptides. In cultured cells, vimentin-1 is more abundant than vimentin-2; in the brain both proteins were present in roughly equal amounts, whereas in the heart only vimentin-1 was found in significant amounts (Fig. 1).

In addition to the presence of two isoforms of each of the known IFPs, it is also possible that there are IFPs unique to *Xenopus*. For example, Charnas *et al.* (1992) described a neuronal IFP (XNIF) that does not correspond to any of the known neuronal IFPs of mammals. Similarly, Hemmati-Brivanlou *et al.* (1992) described a nestin-like protein, which they called tannabin, that is expressed in the growth cones of embryonic neurons. Tannabin appears to be unique in its cellular distribution. Whether XNIF or tannabin have homologs in mammals remains to be determined.[2]

A. The IF System of the *Xenopus* Oocyte

The first clear description of IFs in *Xenopus* was by Perry (1975) who distinguished a class of 10-nm diameter "microfilaments," often associated with desmosomes, from the more common 8-nm diameter "microfilaments" (i.e., actin filaments). This initial study was extended by Gall *et al.* (1983), Franz *et al.* (1983), and Godsave *et al.* (1984a,b) who showed that IFs were present throughout all stages of oogenesis. In the early oocyte (stage I), keratin-type IFs were seen as "sparse" cortical threads. As oogenesis proceeded keratin filaments were found associated with germinal vesicle surface and encircling the mitochondrial cloud, eventually forming a radial system connected to a dense cortical network (Franz *et al.*, 1983; Godsave *et al.*, 1984b; Ryabova *et al.*, 1993). Biochemical studies indicate that the keratin system of the *Xenopus* oocyte is compositionally simple; it consists of a single type II keratin, the homolog of keratin K8 (Franz and Franke, 1986), and two type I keratins, apparently the homologs of keratins K18 and K19 (Franz *et al.*, 1983).[3] This pattern of keratin expression is characteristic of simple epithelial cells.

Using whole-mount immunocytochemistry, it is possible to visualize in detail the organization of the cortical keratin filament network. In the late-

---

[2] Schecter and colleagues have characterized two novel neuronal IF proteins from the goldfish optic nerve, plasticin, and gefiltin (Glasgow *et al.*, 1994); it is currently unknown whether homologous proteins are present in *Xenopus* or other higher vertebrates.

[3] All of these proteins appear to exist in the two psuedoallelic forms.

stage oocyte, this network is organized in a highly asymmetric fashion (Klymkowsky *et al.,* 1987); in the animal hemisphere it forms a disconnected and irregular network, while in the vegetal hemisphere an almost geodesic system of extended filaments is common. The asymmetry in keratin filament organization is also mirrored in other cortical asymmetries (see Dent and Klymkowsky, 1989; Klymkowsky and Karnovsky, 1994), such as the distribution of γ-tubulin (see Chapter 12).

How keratin network asymmetry is established or maintained remains unclear. It is well known that phosphorylation of IFPs can lead to changes in IF organization (see Steinert and Liem, 1990). To test whether the animal–vegetal differences in keratin organization are due to differences in post-translational modification, we injected RNA encoding an epitope-tagged form of *Xenopus* K8 into either the animal or the vegetal hemispheres of the oocyte. Such injected RNAs remain localized to the injected hemisphere over the 1- or 2-day course of the experiment and the translated protein is found to incorporate locally into the keratin network. Two-dimensional gel electrophoresis reveals no obvious difference in the pattern of post-translational modification between protein synthesized in the animal or vegetal hemispheres (J. Bachant, L. Backhus, and M. Klymkowsky, unpublished observations; see Klymkowsky and Karnovsky, 1994). Presumably differences in keratin network organization between the animal and vegetal hemispheres of the oocyte are due to the asymmetrical distribution/ modification of a factor(s) that binds to and modulates keratin filament organization.

Some information as to the nature of this factor has come from studies of the behavior of exogenous vimentin in the oocytes. Godsave *et al.* (1984a) initially reported that vimentin was present in *Xenopus* oocytes and was associated with the interior of the mitochondrial cloud. In our early studies, we also described staining of the mitochondrial cloud by certain (but not all) anti-vimentin antibodies (Dent and Klymkowsky, 1989). We assumed that this was due to the presence of vimentin-2. Excited at the prospect that vimentin-2 could be involved in maintaining the structural integrity of the mitochondrial cloud, we set out to disrupt vimentin organization in the oocyte using the injection of anti-vimentin antibodies and the synthesis of mutated forms of vimentin. First, however, we set out to confirm the presence of vimentin using immunochemical methods. Unfortunately, we were unable to find vimentin-2 in the *Xenopus* oocyte. Our antibody, 14h7 (on deposit at the Developmental Studies Hybridoma bank), recognizes both *X. laevis* vimentins by Western blot (see above) but fails to detect any polypeptide in the 50- to 60-kDa size range in Western blots of early or late-stage *Xenopus* oocytes (Dent and Klymkowsky, 1989); a similar failure to detect vimentin in oocytes has been reported by Franz *et al.* (1983) and Herrmann *et al.* (1989a). On the other hand, when vimentin is synthesized in the oocyte, following the injection of vimentin RNA, the antibody easily detects it (Dent *et al.*, 1992). Newly synthesized, exogenous vimentin does not interact with the mitochondrial cloud (Dent, 1992; Dent *et al.*, 1992) and the assembly dynamics of exogenous vimentin suggest the *de novo* assembly of a vimentin filament network rather than the incorporation of vimentin into a preexisting IF system (Fig. 2). These data lead us to conclude that a nonvimentin polypeptide in the *Xenopus* oocyte exists that cross-reacts with certain anti-vimentin antibodies. The absence of vimentin makes the *Xenopus* oocytes similar to other vertebrate oocytes which express keratins, but not vimentin.

B. Exogenous Vimentin Is Also Asymmetrically Organized in the Oocyte

Vimentin does not copolymerize with the keratins and there does not appear to be a preexisting vimentin network in the oocyte (see above). Therefore, when vimentin RNA is injected into the oocyte it forms a vimentin filament system *de novo* (Fig. 2). Surprisingly, this exogenous vimentin filament system displays the same basic animal–vegetal asymmetry as described previously for the endogenous keratin filament network. Our working hypothesis is that the asymmetric organization of IFs in the cortex of the *Xenopus* oocyte is due to the presence of a factor that actively suppresses the formation of extended IF networks in the animal hemisphere. That the factor is an active "suppressor" is based on the behavior of IFs during oocyte maturation; early in the maturation process, the initially disconnected animal IF system is replaced by an extensive and well-connected IF network (see below) (Dent *et al.*, 1992). Mutational analysis suggests that susceptibility to this factor requires sequences present in the N-terminal head domain of vimentin (Dent *et al.*, 1992). Given the similarity between the behavior of the endogenous keratin filament network, and the exogenous vimentin filament systems created in RNA injection experiments, it seems likely that the same factor can interact with the head of keratins as well.

A rather speculative proposal as to the nature of this factor is based on the characterization of keratin–desmosome interactions. Recently, its has been reported that the desmosomal protein desmoplakin binds to the head domain of type II keratins (Kouklis *et al.*, 1994). Desmoplakin appears to be present in the oocyte and is localized in the cortex; during oocyte maturation, it seems to disappear from the cortex (see Klymkowsky and Karnovsky, 1994). It is possible that free desmoplakin could interact with keratin (and vimentin filaments) and interfere with the formation of extended IF networks (Stappenbeck and Green, 1992; Stappenbeck *et al.*, 1993). It is also known that phosphorylation of desmoplakin can modulate

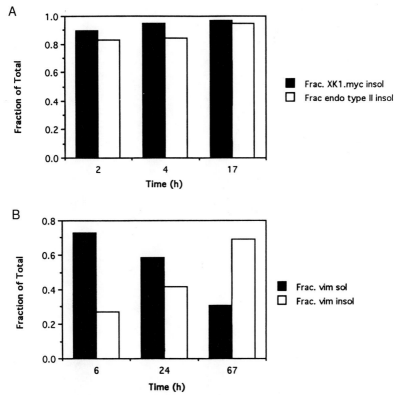

Fig. 2 Fractionation of exogenous keratins and vimentin in the oocyte. Oocytes were injected with RNAs encoding epitope-tagged forms of *Xenopus* keratins (A) or with vimentin (B) and labeled with [^{35}S]-methionine. At various times after injection, the percentage of exogenous keratin in the insoluble fraction was calculated (solid bars in A) and compared with the percentage of insoluble endogenous type II keratin found in uninjected oocytes (open bars in A). The percentage of keratin associated with the insoluble fraction was similar in both keratin RNA-injected and -uninjected oocytes and this percentage remained constant throughout the course of the experiment. In contrast, when a similar analysis was performed on oocytes injected with vimentin RNA (B) we found a clear increase in the percentage of insoluble vimentin (open bars) and a decrease in percentage of soluble vimentin (solid bars) over time. The total amount of radioactivity associated with vimentin (soluble + insoluble) increased throughout a 67-hr labeling period. The behavior of exogenous keratin is consistent with its incorporation into a preexisting keratin filament network, whereas the behavior of exogenous vimentin suggests the *de novo* assembly of a vimentin filament system inside the oocyte. Figure modified from Bachant (1993)

its interaction with keratin filaments (Stappenbeck *et al.*, 1995), which could provide a mechanism for the release of its interaction with IFs during oocyte maturation. This model clearly requires further study.

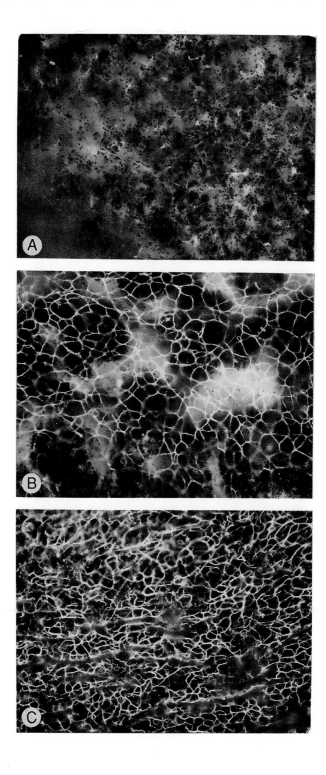

C. Actin Filament, Microtubule, and IF Interactions in the Oocyte

Another hint into the nature of the factor(s) that generates the animal–vegetal asymmetry in keratin network organization comes from a series of studies involving the treatment of oocytes with the antimicrotubule (MT) drug nocodazole, the antiactin filament drug cytochalasin, or both drugs together. Treatment of somatic cells with these drugs can produce profound reorganization of keratin filaments (Knapp *et al.*, 1983a,b). Treatment of oocytes with these drugs produces a dramatic reorganization of animal cortical keratin filaments; in place of the disconnected keratin filaments, a complete and well-connected keratin network appears (Fig. 3). This apparent increase in keratin network organization, in response to anticytoskeletal drugs, is unprecedented and the mechanism that underlies it is completely unclear. Nevertheless, it suggests a close interaction between actin filament, MT, and IF structures in the oocyte.

III. IF Function in the Oocyte

In the absence of direct experimental evidence it is perhaps unwise to discuss the functions of IFs. Knockout mutations in the mouse of the highly conserved vimentin (Colucci *et al.*, 1994) and GFAP (Gomi *et al.*, 1995.; Pekny *et al.*, 1995) genes produce no dramatic (or as yet recognized) phenotypes. These are surprising results, particularly in the case of GFAP, where the protein is a major component of one of the major cell types in the central nervous system. Caution aside, it is a suggestive fact that keratins are a conserved component of the oocyte cytoskeleton in a number of different organisms (for example see Chapter 9 by Gallicano and Capco). Whether they play a functional role or are simply present in a storage form remains unclear.

In the early stage *Xenopus* oocyte, the cage of keratin filaments that surrounds the mitochondrial cloud could be involved in maintaining its integrity. As oogenesis proceeds, the mitochondrial cloud fragments and both mitochondria and associated "germ-plasm" materials move to the

---

Fig. 3 Effects of nocodazole and cytochalasin on keratin filament organization. Oocytes were treated with 10 μg/ml nocodazole (B), 5 μg/ml cytochalasin B (C), or left untreated (A) for 6 hr and then fixed and stained with the anti-keratin antibody 1h5. In the control oocytes, the organization of keratin filaments in the animal hemisphere consists of sparse, scattered filaments. In contrast, treatment with either nocodazole or cytochalasin leads to the appearance of a robust, highly interconnected keratin filament network in the animal hemisphere of the oocyte.

vegetal cortex in an organized manner (see Forristall *et al.,* 1995; Kloc and Etkin, 1995). The molecular switch involved in the fragmentation and migration of the mitochondrial cloud components is unknown. It is possible that changes in IF organization that accompany this process are simply permissive, rather than causal.

Pondel and King (1988) initially proposed that keratin filaments were involved in the anchoring of specific mRNAs to the vegetal cortex of the late-stage oocyte. A number of observations show that this is not the case; the most significant of which are that the disruption of keratin filament organization, mediated by the expression of mutated forms of keratin, does not lead to the solubilization of Vg1 mRNA (Fig. 4) and that Vg1 mRNA can be solubilized without the disruption of the cortical keratin filament network (Klymkowsky *et al.,* 1991). In addition, a number of other vegetally localized RNAs remain associated with the detergent-insoluble "cytoskeletal" fraction even after the complete solubilization of the oocyte's keratin filament system during maturation (see Forristall *et al.,* 1995). Finally, specific mRNAs are found to be associated with a detergent-insoluble fraction in *Drosophila* (Pokrywka and Stephanson, 1994), which appears not to have IFs (see Bartnik and Weber, 1989).

IV. Maturation-Induced Disassembly of Keratin Filaments

The most striking feature of the keratin system of the late-stage *Xenopus* oocyte is its complete disassembly during oocyte maturation. This fact was somehow overlooked by initial studies of the system (see Franz *et al.,* 1983; Gall *et al.,* 1983; Godsave *et al.,* 1984a,b; Gall and Karsenti, 1987), but can be readily observed using either whole-mount immunocytochemistry (Klymkowsky *et al.,* 1987; Klymkowsky and Maynell, 1989) or biochemical fractionation (Pondel and King, 1988; Klymkowsky *et al.,* 1991; Bachant and Klymkowsky, 1995).

The process of keratin filament reorganization during M phase was first observed in cultured cells (see Horwitz *et al.,* 1981; Franke *et al.,* 1982; Lane *et al.,* 1982). In each of these cases, the keratin filament network is transformed into nonfilamentous aggregates as the cell enters M phase; these aggregates reorganize again, back into filaments, as the cell leaves M phase and reenters interphase. In somatic cells, keratins remain in an insoluble form throughout this M-phase reorganization. In the *Xenopus* oocyte, however, keratin filaments disassemble into soluble oligomers (Klymkowsky *et al.,* 1991; Bachant and Klymkowsky, 1995). Our initial studies used sucrose-gradient velocity sedimentation to characterize the soluble form of keratin in the oocyte. During oocyte maturation, we found the appearance of soluble oligomers of varying sizes. In more recent studies,

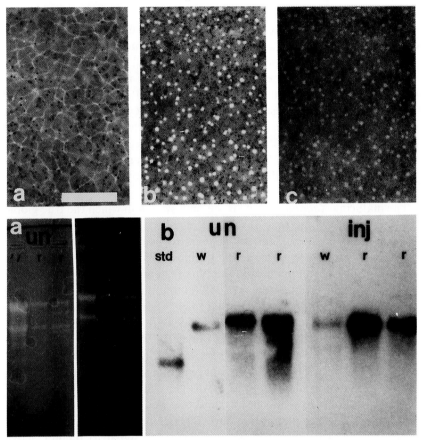

Fig. 4 Disruption of keratin filament organization does not release Vg1 mRNA. To disrupt the oocyte keratin network, we constructed *myc*-tagged and truncated forms of the *Xenopus* keratins XK1 and DG81A. Oocytes were injected with RNA encoding both polypeptides, incubated for 24 hr, and then processed either for whole-mount immunocytochemistry (top) or fractionated to determine whether Vg1 mRNA remained associated with the insoluble fraction (bottom). In the vegetal hemisphere, the normal geodesic system of keratin filaments (a) was lost in mutant keratin RNA-injected oocytes (b and c). Staining with the anti-*myc* antibody reveals the mutant keratin (b), while staining with an anti-keratin antibody reveals the absence of a keratin filament network (c). To determine the effect of disrupting keratin filament organization on Vg1 mRNA insolubility, control and injected oocytes were fractionated and the RNA associated with the insoluble fraction was isolated by Northern blot. In a the ethidium staining of the gel for uninjected (un) and mutant keratin RNA-injected (inj) oocytes are shown. Northern blot analysis (b) with an antisense Vg1 probe revealed that the association of Vg1 mRNA with the insoluble fraction (r) was unaffected. That Vg1 mRNA is concentrated in the insoluble fraction can be seen by comparing the intensity of the signal between RNA isolated from whole oocytes (w) with that of RNA isolated from insoluble residues. Similar amounts of total RNA were loaded for each lane. Figure taken from Bachant (1993).

however, we have used both size-exclusion chromatography and sucrose-gradient sedimentation methods; in these studies we found that the soluble keratin present in the matured oocyte is in fact homogeneous and migrates on size-exclusion column with an apparent molecular weight of ~750 kDa [as opposed to the tetramer, which migrates with an apparent molecular weight of ~ 150 kDa (Coulombe and Fuchs, 1990)]. This species is similar to the soluble keratin present in the prophase oocytes, which also has an apparent molecular weight of 750 kDa on size-exclusion columns (Bachant and Klymkowsky, 1995). Biochemical studies indicate that this oligomeric form is unstable and decays into the tetramer under standard sucrose velocity gradient conditions, which may be why it has not been previously identified.

The maturing *Xenopus* oocyte is the only system in which the true global disassembly of cytoplasmic IFs has been observed. The trigger for the disassembly of keratin filaments appears to be the activation of p34^{cdc2} [maturation-promoting factor (MPF)] kinase (Klymkowsky and Maynell, 1989). During oocyte maturation the type II keratin (K8) becomes hyperphosphorylated (Klymkowsky *et al.,* 1991), but not, apparently, due to the direct action of p34^{cdc2} kinase. Rather, p34^{cdc2} kinase appears to act posttranscriptionally to activate the factors which are required for keratin filament disassembly (Klymkowsky and Maynell, 1989).

To study the process of M-phase reorganization of IFs further, we examined the behavior of both keratin and vimentin filaments in the maturing oocyte. A vimentin filament system was generated in the oocyte (which does not normally possess one) by injecting oocytes with vimentin RNA (see above). During maturation, two distinct processes occur. The first phase runs from exposure to progesterone (the signal to reenter active meiosis) to germinal vesicle breakdown (GVBD). The initially fragmented animal hemisphere keratin (or vimentin) filament network becomes more intact and interconnected (Fig. 5). As oocyte maturation proceeds, however, the behaviors of vimentin and keratin-type IF networks diverge. Vimentin forms a dense and highly interconnected network of filaments and filament bundles that remains intact throughout the maturation process (Dent *et al.,* 1992). In contrast, keratin filaments begin to disappear as the oocyte passes through GVBD. The disappearance of the keratin filament system first begins in the animal hemisphere and eventually extends into the vegetal hemisphere (Klymkowsky and Maynell, 1989). Interestingly, there is often a region around the animal pole (where the oocyte nucleus approaches the cortex) where keratin filament disassembly is delayed. However, by the end of maturation (8–10 hr after exposure to progesterone) keratin filament disassembly is complete; few, if any, intact keratin filaments are visible and the bulk of the keratin is in a soluble form.

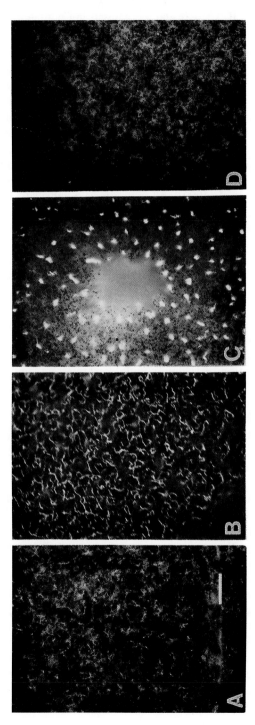

Fig. 5 Changes in keratin organization during oocyte maturation. Illustrated here are the changes seen in keratin filament organization seen during the maturation of the *Xenopus* oocyte. In the prophase oocyte, the keratin network is composed of sparse fibers (A). Shortly after the addition of progesterone, which initiates maturation, a dense, interconnected network of keratin filaments appears (B), which then disassembles, first into aggregates (C) and then into soluble polypeptides (D) (no longer visible by immunofluorescence microscopy).

The difference in the behavior of vimentin and keratin filaments in the maturing *Xenopus* oocytes is intriguing. While both vimentin (Dent *et al.,* 1992) and keratin (Klymkowsky *et al.,* 1991) are hyperphosphorylated during oocyte maturation, only the keratin filament system disassembles. A recent study suggests that the factor that determines whether vimentin filaments remain intact during M phase in somatic cells is the level of $p34^{cdc2}$ kinase activity (Tsujimura *et al.,* 1994). This conclusion would not seem to apply to the *Xenopus* oocyte, however, since vimentin filaments do not disassemble in the metaphase II-arrested oocyte, even though $p34^{cdc2}$ kinase/MPF activity is high in these cells (see Gerhart *et al.,* 1984; Klymkowsky *et al.,* 1991; Dent *et al.,* 1992) and vimentin is a substrate for $p34^{cdc2}$/MPF kinase (Chou *et al.,* 1990; Dent *et al.,* 1992). We therefore conclude that IF disassembly/reorganization during M phase involves other factors in addition to the level of $p34^{cdc2}$ kinase activity.

Thus, the maturation-induced reorganization of keratin filaments is a two-stage process. In the first phase, the factor that suppresses extended IF network formation in the animal hemisphere of the oocyte is inactivated; in the second phase, keratin filaments are disassembled into soluble oligomers. Keratin filament disassembly is accompanied by hyperphosphosporylation of the type II keratin (Klymkowsky *et al.,* 1991). It is clear that the disassembled state must be actively maintained because if protein synthesis is blocked by the addition of cycloheximide, keratins reassemble into extended filaments (Klymkowsky *et al.,* 1991).

An obvious question is: Is the disassembly of keratin filaments physiologically significant? It is interesting to note that the reorganization of IFs during oocyte maturation/early embryogenesis has been described in a wide range of organisms (see Lehtonen *et al.,* 1983; Lehtonen 1985, 1987; Boyle and Ernst, 1989; Plancha *et al.,* 1989, 1991; Schroeder and Otto, 1991; see Chapter 9) and may represent a common transition from the relatively quiescent oocyte to the more dynamic embryo. It could be that the keratin filament disassembly is an epiphenomena; that is, it is an unforeseen (and essentially irrelevant) by-product of some other event(s) occurring during oocyte maturation, e.g., the activation of specific kinases. The second possibility is that keratin filament disassembly is important perhaps as a way of facilitating the remodeling of keratin filaments. In the latter case, the global disassembly of keratin filaments would be similar to the focal disassembly of IFs in the region of the contractile ring of somatic cells (Blose, 1979; Nishizawa *et al.,* 1991). In somatic cells, it appears that the focal disassembly of IFs facilitates the separation of daughter cells; in the oocyte the disassembly of the keratin filament network could simplify the formation of a new embryonic keratin filament system. In this light, it is worth noting that the organization of the embryonic keratin filament system is distinctly different from that of the oocyte (Klymkowsky *et al.,* 1987).

There is a another possible function for the disassembly of the oocyte keratin filament network. In the oocyte, the cortical keratin filament system is connected to a radial system of subcortical filaments (Godsave *et al.,* 1984a,b; Ryabova *et al.,* 1993, 1994); these connections presumably inhibit movements between cortex and the inner cytoplasmic region of the oocyte. In the egg, the establishment of the dorsal–ventral axis of the embryo critically depends on the movement of cortex with respect to the deeper layers of the egg during the first cell cycle (see Gerhart and Keller, 1986; Sive, 1993). Connections between cortical and deeper layers of the egg would therefore interfere with this cortical rotation in the egg (Fig. 6). The disassembly of keratin filaments would remove this possible impediment to cortical rotation.

V. Fertilization and the Reappearance of the Keratin Network

In the egg, the keratin filament system is disassembled. Upon activation of the egg by fertilization, needle prick, or exposure to Ca^{2+} ionophore, the keratin system begins to reappear (see Klymkowsky *et al.,* 1987). The assembly occurs in the presence of cycloheximide and is not dependent on newly synthesized keratins, but appears to involve the reassembly of maternal keratins. Presumably, changes in the level of post-translational modification associated with the progression of the cell outside of meiosis are involved. As the keratin filament system forms, it initially appears as fine isolated fibers; as the microtubule array associated with the cortical rotation

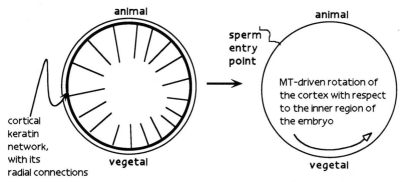

Fig. 6 In the oocyte, keratin forms a cortical system that is connected to radial fibers. In the fertilized egg, microtubules drive the rotation of the cortex with respect to the inner region of the egg. This rotation is critical for the establishment of the dorsal–ventral axis of the embryo. It is possible that the disassembly of the oocyte keratin filament network acts mainly to remove a possible "brake" to this rotation.

of the egg forms and begins to move the cortex (see Chapter 12), the cortical keratin filaments also begin to assume a preferred orientation aligned with the microtubule system. As time proceeds, the keratin filament network begins to form the anastomosing "fishnet" structure associated with the vegetal hemisphere of the embryo. In the animal hemisphere, the keratin system is organized with a finer texture and its organization is much less regular (Klymkowsky *et al.*, 1987; Klymkowsky and Karnovsky, 1994). As cleavage furrows form, they cut through the cortical keratin filament network of the oocyte.

VI. Organization of the Embryonic Keratin Filament System

By the end of the second or third cleavage division, the embryonic keratin filament system is well formed (Klymkowsky *et al.*, 1987). In most cell types, keratin filaments are linked into a supracellular array through interactions with desmosome-type cell–cell junctions. The keratin filament system of the early *Xenopus* embryo, however, is not linked to desmosomes (Perry, 1975), but desmosomal components (e.g., desmosomal cadherins and pla-koglobin) are stored in the oocyte (DeMarais and Moon, 1992; Fouquet *et al.*, 1992; Cordingley *et al.*, 1993; Klymkowsky and Karnovsky, 1994), presumably awaiting an appropriate signal to be assembled into desmosomes. The appearance of desmosomes in *Xenopus* has been rather poorly described. The critical (and essentially only) work on this topic that I know of is the paper by Perry (1975). She found that desmosomes begin to appear before gastrulation and increase with time after that.

The simple epithelial keratins, K8, K18 and K19, appear to be the only IFPs in the early embryo. Zygotic transcription begins in earnest at the midblastula transition (Newport and Kirschner, 1982). At this point, new IFPs begin to appear. The first of these are keratins of the epidermal type (Jonas *et al.*, 1985; Winkles *et al.*, 1985; Miyatani *et al.*, 1986; Fouquet *et al.*, 1988) which are expressed primarily in the superficial cells of the animal hemisphere. These epidermal keratins appear to be homologous to the human epidermal keratins K5, K6, and K14. As the neuroectoderm is induced by the involuting chordomesoderm, the synthesis of epidermal keratins is suppressed in this region of the embryo (Jamrich *et al.*, 1987). There are a number of other changes in keratin expression as the epidermis of the tadpole development and during the course of metamorphosis (see Ellison *et al.*, 1985; Nishikawa *et al.*, 1992).

Two studies indicate that the integrity of the early embryonic keratin filament system is important for normal gastrulation in *Xenopus*. Torpey *et al.* (1992) used antisense oligonucleotides to destroy type II keratin mRNA in the oocyte. Since keratin filaments are obligate heteropolymers, the absence of type II keratin leads to the failure to form keratin filaments. In the resulting embryos, Torpey *et al.* (1992) observed both gastrulation and wound-healing defects.

In our studies, we used the injection of monoclonal anti-keratin antibodies. As in our earlier studies of cultured epithelial cells (Lane and Klymkowsky, 1982; Klymkowsky *et al.*, 1983), injection of anti-keratin antibodies disrupts the organizational integrity of keratin filament systems. In the *Xenopus* embryo injection of the anti-keratin antibody 1h5 disrupts the cortical keratin filament system, but has no effect on gastrulation (Klymkowsky *et al.*, 1992). In contrast, the anti-keratin antibodies AE1 and AE3 appear to interact specifically with subcortical keratin filaments and produce a distinctive gastrulation defect. Embryos injected with these anti-keratin antibodies form a blastopore normally, but the blastopore fails to close (Klymkowsky *et al.*, 1992; Klymkowsky and Karnovsky, 1994). Normally the animal cap of cells thins to a two-cell thick layer during gastrulation; this thinning does not occur in AE3 antibody-injected embryos (A. Karnovsky, unpublished observations). This failure of cellular reorganization could act as a brake on blastopore closure.

Why disruption of keratin filament organization should produce a failure of the animal cap to thin is unclear. Disruption of keratin filament organization in the mouse following disruption of the keratin K8 gene (Baribault and Oshima, 1991; Baribault *et al.*, 1993) suggests that, while many morphogenetic events occur normally, there are also dramatic defects in the behavior of specific epithelial tissues (see Chapter 9 for evidence that cytokeratin 8 is not the prominent form of intermediate filament protein in mouse embryos). For example, placental organization is aberrant and, depending on the genetic background, there can be an extensive hyperplasia of the colon and large intestine (Baribault *et al.*, 1994). The mechanism by which the absence of an intact keratin filament network leads to such a hyperplasia is not clear. A possibility is that the disruption of keratin filament organization leads to changes in intracellular signals arising from interactions with other cellular components, e.g., desmosomes; this altered pattern of signaling may in turn influence other cellular adhesion/cytoskeletal systems.

To address this possibility in the *Xenopus* system, we examined the effects of polypeptides that perturb the keratin–desmosome network. These studies are based on the observation by Kintner (1992) that the isolate cytoplasmic domain of *N*-cadherin acts as a dominant mutation and disrupts normal cellular adhesion when expressed in the embryo. Cadherins are the foundation of both actin filament-associated adherens junctions and IF-associated desmosome-type cell–cell adhesion junctions. "Classical" cadherins are present at adherens junctions, whereas desmosomal cadherins are present at desmosomes. Normal adhesion between cadherins is dependent on the association of cadherin-binding proteins, catenins, with the cytoplasmic tail domain of the cadherin (Gumbiner and McCrea, 1993). In turn, the catenins mediate the interaction between the cadherin and actin filaments. In the case of the adherens junction, the expression of the cytoplasmic domain of N- (Kintner, 1992) or EP-cadherin (Karnovsky and Klymkowsky, 1994) leads to the disruption of normal adhesion between bastomeres, presumably by binding to and sequestering catenins. Similar effects on embryonic integrity have been seen when the level of maternal cadherin is depleted using antisense oligonucleotides (Heasman *et al.,* 1994a).

That a similar approach was also feasible for studying the role of desmosomes was suggested by the work of Troyanovsky *et al.* (1993), who showed that the tail domains of the desmosomal cadherins desmocollin and desmoglein could induce the formation or the disruption of desmosomes in cultured cells. In their study, these desmosomal cadherin tail domains were attached to integral membrane proteins. We reasoned that similar effects could be generated by expressing only the cytoplasmic tail domain. We therefore constructed two "cytoplasmic domain" constructs: one encoding the tail of bovine desmocollin-1b and the other encoding the human desmoglein-1 tail domain. When capped RNA coding for these polypeptides was injected into fertilized eggs, they accumulated and caused defects in gastrulation similar to those produced by AE3 injection (A. Karnovsky *et al.,* manuscript in preparation). In contrast to the phenotypes seen when the transmembrane/cytoplasmic domain of N-cadherin (Kintner, 1992) or the cytoplasmic tail domain of EP-cadherin (Karnovsky and Klymkowsky, 1994) are synthesized in the embryo, cell adhesion appears substantially normal, suggesting that the effects of the desmosomal cadherin tail are not directly due to effects on classical cadherin-based junctions. It remains to be determined whether the effects of the desmosomal cadherin tails on gastrulation are strictly structural, i.e., due to the disruption of desmosome organization or keratin filament attachment, or whether other processes are involved (see below).

VII. Interactions between Adherence and Inductive Systems: The Plakoglobin Connection

It appears that the formation of dorsal mesoderm in *Xenopus* depends on the wnt signaling pathway (Moon, 1993; Moon *et al.,* 1993a,b; Sive, 1993). Wnts (whose name derived from the related proteins wingless (wg) in *Drospholia* and the int-1 protooncogene of mammals) are secreted, short-range signaling polypeptides that presumably act by binding to an as yet unidentified cell surface receptor. In *Xenopus*, the ectopic expression of specific wnts is capable of rescuing axial defects arising from the suppression of dorsal mesoderm induction (induced by UV irradiation during the first cleavage cycle) and of inducing a second neural axis, presumably due to their ability to induce ectopic dorsal mesoderm in the late blastula stage embryo. In *Drosophila*, secretion of wg protein leads to an increase in the cytoplasmic level of the armadillo protein (ARM), which in turn acts to stabilize the expression of the transcription factor engrailed (see Perrimon, 1994; Siegfried *et al.*, 1994). ARM is closely related to the cadherin-binding proteins β-catenin and plakoglobin (γ-catenin) (Peifer *et al.*, 1992). In mammalian cells, exposure to wnt also leads to an increase in the level of soluble β-catenin and plakoglobin (Bradley *et al.*, 1993; Hinck *et al.*, 1994).

That cadherins and their associated proteins play an integral role in inductive signaling in *Xenopus* now appears to be well established (Gumbiner and McCrea, 1993; McCrea *et al.*, 1993; Moon *et al.*, 1993; Brieher and Gumbiner, 1994; Heasman *et al.*, 1994; Klymkowsky and Karnovsky, 1994; Funayama *et al.*, 1995; Karnovsky and Klymkowsky, 1995). Studies by Heasman *et al.* (1994b) demonstrated that the overexpression of cadherin or the downregulation of β-catenin levels, whose interaction with cadherin is required for normal adherens junction formation, both leads to the inhibition of dorsal mesoderm formation. The active signal in the induction of dorsal mesoderm appears to be the ARM-like proteins β-catenin and plakoglobin, since altering β-catenin levels/activity through the injection of anti-β-catenin antibodies (McCrea *et al.*, 1993), the overexpression of β-catenin or plakoglobin proteins (Funayama *et al.*, 1995; Karnovsky and Klymkowsky, 1995), or the expression of a dominant mutant form of the *Xenopus* homolog of zeste-white 3/shaggy kinase (Pierce and Kimelman, 1995; He *et al.*, 1995) leads to dorsal mesoderm differentiation (see Klymkowsky and Parr, 1995). In the context of the intact embryo, the formation of ectopic dorsal mesoderm leads to the induction of secondary neural axes and results in two-headed tadpoles.

How changing levels of ARM-like proteins lead to dorsal mesoderm formation is somewhat more mysterious. In *Drosophila*, increased levels of cytoplasmic ARM appear to be mediated by the suppression of ARM phosphorylation (Peifer *et al.*, 1993; 1994a,b). In the absence of a wg signal, it appears that ARM is phosphorylated by the zeste-white 3 (zw3) gene product; the zw3 gene product is a protein kinase homologous to the glycogen synthetase kinase-3 of mammalian cells. However, the mechanism by which increasing cytoplasmic ARM protein leads to changes in engrailed gene expression remains unclear. It has been proposed that the effect is mediated by the interactions of cytoplasmic ARM with a "wnt effector" (Peifer *et al.*, 1994a; Klymkowsky and Parr, 1995), although there is little evidence as yet to the existence or nature of this molecule.

In our own work, we have been particularly interested in the possibility that changes in keratin filament organization and keratin filament–desmosomal interactions could lead to changes in cell behavior (see above). The observation that overexpression of plakoglobin leads to axis duplication (Karnovsky and Klymkowsky, 1995) provides such a link. Plakoglobin is unique among cadherin-binding proteins in that it interacts with both classical and desmosomal cadherins. At the desmosome, plakoglobin binds to both desmoglein (Korman *et al.*, 1989; Mathur *et al.*, 1994; Troyanovsky *et al.*, 1994) and to the desmocollins (Troyanovsky *et al.*, 1994; P. Cowin, personal communication). By analogy to the interaction between β-catenin and classical cadherins, it appears that plakoglobin interacts with desmosomal cadherins through its central armadillo repeat region (Hülsken *et al.*, 1994; Funayama *et al.*, 1995), a stretch of 13 imperfect ~40 amino acid repeats (Peifer *et al.*, 1994a,b). The expression of this region is both necessary and sufficient to induce axis duplication (Funayama *et al.*, 1995; see Karnovsky and Klymkowsky, 1995).

The effects of overexpressing cadherin (Heasman *et al.*, 1994b) would suggest that the association between catenins and cadherins modulates the signals involved in dorsal mesoderm formation. In our experiments, we have shown that coexpression of the tail domain of the desmosomal cadherin desmoglein-1 suppresses the signaling effects of plakoglobin (Karnovsky and Klymkowsky, 1995). This suggests that the signal involved in the formation of excess dorsal mesoderm either involves the presence of soluble plakoglobin or that plakoglobin must form part of some larger complex to play a role in mesoderm signaling.

More surprisingly is the fact that the exogenous plakoglobin and β-catenin are found concentrated in the nuclei of late-stage blastula cells (Funayama *et al.*, 1995; Karnovsky and Klymkowsky, 1995). The nuclear localization of plakoglobin and β-catenin is seen only under conditions of rather substantial overexpression. It has been shown that when overexpressed, a number of normally cytoplasmic proteins are transported into the nuclei of cells. In polypeptides that lack an intrinsic "nuclear localization sequence," which appears to be the case in both plakoglobin and β-catenin, it is likely that the observed nuclear localization is due to the interaction between these proteins and polypeptides that are themselves normally transported into the nucleus (e.g., transcription factors, nuclear lamins, etc.). We interpret the nuclear localization of β-catenin and plakoglobin as evidence that these proteins interact with nuclear factors, perhaps transcription factors. In the "normal" case, this interaction could block the nuclear transport or activity of these factors. If such factors were active in the control of transcription, changes in the level (or activity) of plakoglobin/ β-catenin could readily alter the pattern of gene expression, which in turn could affect morphogenetic events such as epiboly and convergent extension (see Klymkowsky and Parr, 1995).

VIII. The Appearance of Nonepithelial IFPs

Following gastrulation, a number of nonepithelial IFPs appear in the embryo. The first of these are vimentin and desmin. In the mouse, vimentin expression begins in the mesoderm and most nonepithelial cells express vimentin prior to their differentiation and the expression of differentiation-specific IFPs, such as desmin in muscle and neurofilament proteins in neurons. In *Xenopus*, however, vimentin expression begins later. Vimentin transcription first becomes substantial at stage 14 (early neurula) (Herrmann *et al.*, 1989a) and the protein initially appears within cells of the neural tube (Dent *et al.*, 1989). With time vimentin appears in a number of mesenchymal cell types (Godsave *et al.*, 1986; Dent *et al.*, 1989).

In the nervous system, vimentin expression is followed by the appearance of GFAP, the neurofilament proteins, and the nestin-like protein tannabin (Godsave *et al.*, 1986; Sharpe, 1988; Szaro and Gainer, 1988a,b; Chu and Klymkowsky, 1989; Sharpe *et al.*, 1989; Charnas *et al.*, 1992; Hemmati-Brivanalou *et al.*, 1992). Studies of the role of these IFPs, if any, are few. Szaro *et al.* (1991) used the injection of anti-neurofilament antibodies to examine the role of neurofilaments in neurite extension. In their studies, the presence of anti-neurofilament antibodies blocked neurite extension. However, whether this reflects a normal role for neurofilaments, i.e., that neurite extension depends on the presence of neurofilaments, is not completely clear. In the *quiver* mutant quail, which is effectively a null mutation of the NFL gene, neurite formation appears relatively normal (Ohara *et al.*, 1993). The primary defect seen in these animals, and in transgenic mice with a disruption of their neurofilament system induced by the expression of NFH–*lacZ* chimeric polypeptide (Eyer and Peterson, 1994) is a decrease in the diameter of the largest axons. It is therefore possible that the presence of antibody does more than negate the function of neurofilaments; it may also generate defects related to the presence of abnormal structures within the neural cells. In this light, the observations of Lee *et al.* (1994), Côté *et al.* (1993), and Xu *et al.* (1993) are of interest. They found that expressing human NFH, overexpressing mouse NFL, or expressing a mutated form of mouse NFL leads to the formation of aggregates which in turn lead to motorneuron cell death, a phenotype not seen when the neurofilament system is simply removed.

The functions of vimentin or GFAP in the early embryo are also unclear, particularly given the lack of a overt phenotype following the ablation of these genes in the mouse (see above). It is possible that the effects of these mutations produce mild, but evolutionarily significant, deficits or defects. Alternatively, severe defects may be generated in specific, and as yet unrecognized, aspects of cell/tissue function.

IX. Desmin Organization and Function

We have found *Xenopus* to be particularly useful in the study of desmin organization and function. In most myogenic systems, vimentin is expressed in muscle precursor cells. As these cells begin to differentiate, desmin begins to be expressed and to integrate into the preexisting vimentin filament network. The affinity of desmin and vimentin for one another is high, so this interaction obscures lower-affinity interactions that may be present between desmin and other cellular proteins. As muscle differentiation continues, vimentin synthesis is turned off and the composition of the IF system changes. At the same time, the internal organization of the muscle cell is also changing, with the formation of the contractile apparatus and its attachment to the sarcolemma. By the end of this process, the initial longitudinal IF system of the myoblast has been transformed into the transverse, Z-line-associated network of the mature muscle cell.

In the dorsal myotomal muscle (DMM) of the *Xenopus* tadpole, however, vimentin is never expressed (Dent *et al.,* 1989). Therefore, the organization of desmin, from the time of its first appearance, reflects its interactions with other cellular components. In the DMM, desmin is expressed prior to the segmentation of the somitic mesenchyme (Herrmann *et al.,* 1989b; Cary and Klymkowsky,1994a) and is found concentrated at the tips of the myogenic cells (Cary and Klymkowsky, 1994a). As the somites form through a process of segmentation and rotation (Hamilton, 1969), desmin continues to be found concentrated at the ends of the cells. By the end of somite rotation, desmin is concentrated at the sites of the intersomite junction (ISJ) and is associated with the sarcolemma (Cary and Klymkowsky, 1994a). Desmin eventually associates with Z- lines to form the transverse, sarcolemma-associated IF system characteristic of mature striated muscle. Throughout the process of myogenic differentiation, however, no extensive longitudinal IF network ever forms in DMM cells.

Vimentin and desmin are rather similar in sequence; the differences between them reside primarily in their head domains (Cary and Klymkowsky, 1994b). It is possible to show that the difference in the behavior of vimentin and desmin in muscle cells is due primarily to the nature of their head domains. Vimentin or a chimeric protein consisting of the head domain of vimentin attached to the body of desmin forms a longitudinal filament system when expressed in DMM cells. This type of IF system is similar to that present in other myogenic systems which express vimentin and indicates that the ability to form such longitudinal IF networks is a direct property of vimentin. In contrast, desmin, or a desmin-headed chimera, is found primarily associated with the sarcolemma. This indicates that it is the head of desmin that mediates the interaction between desmin filaments and the sarcolemma. This is an interesting observation in the light of the apparent role of the head of keratin in the interaction with desmoplakin (Kouklis *et al.*, 1994) and data suggesting that the head of vimentin is involved in its interactions with factors in the *Xenopus* oocyte (Dent *et al.*, 1992). However, it is currently unclear what protein(s) mediate the interaction between desmin and the sarcolemma. An intriguing possibility is the lamin-B-like peripheral membrane protein described by Cartaud *et al.* (1995).

To study the role of the desmin filament system in DMM, we used a simple method—the expression of mutated forms of desmin (and vimentin). It is relatively easy to generate dominant mutations in IF proteins by truncation (Albers and Fuchs, 1987; Gill *et al.*, 1990; Wong and Cleveland, 1990) or amino acid changes (see Letai *et al.*, 1993; McLean and Lane, 1995) in key regions of the polypeptide. To study the role of desmin in DMM, we examined the effects of truncated forms of desmin and vimentin; both mutated polypeptides were shown to be capable of disrupting vimentin filament organization in cultured cells (Cary and Klymkowsky, 1995). When expressed in DMM cells, they disrupted normal IF organization as well. As in the case of similar studies on cultured myotubes (Schultheiss *et al.*, 1991), the expression of mutant IF proteins in the DMM had little affect on the general sarcomeric structure of muscle; Z lines formed normally and myofibrils appeared to be aligned longitudinally. The defect observed was localized to the ISJ, the site of myofibril–membrane attachment. Myofibrils are disrupted in this region and the normal morphology of the ISJ membrane is aberrant. Whether these defects are due to the failure of normal ISJ junction morphology to form, or to defects that arise due to a failure of the junction in the course of normal contractile activity remains to be determined. In any case, this work provides the first example of a defect in muscle due to the expression of a mutant IF protein. Given the association of defects in desmin organization with a range of myopathies and cardiomyopathies, it seems likely that disruption of desmin organization leads directly to muscle pathology.

Appendix: Methods

A. General Remarks

Most of the methods used to study IF organization and function in *Xenopus* are rather standard. To adapt them to *Xenopus* requires relatively minor modifications. For example, the immunocytochemical analysis of cortical IF organization is really a simple modification of the methods used to study IF organization in cultured cells (see Klymkowsky *et al.*, 1987). These methods have been described in detail previously (see Klymkowsky and Hanken, 1991).

Although whole-mount staining of oocytes and embryos requires extended incubation times (Dent *et al.,* 1989; Klymkowsky and Hanken, 1991), staining and destaining of cortical structures is much faster. Typically, for such "turbo" staining, we incubate in primary antibody for 1 hr, wash for 5 min, incubate in secondary antibody for 1 hr, and wash for 5 min. To visualize the surface, we routinely use 0.8-mm depression slides, which can be obtained from most general laboratory supply houses. The oocytes are mounted in a airvol-based media (Klymkowsky and Hanken, 1991) and a coverslip is applied to the top; by pressing down on the coverslip, extended regions of the oocyte/embryonic surface are rendered flat and can be viewed using standard confocal epifluorescence microscopes. Photographic exposures of the animal hemisphere are relatively standard; in the vegetal hemisphere the autofluorescence of the yolk requires that exposure times be kept low. Methods of confocal analysis are described in greater detail in Chapter 12.

B. RNA Injection into Oocytes

Concerning the expression of foreign or mutated proteins in *Xenopus* oocytes, the most effective approach for the oocyte or the early embryo is the use of *in vitro* synthesized and capped RNA. Typically, we prepare RNA using a commercially available kit, optimized for the production of capped RNA (e.g., Ambion's mMessage mMachine kit). Our own plasmid system is based on the pSP64T plasmid developed by Krieg and Melton (1984; see also Vize *et al.,* 1991). In pSP64T, the 5' and 3' UTRs of *Xenopus* β-globin have been introduced downstream of a SP6 polymerase promoter. In our version, the SP6 polymerase promoter has been replaced with a T7 promoter (Cary *et al.,* 1994). We have constructed three variants of this plasmid. In one, sequence encoding the "myc" epitope tag is present at the 5' end of the coding sequence. There is also a 3' tagging version of the plasmid and a version without tagging sequence.

We are particularly drawn to the myc tag because the hybridoma cell line that makes the antibody which recognizes it, 9E10 (Evan *et al.,* 1985), can be obtained from the American Type Culture Collection for a modest price (making the antibody inexpensive for the cost conscious); the binding of the antibody to the epitope (MEQKLISEEDLN) is unaffected by aldehyde fixatives, and so it works well at both light and electron microscopic levels (see Cary and Klymkowsky, 1994b, 1995); and the antibody also works well for immunoprecipitation analyzes. The epitope is small, and at least in the case of a number of IF proteins and plakoglobin, does not seem to interfere with function or assembly. I am not aware of anyone who has attempt to determine whether the entire 12 amino acid form is required for 9E10 binding; smaller forms may also work well.

Acknowledgments

I thank the American Cancer Society, the National Science Foundation, and the Muscular Dystrophy Association for their continuing support of our work. I also thank my colleagues, particularly Joe Dent, Jeff Bachant, Robert Cary, and Alla Karnovsky, for their many contributions to the studies described in this work.

References

Albers, K., and Fuchs, E. (1987). The expression of mutant epidermal keratin cDNAs transfected in simple epithelial and squamous cell carcinoma lines. *J. Cell Biol.* **105,** 791–806.

Bachant, J. B. (1993). Assembly and dynamics of keratin intermediate filaments in Xenopus laevis oocytes. Ph.D. Thesis, Univ. of Colorado, Boulder.

Bachant, J. B., and Klymkowsky, M. W. (1995). The identification of a new soluble form of keratin in *Xenopus* oocytes and rabbit reticulocyte lysates. Submitted for publication.

Baribault, H., and Oshima, R. G. (1991). Polarized and functional epithelia can form after the targeted inactivation of both mouse keratin 8 alleles. *J. Cell Biol.* **115,** 1675–1684.

Baribault, H., Price, J., Miyai, K., and Oshima, R. G. (1993). Mid-gestational lethality in mice lacking keratin 8. *Genes Dev.* **7,** 1191–1202.

Baribault, H, Penner, J., Iozzo, R. V., and Wilson-Heiner, M. (1994). Colorectal hyperplasia and inflammation in keratin 8-deficient FVB/N mice. *Genes Dev.* **8,** 2964–2973.

Bartnik, E., and Weber, K. (1989). Widespread occurrence of intermediate filaments in invertebrates: Common principles and diversion. *Eur. J. Cell Biol.* **50,** 17–33.

Blose, S. H. (1979). Ten-nanometer filaments and mitosis: Maintenance of structural continuity in dividing endoethelial cells. *Proc. Natl. Acad. Sci. USA* **76,** 3372–3376.

Boyle, J. A., and Ernst, S. G. (1989). Sea urchin oocytes possess cortical arrays of microfilaments, microtubules and intermediate filaments. *Dev. Biol.* **134,** 72–84.

Bradley, R. S., Cowin, P., and Brown, A. M. (1993). Expression of Wnt-1 in PC12 cells results in modulation of plakoglobin and E-cadherin and increased cellular adhesion. *J. Cell Biol.* **123,** 1857–1865.

Brieher, W. M., and Gumbiner, B. M. (1994). Regulation of C-cadherin function during activin induced morphogenesis of *Xenopus* animal caps. *J. Cell Biol.* **126,** 519–527.

Cartaud, A., Jasmin, B. J., Changeus, J.-P., and Cartaud, J. (1995). Direct involvement of a lamin-B-related (54kDa) protein in the association of intermediate filaments with the postsynaptic membrane of the Torpedo marmorata electrocyte. *J. Cell Sci.* **108,** 153–160.

Cary, R. B., Klymkowsky, M. W., Evans, R. M., Domingo, A., Dent, J. A., and Backhus, L. (1994). Vimentin's tail interacts with actin-containing structures in vivo. *J. Cell Sci.* **107,** 1609–1622.

Cary, R. B., and Klymkowsky, M. W. (1994a). Desmin organization during the differentiation of the dorsal myotome in Xenopus laevis. *Differentiation* **56,** 31–38.

Cary, R. B., and Klymkowsky, M. W. (1994b). Differential organization of desmin and vimentin in muscle is due to differences in their head domains. *J. Cell Biol.* **126,** 445–456.

Cary, R. B., and Klymkowsky, M. W. (1995). Disruption of intermediate filament organization leads to structural defects at the intersomite junction in Xenopus myotomal muscle. *Development* **122,** 1041–1052.

Charnas, L. R., Szaro, B. G., and Gainer, H. (1992). Identification and developmental expression of a novel low molecular weight neuronal intermediate filament protein expressed in Xenopus laevis. *J. Neurosci.* **12,** 3010–3024.

Chou, Y. H., Bischoff, J. R., Beach, D., and Goldman, R. D. (1990). Intermediate filament reorganization during mitosis is mediated by p34cdc2 phosphorylation of vimentin. *Cell* **62,** 1063–1071.

Chu, D. T., and Klymkowsky, M. W. (1989). The appearance of acetylated α-tubulin during early development and cellular differentiation in Xenopus. *Dev. Biol.* **136,** 104–117.

Colucci, G. E., Portier, M. M., Dunia, I., Paulin, D., Pournin, S., and Babinet, C. (1994). Mice lacking vimentin develop and reproduce without an obvious phenotype. *Cell* **79,** 679–694.

Cordingley, H. C., Buxton, R. S., and Magee, A. I. (1993). Investigation into the role of desmosomal cadherins in Xenopus development. *Mol. Biol. Cell* **4,** 438. [Abstract]

Côté, F., Collard, J. F., and Julien, J. P. (1993). Progressive neuronopathy in transgenic mice expressing the human neurofilament heavy gene: A mouse model of anyotrophic lateral sclerosis. *Cell* **73,** 35–46.

Coulombe, P. A., and Fuchs, E. (1990). Elucidating the early stages of keratin filament assembly. *J. Cell Biol.* **111,** 153–169.

DeMarais, A. A., and Moon, R. T. (1992). The armadillo homologs β-catenin and plakoglobin are differentially expressed during early development of Xenopus laevis. *Dev. Biol.* **153,** 337–346.

Dent, J. A. (1992). Expression and organization of vimentin filaments in oocytes and embryos of Xenopus laevis. Ph.D. Thesis, Univ. of Colorado, Boulder.

Dent, J. A., and Klymkowsky, M. W. (1989). "Whole-Mount Analyses of Cytoskeletal Reorganization and Function during Oogenesis and Early Embryogenesis in Xenopus." Academic Press, New York.

Dent, J. A., Polson, A. G., and Klymkowsky, M. W. (1989). A whole-mount immunocytochemical analysis of the expression of the intermediate filament protein vimentin in Xenopus. *Development* **105,** 61–74.

Dent, J. A., Cary, R. B., Bachant, J. B., Domingo, A., and Klymkowsky, M. W. (1992). Host cell factors controlling vimentin organization in the Xenopus oocyte. *J. Cell Biol.* **119,** 855–866.

Ellison, T. R., Mathisen, P. M., and Miller, L. (1985). Developmental changes in keratin patterns during epidermal maturation. *Dev. Biol.* **112,** 329–337.

Evan, G. I., Lewis, G. K., Ramsay, G., and Bishop, J. M. (1985). Isolation of monoclonal antibodies specific for human c-myc proto-oncogene product. *Mol. Cell. Biol.* **5,** 3610–3616.

Eyer, J., and Peterson, A. (1994). Neurofilament-deficient axons and perikaryal aggregates in viable transgenic mice expressing a neurofilament-β-galactosidase fusion protein. *Neuron* **12,** 389–405.

Feng, D. F., and Doolittle, R. F. (1990). Progressive alignment and phylogenetic tree construction of protein sequences. *Methods Enzymol.* **183,** 375–387.

Forristall, C., Pondel, M., Chen, L., and King, M. L. (1995). Patterns of localization and cytoskeletal association of two vegetally localized RNAs, Vg1 and Xcat-2. *Development* **121,** 201–208.

Fouquet, B., Herrmann, H., Franz, J. K., and Franke, W. W. (1988). Expression of intermediate filament proteins during development of Xenopus laevis. III. Identification of mRNAs encoding cytokeratins typical of complex epithelia. *Development* **104,** 533–548.

Fouquet, B., Zimbelmann, R., and Franke, W. W. (1992). Identification of plakoglobin in oocytes and early embryos of Xenopus laevis: Maternal expression of a gene encoding a junctional plaque protein. *Differentiation* **51,** 187–194.

Franke, W. W. (1987). Nuclear lamins and cytoplasmic intermediate filament proteins: A growing multigene family. *Cell* **48,** 3–4.

Franke, W. W., Schmid, E., Grund, C., and Geiger, B. (1982). Intermediate filament proteins in nonfilamentous structures: Transient disintegration and inclusion of subunit proteins in granular aggregates. *Cell* **30,** 103–113.

Franz, J. K., and Franke, W. W. (1986). Cloning of cDNA and amino acid sequence of a cytokeratin expressed in oocytes of Xenopus laevis. *Proc. Natl. Acad. Sci. USA* **83,** 6475–6479.

Franz, J. K., Gall, L., Williams, M. A., Picheral, B., and Franke, W. W. (1983). Intermediate-size filaments in a germ cell: Expression of cytokeratins inoocytes and eggs of the frog *Xenopus Proc. Natl. Acad. Sci. USA* **80,** 6254–6258.

Fuchs, E., and Weber, K. (1994). Intermediate filaments: Structure, dynamics, function, and disease. *Annu. Rev. Biochem.* **63,** 345–382.

Funayama, N., Fagatto, F., McCread, P., and Gumbiner, B. M. (1995). Embryonic axis induction by the armadillo repeat domain of β-catenin: Evidence for intracellular signalling. *J. Cell Biol.* **128,** 959–968.

Gall, L., and Karsenti, E. (1987). Soluble cytokeratins in Xenopus laevis oocytes and eggs. *Biol. Cell.* **61,** 33–38.

Gall, L., Picheral, B., and Gounon, P. (1983). Cytochemical evidence for the presence of intermediate filaments and microfilaments in the egg of Xenopus laevis. *Biol. Cell* **47,** 331–342.

Gerhart, J., and Keller, R. (1986). Region-specific cell activities in amphibian gastrulation. *Annu. Rev. Cell Biol.* **2,** 201–229.

Gerhart, J., Wu, M., and Kirschner, M. (1984). Cell cycle dynamics of an M-phase-specific cytoplasmic factor in Xenopus laevis oocytes and eggs. *J. Cell Biol.* **98,** 1247–1255.

Gill, S. R., Wong, P. C., Monteiro, M. J., and Cleveland, D. W. (1990). Assembly properties of dominant and recessive mutations in the small mouse neurofilament (NF-L) subunit. *J. Cell Biol.* **111,** 2005–2019.

Glasgow, E., Druger, R. K., Fuchs, C., Lane, W. S., and Schechter, N. (1994). Molecular cloning of gefiltin (ON1): Serial expression of two new neurofilament mRNAs during optic nerve regeneration. *EMBO J.* **13,** 297–305.

Godsave, S. F., Anderton, B. H., Heasman, J., and Wylie, C. C. (1984a). Oocytes and early embryos of Xenopus laevis contain intermediate filaments which react with antimammalian vimentin antibodies. *J. Embryol. Exp. Morphol.* **83,** 169–187.

Godsave, S. F., Wylie, C. C., Lane, E. B., and Anderton, B. H. (1984b). Intermediate filaments in the Xenopus oocyte: The appearance and distribution of cytokeratin-containing filaments. *J. Embryol. Exp. Morphol.* **83,** 157–167.

Godsave, S. F., Anderton, B. H., and Wylie, C. C. (1986). The appearance and distribution of intermediate filament proteins during differentiation of the central nervous system, skin and notochord of Xenopus laevis. *J. Embryol. Exp. Morphol.* **97,** 201–223.

Gomi, H., Yokoyama, T., Fujimoto, K., Ikeda, T., Katoh, A., Itoh, T., and Itohara, S. (1995). Mice devoid of the glial fibrillary acidic protein develop normally and are susceptible to scrapie prions. *Neuron* **14,** 29–41.

Gumbiner, B. M., and McCrea, P. D. (1993). Catenins as mediators of the cytoplasmic functions of cadherins *J. Cell Sci.* **(Suppl. 17),** 155–158.

Hamilton, L. (1969). The formation of somites in *Xenopus J. Embryol Exp. Morphol.* **22**, 253–264.

He, X., Saint-Jeannet, J.-P., Woodgett, J. R., Varmus, H. E., and Dawid, I. B. (1995). Glycogen synthase kinase-3 and dorsoventral patterning in Xenopus embryos. *Nature* **374**, 617–622.

Heasman, J., Ginsberg, D., Geiger, B., Goldstone, K., Pratt, T., Yoshida, Noro, C., and Wylie, C. (1994a). A functional test for maternally inherited cadherin in Xenopus shows its importance in cell adhesion at the blastula stage. *Development* **120**, 49–57.

Heasman, J., Crawford, A., Goldstone, K., Garner-Hamrick, P., Gumbiner, B., McCrea, P., Kintner, P., Noro, C. Y., and Wylie, C. (1994b). Overexpression of cadherins and underexpression of β-catenin inhibit dorsal mesoderm induction in early Xenopus embryos. *Cell* **79**, 791–803.

Hemmati-Brivanlou, A., de la Torre, J. R., Holt, C., and Harland, R. M. (1992). A protein expressed in the growth cones of embryonic vertebrate neurons defines a new class of intermediate filament protein. *Neuron* **9**, 417–428.

Herrmann, H., Fouquet, B., and Franke, W. W. (1989a). Expression of intermediate filament proteins during development of Xenopus laevis. I. cDNA clones encoding different forms of vimentin. *Development* **105**, 279–298.

Herrmann, H., Fouquet, B., and Franke, W. W. (1989b). Expression of intermediate filament proteins during development of Xenopus laevis. II. Identification and molecular characterization of desmin. *Development* **105**, 299–307.

Hinck, L., Nelson, W. J., and Papkoff, J. (1994). Wnt-1 modulates cell-cell adhesion in mammalian cells by stabilizing β-catenin binding to the cell adhesion protein cadherin. *J. Cell Biol.* **124**, 729–741.

Horwitz, B., Kupfer, H., Eshhar, Z., and Geiger, B. (1981). Reorganization of arrays of prekeratin filaments during mitosis: Immunofluorescence microscopy with multiclonal and monoclonal prekeratin antibodies. *Exp. Cell Res.* **134**, 281–290.

Hülsken, J., Behrens, J., and Birchmeier, W. (1994). Tumor-supressor gene products in cell contacts: The cadherin–APC–armadillo connection. *Curr. Opin. Cell Biol.* **6**, 711–716.

Jamrich, M., Sargent, T. D., and Dawid, I. B. (1987). Cell-type-specific expression of epidermal cytokeratin genes during gastrulation of Xenopus laevis. *Genes Dev.* **1**, 124–132.

Jonas, E., Sargent, T. D., and Dawid, I. B. (1985). Epidermal keratin gene expressed in embryos of Xenopus laevis. *Proc. Natl. Acad. Sci. USA* **82**, 5413–5417.

Karnovsky, A., and Klymkowsky, M. W. (1994). Expression of the tail domains of desmosomal cadherins and plakoglobin and their effect on gastrulation in Xenopus. *Mol. Biol. Cell.* **5**, 186a. [Abstract]

Karnovsky, A., and Klymkowsky, M. W. (1995). Anterior axis duplication in Xenopus induced by the over-expression of the cadherin-binding protein plakoglobin. *Proc. Natl. Acad. Sci. USA* **92**, 4522–4526.

Kintner, C. (1992). Regulation of embryonic cell adhesion by the cadherin cytoplasmic domain. *Cell* **69**, 225–236.

Kloc, M., and Etkin, L. D. (1995). Two distinct pathways for the localization of RNAs at the vegetal cortex in *Xenopus* oocytes. *Development* **121**, 287–297.

Klymkowsky, M. W. (1995). Intermediate filaments: New proteins, some answers, more questions. *Curr. Opin. Cell Biol.* **7**, 46–54.

Klymkowsky, M. W., and Hanken, J. (1991). Whole-mount staining of Xenopus and other vertebrates. *In "Xenopus laevis: Practical Uses in Cell and Molecular Biology"* (B. K. Kay and H. B. Peng, eds), pp. 419–441. Academic Press, San Diego.

Klymkowsky, M. W., and Karnovsky, A. (1994). Morphogenesis and the cytoskeleton: Studies of the Xenopus embryo. *Dev. Biol.* **165**, 372–384.

Klymkowsky, M. W., and Maynell, L. A. (1989). MPF-induced breakdown of cytokeratin filament organization in the maturing Xenopus oocyte depends upon the translation of maternal mRNAs. *Dev. Biol.* **134,** 479–485.

Klymkowsky, M. W., and Parr, B. (1995). A glimpse into the body language of cells: the intimate connection between cell adhesion and gene expression. *Cell,* in press.

Klymkowsky, M. W., Shook, D. R., and Maynell, L. A. (1992). Evidence that the deep keratin filament system of the Xenopus embryo acts to ensure normal gastrulation. *Proc. Natl. Acad. Sci. USA* **89,** 8726–8740.

Klymkowsky, M. W., Maynell, L. A., and Nislow, C. (1991). Cytokeratin phosphorylation, cytokeratin filament severing and the solubilization of the maternal mRNA Vg1. *J. Cell Biol.* **114,** 787–797.

Klymkowsky, M. W., Bachant, J. B., and Domingo, A. (1989). Functions of intermediate filaments. *Cell Motil. Cytoskeleton* **14,** 309–331.

Klymkowsky, M. W., Maynell, L. A., and Polson, A. G. (1987). Polar asymmetry in the organization of the cortical cytokeratin system of Xenopus laevis oocytes and embryos. *Development* **100,** 543–557.

Klymkowsky, M. W., Miller, R. H., and Lane, E. B. (1983). Morphology, behavior, and interaction of cultured epithelial cells after the antibody-induced disruption of keratin filament organization. *J. Cell Biol.* **96,** 494–509.

Knapp, L. W., O'Guin, W. M., and Sawyer, R. H. (1983a). Drug-induced alterations of cytokeratin organization in cultured epithelial cells. *Science* **219,** 501–503.

Knapp, L. W., O'Guin, W. M., and Sawyer, R. H. (1983b). Rearrangement of thekeratin cytoskeleton after combined treatment with microtubule and microfilament inhibitors. *J. Cell Biol.* **97,** 1788–1794.

Krieg, P. A., and Melton, D. A. (1984). Functional messenger RNAs are produced by SP6 *in vitro* transcription of cloned cDNAs. *Nucleic Acids Res.* **12,** 7057–7070.

Kobel, H. R., and Pasquier, L. D. (1986). Genetics of polypoid Xenopus. *Trends Genet.* **3,** 310–315.

Korman, N. J., Eyre, R. W., Kalus-Kovtun, V., and Stanley, J. R. (1989). Demonstration of an adhering-junction molecule (plakoglobin) in the autoantigens of pemphigus foliaceus and pemphigus vulgaris. *N. Engl. J. Med.* **321,** 631–635.

Kouklis, P. D., Hutton, E., and Fuchs, E. (1994). Making a connection: Direct binding between keratin intermediate filaments and desmosomal proteins. *J. Cell Biol.* **127,** 1049–1060.

Lane, E. B., and Klymkowsky, M. W. (1982). Epithelial tonofilaments: Investigating their form and function using monoclonal antibodies. *Cold Spring Harbor Symp. Quant. Biol.* **46,** 387–402.

Lane, E. B., Goodman, S. L., and Trejdosiewicz, L. K. (1982). Disruption of the keratin filament network during epithelial cell division. *EMBO J.* **1,** 1365–1372.

Lee, M. K., Marszalek, J. R., and Cleveland, D. W. (1994). A mutant neurofilament subunit causes massive, selective motor neuron death: Implications for the pathogenesis of human motor neuron disease. *Neuron* **13,** 975–988.

Lehtonen, E. (1985). A monoclonal antibody against mouse oocyte cytoskeleton recognizing cytokeratin-type filaments. *J. Embryol. Exp. Morphol.* **90,** 197–209.

Lehtonen, E. (1987). Cytokeratins in oocytes and preimplantation embryos of the mouse. *Curr. Topics Dev. Biol.* **22,** 153–173.

Lehtonen, E., Lehto, V. P., Vartio, T., Badley, R. A., and Virtanen, I. (1983). Expression of cytokeratin polypeptides in mouse oocytes and preimplantation embryos. *Dev. Biol.* **100,** 158–165.

Letai, A., Coulombe, P. A., McCormick, M. B., Yu, Q. C., Hutton, E., and Fuchs, E. (1993). Disease severity correlates with position of keratin point mutations in patients with epidermolysis bullosa simplex. *Proc. Natl. Acad. Sci. USA* **90,** 3197–3201.

Liem, R. K. (1993). Molecular biology of neuronal intermediate filaments. *Curr. Opin. Cell Biol.* **5**, 12–16.

Mathur, M., Goodwin, L., and Cowin, P. (1994). Interactions of the cytoplasmic domain of the desmosomal cadherin Dsg1 with plakoglobin. *J. Biol. Chem.* **269**, 14075–14080.

McCrea, P. D., Brieher, W. M., and Gumbiner, B. M. (1993). Induction of a secondary body axis in Xenopus by antibodies to β-catenin. *J. Cell Biol.* **123**, 477–484.

McLean, W. H. I., and Lane, E. B. (1995). Intermediate filaments in disease. *Curr. Opin. Cell Biol.* **7**, 118–125.

Miyatani, S., Winkles, J. A., Sargent, T. D., and Dawid, I. B. (1986). Stage-specific keratins in Xenopus laevis embryos and tadpoles: The XK81 gene family. *J. Cell Biol.* **103**, 1957–1965.

Moon, R. T. (1993). In pursuit of the functions of the Wnt family of developmental regulators: Insights from Xenopus laevis. *Bioessays* **15**, 91–97.

Moon, R. T., Christian, J. L., Campbell, R. M., McGrew, L. L., DeMarais, A. A., Torres, M., Lai, C. J., Olson, D. J., and Kelly, G. M. (1993a). Dissecting Wnt signalling pathways and Wnt-sensitive development processes through transient misexpression analyses in embryos of Xenopus laevis. *Dev. Suppl.* **1993**, 85–94.

Moon, R. T., DeMarais, A., and Olson, D. J. (1993b). Responses to *Wnt* signals in vertebrate embryos may involve changes in cell adhesion and cell movement. *J. Cell Sci. Suppl.* **17**, 183–188.

Newport, J., and Kirschner, M. (1982). A major developmental transition in early Xenopus laevis embryos: 1. Characterization and timing of cellular changes at the midblastula stage. *Cell* **30**, 675–686.

Nishikawa, A., Shimizu-Nishikawa K., and Miller, L. (1992). Spatial, temporal, and hormonal regulation of epidermal keratin expression during development of the frog, Xenopus laevis. *Dev. Biol.* **151**, 145–153.

Nishizawa, K., Yano, T., Shibata, M., Ando, S., Saga, S., Takahashi, T., and Inagaki, M. (1991). Specific localization of phosphointermediate filament protein in the constricted area of dividing cells. *J. Biol. Chem.* **266**, 3074–3079.

Ohara, O., Gahara, Y., Miyake, T., Teraoka, H., and Kitamura, T. (1993). Neurofilament deficiency in quail caused by nonsense mutation in neurofilament-L gene. *J. Cell Biol.* **121**, 387–395.

Peifer, M., Berg, S., and Reynolds, A. B. (1994). A repeating amino acid motif shared by proteins with diverse cellular roles. *Cell* **76**, 789–791.

Peifer, M., McCrea, P. D., Green, K. J., Wieschaus, E., and Gumbiner, B. M. (1992). The vertebrate adhesive junction proteins β-catenin and plakoglobin and the Drosophila segment polarity gene armadillo form a multigene family with similar properties. *J. Cell Biol.* **118**, 681–691.

Peifer, M., Orsulic, S., Pai, L.-M., and Loureiro, J. (1993). A model system for cell adhesion and signal transduction in Drosophila. *Development* **(Suppl.)**, 163–176.

Peifer, M., Pai, L. M., and Casey, M. (1994a). Phosphorylation of the *Drosophila* adherens junctions protein armadillo: Roles for wingless signal and zest-white 3 kinase. *Dev. Biol.* **166**, 543–556.

Peifer, M., Sweeton, D., Casey, M., and Wieschaus, E. (1994b). Wingless signal and zeste-white 3 kinase trigger opposing changes in the intracellular distribution of armadillo. *Development* **120**, 369–380.

Pekny, M., Leveen, P., Pekna, M., Eliasson, C., Westermark, B., and Betsholtz, C. (1995). Mice deficient for glial fibrillary acidic protein (GFAP) generated by gene targeting. *J. Cell. Biochem. Suppl.* **19B**, 139. [Abstract]

Perrimon, N. (1994). The genetic basis of patterned baldness in Drosophila. *Cell* **76**, 781–784.

Perry, M. M. (1975). Microfilaments in the external surface layer of the early amphibian embryo. *J. Embryol. Exp. Morphol.* **33**, 127–146.

Pierce, S. B., and Kimelman, D. (1995). Regulation of Spemann organizer formation by the intercellular kinase Xgsk-3. *Development* **121**, 755–765.

Plancha, C. E., Carmo-Fonseca, M., and David-Ferreira, J. (1989). Cytokeratin filaments are present in golden hamster oocytes and early embryos. *Differentiation* **42**, 1–9.

Plancha, C. E., Carmo-Fonseca, M., and David-Ferreira, J. (1991). Cytokeratin in early hamster embryogenesis and parthenogenesis: Reorganization during mitosis and association with clusters of interchromatinlike granules. *Differentiation* **48**, 67–74.

Pokrywka, N. J., and Stephanson, E. C. (1994). Localized RNAs are enriched in cytoskeletal extracts of Drosophila oocytes. *Dev. Biol.* **166**, 210–219.

Pondel, M. D., and King, M. L. (1988). Localized maternal mRNA related to transforming growth factor β mRNA in concentrated in a cytokeratin-enriched fraction from Xenopus oocytes. *Proc. Natl. Acad. Sci. USA* **85**, 7612–7616.

Ryabova, L. V., Virtanen, I., Lehtonen, E., Wartiovaara, J., and Vassetzky, S. G. (1993). Morphology of the keratin cytoskeleton of Xenopus oocytes as studied using heterologous monoclonal antibodies. *Russian J. Dev. Biol.* **24**, 364–372.

Ryabova, L. V., Vassetzky, S. G., and Capco, D. G. (1994). Development of cortical contractility in the Xenopus laevis oocyte mediated by reorganization of the cortical cytoskeleton: A model. *Zygote* **2**, 263–271.

Schroeder, T. E., and Otto, J. J. (1991). Snoods: A periodic network containing cytokeratin in the cortex of starfish oocytes. *Dev. Biol.* **144**, 240–247.

Schultheiss, T., Lin, Z. X., Ishikawa, H., Zamir, I., Stoeckert, C. J., and Holtzer, H. (1991). Desmin/vimentin intermediate filaments are dispensable for many aspects of myogenesis. *J. Cell Biol.* **114**, 953–966.

Sharpe, C. R. (1988). Developmental expression of a neurofilament-M and two vimentin-like genes in Xenopus laevis. *Development* **103**, 269–277.

Sharpe, C. R., Pluck, A., and Gurdon, J. B. (1989). XIF3, a Xenopus peripherin gene, requires an inductive signal for enhanced expression in anterior neural tissue. *Development* **107**, 701–714.

Siegfried, E., Wilder, E. L., and Perrimon, N. (1994). Components of wingless signalling in Drosophila. *Nature* **367**, 76–80.

Sive, H. L. (1993). The frog prince-ss: A molecular formula for dorsoventral patterning in *Xenopus*. *Genes Dev.* **7**, 1–12.

Stappenbeck, T. A., Lamb, J. A., Corcoran, C. M., and Green, K. J. (1995). Phosphorylation of the desmoplakin C-terminus negatively regulates its interaction with keratin intermediate filament networks. *J. Biol. Chem.* **269**, 29351–29354.

Stappenbeck, T. S., Bornslaeger, E. A., Corcoran, C. M., Luu, H. H., Virata, M. L., and Green, K. J. (1993). Functional analysis of desmoplakin domains: Specification of the interaction with keratin versus vimentin intermediate filament networks. *J. Cell Biol.* **123**, 691–705.

Stappenbeck, T. S., and Green, K. J. (1992). The desmoplakin carboxyl terminus coaligns with and specifically disrupts intermediate filament networks when expressed in cultured cells. *J. Cell Biol.* **116**, 1197–1209.

Steinert, P. M., and Liem, R. K. (1990). Intermediate filament dynamics. *Cell* **60**, 521–523.

Szaro, B. G., and Gainer, H. (1988a). Identities, antigenic determinants, and topographic distributions of neurofilament proteins in the nervous systems of adult frogs and tadpoles of Xenopus laevis. *J. Comp. Neurol.* **273**, 344–358.

Szaro, B. G., and Gainer, H. (1988b). Immunocytochemical identification of non-neuronal intermediate filament proteins in the developing Xenopus laevis nervous system. *Brain Res.* **471**, 207–224.

Szaro, B. G., Grant, P., Lee, V. M., and Gainer, H. (1991). Inhibition of axonal development after injection of neurofilament antibodies into a Xenopus laevis embryo. *J. Comp. Neurol.* **308,** 576–585.

Torpey, N., Wylie, C. C., and Heasman, J. (1992). Function of maternal cytokeratin in Xenopus development. *Nature* **357,** 413–415.

Troyanovsky, S. M., Eshkind, L. G., Troyanovsky, R. B., Leube, R. E., and Franke, W. W. (1993). Contributions of cytoplasmic domains of desmosomal cadherins to desmosome assembly and intermediate filament anchorage. *Cell* **72,** 561–574.

Troyanovsky, S. M., Troyanovsky, R. B., Eshkind, L. G., Krutovskikh, V. A., Leube, R. E., and Franke, W. W. (1994a). Identification of the plakoglobin-binding domain in desmoglein and its role in plaque assembly and intermediate filament anchorage. *J. Cell Bio.* **127,** 151–160.

Troyanovsky, S. M., Toryanovsky, R. B., Eshkind, L. G., Leube, R. E., and Franke, W. W. (1994b). Identification of amino acid sequence motifs in desmocollin, a desmosomal glycoprotein, that are required for plakoglobin binding and plaque formation. *Proc. Natl. Acad. Sci. USA* **91,** 10790–10794.

Tsujimura, K., Ogawara, M., Takeuchi, Y., Imajoh-Ohmi, S., Ha, M. H., and Inagaki, M. (1994). Visualization and function of vimentin phosphorylation by cdc2 kinase during mitosis. *J. Biol. Chem.* **269,** 31097–31106.

Vize, P. D., Melton, D. A., Hemmati-Brivanlou, A., and Harland, R. M. (1991). Assays for gene function in developing Xenopus embryos. *In* "*Xenopus laevis: Practical Uses in Cell and Molecular Biology*" (B. K. Kay and H. B. Peng, eds.), pp. 367–387. Academic Press, San Diego.

Winkles, J. A., Sargent, T. D., Parry, D. A., Jonas, E., and Dawid, I. B. (1985). Developmentally regulated cytokeratin gene in Xenopus laevis. *Mol. Cell. Biol.* **5,** 2575–2581.

Wong, P. C., and Cleveland, D. W. (1990). Characterization of dominant and recessive assembly-defective mutations in mouse neurofilament NF-M. *J. Cell Biol.* **111,** 1987–2003.

Xu, Z., Cork, L. C., Griffin, J. W., and Cleveland, D. W. (1993). Increased expression of neurofilament subunit NF-L produces morphological alterations that resemble the pathology of human motor neuron disease. *Cell* **73,** 23–33.

Index

A

Actin, *see also* F-actin
  in anural developers, 266
  bundles
    equatorial, 210–211
    ring, 207–208, 222
  cables, 390–391
  component of unfertilized egg,
    344–348
  cortical, in *Xenopus* oocytes, 435–436,
    443–444, 446–447
  network
    in egg shape change, 2
    endoplasmic, disruption, 212–214
  nonfilamentous, 107–108, 116–118,
    347–348
  organization changes, fertilization-
    induced, 348–352
  polymerization, and sperm incorporation,
    349–350
Actin-binding proteins
  biochemical preparation, 123–126
  in egg cortex, 437–438
  function in actin cytoskeleton, 175–188
Actin cap
  apical orientation, 170–173
  localization of actin-binding proteins,
    182–184
  role of 13D2 protein, 187–188
Actin cytoskeleton
  cortical, developmental role, 204–221
  dynamics during egg fertilization,
    104–107
  role in early *Drosophila* development,
    167–193
  subcortical, developmental role,
    223–225
Actin filament-binding proteins, 124–125
Actin filament-capping and -severing
    proteins, 108–111
Actin filament cross-linking proteins,
    111–115

Actin filaments
  bundles
    in early embryo, 170
    in marginal cell cortex, 368
  contractile arcs, 218–220
  endoplasmic population, 105
  and epiboly, 365–368
  forming cortical network, 281–282
  interactions with microtubules and
    intermediate filaments, 463
  organization
    mammalian species, 327–328
    rodent, 322–325
  polymerization, 108–109, 114–118
  reorganization, 211–213
  visualization, 339
α-Actinin, egg, enrichment in cortical
    region, 113
Actin lattice
  contractions, 215
  polar cortical, during early cleavage,
    221–223
Actin–membrane cross-linking proteins,
    111–115
Actin–membrane cytoskeleton
  dynamics during early embryogenesis,
    115–123
  reorganization, 102–104
Actin monomer-binding proteins, 107–108
Actolinkin, complex with actin, 109
Actomyosin, based cortical contraction, 446
Aequorin, coinjection with fluorescein, 361
Amphibians, *see also Xenopus laevis*
  microtubule assembly regulation,
    412–413
  oogenesis and early development,
    385–405, 417–419
Anillin, nuclear localization, 184
Animal–vegetal axis
  amphibian oocyte, 385–386
  asymmetry of keratin and vimentin
    organization, 459–460
  built-in polarity, 257